This volume provides a wide-ranging review of the variety and specificity of receptors, including those for neurotransmitters, hormones, growth factors, oncogenes and antigens, and looks in depth at the fundamental features of multi-subunit complexes and their vital role in co-ordinating and controlling the functions of the body. These receptors which are the sites of action of many types of drugs, are constituted of single or multiple protein subunits. They may be organised as ligand-operated ion channels in a cell membrane or they may be concerned with enzyme activation to generate second messengers or DNA transcription following a change in the association–dissociation state of the receptor complex.

Advances at the molecular level have enabled scientists to relate the structure of these receptor subunits to their specific mode of action and their function and this, in turn, is leading the way forward to exciting new research with important practical applications for the treatment of disease. The 14 chapters in this volume summarise these advances across a wide range of disciplines, and the resulting publication will therefore be of interest to research scientists in endocrinology, immunology, pharmacology, oncology and neurobiology.

RECEPTOR SUBUNITS AND COMPLEXES

RECEPTOR SUBUNITS AND COMPLEXES

Edited by
A. BURGEN AND E. A. BARNARD

CAMBRIDGE
UNIVERSITY PRESS

1992.

Published by the Press Syndicate of the University of Cambridge
The Pitt Building, Trumpington Street, Cambridge CB2 1RP
40 West 20th Street, New York, NY 10011-4211, USA
10 Stamford Road, Oakleigh, Victoria 3166, Australia

First published 1992

Printed in Great Britain at the University Press, Cambridge

A catalogue record of this book is available from the British Library

Library of Congress cataloguing in publication data
Receptor subunits and complexes/A. S. V. Burgen and E. A. Barnard.
    p.   cm.
Includes index.
ISBN 0 521 36612 7
1. Cell receptors – Structure–activity relationships. 2. Receptor-ligand complexes.
I. Burgen, A.S.V. II. Barnard, Eric A., 1927
QH603.C43R423 1992
574.87 – dc20   91-28581 CIP

ISBN 0 521 36612 7 hardback

UP

Contents

Contributors

E. A. Barnard
MRC Molecular Neurobiology Unit, Hills Road, Cambridge CB2 2QH, UK

J. S. Bonifacino
Cell Biology and Metabolism Branch, National Institute of Child Health and Human Development, Bethesda, MD 20892, USA

H. R. Bourne
Department of Pharmacology, Medicine and the Cardiovascular Research Institute, University of California, San Francisco, CA 94143-0480, USA

A. Burgen
Department of Pharmacology, University of Cambridge, Tennis Court Road, Cambridge CB2 1QJ, UK

M. G. Caron
Howard Hughes Medical Institute, Departments of Biochemistry, Cell Biology and Medicine (Cardiology), Duke University Medical Center, Durham, NC 27710, USA

M. J. Fry
Ludwig Institute for Cancer Research, 91 Ridinghouse Street, London W1P 8BT, UK

C. K. Glass
Eukaryotic Regulatory Biology Program, Center for Molecular Genetics, School of Medicine, M-013, University of California, San Diego, La Jolla, CA 92093-0613, USA

M. Hnatowich
Howard Hughes Medical Institute, Departments of Biochemistry, Cell Biology and Medicine (Cardiology), Duke University Medical Center, Durham, NC 27720, USA

J. M. Holloway
Center for Molecular Genetics and Department of Chemistry, School of Medicine M-013, University of California, San Diego, La Jolla, CA 92093-613, USA

L. N. Johnson
Laboratory of Molecular Biophysics, University of Oxford, Rex Richards Building, Oxford OX1 3QU, UK

R. D. Klausner
Cell Biology and Metabolism Branch, National Institute of Child Health and Human Development, Bethesda, MD 20892, USA

R. J. Lefkowitz
Howard Hughes Medical Institute, Departments of Biochemistry, Cell Biology and Medicine (Cardiology), Duke University Medical Center, Durham, NC 27710, USA

A. Maelicke
Institute of Physiological Chemistry and Pathobiochemistry, Johannes Gutenberg University Medical School, D-6500 Mainz, Germany

R. Tyler Miller
Department of Medicine, University of Texas Health Sciences Center, 5323 Harry Hines Blvd., Dallas, Tx 75235, USA

T. P. Misko
Monsanto Corporation, Molecular Genetics and Mammalian Biology Group, St Louis, MI 63198, USA

B. F. O'Dowd
Addiction Research Foundation, 33 Russell Street, Toronto, Canada

G. Panayotou
Ludwig Institute for Cancer Research, 91 Ridinghouse Street, London W1P 8BT, UK

N. C. Price
Department of Molecular and Biological Sciences, University of Stirling, Stirling FK9 4LA, UK

M. J. Radeke
Department of Biology, University of California, Santa Barbara, Santa Barbara, CA 93106, USA

M. G. Rosenfeld
University of California, 9500 Silman Drive San Diego, La Jolla, CA 92093–0648 USA

M. Schimerlik
Department of Biochemistry and Biophysics, Oregon State University, Carvallis, OR 97331-6503, USA

E. M. Shooter
Department of Neurobiology, Standord University School of Medicine, Stanford, CA 94305-5401, USA

K. Siddle
Department of Clinical Biochemistry, University of Cambridge, Addenbrooke's Hospital, Cambridge CB2 2QR, UK

R. D. Vale
Department of Pharmacology, University of California, San Francisco, San Francisco, CA 94135, USA

M. D. Waterfield
Ludwig Institute for Cancer Research, 91 Ridinghouse Street, London W1P 8BT, UK

A. M. Weissman
Experimental Immunology Branch, National Cancer Institute, Bethesda, MD 20892, USA

1

Introduction

ARNOLD BURGEN

The concept of a receptor was invented by Langley to account for the specificity of drug action. It was seen as a structure in the biological system that recognised the drug, and that, in some way, interaction between the two was responsible for the drug's effects. The recognition process needed to be specific in order to recognise a unique ligand, or class of ligand, but not others. Gradually it was realised that drug receptors were, in most cases, the site of action of endogenous regulators whether called transmitters, hormones, growth controllers, etc, for largely historical or topological reasons.

It is now accepted that such regulating systems are fundamental to the orderly operation of all living systems, controlling cell division and differentiation and being responsible for regulation of higher order events such as heart rate and respiratory movement, and, ultimately, perception and thought. All these regulatory processes are either actual or potential sites of drug action and, indeed, the range of available agents that interact with regulatory systems is growing rapidly.

The approaches to exploring receptors vary in the different branches of biology largely for historical reasons. In some branches, development of a wide range of synthetic ligands has had a major influence, in others genetics and mutation of the receptor have been prominent. In yet others, some aspects of the integration into the cell machinery has been prominent. Despite these divergent emphases, the principles of receptor action are simple and common to them all. One of the main reasons for bringing examples of major classes of receptors together in one place is to illuminate these common problems and the essential unity of the concept by juxtaposition. There are also fundamental problems of protein machinery that underlie the operation of all these receptors which are dealt with in Chapters 2–4.

The idea of drug receptors has been turned from an operational concept into a phenomenon that can be studied at the molecular level by the development of ligand binding methods and, more recently, by the isolation of receptor molecules and direct studies of their structure. The isolation of receptors from tissues was difficult both because of their low abundance and because of their integration in cell membranes. The breakthrough came from the acetylcholine (nicotinic) receptor in the electric organ of elapid fishes. These are specialised structures for generating massive external electric currents by the synchronised action of receptors. For this reason, the organ has evolved so that its membrane protein is predominantly receptor protein, and, since the organs are large, it provides an abundance of material which was relatively easily purified to homogeneity. Because of these unique advantages, more is known about the properties of this receptor than about any other. For most other receptors, the difficulty of isolation and the low abundance has meant that detailed studies have had to await the development of gene expression methods. The use of these has revealed an unexpected multiplicity of genes for receptors, so that it is now a generality that each drug receptor turns out to represent a group of closely related receptors that may be differentially expressed in different cells or even co-expressed in the same cell.

What are the essential features of a receptor? First, it must recognise the appropriate ligand or group of ligands; this involves a more or less complex binding process which is selective and hence involves specific molecular features of the ligand and a complementary binding site on the receptor. Since the receptor is the medium for transmitting information, it has an effector function, which means that it must have a second binding site to which an effector molecule can be bound. The binding of an activating ligand at the 'drug binding site' then leads to some action at the effector site – for this to happen there needs to be coupling between the two sites. Many drug receptors are integrated in the boundary membrane of cells, the receptor site being located on the outer face of the membrane and the effector site on the inner surface. The coupling process is then transmembrane and is a means of communicating extra-cellular signals to the cell interior.

Receptor activation and the coupling process are both dependent on conformation change.

Most receptors are proteins. The final conformation that an unfolded peptide adopts depends on the generation of various substructure motifs such as helices and sheets dependent on local residue interaction and the production of a global tertiary conformation also involving interactions between non-contiguous residues. The final conformation is that accessible state of lowest free energy and highest entropy. This state may be unique or there may be several conformations of similar free energy content which may co-exist and be populated accordingly; the energy barrier between these states will determine whether or not they can readily interconvert. It will be recalled that few proteins are indefinitely stable in solution and become denatured spontaneously without covalent change; evidently in these instances the denatured state is more stable and the active state is essentially metastable. However, some proteins can be denatured and fully renatured quite simply. The lesson is that, while a receptor protein may exist in one conformation, it is potentially capable of existing in other conformations.

What happens when a ligand is bound to the protein? The global free energy of the ligand–protein complex is different from that of the free protein and the complex *may* relax into a new conformation of lower free energy that is now accessible and for this conformation change to produce effects it is necessary that changes in conformation are also produced at the effector site. This process is usually referred to as the ligand producing a *conformation induction*. If the binding of the ligand does not favour the generation of the active conformation, there are other possibilities. One is that the ligand stabilises the existing conformation (i.e. increases the depth of the conformational energy well) or that it favours an alternative conformation that is not coupled to an active state of the effector site. In these latter cases the ligand will act as an antagonist.

An alternative to conformation induction may be found where the unliganded receptor exists as an interconverting equilibrium mixture of two conformations. If a ligand binds uniquely (or selectively) to one of the conformations, the equilibrium will shift in its favour, thus producing *conformation selection*. There are kinetic differences between these two mechanisms. In the first, the generation of activity is dependent on the binding constant for the resting conformation and the rate of transition into the active form. In the second mechanism, the degree of possible activation depends on the unliganded equilibrium between conformations. For there to be the possibility of a large change in the amount of the active conformation, it must be in low abundance in the absence of ligand. Since it also implies that there are active molecules in the absence of ligand, it includes the possibility of antagonists that have actions opposite to agonists. Neither of these phenomena has been clearly demonstrated. In the case of the nicotinic receptor, spontaneous activity has not been detected and the abundance of the active state in the absence of ligand is less than 10^{-3}. In other cases the limit has been set even lower. Thus, while conformation selection was the original mechanism described for allosteric control of enzymes, it is not apparently the commonest mode of receptor action. Ideally, conformation changes produced by drugs should be delineated by structural methods such as X-ray crystallography and three-dimensional NMR. So far this has not been possible for technical reasons, so that the evidence for conformation change has come from binding parameters and kinetic methods, the latter notably in the nicotinic receptor. However, much has been discovered about conformational changes occurring in single proteins and multi-subunit proteins as a result of ligand binding ranging from the classic oxygen binding by haemoglobin to a recent example of the effects of coenzymes, substrates and inhibitors on dihydrofolate reductase. (Bystroff & Kraut, 1991). These problems are covered in detail in Chapters 2 and 3. At first sight, it may seem surprising to see that the conformation changes produced by ligands are rather limited and that the general structure of proteins is little affected, but this is consistent with the view that much of the structure is determined by local interactions between successive residues and not by global energy considerations. Secondly, since specific interaction of ligands with the receptor will be very sensitive to quite small distance changes (notably van der Waal's attractive and repulsive components), such changes are perfectly adequate to account for the alterations in the ligand binding.

In a multi-subunit protein, the effect of a ligand-induced conformation at one subunit may have a number of possible effects on the other proteins in the complex. At its simplest, it may not affect the other proteins at all, which are thus just providing a framework for the assembly. The effector in the ligand-binding unit in this case is not dependent directly on a protein interaction. A second possibility is that the conformation change in the ligand-binding unit affects the interfacial region between some of the subunits and this can result in the binding energy of the subunits changing. This leads to either dissociation or association of subunits. Yet another possible outcome is that the changed conformation of the ligand-binding subunit at the interface with an associated subunit leads to a change in the conformation of that subunit, so that we have a transmitted conformation change between subunits. We will see examples of all these in later chapters.

One very active area of research on multi-subunit receptors concerns the specificity of assembly. For example, it was established many years ago that the nicotinic receptor in the electric organ was constituted by four types of subunit of rather similar size in which one subunit, α, was the exclusive acetylcholine binding subunit and was represented twice and the other three subunits represented singly, giving an overall composition of $\alpha_2\beta\gamma\delta$ and an unusual pentameric symmetry. More recently, it has been found that the nicotinic receptors in autonomic ganglia and in the central nervous system are distinct structural types and each contains only two types of subunit. At first, the evidence seemed to suggest that, in these receptors, the symmetry was tetrameric and could be regarded as $\alpha_2\beta_2$. More recently, a study by Cooper *et al.* (1991) using mutated subunits with altered conductances has revealed that the ganglionic nicotinic receptor is also pentameric with a $\alpha_2\beta_3$ composition. A very general current problem is how multi-subunit receptors are assembled in any regular fashion when there is an abundance of related subtypes of the basic units present in the cell. Some experiments with mixed enzyme subunits show that sorting of subunits and the avoidance of hybrid production is efficient. In various systems described in this book, the same problem is addressed and the answers are by no means so clear; hybrid formation seems to occur but it remains unclear how common or important it is.

At present, the concept is that receptors fall into a rather small number of basic patterns that may be termed receptor families. Within a given family, the assemblage is similar and the component proteins show many common features such as a considerable degree of sequence homology and a common pattern of hydrophobic sequences. Likewise, the effector systems controlled by the receptors are few in number. By contrast, the number of regulator ligands is large and correspondingly so is the number of primary receptors. This number is expanded still further by the existence of receptor subtypes, which still respond to the type ligand but may have significantly different operational parameters and have differential selectivity for drugs.

There are two main groups of receptors that respond by change in the binding of the receptor subunit to other subunits. The first is characterised by the steroid receptors. These are associated in the resting state with a chaperone, the heat shock protein (hsp 90). Interaction with a steroid causes dissociation of the complex followed by self-association of the receptor subunit to form a dimer which then binds to DNA (see DeMarzo *et al.*, 1991). In G protein receptors, the association of receptor with G protein components is regulated by the agonist and by guanine nucleotides interacting with the G_α subunit. In the absence of the appropriate ligand, the G subunits are associated with the receptor but in the active state the separated G_α subunit is the effector (Bourne *et al.*, 1991; Birnbaumer, 1990).

Examples of non-dissociating receptors are the nicotinic, GABA and glycine receptors in which ligand interaction leads to the opening of a channel in the cell membrane through which various classes of inorganic ions can move. A quite distinct type is the insulin receptor in which interaction with the α unit in the cell exterior leads to activation of phosphokinase on the intracellular part of the β subunit. However, since the α and β subunits are covalently joined by a disulphide, in this respect this system strains the definition of subunit, but, since pairs of both subunits are found in the receptor, all is well.

Yet another class is that of the antigen receptors whose response is to cause an invagination of the receptor into the cell interior. This is a process that proceeds with many, perhaps all, receptors as part of their turnover and regulation of receptor number. For the antigen receptor this is the primary reaction.

The development of drugs to interact with a receptor system can follow several strategies. The first is to simulate the actions of the endogenous regulator in some advantageous fashion, for instance, making administration simpler (cp. morphine and the endogenous opioid pepides) or giving a prolonged or abbreviated duration of action. Equally, one might interfere with the endogenous action by some form of antagonism. The simplest view of antagonism is that of an agent which interacts with the agonist site of the receptor without leading to the conformation change that activates an effector and simply competes with the endogenous agonist. This may occur through binding in the same conformation as the unliganded state but there are other possibilities. For instance, the conformation changes produced by agonists at the nicotinic receptor can relax into an inactivated state. This is responsible for the fade of the response on continued application of the agonist and is also responsible for the neuromuscular block that attends the action of cholinesterase inhibitors which interfere with the removal of synaptically released acetylcholine. Suxamethonium represents a drug with a high affinity for the inactivated state of this receptor and blocks in this way.

Strictly, these represent non-competitive antagonists but the distinction from classical antagonists is not always easy to establish. The existence of a subunit structure offers the possibility of perturbing the system by interaction other than at the agonist binding domain. The NMDA receptor is excited by glutamate and aspartate and inactivates rapidly and extensively with a time course of less than a second. Glycine, which is not a direct activator of this receptor, drastically reduces the rate of desensitisation and stabilises the active state apparently by acting at a distinct binding domain (Mayer *et al.*, 1989). Incidentally, the NMDA (and the glycine) receptors can have multiple open states implying more than a single active conformation (Hamill *et al.*, 1983).

In the GABA receptor, the agonist binding site is in the α receptor subunit but some drugs that modify the receptor action bind onto the β subunit. The GABA receptor, when activated, opens a chloride channel. Diazepam increases the frequency of channel openings within a burst without altering the duration of the burst, indicating a shorter excited lifetime (Study & Barker 1981). On the other hand, pentobarbitone decreases the frequency of bursts but increases their duration, apparently by changing the kinetics of GABA binding. As the details of receptor action are explored in more detail, the range and subtlety of the ways in which receptor action can be modulated will become apparent.

Our knowledge about biological regulators and their receptors, although impressive, is very far from complete and we can expect both new systems and new concepts to develop. Nevertheless, the cross-section of receptors discussed in this book provides a broad view of the current position.

References

Birnbaumer, L. (1990). G Proteins in signal transduction. *Annual Reviews in Pharmacology and Toxicol.*, **30**, 675–705.

Bourne, H. R., Sanders, D. A. & McCormick, F. (1991). The GTPase superfamily: conserved structure and molecular mechanism. *Nature, London*, **349**, 117–27.

Bystroff, C. & Kraut, J. (1991). Crystal structure of unliganded *Escherichia coli* dihydrofolate reductase. Ligand-induced conformational changes and cooperativity in binding. *Biochemistry*, **30**, 2227–39.

Cooper, E., Couturier, S. & Ballivet, M. (1991). Pentameric structure and subunit stoichiometry of a neuronal nicotinic acetylcholine receptor. *Nature, London*, **350**, 235–8.

DeMarzo, A. M., Beck, C. A., Onate, S. A. & Edwards, D. P. (1991). Dimerization of mammalian progesterone receptors occurs in the absence of DNA and is related to the release of the 90 kD heat shock protein. *Proceedings of the National Academy of Sciences, USA*, **88**, 72–6.

Hamill, O. P., Bormann, J. & Sakmann, B. (1983). Activation of multiple conductance state chloride channels in spinal neurones by glycine and gaba. *Nature, London*, **305**, 805–8.

Mayer, M. L., Vyklicky, L. & Clements, J. (1989). Regulation of NMDA receptor desensitisation in mouse hippocampal neurones by glycine. *Nature, London*, **338**, 425–7.

Study, R. E. & Barker, J. L. (1981). Diazepam and pentobarbital: fluctuation analysis reveals different mechanisms for potentiation of γ-aminobutyric acid responses in cultured central neurones. *Proceedings of the National Academy of Sciences, USA*, **78**, 7180–4.

2

Folding and assembly of multi-subunit proteins

NICHOLAS C. PRICE

Introduction

The purpose of this chapter is to summarise the present state of knowledge regarding the mechanism(s) by which multi-subunit proteins acquire their native three-dimensional structure. Most of the information comes from studies of the refolding and reassembly of denatured and dissociated proteins, and the extent to which conclusions from these studies can be applied to folding and assembly *in vivo* is not yet established. It should also be noted that the majority of studies refer to well-characterised soluble proteins, and the processes involved in the assembly of membrane-bound proteins such as multi-subunit receptors are not yet so clearly elucidated. From these studies, certain conclusions have been reached regarding the mechanisms of folding and assembly of proteins and the reader is referred to a number of excellent detailed reviews (Creighton, 1978; Kim & Baldwin, 1982; Ghélis & Yon, 1982; Jaenicke, 1982, 1987; Tsou, 1988).

This chapter begins with a general discussion of the folding problem and compares aspects of the process *in vivo* and *in vitro*. The various methods which have been used to study folding are then outlined. Results obtained with monometric proteins (both single- and multi-domain) are then described; these provide the foundation for the study of multi-subunit proteins. In such multi-subunit proteins, the processes of folding and association must be coordinated and it has generally been observed that the overall pathway involves the formation of 'structured' intermediates which may possess biological activity under appropriate circumstances.

There are some well-documented examples of multi-subunit proteins, such as glutamate dehydrogenase and ribulose bisphosphate carboxylase–oxygenase which have not so far been shown to be capable of regaining biological activity after unfolding. The recent demonstration that 'chaper-

9

one' proteins assist in the assembly of ribulose bisphosphate carboxylase–oxygenase in the chloroplast may offer clues as to how such large proteins fold and assemble *in vivo* (Hemmingsen *et al.*, 1988) (see pp. 27–32).

The need to understand the mechanism of protein folding has been given fresh impetus by recent developments in recombinant DNA technology. Overexpression of eukaryotic genes in *E. coli* leads, in a number of cases, to the formation of inclusion bodies in which the desired protein is in an insoluble, highly aggregated, form. Recovery of the active protein is usually attempted by denaturation and refolding under carefully controlled conditions; this can often represent the most difficult and unpredictable part of the entire protein production process (Marston, 1986). The development of the elegant technique of site-directed mutagenesis (Rossi & Zoller, 1987) allows, in principle, the replacement of an amino acid at any position in a polypeptide chain by any of the other 19 amino acids. It is thus possible to generate mutant proteins in which the factors affecting biological activity can be studied in a systematic fashion (Fersht, 1987). However, the rules governing the folding of peptide chains must be understood so that the effects of the amino acid replacements on the folding process and hence on the three-dimensional structure of the folded protein can be predicted. Site-directed mutagenesis can also be used to study the involvement of particular amino acids in the folding process itself. The results obtained with the monomeric proteins bovine pancreatic trypsin inhibitor (Goldenberg, 1988) and dihydrofolate reductase (Perry *et al.*, 1987) illustrate the power of the technique in examining details of the folding process in cases where the general features of the process have been deduced by other techniques.

The folding problem

Structure prediction methods

Considerable effort has been expended in attempts to calculate or predict the three-dimensional structure of a protein purely on the basis of its amino acid sequence. A summary of the approaches involved has been given by Jaenicke (1987). *Ab initio* calculations of the most stable structure (i.e. of lowest free energy) have proved of value only for short oligopeptides. For most proteins of biological interest containing over, say, 100 amino acids, a more fruitful approach has been to develop 'semi-empirical' methods in which proteins of known three-dimensional structure are examined in order to categorise amino acids or short sequences of amino acids according to the types of secondary structure in which they are most likely to occur. Predictive algorithms can then be devised and applied to predict secondary structure elements in other proteins. In a detailed survey of three of the most widely used such algorithms (those due to Lim, 1974, Chou & Fasman, 1978 and Garnier *et al.*, 1978), it was found that no method gave better than about 60 % correct prediction of secondary structure (Kabsch & Sander, 1983). More recently, a consensus approach has been used in which several prediction methods are applied to a given amino acid sequence; a structural feature is classed as 'predicted' if several (at least half) methods agree (Sawyer *et al.*, 1988). It is clear that very similar arrangements of secondary structural elements in proteins, such as the α-helix/β-sheet motif characteristic of nucleotide-binding domains, can be generated by a variety of amino acid sequences. Thus a 'folding code', which is based on relatively short range amino acid sequences alone, must be highly degenerate, and indeed may even not exist at all (Jaenicke, 1987). The fundamental difficulties with structure prediction methods arise from the complexities of protein molecules and uncertainties about the forces within the protein and between the protein and the solvent which determine the most stable folded structure. The use of 'neural network' (pattern learning) methods for structure prediction has been described (Holley & Karplus, 1989). Such methods have been claimed to give more reliable predictions than the earlier procedures.

Studies of protein folding **in vivo**

Ideally, the folding and assembly of proteins should be studied during the translation of mRNA on the ribosomes. However, there are obvious inherent difficulties in achieving perfect synchronisation of the translation process and, indeed, many of the physical methods for studying the conformation of proteins cannot be directly applied to protein synthesising systems and/or require relatively large quantities of well-characterised material. Some aspects of the folding process have been studied *in vivo*; thus disulphide bonds have been shown to occur in nascent polypeptide chains of serum albumin (Peters & Davidson, 1982) and of immuno-globulins (Bergman & Kuehl, 1979*a, b*). The formation of such bonds in these secreted proteins implies that folding of the polypeptide chain has already occurred so as to bring the correct pairs of cysteine side chains into juxtaposition. It has also been shown that nascent polypeptide chains of β-galactosidase can cross-react with antibodies raised against the completed protein (Hamlin & Zabin, 1972) again implying that folding to generate the required epitope(s) has occurred while the growing polypeptide chain is still bound to the ribosome. Other work has indicated that β-galactosidase chains may fold and even assemble to generate the catalytically active tetramer while still associated with the ribosomes (Zipser, 1963; Kiho & Rich, 1964).

While these studies have indicated that folding can occur during biosynthesis, there is little indication of the extent or rate of such folding. In order to examine the folding in more detail, it is necessary to use physical techniques which can monitor protein conformation and this can most conveniently be done by studying the process *in vitro*, as described in the next section.

Studies of protein folding **in vitro**

The *in vitro* method involves the unfolding and dissociation of a completed protein by treatment with denaturing agents (such as guanidinium chloride or urea), followed by removal of the agent by dilution or dialysis. (In some cases it is desirable to retain some secondary and tertiary structure in a dissociated protein. The refolding and reassembly processes can be studied by a variety of methods.

A comparison of folding **in vivo** and **in vitro**

Although *in vitro* studies of protein folding have yielded a considerable amount of information, it is important to investigate the extent to which the results can be used to understand folding *in vivo*. The two principal criteria which have been used in such comparisons involve the *rate* of the process and the *final product* of the process. The rate of folding *in vitro* for many proteins is fast enough (within seconds or minutes) to be compatible with estimates of the rate of folding of nascent polypeptide chains *in vivo*. Thus, for instance, the monomeric fructose 1,6-bisphosphate aldolase from *Staphylococcus aureus* regains activity after denaturation in guanidinium chloride within 10 seconds (Rudolph et al., 1983), and myoglobin refolds within this time after unfolding at low pH (Shen & Hermans, 1972). These times can be compared with estimates of 30 second or less for the folding of proteins *in vivo* (Tsou, 1988). In other cases, however, the refolding of proteins *in vitro* is much slower, requiring several minutes or hours, and would appear to be incompatible with the rate of folding *in vivo*. (Even this slow rate of folding is much faster than the estimated 10^{50} years it would take a polypeptide chain of 100 amino acids to find its most stable conformation by a random search pathway (Karplus & Weaver, 1976). The folding pathway appears to be directed. The refolding of those proteins where disulphide bond formation is involved (e.g. ribonuclease) can be accelerated by addition of protein disulphide isomerase, an enzyme found in the microsomal fraction of secretory tissues such as liver and pancreas (Freedman, 1984). There is now strong evidence that the isomerase is involved in the *in vivo* folding of proteins such as immuno-globulins (Roth & Pierce, 1987) and procollagen (Koivu & Myllylä, 1987) and in the co-translational formation of disulphide bonds in γ-gliadin, a wheat storage protein (Bulleid & Freedman, 1988). Catalysis by prolyl isomerase is also likely to accelerate the *in vivo* folding of those proteins (e.g. immunoglobulin light chain) where *cis* ⇌ *trans* isomerisation of peptide bonds of proline has been implicated as a slow step in protein folding (Lang *et al.*, 1987). It has been demonstrated that prolyl isomerase is identical with the cyclosporin-A binding protein, cyclophilin, although the functional significance of this identity is unclear (Fischer *et al.*, 1989; Takahashi *et al.* 1989). The slow refolding of other proteins *in vitro* may be due to the process of 'domain pairing', i.e. the correct alignments of individual folded units; it is likely that *in vivo* this process is made more efficient by co-translational folding. There is also recent evidence that co-translational glycosylation could serve to promote the correct folding and assembly of proteins *in vitro* .

The final product of refolding *in vitro* has been shown generally to be identical with the original native protein, by criteria such as biological activity, conformation and relative molecular mass (Jaenicke, 1987). However, in most cases, the yield in the refolding process is less than 100 %, indicating that side reactions, such as formation of aggregates, occur. This point is discussed further on pp. 14–15.

Is the completed polypeptide chain required for correct folding?

The studies mentioned previously have shown that nascent polypeptide chains can acquire at least a certain degree of three-dimensional structure.

There is also considerable evidence to show that fragments of proteins can fold to form stable, compact structures similar to those found in the completed proteins (Wetlaufer, 1981; Wright *et al.*, 1988). Thus the peptide corresponding to amino acids 1–45 of adenylate kinase binds MgATP with an affinity comparable to that of the complete enzyme (194 amino acids). NMR studies suggest that the peptide has a very similar conformation to that of the corresponding part of the enzyme (Fry *et al.*, 1985). On the other hand, there are some well-documented cases in which the entire polypeptide chain appears to be required for correct folding. Removal of the C-terminal 23 amino acids of Staphylococcal nuclease yields a large fragment (1–126) which has little or no ordered structure (Taniuchi & Anfinsen, 1969). Similarly, removal of the C-terminal four amino acids from ribonuclease leads to a derivative (1–120) which is unable to refold correctly after denaturation and reduction of the disulphide bonds (Taniuchi, 1970). In other cases, the removal of C-terminal amino acids has little effect on the folding process; thus the removal of 17 amino acids from bacteriorhodopsin has no effect on the refolding of the protein or its ability to be incorporated into vesicles with full proton-trans-locating activity (Liao & Khorana, 1984).

On balance, therefore, it seems that the available evidence favours the hypothesis that proteins acquire structure during biosynthesis of the polypeptide chain, although in some cases the final C-terminal amino acids may be required before the correct structure is finally specified. The rate of addition of amino acids during translation of mRNA has been estimated to be about 5–10 amino acids per second (Darnell *et al.*, 1986). There is thus ample time during biosynthesis to allow the formation of folded structures, which in the case of a small protein such as α-lactalbumin can occur within 20 ms (Gilmanshin & Ptitsyn, 1987).

The efficiency of folding

As mentioned on p. 13, folding of proteins *in vitro* is often less than 100% efficient. The major side reaction appears to be the formation of aggregates; these arise from the involvement of groups of amino acids in *intermolecular* contacts. In the native protein such groups would be involved in *intramolecular* (or *interdomain*) contacts. Aggregation is generally found to be more significant in proteins of larger subunit mass, because of the greater number of folded domains in such molecules. In monomeric multi-domain proteins such as octopine dehydrogenase (Teschner *et al.*, 1987) the recovery of biological activity is often slow and incomplete because of the failure to coordinate the processes of polypeptide chain folding (i.e. domain formation) and domain pairing. In the refolding of multi-subunit proteins, folded or partially folded monomers may associate to give 'wrong' aggregates. The extent of aggregation may be reduced by lowering the concentration of protein during refolding, since the reactions leading to aggregation are generally of higher kinetic order than those leading to formation of the correct multi-subunit protein. However, when the concentration of protein is too low during refolding, additional complications can arise from the adsorption or instability of the protein or intermediates in the refolding pathway. The technological aspects of refolding of unfolded proteins have been discussed (Jaenicke, 1987).

It is generally assumed that folding *in vivo* is a much more efficient process, and that the problems of aggregation which occur *in vitro* are avoided by the occurrence of co-translational folding. Domain structures which could then interact would thus be formed in a temporal sequence. Indeed, it has been proposed that, in some cases, gene sequences may have evolved so that the translation process may 'pause' at certain points to allow the correct folding to occur; such pauses would be most likely where a cluster of rare codons occurs in the mRNA (Purvis *et al.*, 1987; but see McNally *et al.*, 1989). In at least one case, however, there is evidence that folding *in vivo* is not 100 % efficient. The folding and association of the newly synthesised polypeptide chains of the tail spike endorhamnosidase of *Salmonella* phage P22 to yield trimers is 90 % efficient at 27 °C but only 15 % efficient at 42 °C. The data are consistent with a labile intermediate in the folding or assembly process; from its immunological properties this intermediate appears to be largely unfolded (Goldenberg *et al.*, 1982, 1983). It is a matter for speculation as to how general is the lack of complete folding *in vivo*, but it could be connected with the occurrence of the various intracellular enzyme systems for protein degradation.

Methods for studying folding *in vitro*

The unfolding of proteins

The unfolding of proteins is achieved most conveniently by the use of denaturing agents such as guanidinium chloride or urea or by extremes of pH. Experimental evidence indicates that the extent of unfolding can vary for different denaturing conditions (Ghélis & Yon, 1982; Jaenicke, 1987), ranging from local distortions of the polypeptide backbone to complete randomisation of the polypeptide chain (i.e. random coil formation). Concentrated (6 mol dm^{-3}) solutions of guanidinium chloride generally lead to complete unfolding of globular proteins as shown by NMR, infra-

Table 2.1 *Techniques which can be used to study the refolding of proteins*

Property studied	Technique	Reference
Biological activity	Kinetics of reactivation	Chan *et al.* (1973)
	Kinetics of ligand binding	Garel & Baldwin (1973)
Protein conformation	Fluorescence	Teipel & Koshland (1971)
	Circular dichroism	Labhardt (1986)
	Hydrogen–tritium exchange	Kim (1986)
	Hydrogen–deuterium exchange (followed by NMR analysis)	Roder *et al.* (1988) Udgaonkar & Baldwin (1988)
	Chemical modification of amino acid side chains	Ghélis (1980)
	Susceptibility to proteolysis	Girg *et al.* (1981)
	Trapping of intermediates in disulphide-bonded proteins	Creighton (1978)
State of association	Chemical cross-linking	Hermann *et al.* (1981)
	Hybridisation	Bothwell & Schachman (1980*a*, *b*)
	Separation of intermediates	Girg *et al.* (1983)

red spectroscopy and viscosity studies (Jaenicke, 1987). Disulphide bonds in proteins can be broken by addition of 2-mercaptoethanol or dithiothreitol. It is also advisable to add these reducing agents during unfolding of proteins containing cysteine side chains in order to prevent oxidative damage. Dithiothreitol has also been shown to remove traces of heavy metal ions which could catalyse oxidation by oxygen (Bickerstaff *et al.*, 1980).

In some cases it is desirable to bring about only limited unfolding. Thus in the case of multi-subunit proteins it is possible to focus on the association rather than the folding steps by using conditions which cause dissociation but not the total loss of secondary and tertiary structure.

The refolding of unfolded proteins

The refolding process is initiated by removal of the structure-perturbing agent, e.g. by dilution or dialysis, or by a pH-jump. The reoxidation of reduced proteins can be brought about efficiently by mixtures of oxidised

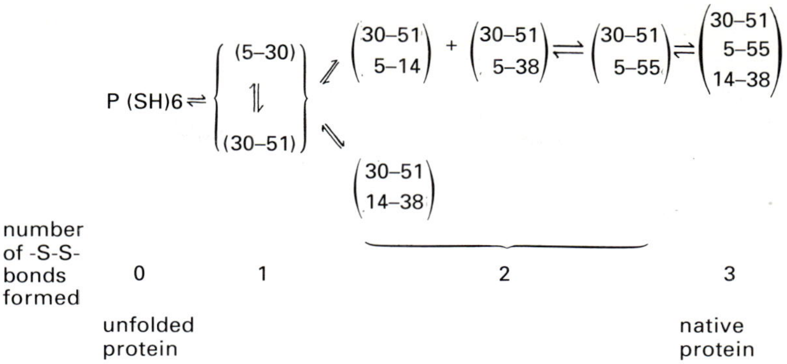

$$P\,(SH)6 \rightleftharpoons \left\{ \begin{array}{c} (5\text{-}30) \\ \Updownarrow \\ (30\text{-}51) \end{array} \right\} \begin{array}{c} \nearrow \\ \\ \searrow \end{array} \begin{array}{c} \begin{pmatrix} 30\text{-}51 \\ 5\text{-}14 \end{pmatrix} + \begin{pmatrix} 30\text{-}51 \\ 5\text{-}38 \end{pmatrix} \rightleftharpoons \begin{pmatrix} 30\text{-}51 \\ 5\text{-}55 \end{pmatrix} \rightleftharpoons \begin{pmatrix} 30\text{-}51 \\ 5\text{-}55 \\ 14\text{-}38 \end{pmatrix} \\ \\ \begin{pmatrix} 30\text{-}51 \\ 14\text{-}38 \end{pmatrix} \end{array}$$

number of -S-S- bonds formed			
0	1	2	3
unfolded protein			native protein

Fig. 2.1. The pathway of refolding of reduced bovine pancreatic trypsin inhibitor (Creighton, 1978). Intermediates with the designated disulphide bonds were trapped by reaction of remaining cysteine thiol groups with iodoacetate.

and reduced glutathione (Rudolph & Fuchs, 1983). A number of techniques (see Table 2.1) can be used to monitor the refolding; each of these gives information about particular aspects of the process and the results have to be taken together to build up a model for refolding. It is necessary to ensure that the time scale of any technique is compatible with the rate of the process being monitored; many of the techniques are suitable for processes in the seconds–minutes time range. Very rapid (< 1 s) steps in folding can be monitored by stopped-flow procedures using circular dichroism or fluorescence (Gilmanshin & Ptitsyn, 1987). The extent to which the information on the refolding pathway can be interpreted will, of course, be much greater in those cases where the structure of the native protein has already been determined by X-ray crystallography.

The folding of monomeric proteins

Single domain proteins
Bovine pancreatic trypsin inhibitor The refolding of this small protein (58 amino acids) after reduction of the disulphide bonds has been extensively studied by Creighton (1978). Intermediates were separated after remaining free cysteine groups had been quenched by reaction with iodoacetate and a pathway proposed (Fig. 2.1). The pathway shows the pattern

of disulphide bond formation resulting from folding of the polypeptide chain so as to bring the appropriate pairs of cysteine side chains into juxtaposition. Further details of the folding pathway have been obtained by the use of site-directed mutagenesis in which Cys 14 and Cys 38 are both converted to Ser (Goldenberg, 1988). The kinetics of folding of the mutant protein were not affected, consistent with the proposed pathway which involves rearrangements of intermediates in which Cys 14 and Cys 38 are paired in non-native arrangements. The conformations of the trapped intermediates have been examined by NMR (States *et al.*, 1987). These studies show that many conformational features of the native protein including the central β-sheet fold are present in the intermediate containing just one of the three normal disulphide bonds (30–51). Intermediates containing two of the correct disulphide bonds (e.g. 5–55, 30–51) have structures very similar to that of the native protein. Thus a more complete description of the folding process can now be proposed.

Bovine pancreatic ribonuclease The refolding of reduced ribonuclease (124 amino acids) has been studied by the quenching technique described for bovine pancreatic trypsin inhibitor. In this case, the pathway involves a high energy barrier separating the native conformation with four disulphide bonds from the other species (Creighton, 1978):

$$R \rightleftharpoons I \rightleftharpoons II \rightleftharpoons III \rightleftharpoons IV$$
$$\Updownarrow$$
$$N$$

where R and N are fully reduced and native protein respectively; and I, II, III and IV represent the large numbers of intermediates with one, two, three and four disulphide bonds respectively. These intermediates appear to have relatively little ordered structure.

The refolding of reduced ribonuclease is thus rather slow *in vitro*, but can be accelerated by addition of protein disulphide isomerase.

The refolding of non-reduced ribonuclease (i.e. with the disulphide bonds intact) after denaturation in guanidinium chloride is much faster than that of the reduced enzyme. At equilibrium the unfolded non-reduced enzyme consists of a mixture of $\approx 20\%$ of a fast-folding species ((U_F) and $\approx 80\%$ of a slow-folding species (U_s) (Garel & Baldwin, 1973). The two species appear to differ in the conformation of the peptide bonds to one or more of the four prolines in the enzyme. Thus most, if not all, of the U_s molecules contain a *trans* Pro 93; this bond has to isomerise to the *cis* form

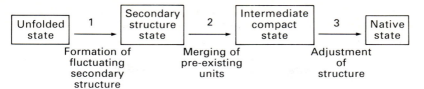

Fig. 2.2. A general model for folding of monomeric proteins. Stages 1, 2 and 3 involve predominantly hydrogen bonds, hydrophobic forces and van der Waals forces respectively.

(in which it occurs in the native enzyme) before rapid folding can occur (Schmid *et al.*, 1986). The isomerisation, and hence the folding process, can be accelerated by prolyl isomerase. By using two-dimensional NMR techniques to monitor the exchange of protons between the peptide chain amide groups and solvent, it has been shown that an early intermediate in the folding of non-reduced ribonuclease has a significant amount of stable secondary structure (Udgaonkar & Baldwin, 1988).

Multi-domain proteins

Many proteins consist of multiple folding units or 'domains', the existence of which can be inferred from X-ray crystallography, limited proteolysis or comparisons of amino acid sequences. The folding mechanisms of monomeric multi-domain proteins are intermediate between those of monomeric single domain proteins and multi-subunit proteins. Proteins which have been studied extensively include penicillinase, phospho-glycerate kinase, α-lactalbumin and thermolysin (Jaenicke, 1987).

Penicillinase consists of three domains which can be identified by proteolysis. During refolding after denaturation in guanidinium chloride, the native secondary structure is formed rapidly whereas the recovery of the native tertiary structure is slow, in accordance with the general model for refolding. Treatment of the enzyme with cyanogen bromide produces three fragments of approximately equal size which can associate with each other. Although this complex does not exhibit enzyme activity, it does possess the secondary structure and immunological properties of the native protein. The complex also undergoes reversible unfolding in guanidinium chloride, showing the importance of domain pairing in the overall folding mechanism (Adams *et al.*, 1980).

A general model for the folding of monomeric proteins

Taking account of the finding that secondary structure is formed very rapidly during the refolding of proteins, it is possible to propose a general outline model for protein folding. In this model the formation of the final native structure involves the merging and rearrangement with respect to each other of the rapidly formed folded units ('domain pairing') (Fig. 2.2). The intermediate compact state has been termed a 'molten globule'. This globule is compact, possesses a dense interior and does not undergo temperature melting. It has a high secondary structure content, but the amino acid side chains are more mobile and less tightly packed than in the native protein (Ptitsyn, 1987).

The later 'domain pairing' can be slow and in some cases relatively inefficient, leading to the lack of 100% recovery of native biological activity. Tandon & Horowitz (1988) have shown that certain non-ionic detergents (Triton X-100 and dodecyl-β-D-maltoside) can assist in the refolding of certain monomeric enzymes such as rhodanese and adenosine deaminase after unfolding in guanidinium chloride. It is proposed that these detergents could act by stabilising transient exposed hydrophobic areas of the protein which are subsequently involved in 'domain pairing'. In this way incorrect pairing which might lead to the formation of wrong aggregates could be avoided. However, these observations cannot be extended to all proteins (Tandon & Horowitz, 1988).

The folding and assembly of multi-subunit proteins

Structured intermediates

The study of the folding of multi-subunit proteins is inherently more complex than that of the monomeric systems, because of the occurrence of both folding and association reactions. It is, however, often possible to focus attention on the association steps by using denaturing conditions which cause dissociation but do not lead to the total loss of native secondary and tertiary structure. These 'structured intermediates' can be formed by treatments such as cold inactivation, application of high hydrostatic pressure, or by the addition of stabilising ions (Jaenicke, 1987). Thus in the case of lactate dehydrogenase addition of 1 mol dm^{-3} Na$_2$SO$_4$ at low pH (2.3) prevents complete unfolding of the dissociated subunits (Hermann *et al.*, 1981).

Kinetics of reactivation

The concentration dependence of the rate of regain of biological activity can be used to indicate whether or not association processes are involved (folding reactions would be first-order; association reactions generally second-order). Thus in the case of the dimeric enzyme triose-phosphate isomerase, it was found that at low concentrations of enzyme ($< 0.1 \, \mu g$ cm^{-3}) reactivation after unfolding in guanidinium chloride followed second-order kinetics (Waley, 1973). This indicates that the folded monomers could not possess appreciable activity and that the rate-limiting step was association to produce an active dimer. This conclusion was confirmed and extended by Zabori *et al.* (1980). By contrast, the reactivation of the dimeric enzyme creatine kinase after unfolding in 8 mol dm^{-3} urea obeys first-order kinetics indicating in this case that the folded monomers possess enzyme activity (Grossman *et al.*, 1981). The conclusions from these types of experiments can be checked against the results of the matrix-binding technique (Chan & Mosbach, 1976) in which isolated subunits can be prepared attached covalently to a matrix such as Sepharose. (It should be noted, however, that there are a number of problems in this latter approach (Zabori *et al.*, 1980).)

Population analysis during refolding

An analysis of the kinetics of reactivation of dimeric proteins is straightforward since only one type of association step is involved. However, with more complex multi-subunit proteins such as tetramers or hexamers, it is difficult to assign such associative steps with certainty. For instance, in the case of the tetrameric enzyme fructose-1,6-bisphosphate aldolase from rabbit muscle, it was shown that an association step was involved in the regain of full activity. However, it was not possible on this basis to decide whether the fully active species was the dimer or the tetramer (Rudolph *et al.*, 1977*b*).

In order to analyse the kinetics of the association processes in more detail, and to assign properties to the individual species involved, it is necessary to obtain information on the population distribution during refolding, or to be able to isolate the intermediates involved (as in the case of lactate dehydrogenase.

The population distribution during refolding can be analysed by the elegant technique of chemical cross-linking developed by Jaenicke and his colleagues (see Fig. 2.3). A number of conditions must be fulfilled in this

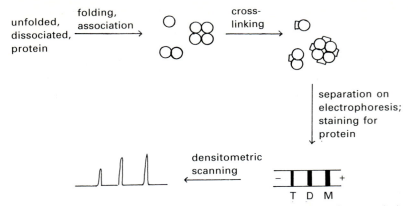

Fig. 2.3. The cross-linking technique to determine the population distribution during refolding of a tetrameric enzyme (Hermann *et al.*, 1981). After rapid cross-linking, the cross-linked species are separated on polyacrylamide gel electrophoresis in the presence of sodium dodecyl-sulphate (M, D and T represent monomer, dimer and tetramer respectively). The gel is scanned densitometrically after staining for protein.

experiment. Cross-linking must be quantitative and rapid compared with the rates of association reactions. Intermolecular cross-linking must be negligible and the reaction must not interfere with the refolding process. Detailed investigations have shown that glutaraldehyde is the reagent of choice for cross-linking (Hermann *et al.*, 1981), and the application of the technique in studies on the refolding of lactate dehydrogenase and phosphoglycerate mutase is described here. The applicability of the technique must be checked in each case; for some proteins such as mitochondrial malate dehydrogenase or rabbit muscle phosphoglycerate mutase cross-linking is not quantitative, presumably because of an unfavourable distribution of lysine side chains.

Hybridisation can also be used to determine the population distribution. In this case an excess of modified or mutant subunits is added to samples taken from the refolding mixture; these added subunits will then form hybrids with 'incomplete' intermediates. The various species formed can be separated and quantitated by gel electrophoresis. In these experiments it is necessary to show that the modified subunits do not affect the assembly mechanism, e.g. by subunit exchange (Bothwell & Schachman, 1980 *a*, *b*).

Results obtained on multi-subunit proteins

In this section, the results obtained from studies of the refolding of a number of multi-subunit proteins are summarised. Further details can be found in the reviews by Jaenicke (1982, 1987). In the next section the main conclusions from this work are outlined.

Lactate dehydrogenase The refolding of the tetrameric lactate dehydrogenase from pig heart or pig skeletal muscle after unfolding by guanidinium chloride has been studied by glutaraldehyde cross-linking and by reactivation (Jaenicke, 1987). The overall pathway is of the type:

$$4M' \xrightarrow{k_1} 4M \underset{}{\overset{\text{fast}}{\rightleftharpoons}} 2D \xrightarrow{k_2} T$$

where M′ and M represent different conformations of monomer respectively, D the dimer and T the tetramer.

The monomer $\rightleftharpoons$ dimer equilibrium is rapid ($K = 3 \times 10^8$ (mol dm^{-3})$^{-1}$), occurring at a rate close to that of a diffusion controlled reaction. The values of k_1 and k_2 are 0.8×10^{-3} s^{-1} and 3.0×10^4 (mol dm^{-3})$^{-1}$ s^{-1} respectively for the skeletal muscle enzyme. At sufficiently low concentration ($\leqslant 1$ µg cm^{-3}) the association of dimers to tetramer becomes very slow, effectively reaching completion only after 24 hours. Reactivation parallels the formation of tetramers. Circular dichroism measurements suggest that the formation of M′ in the above scheme is preceded by one or more rapid folding steps so that $\geqslant 75\%$ of the native secondary structure is regained within 10 seconds.

If the enzyme is unfolded at low pH (2.3) in the presence of 1 mol dm^{-3} Na$_2$SO$_4$, the slow M′ → M transition is bypassed and effectively only the 4M $\rightleftharpoons$ 2D → T part of the refolding process is observed.

The dimeric intermediate, D, is inactive in the standard enzyme assay. However, when assays are performed in the presence of 1.5 mol dm^{-3} (NH$_4$)$_2$SO$_4$, the kinetics of reactivation show that the dimer has considerable activity (approximately 50% of the tetramer). This conclusion has been substantiated by isolation of the intermediate (Girg *et al.*, 1983). Addition of thermolysin during refolding causes cleavage of peptide bonds in the flexible N-terminal 'arm' of the polypeptide chain. Since this 'arm' is involved in the association between dimers, the formation of tetramer is prevented. The isolated dimers show activity in the presence of (NH$_4$)$_2$SO$_4$ and possess spectroscopic properties similar to the native enzyme (Opitz *et al.*, 1987).

Phosphoglycerate mutase The refolding of the tetrameric enzyme from bakers' yeast after denaturation by guanidinium chloride has been monitored by glutaraldehyde cross-linking, by reactivation and by susceptibility to proteolysis (Hermann *et al.*, 1983, 1985; Johnson & Price, 1986). The overall pathway can be summarised as:

$$4M' \xrightarrow{\text{fast}} 4M \underset{k_{-1}}{\overset{k_1}{\rightleftharpoons}} 2D \xrightarrow{k_2} T$$

where k_1, k_{-1} and k_2 have values 6.25×10^3 (mol dm^{-3})$^{-1}$ s^{-1}, 6×10^{-3} s^{-1} and 2.75×10^4 (mol dm^{-3})$^{-1}$ s^{-1} respectively.

As in the case of lactate dehydrogenase, regain of native secondary structure is largely complete within the time of manual mixing and recording. The intermediates M and D each possess 35 % of the activity of the tetramer; however this activity, unlike that of the tetramer, is very sensitive to proteases such as trypsin, chymotrypsin and thermolysin. These proteases also cause extensive degradation of the polypeptide chains of the M and D species showing that unlike the tetramer they must possess relatively 'loose' structures. It is interesting to note that recently a monomeric form of phosphoglycerate mutase has been isolated from the fission yeast *Schizosaccharomyces pombe* (Johnson & Price, 1987). This monomeric enzyme refolds very rapidly after unfolding in guanidinium chloride.

Aspartokinase-homoserine dehydrogenase The bifunctional enzyme aspartokinase-homoserine dehydrogenase is a tetramer and possesses a well-characterised multi-domain structure (Dautry-Varsat & Garel, 1981). The aspartokinase and homoserine dehydrogenase activities are located towards the N-terminal and C-terminal parts of the polypeptide chain respectively. By proteolytic and genetic techniques it has been shown that the aspartokinase fragment is monomeric, whereas a dimeric structure is required for the expression of homoserine dehydrogenase activity. From studies of refolding of the fragments, and of the intact polypeptide chain after unfolding by guanidinium chloride, the following pathway has been proposed (Dautry-Varsat & Garel, 1981; Vaucheret *et al.*, 1987):

$$4M' \xrightarrow{k_1} 4M \xrightarrow{k_2} 2D \xrightarrow{\text{fast}} T$$

where k_1 and k_2 are 6×10^{-4} s^{-1} and 7×10^4 (mol dm^{-3})$^{-1}$ s^{-1} respectively.

In the scheme, M shows aspartokinase activity, D shows aspartokinase and homoserine dehydrogenase activities and T shows both activities and regulation by the allosteric effector, threonine.

In the refolding of this bifunctional enzyme, formation of aggregates is a major competing reaction (Vaucheret *et al.*, 1987). The yield of aspartokinase or homoserine dehydrogenase activity declines steadily as the concentration of enzyme during refolding increases from $\approx 100\%$ at 5 nmol dm^{-3} to $\approx 10\%$ at 100 nmol dm^{-3}. Thus the 'branch point' for formation of native enzyme or large aggregates must be before the domain pairing step which is required for the expression of aspartokinase activity. All the protein which has reached the 'correct' monomer structure (for expression of aspartokinase activity) will proceed further to form the native tetramer.

Apoferritin This iron-binding protein represents a considerable increase in complexity from the earlier examples in this section. The protein from horse spleen consists of 24 subunits each of relative molecular mass 18 500 which associate to form a near spherical hollow shell; the iron core is not required for the assembly process. Following unfolding under extreme conditions (6 mol dm^{-3} guanidinium chloride at high or low pH) the native protein can be reassembled on removal of the denaturing agent. Cross-linking and circular dichroism have been used to monitor the reassembly process, and a pathway involving the formation of intermediate dimers, trimers, hexamers and dodecamers has been proposed (Jaenicke, 1987):

$$24M' \rightarrow 24M_1 \rightleftharpoons 8M_2 + 8M_1 \rightleftharpoons 8M_3 \rightleftharpoons 4M_6 \rightleftharpoons 2M_{12} \rightleftharpoons M_{24}$$

Pyruvate dehydrogenase complex The multienzyme complexes from *E. coli* and pig heart (of $M_r \approx 5 \times 10^6$) can be reassembled after separation of the individual polypeptide chains or after dissociation and unfolding. The larger ($M_r \approx 10^7$) complex from *Bacillus stearothermophilus* cannot be dissociated into functional protomers, because of the instability of the pyruvate decarboxylase component. However, after denaturation of all components in stoichiometric amounts, partial reassembly can be observed (Jaenicke & Perham, 1982). The regain of overall activity is found to obey first-order kinetics, with association steps becoming significant only at very low enzyme concentrations. The nature of the rate-limiting step remains obscure; however, neither the folding of the lipoamide dehydrogenase (E3) nor the reshuffling of the lipoamide acetyl transferase (E2) core can be crucial.

Conclusions regarding the assembly of multi-subunit proteins

Overall mechanism The different examples described in the last section share a number of common features. One or more first-order folding steps occur to generate monomeric intermediates which can then associate to form dimers and larger assemblies. It is also possible that these dimers (and other intermediates) may undergo kinetically significant folding steps in order to develop the correct interfaces for further association. A general mechanism would thus involve a sequence of folding and association steps:

$$nM' \rightarrow nM \rightarrow \tfrac{n}{2}D' \rightarrow \tfrac{n}{2}D \rightarrow \tfrac{n}{4}T' \rightarrow \tfrac{n}{4}T \rightarrow \text{etc}$$

where M', D', T' represent different conformations of the monomer (M), dimer (D) and tetramer (T) respectively.

Although the rate constants for the various steps (and hence the nature of the rate-determining step) will vary between different proteins, it does seem clear that, in all cases, there is an initial folding step in which a considerable proportion of the native secondary structure of the protein is generated. This folding step occurs quickly (within a few seconds), and in some cases the folded monomer may possess biological activity. In a number of cases the rapidly formed folded monomer must undergo further slow rearrangement ('domain-pairing') before association can occur.

Jaenicke (1987) has made a compilation of the rate constants for the association steps during the assembly of a number of multi-subunit enzymes. The second-order rate constants are in the range $10^4 - 10^5$ (mol dm^{-3})$^{-1}$ s^{-1}. Comparison with the calculated rate constant for diffusion controlled association of proteins (approximately 10^8 (mol dm^{-3})$^{-1}$ s^{-1} (Koren & Hammes, 1976)) suggests that the association reactions in the assembly process are subject to a high degree of steric restriction, consistent with specific interactions at preformed 'contact' sites.

Specificity of association In the earlier discussion of multi-domain monomeric proteins it was noted that the formation of the native structure requires precise correlation (pairing) between the individual folded units (domains). Similar considerations must also apply to the association reactions between subunits, since *in vivo* the assembly of a multi-subunit protein occurs in the presence of subunits of other proteins. Evidence for the specificity of the association reactions was obtained by Cook & Koshland (1969) who studied the unfolding and refolding of mixtures of multi-subunit proteins. There was no evidence for the formation of hybrid proteins or for the influence of one protein on the refolding of another. This conclusion was substantiated in more definitive experiments by Gerl *et al.* (1985) who studied the refolding of two dimeric proteins of similar tertiary structures and whose mechanisms of folding and assembly were similar (pig mitochondrial malate dehydrogenase and D-lactate dehydrogenase from *Limulus polyphemus*). Again, no evidence for hybrid proteins was obtained, indicating that the association between subunits is highly specific. The specificity of association has also been observed in studies on the assembly of enterotoxins which possess an AB_5 quaternary structure. The A subunit appears to play a coordinating role in the assembly process (Hardy *et al.*, 1988).

Effects of ligands Studies of the effects of ligands or cofactors on the refolding of multi-subunit enzymes have often been confused and contradictory. In general, it would be expected that ligands could stabilise intermediates or the final product of the pathway, thereby shifting the equilibrium towards the native state and hence increasing the yield of native material. Since the binding of ligand with a high affinity (to act as a 'primer') requires a conformation which is at least similar to that of the native state, ligands should have little effect on the *rate* of the refolding process (for a fuller discussion of this, see Jaenicke, 1987).

Deal (1969) reported that NAD^+ was required for the refolding of the tetrameric glyceraldehyde-3-phosphate dehydrogenase. Later work (Rudolph *et al.*, 1977a; Krebs *et al.*, 1979) showed that refolding could be achieved in the absence of NAD^+ or NADH, although these ligands accelerated the process by acting on a rate-determining step late in the pathway. The lack of effects of NAD^+ or NADH on the refolding of other multi-subunit dehydrogenases (malate dehydrogenase, lactate dehydrogenase and the pyruvate dehydrogenase complex) suggests that any 'nucleating' effects of these ligands cannot be a general phenomenon.

Side reactions As mentioned previously in a number of cases refolding of multi-subunit proteins from their isolated subunits proceeds to only a very limited extent. For instance, all attempts to refold the hexameric glutamate dehydrogenase from a variety of sources have failed to regenerate enzyme activity (Müller & Jaenicke, 1980; West & Price, 1988). In other cases the yield of active proteins is highly concentration dependent, declining at high concentrations of protein, e.g. pyruvate kinase (Price & Stevens, 1983) and aspartokinase–homoserine dehydrogenase (Vaucheret *et al.*, 1987). The principal side reaction in these cases appears to be formation of 'wrong' aggregates; this is favoured at high concentrations because the association reactions leading to aggregate formation are generally of higher kinetic order than those leading to formation of the native protein.

In vivo it is likely that the occurrence of co-translational folding of the polypeptide chain would prevent the incorrect inter-actions responsible for aggregate formation in *in vitro* experiments. The diminution or elimination of slow 'domain pairing' steps would accelerate the formation of active protein. Specialised binding proteins ('chaperone' proteins) may also be involved in the assembly process *in vivo*. (It should be remembered that *in vivo* folding and assembly may not always be 100 % efficient.)

The folding and assembly of translocated proteins

Many proteins are located in intracellular compartments distinct from their site of translation, or are secreted from cells. In these cases, there are a number of possible steps (translocation, processing, folding and assembly) which may occur between translation and the arrival of the mature assembled protein at its correct destination. This section will describe some of the evidence on which our understanding of these processes is based. A review of some aspects of this topic has been presented by Pfeffer & Rothman (1987).

Signals for translocation

The studies of precursors of secreted proteins, such as immunoglobulins, albumin and lysozyme, led to the development of the 'signal hypothesis', in which a 'signal sequence' of amino acids directs the nascent polypeptide chain across the endoplasmic reticulum membrane. The signal sequence (15–35 amino acids in length) contains a positively charged amino acid near the N-terminus and a 'core' of hydrophobic amino acids; it is recognised by a signal recognition particle which helps to 'thread' the nascent chain through the membrane. The signal sequence is subsequently removed by the action of a signal peptidase located in the lumen of the endoplasmic reticulum, where the polypeptide may undergo further modification such as disulphide bond formation or glycosylation prior to secretion or delivery to lysosomes or the plasma membrane (Walter *et al.*, 1984).

Analogous signals exist to direct precursors of other proteins to their correct destination in mitochondria or chloroplasts. In the case of mitochondria the overall features of the signal (or 'targeting') sequences for the four possible destinations (matrix, inner membrane, outer membrane, inter-membrane space) have been elucidated (Hurt & van Loon, 1986). The amino terminal sequence (approximately 15 amino acids) of a precursor targeted for the matrix, for instance, contains a periodic arrangement of positively charged amino acids and is generally devoid of negatively charged amino acids. (The precursor for glutamate dehydrogenase, a matrix enzyme, seems to be an exception since it does contain two negative amino acids in the targeting sequence (Amuro *et al.* (1988).) If the 'matrix-targeting' sequence is followed by a 'stop-transport' sequence (a long, uninterrupted stretch of uncharged amino acids followed by a cluster of positively charged amino acids), transport across the outer or inner membrane is prevented and the protein becomes an outer membrane protein or a protein exposed to or released into the inter-membrane space. Most of the target sequence can be removed by proteolytic cleavage which can occur in one or two steps (Roise & Schatz, 1988; Pfanner *et al.*, 1988*a*). The translocation of proteins into, or across, the inner mitochondrial membrane requires a membrane potential and nucleoside triphosphates (see the next section).

State of proteins during translocation

The models proposed for translocation and processing envisage that proteins cross the membrane in a conformation such that signal or target sequences can be recognised and in which proteolytic cleavage sites are exposed. A number of different types of experiment have shown that proteins are indeed unfolded to a large extent during translocation (Bychkova *et al.*, 1988).

Schleyer & Neupert (1985) showed that, during translocation of the precursors of the mitochondrial matrix proteins (such as the β subunit of F1 ATPase), the amino-terminal 'targeting' sequence could be removed by the protease located in the matrix while a major proportion of the polypeptide chain was still present at the outer face of the outer membrane and susceptible to externally added proteinase K. Clearly, in this case, the protein is in an unfolded, extended conformation. Pfanner *et al.* (1988*b*) showed that the requirement for nucleoside triphosphates (ATP or GTP) for the import into mitochondria is to keep the precursor proteins in an unfolded conformation. In this experiment, the import of native and unfolded forms of porin (an outer membrane protein) was studied; only the native form required the triphosphate for import.

In an ingenious experiment, Eilers & Schatz (1986) studied the uptake into mitochondria of a fusion protein in which the residues 1–22 of the presequence of the imported protein cytochrome oxidase subunit IV were added to the N-terminus of the cytosolic enzyme dihydrofolate reductase. The resulting fused protein is imported and proteolytically processed by isolated yeast mitochondria or by mitochondria in living yeast cells. Methotrexate, which acts as a competitive inhibitor towards folate, specifically blocks the import of the fusion protein, almost certainly by binding tightly to the dihydrofolate reductase component and preventing the unfolding of the protein which is a prerequisite for transport.

If proteins are translocated in a partially or completely unfolded form, association to generate multi-subunit proteins can only occur after translocation since, as we have seen, association of subunits is highly specific. The uptake of aspartate aminotransferase into mitochondria was shown by cross-linking experiments to occur via the monomeric form; association to generate the dimeric enzyme would then occur subsequently (O'Donovan *et al.*, 1984).

A number of reports have also shown that unfolded or non-assembled proteins are retained in the lumen of the endoplasmic reticulum (Pfeffer & Rothman, 1987). For instance, the haemagglutinin of influenza virus is translocated co-translationally across the endoplasmic reticulum membrane. The folding and assembly of the protein into the mature trimeric unit could be monitored by various methods including susceptibility to proteolysis by trypsin, and cross-linking. It was found that the folding and assembly took approximately 7–10 minutes and was completed before the protein was transported to the Golgi apparatus. Mutants of the haemagglutinin which could not be transported were correlated with the accumulation of unfolded monomeric forms of the protein; these forms were found to be associated with a binding protein which had been previously shown to be capable of binding to the heavy chains of immunoglobulins (Gething *et al.*, 1986). The role of these types of binding proteins in the assembly of multi-subunit proteins is discussed later.

From the results of studies of the refolding of native (i.e. glycosylated) and non-glycosylated invertase from yeast, it has been suggested that co-translational glycosylation in the lumen of the endoplasmic reticulum may act as a further means of control of folding by preventing aggregation of partially structured protein chains and thus allowing proper refolding in the presence of high concentrations of other polypeptide chains (Schülke & Schmid, 1988). The wider applicability of this conclusion remains to be tested.

Chaperone proteins

As mentioned above, a protein which binds to the heavy chains of immunoglobins has been detected in the lumen of endoplasmic reticulum of lymphoid cells. The function of this binding protein is to assist the assembly of immunoglobulin molecules by binding to the hydrophobic surfaces of newly synthesised heavy chains and thereby prevent aggregation of these chains or association with other proteins in the endoplasmic reticulum. Addition of ATP causes the release of heavy chains from the binding protein *in vitro* (and presumably also *in vivo*), allowing any light chains to associate with the heavy chains in a controlled fashion (Munro & Pelham, 1986). It has also been suggested that the binding protein might serve to disrupt any aggregates of the heavy chains which may have formed in the lumen of the endoplasmic reticulum.

It is now clear that the immunoglobulin heavy chain binding protein is just one example of a number of proteins whose function is to direct folding and assembly processes *in vivo*; these proteins have been termed 'chaperone proteins' (Ellis, 1987; Hemmingsen *et al.*, 1988; Ellis & Hemmingsen, 1989). The heat shock protein hsp70 produced in mammalian cells shows 60% identity with the heavy chain binding protein, and may be involved in binding to, and promoting the disaggregation of, nuclear proteins which become insoluble on heat shock (Ellis, 1987). However, one of the best characterised 'chaperone proteins' participates in the assembly of ribulose bisphosphate carboxylase–oxygenase (Ellis, 1987; Hemmingsen *et al.*, 1988) which catalyses the CO_2-fixation step in photosynthesis. The enzyme from higher plants has the subunit structure L_8S_8 where the large (L) and small (S) chains are of M_r 53000 and 15000 respectively and are encoded by chloroplast and nuclear genes respectively. The precursor S chain is proteolytically processed as it is translocated across the chloroplast membrane. The L subunit is formed by processing of a precursor within the chloroplast. All attempts to reconstitute an active enzyme *in vitro* from the isolated mature L and S chains have proved unsuccessful. A binding protein has been isolated which is involved in the assembly process *in vivo*. The protein (of subunit structure $\alpha_6\beta_6$) is found to be associated with newly synthesised L subunits in chloroplast extracts and keeps the latter subunits in solution. By contrast, L subunits synthesised from genes cloned in *E. coli* form large insoluble aggregates (Roy & Cannon, 1988). The controlled release of L subunits from the complex with the binding protein would permit correct association with S subunits to form the final (L_8S_8) product. It is also possible that the binding protein could serve to store S subunits to compensate for irregularities in supplies of the S subunit. As in the case of the immunoglobulin heavy chain binding protein, release of bound L subunits from the complex appears to depend on the addition of ATP, although the equilibria involved are complex (Roy & Cannon, 1988).

Correct post-translational assembly of the head proteins of bacteriophages lambda and T4 and of the tail proteins of bacteriophage T5 requires the involvement of the *E. coli* gene products, *gro*L and *gro*ES (Hemmingsen *et al.*, 1988). The predicted amino acid sequence of the *E. coli groEL* protein is very similar (46% identity) to that of the α subunit of the chloroplast binding protein involved in assembly of ribulose bisphosphate carboxylase–oxygenase. On this basis, it has been concluded that the two proteins are homologous in evolutionary terms. In addition the heat shock protein hsp60 which appears to be related both structurally and immunologically to the *groEL* protein has been found in mitochondria from a wide variety of organisms, indicating that it might well be involved in the folding and assembly of imported proteins (Hemmingsen *et al.*, 1988). In support of this idea it has been shown that hsp60 is required for the correct assembly of various oligomeric proteins such as F1 ATPase and the Rieske Fe/S protein imported into yeast mitochondria (Cheng *et al.*, 1989). The folding of proteins has been shown to occur at the surface of hsp60 in an ATP-dependent manner, followed by release of the bound polypeptide (Ostermann *et al.*, 1989).

Conclusion

This chapter has dealt with current ideas on the folding and assembly of multi-subunit proteins. The *in vitro* studies of refolding and reassembly of completed polypeptide chains can be used to give a guide to the processes *in vivo*. From the studies *in vitro* it is clear that, as the complexity of the polypeptide chain increases, so do the possibilities of incorrect domain pairing both within and between polypeptide chains. *In vivo*, the occurrence of co-translational folding, the influence of post-translational modifications and the involvement of 'chaperone proteins' can all play a part in directing the folding and assembly of multi-subunit proteins. It is a task for future work to establish the contribution of these effects to the formation of the final product.

References

Adams, B., Burgess, R. J., Carrey, E. A., Mackintosh, I. R., Mitchinson, C., Thomas, R. N. & Pain, R. H. (1980). The role of folding units in the kinetic folding of globular proteins. In: *Protein Folding* ed. R. Jaenicke, pp. 447–67. Amsterdam; Elsevier/North Holland.

Amuro, N., Yamaura, M., Goto, M. & Okazaki, T. (1988). Molecular cloning and nucleotide sequence of the cDNA for human liver glutamate dehydrogenase precursor. *Biochemical and Biophysical Research Communications*, **152**, 1395–400.

Bergman, L. W. & Kuehl, W. M. (1979*a*). Formation of intermolecular disulphide bonds on nascent immunoglobulin polypeptides. *Journal of Biological Chemistry*, **254**, 5690–4.

Bergman, L. W. & Kuehl, W. M. (1979*b*). Formation of an intrachain disulphide bond on nascent immunoglobulin light chains. *Journal of Biological Chemistry*, **254**, 8869–76.

Bickerstaff, G. F., Paterson, C. & Price, N. C. (1980). The refolding of denatured rabbit muscle creatine kinase. *Biochimica et Biophysica Acta*, **621**, 305–14.

Bothwell, M. A. & Schachman, H. K. (1980*a*). Equilibrium and kinetic studies of the association of catalytic and regulatory subunits of aspartate transcarbamoylase. *Journal of Biological Chemistry*, **255**, 1962–70.

Bothwell, M. A. & Schachman, H. K. (1980*b*). A model for the assembly of aspartate transcarbamoylase from catalytic and regulatory subunits. *Journal of Biological Chemistry*, **255**, 1971–7.

Bulleid, N. J. & Freedman, R. B. (1988). Defective co-translational formation of disulphide bonds in protein disulphide-isomerase-deficient microsomes. *Nature*, London, **335**, 649–51.

Bychkova, V. E., Pain, R. H. & Ptitsyn, O. B. (1988). The molten globule state is involved in the translocation of proteins across membranes? *FEBS Letters*, **238**, 231–4.

Chan, W. W.-C., Mort, J. S., Chong, D. K. K. & Macdonald, P. D. M. (1973). Studies on protein subunits. III. Kinetic evidence for the presence of active subunits during the renaturation of muscle aldolase. *Journal of Biological Chemistry*, **248**, 2778–84.

Chan, W. W.-C. & Mosbach, K. (1976). Effects of subunit interactions on the activity of lactate dehydrogenase studied in immobilised enzyme systems. *Biochemistry*, **15**, 4215–22.

Cheng, M. Y., Hartl, F.-U., Martin, J., Pollock, R. A., Kalousek, F., Neupert, W., Hallberg, E. M., Hallberg, R. L. & Horwich, R. L. (1989). Mitochondrial heat-shock protein hsp60 is essential for assembly of proteins imported into yeast mitochondria. *Nature, London*, **337**, 620–5.

Chou, P. Y. & Fasman, G. D. (1978). Prediction of the secondary structure of proteins from their amino acid sequence. *Advances in Enzymology*, **47**, 45–148.

Cook, R. A. & Koshland, D. E. Jr. (1969). Specificity in the assembly of multisubunit proteins. *Proceedings of the National Academy of Sciences, USA*, **64**, 247–54.

Creighton, T. E. (1978). Experimental studies of protein folding and unfolding. *Progress in Biophysics and Molecular Biology*, **33**, 231–97.

Darnell, J. E., Lodish, H. & Baltimore, D. (1986). *Molecular Cell Biology*, p. 121. New York: Scientific American Books.

Dautry-Varsat, A. & Garel, J.-R. (1981). Independent folding regions in aspartokinase-homoserine dehydrogenase. *Biochemistry*, **20**, 1396–401.

Deal, W. C. Jr. (1969). Metabolic control and structure of glycolytic enzymes. IV. Nicotinamide-adenine dinucleotide dependent *in vitro* reversal of dissociation and possible *in vivo* control of yeast glyceraldehyde 3-phosphate dehydrogenase synthesis. *Biochemistry*, **8**, 2795–805.

Eilers, M. & Schatz, G. (1986). Binding of a specific ligand inhibits import of a purified precursor protein into mitochondria. *Nature*, London, **322**, 228–32.

Ellis, R. J. (1987). Proteins as molecular chaperones. *Nature*, London, **328**, 378–9.

Ellis, R. J. & Hemmingsen, S. M. (1989). Molecular chaperones: proteins essential for the biogenesis of some macromolecular structures. *Trends in Biochemical Sciences*, **14**, 339–42.

Fersht, A. R. (1987). Dissection of the structure and activity of the tyrosyl-tRNA synthetase by site-directed mutagenesis. *Biochemistry*, **26**, 8031–7.

Fischer, G., Wittman-Liebold, B., Lang, K., Kiefhaber, T. & Schmid, F. X. (1989). Cyclophilin and peptidyl-prolyl *cis–trans* isomerase are probably identical proteins. *Nature*, London, **337**, 476–8.

Freedman, R. B. (1984). Native disulphide bond formation in protein biosynthesis: evidence for the role of protein disulphide isomerase. *Trends in Biochemical Sciences*, **9**, 438–41.

Fry, D. C., Kuby, S. A. & Mildvan, A. S. (1985). NMR studies of the MgATP binding site of adenylate kinase and of a 45-residue peptide fragment of the enzyme. *Biochemistry*, **24**, 4680–94.

Garel, J.-R. & Baldwin, R. L. (1973). Both the fast and slow refolding reactions of ribonuclease A yield native enzyme. *Proceedings of the National Academy of Sciences, USA*, **70**, 3347–51.

Garnier, J., Osguthorpe, D. J. & Robson, B. (1978). Analysis of the accuracy and implications of simple methods for predicting the secondary structure of globular proteins. *Journal of Molecular Biology*, **120**, 97–120.

Gerl, M., Rudolph, R. & Jaenicke, R. (1985). Mechanism and specificity of reconstitution of dimeric lactate dehydrogenase from *Limulus polyphemus*. *Biological Chemistry Hoppe-Seyler*, **366**, 447–54.

Gething, M.-J., McCammon, K. & Sambrook, J. (1986). Expression of wild-type and mutant forms of influenza haemagglutinin: the role of folding in intracellular transport. *Cell*, **46**, 939–50.

Ghélis, C. (1980). Transient conformational states in proteins followed by differential labelling. *Biophysical Journal*, **32**, 503–14.

Ghélis, C. & Yon, J. (1982). *Protein Folding*. New York, Academic Press.

Gilmanshin, R. I. & Ptitsyn, O. B. (1987). An early intermediate of refolding α-lactalbumin forms within 20 ms. *FEBS Letters*, **223**, 327–9.

Girg, R., Rudolph, R. & Jaenicke, R. (1981). Limited proteolysis of porcine-muscle lactic dehydrogenase by thermolysin during reconstitution yields dimers. *European Journal of Biochemistry*, **119**, 301–5.

Girg, R., Jaenicke, R. & Rudolph, R. (1983). Dimers of porcine skeletal muscle lactate dehydrogenase produced by limited proteolysis during reassociation are enzymatically active in the presence of stabilizing salt. *Biochemistry International*, **7**, 433–41.

Goldenberg, D. P. (1988). Kinetic analysis of the folding and unfolding of a mutant form of bovine pancreatic trypsin inhibitor lacking the cysteine-14 and -38 thiols. *Biochemistry*, **27**, 2481–9.

Goldenberg, D. P., Berget, P. B. & King, J. (1982). Maturation of the tailspike endorhamnosidase of *Salmonella* phage P22. *Journal of Biological Chemistry*, **257**, 7864–71.

Goldenberg, D. P., Smith, D. H. & King, J. (1983). Genetic analysis of the folding pathway for the tail spike proteins of phage P22. *Proceedings of the National Academy of Sciences, USA*, **80**, 7060–4.

Grossman, S. H., Pyle, J. & Steiner, R. J. (1981). Kinetic evidence for active monomers during the reassembly of denatured creatine kinase. *Biochemistry*, **20**, 6122–8.

Hamlin, J. & Zabin, I. (1972). β-galactosidase: immunological activity of ribosome-bound growing polypeptide chains. *Proceedings of the National Academy of Sciences, USA*, **69**, 412–16.

Hardy, S. J. S., Holmgren, J., Johansson, S., Sanchez, J. & Hirst, T. R. (1988). Coordinated assembly of multisubunit proteins: oligomerization of bacterial enterotoxins *in vivo* and *in vitro*. *Proceedings of the National Academy of Sciences, USA*, **85**, 7109–13.

Hemmingsen, S. M., Woolford, C., van der Vries, S. M., Tilly, K., Dennis, D. T., Georgopoulos, C. P., Hendrix, R. W. & Ellis, R. J. (1988). Homologous plant and bacterial proteins chaperone oligomeric protein assembly. *Nature*, London, **333**, 330–4.

Hermann, R., Jaenicke, R. & Rudolph, R. (1981). Analysis of the reconstitution of oligomeric enzymes by cross-linking with glutaraldehyde: kinetics of reassociation of lactic dehydrogenase. *Biochemistry*, **20**, 5195–201.

Hermann, R., Rudolph, R., Jaenicke, R., Price, N. C. & Scobbie, A. (1983). The reconstruction of denatured phosphoglycerate mutase. *Journal of Biological Chemistry*, **258**, 11014–19.

Hermann, R., Jaenicke, R. & Price, N. C. (1985). Evidence for active intermediate during the reconstitution of yeast phosphoglycerate mutase. *Biochemistry*, **24**, 1817–21.

Holley, L. H. & Karplus, M. (1989). Protein secondary structure prediction with a neural network. *Proceedings of the National Academy of Sciences, USA*, **86**, 152–6.

Hurt, E. C. & van Loon, A. P. G. M. (1986). How proteins find mitochondria and intramitochondrial compartments. *Trends in Biochemical Sciences*, **11**, 204–7.

Jaenicke, R. (1982). Folding and association of proteins. *Biophysics of Structure and Mechanism*, **8**, 231–56.

Jaenicke, R. (1987). Folding and association of proteins. *Progress in Biophysics and Molecular Biology*, **49**, 117–237.

Jaenicke, R. & Perham, R. N. (1982). Reconstitution of the pyruvate dehydrogenase multienzyme complex from *Bacillus stearothermophilus*. *Biochemistry*, **21**, 3378–85.

Johnson, C. M. & Price, N. C. (1986). The susceptibility towards proteolysis of intermediates during the renaturation of yeast phosphoglycerate mutase. *Biochemical Journal*, **236**, 617–20.

Johnson, C. M. & Price, N. C. (1987). Denaturation and renaturation of the monomeric phosphoglycerate mutase from *Schizosaccharomyces pombe*. *Biochemical Journal*, **245**, 525–30.

Kabsch, W. & Sander, C. (1983). How good are predictions of protein secondary structure? *FEBS Letters*, **155**, 179–82.

Karplus, M. & Weaver, D. L. (1976). Protein-folding dynamics. *Nature*, London, **260**, 404–6.

Kiho, Y. & Rich, A. (1964). Induced enzyme formed on bacterial polyribosomes. *Proceedings of the National Academy of Sciences, USA*, **51**, 111–8.

Kim, P. S. (1986). Amide proton exchange as a probe of protein folding pathways. *Methods in Enzymology*, **131**, 136–56.

Kim, P. S. & Baldwin, R. L. (1982). Specific intermediates in the folding reactions of small proteins and the mechanism of protein folding. *Annual Review of Biochemistry*, **51**, 459–89.

Koivu, J. & Myllylä, R. (1987). Interchain disulphide bond formation in types I and II procollagen. *Journal of Biological Chemistry*, **262**, 6159–64.

Koren, R. & Hammes, G. G. (1976). A kinetic study of protein–protein interactions. *Biochemistry*, **15**, 1165–71.

Krebs, H., Rudolph, R. & Jaenicke, R. (1979). Influence of coenzyme on the refolding and reassociation *in vitro* of glyceraldehyde-3-phosphate dehydrogenase from yeast. *European Journal of Biochemistry*, **100**, 359–64.

Labhardt, A. M. (1986). Folding intermediates studied by circular dichroism. *Methods in Enzymology*, **131**, 126–35.

Lang, K., Schmid, F. X. & Fischer, G. (1987). Catalysis of protein folding by prolyl isomerase. *Nature*, London, **329**, 268–70.

Liao, M.-J. & Khorana, H. G. (1984). Removal of the carboxyl-terminal peptide does not affect refolding or function of bacteriorhodopsin as a light-dependent proton pump. *Journal of Biological Chemistry*, **259**, 4194–9.

Lim, V. I. (1974). Algorithms for prediction of α-helical and β-structural regions in globular proteins. *Journal of Molecular Biology*, **88**, 873–94.

Marston, F. A. O. (1986). The purification of eukaryotic polypeptides synthesised in *Escherichia coli*. *Biochemical Journal*, **240**, 1–12.

McNally, T., Purvis, I. J., Fothergill-Gilmore, I. A. & Brown, A. J. P. (1989). The yeast pyruvate kinase gene does not contain a string of non-preferred codons: revised nucleotide sequence. *FEBS Letters*, **247**, 312–16.

Müller, K. & Jaenicke, R. (1980). Denaturation and renaturation of bovine liver glutamate dehydrogenase after dissociation in various denaturants. *Zeitschrift für Naturforschung*, **35c**, 222–8.

Munro, S. & Pelham, H. R. B. (1986). An hsp70-like protein in the ER: identity with the 78 kd glucose-regulation protein and immunoglobulin heavy chain binding protein. *Cell*, **46**, 291–300.

O'Donovan, K. M. C., Doonan, S., Marra, E. & Passaella, S. (1984). Import of aspartate aminotransferase into mitochondria. *Biochemical Society Transactions*, **12**, 444–5.

Opitz, U., Rudolph, R., Jaenicke, R., Ericsson, L. & Neurath, H. (1987). Proteolytic dimers of porcine muscle lactate dehydrogenase: characterisation, folding and reconstitution of the truncated and nicked polypeptide chain. *Biochemistry*, **26**, 1399–406.

Ostermann, J., Horwich, A. L., Neupert, W. & Hartl, F.-U. (1989). Protein folding in mitochondria requires complex formation with hsp60 and ATP hydrolysis. *Nature*, London, **341**, 125–30.

Perry, K. M., Onuffer, J. J., Touchette, N. A., Herndon, C. S., Gittelman, M. S., Matthews, C. R., Chen, J.-T., Mayer, R. J., Taira, K., Benkovic, S. J., Howell, E. E. & Kraut, J. (1987). Effect of single amino acid replacements on the folding and stability of dihydrofolate reductase from *Escherichia coli*. *Biochemistry*, **26**, 2674–82.

Peters, T. Jr. & Davidson, L. K. (1982). The biosynthesis of rat serum albumin. *Journal of Biological Chemistry*, **257**, 8847–53.

Pfanner, N., Pfaller, R. & Neupert, W. (1988a). How finicky is mitochondrial protein import? *Trends in Biochemical Sciences*, **13**, 165–7.

Pfanner, N., Pfaller, R., Kleene, R., Ito, M., Tropschug, M. & Neupert, W. (1988b). Role of ATP in mitochondrial protein import. *Journal of Biological Chemistry*, **263**, 4049–51.

Pfeffer, S. R. & Rothman, J. E. (1987). Biosynthetic protein transport and sorting by the endoplasmic reticulum and Golgi. *Annual Review of Biochemistry*, **56**, 829–52.

Price, N. C. & Stevens, E. (1983). The refolding of denatured rabbit muscle pyruvate kinase. *Biochemical Journal*, **209**, 763–70.

Ptitsyn, O. R. (1987). Protein folding: hypotheses and experiments. *Journal of Protein Chemistry*, **6**, 273–93.

Purvis, I. J., Bettany, A. J. E., Santiago, T. C., Coggins, J. R., Duncan, K., Eason, R. & Brown, A. J. P. (1987). The efficiency of folding of some proteins is increased by controlled rates of translation *in vivo*. A hypothesis. *Journal of Molecular Biology*, **193**, 413–17.

Roder, H., Elöve, G. A. & Englander, S. W. (1988). Structural characterization of folding intermediates in cytochrome *c* by H-exchange labelling and proton NMR. *Nature*, London, **335**, 700–4.

Roise, D. & Schatz, G. (1988). Mitochondrial presequences. *Journal of Biological Chemistry*, **263**, 4509–11.

Rossi, J. & Zoller, M. (1987). Site-specific and regionally directed mutagenesis of protein-encoding sequences. In *Protein Engineering*, ed. D. L. Oxender and C. F. Fox, pp. 51–63. New York: Alan R. Liss, Inc.

Roth, R. A. & Pierce, S. B. (1987). *In vivo* cross-linking of protein disulphide isomerase to immunoglobulins. *Biochemistry*, **26**, 4179–82.

Roy, H. & Cannon, S. (1988). Ribulose bisphosphate carboxylase assembly. What is the role of the large subunit binding protein? *Trends in Biochemical Sciences*, **13**, 163–5.

Rudolph, R. & Fuchs, I. (1983). Influence of glutathione on the reactivation of enzymes containing cysteine or cystine. *Hoppe-Seylers Zeitschrift für physiologische Chemie*, **364**, 813–20.

Rudolph, R., Bohrer, M. & Fischer, S. (1983). Physicochemical characterisation of a fast refolding monomeric class I fructose-1,6-bisphosphate aldolase from *Staphylococcus aureus*. *European Journal of Biochemistry*, **131**, 383–6.

Rudolph, R., Heider, I. & Jaenicke, R. (1977*a*). Mechanism of reactivation and refolding of glyceraldehyde-3-phosphate dehydrogenase from yeast after denaturation and dissociation. *European Journal of Biochemistry*, **81**, 563–70.

Rudolph, R., Westhof, E. & Jaenicke, R. (1977*b*). Kinetic analysis of the reactivation of rabbit muscle aldolase after denaturation with guanidine-HCl. *FEBS Letters*, **73**, 204–6.

Sawyer, L., Fothergill-Gilmore, L. A. & Freemont, P. S. (1988). The predicted secondary structures of class I fructose-bisphosphate aldolases. *Biochemical Journal*, **249**, 789–93.

Schleyer, M. & Neupert, W. (1985). Transport of proteins into mitochondria: translocational intermediates spanning contact sites between outer and inner membranes. *Cell*, **43**, 339–50.

Schmid, F. X., Grafl, R., Wrba, A. & Beintema, J. J. (1986). Role of proline peptide bond isomerisation in unfolding and refolding of ribonuclease. *Proceedings of the National Academy of Sciences, USA*, **83**, 872–6.

Schülke, N. & Schmid, F. X. (1988). Effect of glycosylation on the mechanism of renaturation of invertase from yeast. *Journal of Biological Chemistry*, **263**, 8832–7.

Shen, L. L. & Hermans, J. Jr. (1972). Kinetics of conformation change of sperm-whale myoglobin. I. Folding and unfolding of metmyoglobin following pH jump. *Biochemistry*, **11**, 1836–41.

States, D. J., Creighton, T. E., Dobson, C. M. & Karplus, M. (1987). Conformation of intermediates in the folding of the pancreatic trypsin inhibitor. *Journal of Molecular Biology*, **195**, 731–9.

Takahashi, N., Hayano, T. & Suzuki, M. (1989). Peptidyl-prolyl *cis–trans* isomerase is the cyclosporin A-binding protein cyclophilin. *Nature*, London, **337**, 473–5.

Tandon, S. & Horowitz, P. (1988). The effects of lauryl maltoside on the reactivation of several enzymes after treatment with guanidinium chloride. *Biochimica et Biophysica Acta*, **955**, 19–25.

Taniuchi, H. (1970). Formation of randomly paired disulphide bonds in des-(121–124)-ribonuclease after reduction and reoxidation. *Journal of Biological Chemistry*, **245**, 5459–68.

Taniuchi, H. & Anfinsen, C. B. (1969). An experimental approach to the study of folding of Staphylococcal nuclease. *Journal of Biological Chemistry*, **244**, 3864–75.

Teipel, J. W. & Koshland, D. E. Jr. (1971). Kinetic aspects of conformational changes in proteins. II. Structural changes in reactivation of denatured proteins. *Biochemistry*, **10**, 798–805.

Teschner, W., Rudolph, R. & Garel, J.-R. (1987). Intermediates on the folding pathway of octopine dehydrogenase from *Pecten jacobaeus*. *Biochemistry*, **26**, 2791–6.

Tsou, C.-L. (1988). Folding of the nascent peptide chain into a biologically active protein. *Biochemistry*, **27**, 1809–12.

Udgaonkar, J. B. & Baldwin, R. L. (1988). NMR evidence for an early framework intermediate on the folding pathway of ribonuclease A. *Nature, London*, **335**, 694–9.

Vaucheret, H., Signon, L., Le Bras, G. & Garel, J.-R. (1987). Mechanism of renaturation of a large protein, aspartokinase-homoserine dehydrogenase. *Biochemistry*, **26**, 2785–90.

Waley, S. G. (1973). Refolding of triose phosphate isomerase. *Biochemical Journal*, **135**, 165–72.

Walter, P., Gilmore, R. & Blobel, G. (1984). Protein translocation across the endoplasmic reticulum. *Cell*, **38**, 5–8.

West, S. M. & Price, N. C. (1988). The unfolding and refolding of glutamate dehydrogenases from bovine liver, baker's yeast and *Clostridium symbiosum*. *Biochemical Journal*, **251**, 135–9.

Wetlaufer, D. B. (1981). Folding of protein fragments. *Advances in Protein Chemistry*, **34**, 61–92.

Wright, P. E., Dyson, H. J. & Lerner, R. A. (1988). Conformation of peptide fragments of proteins in aqueous solution: implications for initiation of protein folding. *Biochemistry*, **27**, 7167–75.

Zabori, S., Rudolph, R. & Jaenicke, R. (1980). Folding and association of triose phosphate isomerase from rabbit muscle. *Zeitschrift für Naturforschung*, **35c**, 999–1004.

Zipser, D. (1963). Ribosome-bound β-galactosidase of *Escherichia coli*. *Journal of Molecular Biology*, **7**, 739–51.

3
Allosteric proteins

L. N. JOHNSON

Introduction

The inclusion of a chapter on allosteric enzymes in a book on receptors perhaps requires some justification. In simple terms, ligand binding to a cell surface receptor triggers changes at the intracellular region that result in one of three types of events. Ligand–receptor binding can lead to protein–protein association as in the binding of the G proteins to β-adrenergic receptors or rhodopsin; or to increased enzyme activity as in tyrosine kinase activity stimulated by the binding of insulin or epidermal growth factor to their respective receptors; or to changes in selective ion permeability arising from channel opening as in the acetylcholine receptor. Each response is characterised by long-range communication. Commonly, the activity of cell receptors is correlated with changes in association of protein molecules in response to ligand binding. Thus long-range communication between sites may be mediated by subunit–subunit interactions that are similar to those involved in the transmission of information in allosteric enzymes. Up to 1987, haemoglobin was the only protein for which detailed structural information on the allosteric transitions was available. In 1988, high resolution refined structures were reported for liganded and unliganded complexes of three further proteins: aspartate carbamoyltransferase, phosphofructokinase and glycogen phosphorylase. The new structures have greatly increased the basic knowledge of site–site communication. This chapter will describe these advances with special reference to those features that may be relevant for signal transduction. In addition, there is one example, glycogen phosphorylase, of an enzyme that is controlled by reversible phosphorylation, a phenomenon of relevance to receptor action in connection with autophosphorylation and with cellular events triggered by phosphorylation of exogenous substrates. A masterly comprehensive review of allosteric proteins has recently appeared (Perutz, 1989).

To a first approximation, all four proteins conform to the Monod, Wyman and Changeux (1965) model (MWC model) for allosteric transitions. They all fulfil the basic assumptions of the unifying theory. All four proteins are oligomers, with the individual subunits associated about at least one axis of symmetry. The symmetry of the binding sites is the same as the symmetry of the oligomeric structure. Each enzyme exhibits a definite tertiary structure which is associated with a specific quaternary structure. Two (or at least two) quaternary structures are accessible to these enzymes and these states differ in the distribution and/or energy of the inter-subunit interactions. As a result, the affinity of one (or several) of the stereospecific sites is altered when a transition occurs from one to another state. The two states are designated T and R. The T state exhibits low affinity for substrate and allosteric activators and a high affinity for inhibitors. The R state is stabilised by, and exhibits high affinity for, substrates and activators and low affinity for inhibitors. In general, symmetry is conserved as the enzymes switch from one state to another, but some small deviations from symmetry have been noted for T state aspartate carbamoyltransferase. The MWC model accounts for both homotropic and heterotropic effects and for their interdependence, and predicts these interactions solely on the basis of symmetry considerations.

In the MWC terminology, L is the equilibrium constant that describes the ratio between the concentrations of the T and R state molecules in the absence of ligand ($L = T_o/R_o$); c is the ratio of the dissociation constants of ligand for the R and T states ($c = K_R/K_T$). Homotropic effects refer to interactions between sites that bind the same ligand; heterotropic effects refer to interactions between sites that bind different ligands. Some values for these constants are given in Table 3.5 and are discussed further on pp. 82–3. The X-ray studies define stable structural states amenable to crystallisation. As yet, the crystallographic analysis cannot provide information on the pathway between these states.

The MWC model contains no proposals about the nature of the binding sites nor the changes in protein tertiary structure needed to accomplish the change in affinity. Examination of the known structures of the four allosteric proteins reveals definite changes in tertiary structure that are associated with ligand binding. The ligand binding sites are located either directly at a subunit–subunit interface or at a site connected to the interface so that binding is closely associated with changes at the interface. In this respect the enzymes show properties envisaged in the sequential model of Koshland (Koshland, Nemethy & Filmer, 1966; Koshland, 1969) and the induced fit theory of enzyme action (Koshland, 1958; Koshland & Neet, 1969).

There are many examples of changes in tertiary structure induced by ligand binding to proteins which are not allosterically regulated. Firstly, most protein structures are organised into domains with the catalytic site located at the domain interface. Structural changes often occur through domain movements (Janin & Wodak, 1983; Lesk & Chothia, 1984; Bennett & Huber, 1984). For example, in many two substrate reactions, the substrates need to be brought together and removed from bulk solvent in order for reaction to occur. In hexokinase, citrate synthase, liver alcohol dehydrogenase and glyceraldehyde 3-phosphate dehydrogenase (Steitz *et al.*, 1981; Wiegand *et al.*, 1984; Eklund *et al.*, 1981; Sharzynski & Wonacott, 1988) this is achieved by each substrate binding to a separate domain of the enzyme and triggering domain closure. Secondly, structural changes leading to correct substrate recognition sites and correct orientation of groups to stabilise the transition state may be brought about by mobile to rigid transformations as for example in the activation of the serine proteinases from their zymogen precursors (Huber & Bode, 1978). Thirdly, the recent studies with the Trp repressor/operator system has demonstrated how binding of a small ligand (L-tryptophan) can trigger structural changes remote from the binding site so that the dimer molecule is activated to bind operator DNA (Zhang *et al.*, 1987; Otwinowski *et al.*, 1988; Marmorstein & Sigler, 1988). The X-ray studies show that L-trp induces large tertiary, but no quaternary, structural changes that enable the flexible reading heads of the helix–turn–helix motifs to bind to successive grooves of the DNA.

The four examples of allosteric protein structures show how ligand-induced tertiary structural changes are correlated with quaternary structural changes. These may provide a model for receptor response. In the β-adrenergic receptor, association of agonist at an extracellular site stimulates the association of the G protein at an intracellular site. In turn, the agonist–receptor–G protein complex favours a change in the G protein guanine nucleotide binding site from closed (in the absence of agonist-receptor) to open, allowing GTP/GDP exchange. Likewise, the guanine nucleotide site affects the agonist binding site promoting a high affinity agonist site when GDP is dissociated and a low affinity site when GDP is associated in the receptor–G protein complex (evidence reviewed by Gilman, 1987). It seems plausible that the effects between distant binding sites are mediated through changes at the receptor/G protein interface. In the epidermal growth factor receptor, binding of the hormone at the extracellular domain results in activation of the intracellular tyrosine

Table 3.1. *Crystal structures of haemoglobin*

State	Conditions	Space group and unit cell dimensions	Resolution (and crystallographic R factor)	Reference
T	Human deoxy (ammonium sulphate)	$P2_1$ a = 63.2 Å b = 83.6 Å c = 53.8 Å β = 99.3 Å asymmetric unit tetramer	1.74 Å (0.16)	Fermi *et al.*, 1984
R	Human oxy (3 M phosphate buffer pH 8.5)	$P4_12_12$ a = b = 53.7 Å c = 193.8 Å asymmetric unit one $\alpha\beta$ dimer	2.1 Å (0.223)	Shanaan, 1983

kinase domain but the two domains are linked only through a single membrane spanning helix. The allosteric oligomerisation model (Schlessinger, 1988) suggests that, instead of transmission through the membrane region, activation is triggered by association of subunits that bring neighbouring cytoplasmic domains into proximity. In the acetylcholine receptor, the ring of five homologous membrane spanning subunits (Changeux *et al.*, 1984, 1987) responds to chemical stimuli by opening a water filled channel that is situated within the membrane at a distance of about 60 Å from the ligand-binding site (Toyoshima & Unwin, 1988). In this instance, a ligand-induced change resulting in helix/helix movements at the subunit interfaces could provide a mechanism for pore opening (Giraudat *et al.*, 1987; Unwin *et al.*, 1988). The intimate connection between induced local tertiary structural changes with changes at the subunit interfaces that are observed with allosteric proteins provides some clues as to how signal transduction might be effected with receptor molecules.

Enzymes

Haemoglobin

Haemoglobin, an honorary enzyme, has been until recently the only structural model for an allosteric protein. The fundamental work of Perutz (1970) led to a description of the interactions that stabilise the T (deoxy)

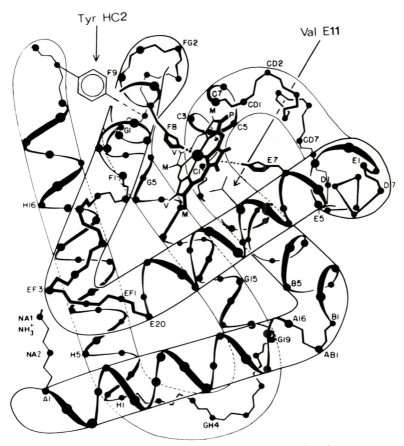

Fig. 3.1. A schematic diagram of the secondary and tertiary structure of haemoglobin subunits showing α carbons and coordination of the haem. The diagram shows the proximal histidine F8 linked to the haem iron, the distal residues, His E7 and Val E11, and also Tyr HC2, which are important in the haemoglobin mechanism. (From Perutz, 1987.)

and R (oxy) states. Refinements of these structures (Table 3.1) have substantiated these proposals and allowed further details of the basic model to be described (Baldwin & Chothia, 1979; Imai, 1982; Dickerson & Geiss, 1983; Perutz, 1987). The association of the two α and β subunits to form a tetramer provides a mechanism for decreasing the oxygen affinity of the haem groups by about 100-fold relative to the affinity of the free subunits. By responding in conformation between the T and R states, the protein is able to react to small changes in oxygen concentration encountered between the lungs and the tissues.

Each subunit (141 amino acids α chains and 146 amino acids β chains) is

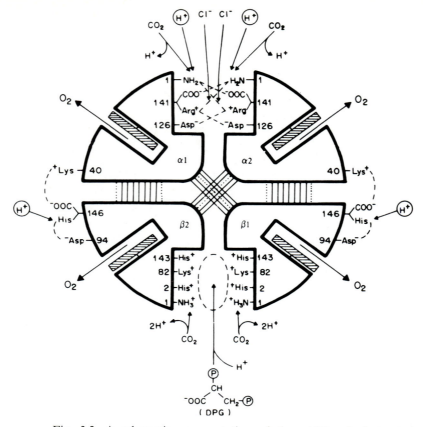

Fig. 3.2. A schematic representation of the additional electrostatic interactions that characterise T state haemoglobin. Arrows indicate the binding sites for non-haem ligands and also the direction of their reactions when oxygen molecules leave the haem groups. The numbers attached to the amino acid residues are the sequence numbers. Circled protons are those binding to the Bohr groups. Salt bridges, van der Waals contacts and hydrogen bonds are indicated by broken line, solid line, and dotted line, respectively. (From Imai, 1982.) Latest results indicate that the carboxylate of the C-terminal residue Arg 141 (α1) is linked to the side chain of Lys 127(α2) (Perutz, 1989).

composed of 8 α helices, labelled A to H with the haem group embedded in a hydrophobic pocket and linked to the protein by a bond between the iron and His F8 (Fig. 3.1). In the tetramer, the main interactions are across the $\alpha_1\beta_1$ and $\beta_1\beta_2$ interfaces with few contacts between like subunits. Caged-in water molecules form an essential part of the interactions. There are marked changes in quaternary structure between T and R states. In the T

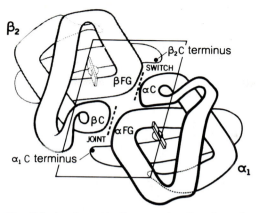

Fig. 3.3. A schematic diagram showing the $\alpha_1\beta_2$ contact joint and switch at the subunit interface of haemoglobin. (From Dickerson & Geiss, 1983.)

state, the cavity around the central two-fold symmetry axis has the right dimension to accommodate the allosteric inhibitor 2,3-diphosphoglycerate (DPG) which binds with 1:1 stoichiometry making strong ionic interactions with basic groups at the $\beta_1\beta_2$ interface (Fig. 3.2). In the R state, the cavity is narrowed so that DPG is excluded.

The T state quaternary structure is stabilised by salt links. At the $\alpha_1\alpha_2$ interface, the carboxylate group and the side chain of the terminal residue Arg 141 are linked to the side chains of Lys 127 and Asp 126, respectively, of the other α chain (Fig. 3.2). The contact between arginine $(\alpha_1)/\alpha$-amino (α_2) is stabilised by a bridging chloride ion. At the $\alpha_1\beta_2$ interface, the C terminal carboxylate and side chain of His 146 (β_2) are linked to Lys 40 (α_1) and to Asp 94 (β_2) respectively. At physiological pH, hydrogen ions, chloride ions and carbon dioxide decrease the oxygen affinity by lowering the oxygen affinity of the T state and retarding the T to R transition. Substantial experimental evidence supports the structural proposals that every one of these heterotropic ligands acts by fortifying the salt bridges of the T structure (Perutz, 1987) (Fig. 3.2).

On oxygenation, the subunits rotate relative to each other by about 15°. The haems change their angles of tilt by a few degrees and the helical regions at the $\alpha_1\beta_2$ interface move by 2 to 3 Å relative to each other (Fig. 3.3). Helix $C\beta_2$ which is in contact with the $FG\alpha_1$ corner forms a ball and socket joint around which the dimers turn, while helix $C\alpha_1$ slides relative to $FG\beta_2$. Together, the two contacts form a two-way switch that ensures that the $\alpha\beta$ dimers click back and forth between two stable positions, T and R.

Fig. 3.4. A stereo view of the oxygen binding sites of the α subunits of human haemoglobin. The oxy and deoxy structures are represented by full and broken lines, respectively. The proximal histidine F8 moves towards the haem as the iron moves into the plane of the haem to take up oxygen. The distal histidine E7 moves to form a hydrogen bond with the oxygen. (From Perutz, 1987.)

The two quaternary structural states are maintained even in hybrid structures such as α(oxy)β(deoxy) (Brzowowski *et al.*, 1984) or liganded T state haemoglobin (Liddington *et al.*, 1988) and no intermediate quaternary states have been observed. On oxygenation, the salt bridges that constrain the α and β chains in the T state are broken.

The changes in quaternary structure are triggered by relatively small changes in tertiary structure that are triggered by oxygen binding to the haem iron. In the T state, the iron is 5-coordinated, high spin state and displaced from the mean plane of the porphyrin by 0.58 Å (α chains) and 0.50 Å (β chains). Access to the haem pocket is barred by the distal His E7 and Phe CD1 in both chains and by Val E11 in the β chains. In order to admit oxygen, these residues have to move, resulting in concerted movements of helices A and E relative to the haem. Further, when iron combines firmly with oxygen, it shifts to the low spin state and moves into the plane of the porphyrin taking the proximal histidine and helix F with it. This is the vital movement that triggers the T to R transition (Fig. 3.4). Helix F transmits the movements from the iron to the FG segments. As a result, the FG segment of α_1 moves 2.5 Å closer to the FG segment of β_1 and there is a change of packing at the subunit interface.

In terms of allosteric theory, the model that emerges from the X-ray studies conforms well to the MWC model with changes in tertiary structure firmly linked with quaternary structural changes. A structural understanding of the restraints on the T state structure provides a pleasing explanation of how seemingly disparate ligands (H^+, Cl^-, DPG and CO_2) can exert their effects by binding at sites distinct from the oxygen binding site. Within this framework, there are some deviations from the MWC model in that the T state structures are able to exhibit a variation in affinity for oxygen (K_T) as a function of allosteric ligand concentration (Table 3.5). These T state structures presumably differ by subtle changes in tertiary structure that are governed by the strong restraints on the T state quaternary structure (Perutz, 1987).

Aspartate carbamoyltransferase

Aspartate carbamoyltransferase (EC 2.1.3.2) is probably the most studied regulatory protein, after haemoglobin, and many of the proposals for the roles of individual side chains have been tested by site-directed mutagenesis (Kantrowitz & Lipscomb, 1988). The enzyme catalyses the formation of carbamoyl-L-aspartate and phosphate from carbamoyl phosphate and L-aspartate, the first step in pyrimidine synthesis for many prokaryotes (Fig. 3.5). The binding of substrates and activity is regulated homotropically

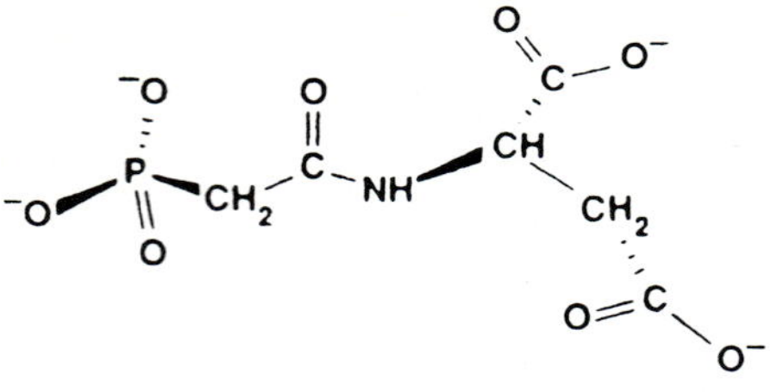

N-phosphonacetyl-L-aspartate (PALA)

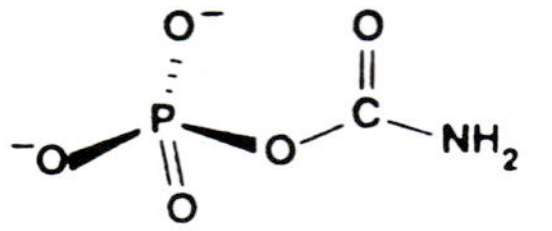

Carbamyl phosphate

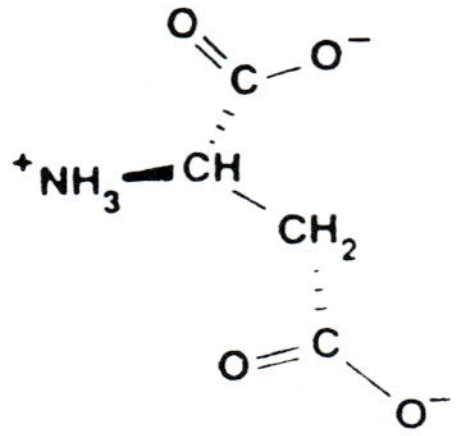

L-aspartate

Fig. 3.5. The structures of the two substrates of the reaction catalysed by aspartate carbamoyltransferase and PALA, the stable tightly bound inhibitor. (From Kantrowitz & Lipscomb, 1988.)

(Gerhart & Pardee, 1962; Bethell *et al.*, 1968) and heterotropically by nucleotide triphosphates. CTP, the final product of the pyrimidine pathway, is an allosteric inhibitor and ATP, the product of a parallel pathway of purine synthesis, is an activator of catalysis.

The enzyme is a dodecamer with total molecular weight 310000 and is composed of two types of chains. The catalytic chains (C), M_r 33000, bind substrates and the regulatory chains (R), M_r 17000 bind nucleoside triphosphate effectors. The chains are combined in the ratio of two catalytic trimers ($2C_3$) and three regulatory dimers ($3R_2$) to form a C_6R_6 complex with dihedral symmetry D_3 (Fig. 3.6). The two catalytic trimers are arranged in an almost eclipsed conformation with their three-fold axis parallel to a crystallographic three-fold axis. The separation of the catalytic sites between the C1 and C2 subunits is 32 Å in both the T and R states. The separation between the catalytic sites of the C1 and C4 subunits is 40 Å in the T state and 50 Å in the R state. The three regulatory dimers interdigitate between the catalytic subunits and are related by a non-crystallographic two-fold rotation axis. The regulatory sites R1 to R6 are separated by 20 Å and the distance between a regulatory site and catalytic site is 60 Å. Aspartate carbamoyltransferase has been studied by

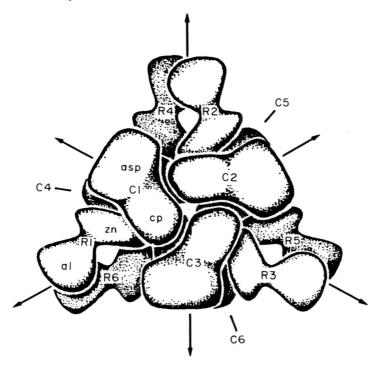

Fig. 3.6. A schematic view of the aspartate carbamoyltransferase dodecamer viewed along the three-fold axis. Arrows indicate the molecular two-fold axes. C, catalytic chain; R, regulatory chain; asp, aspartate (equatorial) domain; cp, carbamyl phosphate (polar) domain; zn, zinc domain; al, allosteric effector domain. (From Krause *et al.*, 1987.)

many different techniques and to a first approximation its kinetic and physical properties conform with the MWC model (e.g. Foote & Schachman, 1985; Schachman, 1988), although some kinetic properties associated with binding to the regulatory subunit are better explained by the sequential model (for review see Hervé, 1988). Sedimentation experiments had indicated that the R state is significantly enlarged relative to the T state (Gerhart & Schachman, 1968; Howlett & Schachman, 1977). X-ray structural studies by Lipscomb and his colleagues have shown the molecular basis for this enlargement and provided a detailed stereochemical mechanism for homotropic effects.

High resolution structures have been solved for the unligated enzyme (T state), the CTP-inhibited enzyme (T state) and the bisubstrate analogue complexed enzyme (R state) (Table 3.2). The two T state structures are

Table 3.2. *Crystal structures of aspartate carbamolytransferase*

State	Conditions	Space group and unit cell dimensions	Resolution (and crystal-lographic R factor)	Reference
T	Unligated	P321 $a = b = 122.1$ Å $c = 142.2$ Å asymmetric unit R_2C_2	2.6 Å (0.24)	Monaco *et al.*, 1978 Ke *et al.*, 1984
T	0.36 mM CTP pH 5.8	P321 $a = b = 121.8$ Å $c = 142.1$ Å asymmetric unit R_2C_2	2.6 Å (0.155)	Honzatko *et al.*, 1982 Ke *et al.*, 1984 Kim *et al.*, 1987
R	1 mM PALA pH 5.9	P321 $a = b = 122.1$ Å $c = 156.2$ Å asymmetric unit R_2C_2	2.4 Å (0.165)	Krause *et al.*, 1987 Ke *et aql.*, 1988

very similar (Ke *et al.*, 1984). The conformation of the individual subunits is shown in Fig. 3.7. The enzyme exhibits α/β topology for both subunits. The catalytic subunit is composed of two domains termed the equatorial (or aspartate) domain and the polar (or carbamoyl phosphate) domain which is situated close to the three-fold axis. The regulatory subunit is also divided into two domains. The zinc domain is composed of mostly β sheet with zinc chelated to four cysteines in a loop which is close to the polar domain of the catalytic subunit. The allosteric (or CTP binding) domain is on the periphery of the molecule.

Structural studies on the R state were initially hampered by substrate turnover and the instability of carbamoyl phosphate. The problem was solved with the bisubstrate analogue *N*-(phosphonacetyl)-L-aspartate (PALA), a strong inhibitor ($K_i = 2.7 \times 10^{-8}$ M) that resembles both carbamoyl phosphate and aspartate (Fig. 3.5). Evidence that PALA is a good model for the substrates has come from a crystal structure determination of the enzyme in the presence of carbamoyl phosphate and succinate (a competitive inhibitor of aspartate) which showed similar protein–ligand contacts to those observed in the PALA complex (Gouaux & Lipscomb, 1988).

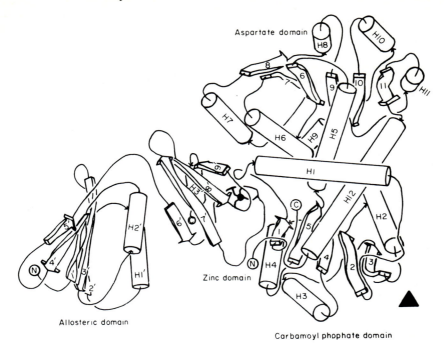

Fig. 3.7. A schematic view of the RC subunit of aspartate carbamoyl-transferase. Helices are shown as cylinders and numbered H1 to H12 in the catalytic subunit and H1′ to H3′ in the regulatory subunit. β strands are shown as arrows and are numbered 1 to 11 in the catalytic subunit and 1′ to 9′ in the regulatory subunit. (From Krause *et al.*, 1987.)

The catalytic site is situated at the domain interface and at the interface between adjacent catalytic subunits of the catalytic trimer (e.g. C1 and C2 Fig. 3.6). Residues required for maximal activity are contributed by these adjacent subunits. In the R state structure (Ke *et al.*, 1988), PALA binds to each of the six catalytic subunits with the phosphonate group interacting with groups from the polar domain and with two residues from the catalytic subunit adjacent in the trimer, Ser 80′ and Lys 84′ (loop between S3 and H3), where superscript prime denotes a symmetry related subunit (Fig. 3.8). The aspartate moiety extends into the equatorial domain with the carboxylate groups stabilised by several arginine residues: the α-carboxylate interacts with Arg 167 (helix H6); the β-carboxylate interacts with Arg 229 and Gln 231 (S9–H9 loop). Both Arg 105 and Lys 84′ bridge the phosphonate and the carboxylates of the bisubstrate. The other polar groups of the bisubstrate are stabilised by hydrogen bonds that include a contact between His 134 (start of helix H5) to the carbonyl oxygen of the

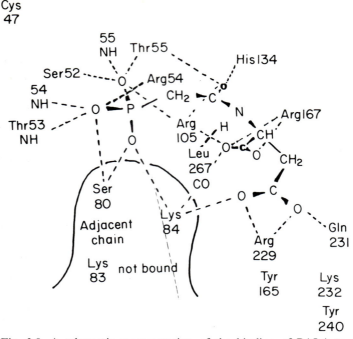

Fig. 3.8. A schematic representation of the binding of PALA to the catalytic site of aspartate carbamoyltransferase. Hydrogen bonds and salt links are indicated. Lys 84 and Ser 80 come from an adjacent subunit within the same catalytic trimer. Residues studied by other methods (Cys 47, Lys 83, Tyr 165, Lys 232 and Tyr 240) are too far away to make direct interactions. (From Krause *et al.*, 1987.)

bisubstrate (Fig. 3.8). In this conformation, the substrate analogue is trapped in the catalytic site with very restricted access to bulk solvent although the aspartate moiety is in contact with bound water molecules (Krause *et al.*, 1987). Transition from T to R results in closure of the domain interface of the catalytic subunits. In the T state the catalytic site is open and filled with clusters of water molecules. The phosphate recognition site is still effective since pyrophosphate has been observed to bind to T state crystals (Honzatko *et al.*, 1982).

Conversely, in the R to T transition the domain interface of the regulatory subunits is closed. The allosteric inhibitor CTP binds to the six regulatory subunits but the binding to the R1 (and its symmetry related equivalents) is stronger than to R6 (and its equivalents) (Kim *et al.*, 1987). Exceptionally for an allosteric effector, the site is located entirely within one subunit and is not at a subunit interface. There are good contacts between the base and ribose of the CTP molecule and the enzyme that involves amino acid residues from the N terminal portion of the chain. The phosphate groups do not make strong interactions, and this may be a result of incomplete association of the inhibitor at pH 5.8. Conformational heterogeneity is observed for the triphosphates bound at the R1 and R6 subunits and there are violations of the non-crystallographic two-fold axis which affect the water structure and residues at the subunit interfaces (e.g. Arg 130) and the catalytic sites (e.g. Arg 54). It is not yet certain whether these violations are a consequence of crystal packing or if they relate to aspects of negative cooperativity for CTP binding that could be better accommodated in a sequential model for heterotropic allosteric effects.

The major changes in tertiary structure induced on binding PALA are a movement in α carbon positions of up to 8 Å in the 240s loop (residues 225 to 245) of the catalytic chains. The accompanying closure of the catalytic site is about 2 Å (residues 50 to 55) and shifts of up to 5 Å are seen for residues 70 to 75 that affect the positions of the 80s loop resulting in a shift of Lys 84 side chain by 10 Å. Overall the changes in tertiary structure are not excessive (root mean square differences in Cα positions of about 1 to 1.3 Å between T and R structures (Ke *et al.*, 1988)) but shifts of individual side chains result in changes in ionic contacts within subunits and large changes in quaternary structure (Kraus *et al.*, 1987). The major changes in quaternary structure on transition from T to R involve a rotation of each catalytic trimer by 5° about the three-fold axis (and since the rotation for one trimer is opposite to that of the other the net reorientation is 10°) and a separation of the two trimers along the three-fold axis by 12 Å. The regulatory dimers rotate 15° about their respective two fold axes so as to conserve the C–R–R–C interactions where the two C subunits are in different trimers (e.g. C1–R1–R6–C6) (Fig. 3.9). The combined rotations achieved by the catalytic trimers and regulatory dimers about their respective axes has appropriately been termed a differential gear mechanism by Perutz (1989); the rotation of one unit is linked to the rotation of the other units. In the T state, the 240s loops stack neatly side by side at the C1–C4 interface. In the R state, they stack nearly on top of one another forcing the two catalytic trimers apart. The movement of the 240s loop is triggered by the shift of Arg 229. As Arg 229 moves to make contact with PALA, it breaks its contacts with Glu 272 and makes a new contact with Glu 233, a feature that requires movement of the 240s loop. Despite the large magnitude of the quaternary structural change, there is evidence that at least part of this change can be accomplished in the crystal lattice (Gouaux & Lipscomb, 1989).

The importance of many groups in crucial interactions has been demonstrated by single mutant and by site-directed mutagenesis studies

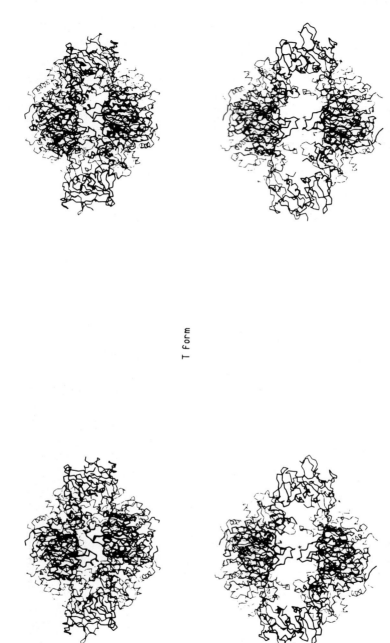

T form

Fig. 3.9. Stereo α-carbon tracings of the polypeptide chains of the entire aspartate carbamoyltransferase molecule in the T and R states. The view is perpendicular to the three-fold axis which lies vertically in the plane of the paper. One of the two-fold axes projects out from the centre of the enzyme towards the viewer. The conspicuous loop nearest the centre of the diagram is the 240s loop. Behind the 240s loop lies the central cavity of the molecule. (From Krause *et al.*, 1987.)

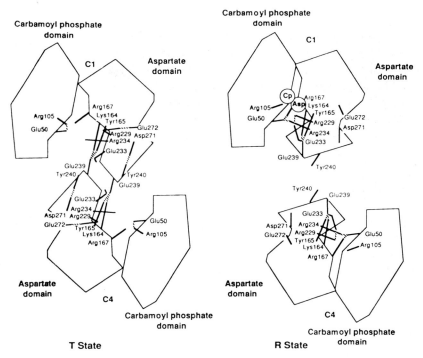

Fig. 3.10. A model for the mechanism of homotropic cooperativity in aspartate carbamoyltransferase. The two extreme conformations of a C1–C4 pair are shown schematically in the T state (left) and R state (right). The various interactions shown are repeated in the C2–C5 and C3–C6 pairs by operation of the three-fold axis of symmetry. Upon aspartate binding in the presence of carbamoyl phosphate, the aspartate domain moves towards the carbamoyl phosphate domain, which results in closure of the catalytic site. The 240s loops of C1 and C4 undergo large changes from being side by side in the T state to almost on top of each other in the R state. The separation of the subunits leads to elongation of the molecular shape. The ionic interactions that are broken and made following these changes in tertiary and quaternary structure are shown in the diagram and discussed in the text. For example the inter-domain linkage between Glu 50 and Arg 234 stabilises the R state whereas the link between Tyr 240 and Asp 271 stabilises the T state. The salt link between Glu 233 and Arg 229 correctly positions the latter residue to interact with the β carboxylate of aspartate in the R state. (From Kantrowitz & Lipscomb, 1988.)

(Ladjimi *et al.*, 1988; Robey *et al.*, 1986; Kantrowitz & Lipscomb, 1988). The T to R transition involves changes in tertiary structure that result in a switching of ionic contacts (Fig. 3.10). In the T conformation, Glu 50 binds to Arg 105 (interdomain contact), Arg 234 hydrogen bonds to main chain oxygen of Leu 235' in the adjacent subunit, and His 134 and Arg 167 do not make significant contacts. In the R state, on binding of PALA the inter-domain contacts are strengthened. Arg 105 binds to the phosphonate of PALA and Glu 50 then interacts with Arg 167 and Arg 234. Arg 167 and His 134 bind to PALA. The mutants Glu 50 to Ser or Arg 234 to Ser exhibit properties of a T state enzyme which has low affinity for aspartate and little cooperativity. Evidently, the ionic link between Glu 50 and Arg 234 is an important factor in stabilising the R state conformation although neither group interacts directly with substrate. The link between Glu 233 and Arg 229 which exists only in the R state after domain closure appears to hold Arg 229 in the proper orientation for catalysis. Conversion of Glu 233 to Ser results in an enzyme with 80-fold lower activity (Kantrowitz & Lipscomb, 1988).

Domain closure necessitates a change in the 240s loop that results in major quaternary structure changes at the C1–C4 interface. The numerous hydrogen bonds and van der Waals contacts of the T state are replaced by a limited set of contacts in the R state. For example, the inter-subunit contact Glu 239 (C1) to Lys 164 (C4) and Tyr 165 (C4) in the T state is broken and replaced by an intra-subunit link between Glu 239 and Lys 164 (both within C1) in the R state (Fig. 3.10). The Glu 239 to Gln mutant results in an enzyme which has properties of the R state showing that interactions between upper and lower trimers are critical for stability of the T state. Likewise, mutagenesis experiments have shown that the intra-domain contact between Tyr 240 and Asp 271 is important but not critical for stabilising the T state. In addition to these changes at the C1–C4 interface, there are also rearrangements at the C1–R1, C1–R4 and C1–C2 interfaces. Overall, the contacts between C1 and R1 are strengthened and those between C 1 and R4 and C1 and C4 are significantly weakened on the transition from T to R state. These changes have provided the basis for understanding homotropic cooperativity between the catalytic sites.

It is proposed that the binding of aspartate in the presence of carbamoyl phosphate leads to closure of the two domains of the catalytic subunits and triggers changes in the 240s loop and the 80s loop. This allows access of Arg 229 to the β carboxylate of aspartate and Lys 84' to the phosphate. The changes in tertiary structure lead to a high affinity aspartate site and to increased catalysis but can only be accomplished at the expense of quaternary structure changes. The C1–C4 and C1–R4 interfaces are disrupted, the trimers separate and the equatorial domains move relative to the polar domains. In a concerted mechanism, the catalytic sites for all the other subunits are locked by salt links in the R conformation, thus creating a high-affinity, high-activity state.

Table 3.3. *Crystal structures of phosphofructokinase*

State	Source	Conditions	Space group and unit cell dimensions	Resolution (and crystallographic R factor)	Reference
T	*B. stearo-thermophilus*	10 mM 2-phospho-glycolate	$P2_12_11_1$ a = 131.5 Å b = 114.1 Å c = 96.1 Å asymmetric unit tetramer	7 Å 2.5 Å	Evans *et al.*, 1986 P. R. Evans & T. Schirmer (unpublished observations)
R	*B. stearo-thermophilus*	~ 2 M potassium phosphate 2 mM Fru6P	I222 a = 122.5 Å b = 84.1 Å c = 61.5 Å asymmetric unit monomer	2.4 Å	Evans & Hudson (1979) Evans *et al.*, 1981
R	*E. coli*	1 mM Fru6P* 5–10 mM ADP 5 mM $MgCl_2$ in presence of 51% 2-methyl-2,-4-pentanediol	$P2_12_12$ a = 112.3 Å b = 85.4 Å c = 77.1 Å asymmetric unit dimer	2.4 Å (0.165)	Shirakihara & Evans 1988
R	*E. coli*	14% w/v polyethylene glycol 6000 1.0 M NaCl Tri-Cl pH 7.7 No ligands	C2 a = 177.0 Å b = 66.4 Å c = 154.0 Å β = 118.8Å asymmetric unit two half-tetramers	2.4 Å (0.168)	Rypniewski & Evans (1989)

* Crystallographic analysis showed Fru6P had been converted to Fru1,6P.

A number of proposals exist for possible catalytic mechanisms, the most recent of which (Gouaux *et al.*, 1987) is based on molecular model building from the PALA complex. In this mechanism, the α-amino group of aspartate (Fig. 3.5) is deprotonated by a base (possibly His 134). Attack by the amino group on the carbonyl carbon of carbamoyl phosphate leads to a tetrahedral intermediate which is stabilised by interactions of the carbonyl oxygen with His 134, Arg 105 and Thr 55. Intramolecular proton transfer from the amino group to the phosphate could facilitate release of the phosphate and formation of product could also be favoured by the

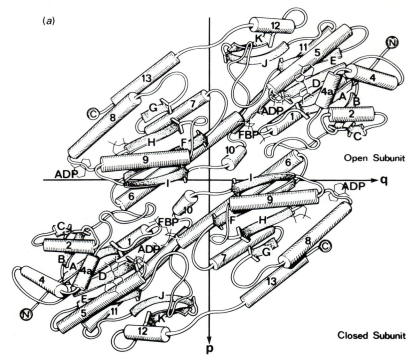

(*a*)

Open Subunit

Closed Subunit

Fig. 3.11. (*a*) For legend see facing page.

interaction of Lys 84′ (from the adjacent subunit) and Arg 54. Substitution of Lys 84 by Gln leads to a 4000-fold reduction in activity relative to wild type and substitution of His 134 by Ala to a 20-fold reduction in activity (Robey *et al.*, 1986).

The heterotropic regulation of catalysis by nucleoside triphosphates is less well understood. The changes in the orientation of the two domains of the regulatory subunits which adopt a more open conformation in the R state and the changes at the R–C interfaces are obviously important but an understanding of their significance requires further work.

Phosphofructokinase

Phosphofructokinase (EC 2.7.1.11) is a key regulatory element in glycolysis. It catalyses the phosphorylation of fructose 6-phosphate (Fru6P) by ATP to form fructose 1,6-biphosphate (Fru1,6P).

$$\text{Fru6P} + \text{ATP.Mg}^{2+} \rightleftharpoons \text{Fru1,6P} + \text{ADP.Mg}^{2+}$$

(b)

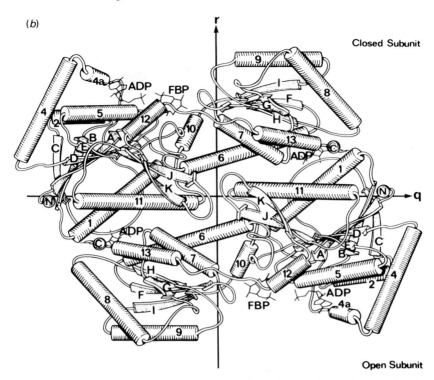

Fig. 3.11. Schematic representation of the phosphofructokinase molecule. Helices (numbered 1 to 13) are represented by cylinders and β sheet strands (labels A to K) by arrows. (a) View of two subunits related by the molecular r axis: the other two subunits of the tetramer are behind. (b) A view of two subunits along the molecular p axis: the other two subunits are behind. (From Shirakihara & Evans, 1988.)

Phosphofructokinase from various organisms is subject to feedback control so the activity is sensitive to the energy needs of the cell. Protein crystallographic studies are well advanced on the bacterial enzymes which exhibit similar but simpler allosteric properties to the homologous mammalian enzymes (Poorman *et al.*, 1984). The bacterial enzymes are tetramers with subunit molecular weight around 35 000. Early kinetic studies (Blangy *et al.*, 1968) indicated that the enzyme is well described by the MWC model. The enzyme acts solely as a K system with both T and R states showing the same catalytic rate but different affinities for the principal substrate Fru6P. The enzyme exhibits no cooperativity for ATP, but is allosterically activated by ADP and other diphosphonucleosides and inhibited by phosphoenolpyruvate (PEP).

Four different crystal structures of phosphofructokinase from *Escherichia coli* and *Bacillus stearothermophilus* have been solved by P. R. Evans and his colleagues (Table 3.3.). These show three distinct conformations associated with catalysis, co-operativity of substrate binding and binding of allosteric effectors. Each subunit consists of two domains and each domain is organised around a mostly parallel β sheet sandwiched between α helices (Fig. 3.11). The catalytic site is in a cleft between the domains and at the subunit interface. The substrate ATP (or ADP) site is wholly within one subunit but the 6-phosphate site of the substrate Fru6P (or Fru1,6P) is at the subunit interface. The allosteric effector site binds both the activator ADP and the inhibitor PEP and is located 35 Å from the catalytic site and at a second subunit interface (Fig. 3.11).

The four subunits of the tetramer are related by 222 symmetry and the three dyad axes have been labelled p, q, r. The separations of the catalytic sites about the p, q and r axes are 47 Å, 52 Å and 45 Å, respectively. The allosteric site is located on the p interface which is the most extensive of the three interfaces. This interface is formed by packing between helices $\alpha 11$ to $\alpha 6'$, $\alpha 7'$ and $\alpha 13'$ and between the bend βJ and βK to $\beta J'$ and $\beta K'$. On the T to R transition there is relatively little movement at this interface and the molecule responds as a dimer with the greatest shifts at the r interface, the interface that carries the substrate. At the r interface the C terminal end of helix $\alpha 6$ and the loop $\alpha 6/\beta F$ pack against the loop between $\beta C'$ and $\alpha 4'$ (Fig. 3.11(a)) with many hydrogen bonds between ionisable residues that also interact with the substrate. In addition, there are side chain interactions across the dyad between Thr 145, Glu148 and Arg 152 (from successive turns of the $\alpha 6$ helix) and their symmetry counterparts and between the end of strand βI and its symmetry counterpart. The q interface exhibits negligible contacts between subunits and there is a solvent-filled hole approximately 7 Å in diameter through the centre of the molecule.

The most detailed information on the binding of substrates to the R state of the enzymes have come from the work with *E. coli* phosphofructokinase complexed with the products, Fru1,6P and ADP/Mg^{2+} (Shirakihara & Evans, 1988). Two slightly different conformations at the catalytic site were observed. In two of the subunits the site was 'closed' and the Mg^{2+} ion was found to bridge across the 1-phosphate of one substrate to the β-phosphate of ADP (Fig. 3.12). In the other two subunits the site was 'open' with the Mg^{2+} firmly associated with the ADP and the two substrates were further apart by 1.5 Å. These conformations represent two successive stages along the reaction pathway in which closure of the subunit is required to bring the reactants together. The changes are probably not associated with

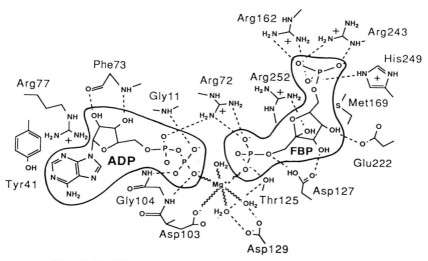

Fig. 3.12. The catalytic site of phosphofructokinase showing the interactions with the products Fru1,6P and ADP in the 'closed' subunits where the magnesium ion bridges the phosphates of the products. Arg 162 and Arg 243 come from an adjacent subunit related by the r axis dyad. (From Shirakihara & Evans, 1988.)

cooperativity since the product Fru1,6P is in the same position at each catalytic site and the subunit interfaces are not changed. In the closed conformation, the constellation of groups surrounding the substrate provide compelling evidence for a catalytic mechanism that involves the key residue Asp 127. It is proposed that Asp 127 acts as a base to abstract the proton from the 1-OH of Fru6P, thus allowing attack by the terminal phosphate of ATP. The pentacoordinate transition state formed after attack has begun but before the leaving group has left is stabilised by interactions of the phosphate with Arg 171, Arg 72, Thr 125 and the peptide nitrogen of Gly 11 (Fig. 3.12). Changing Asp 127 to Ser in a site-directed mutagenesis experiment resulted in a reduction of catalytic rate by 18 000 in the forward direction and 3100 in the reverse direction with no change in the Michaelis constant for Fru6P and a decrease in the Michaelis constant for Fru1,6P by 45 (Hellinga & Evans, 1987).

The contacts to ADP/Mg^{2+} at the allosteric effector site are shown in Fig. 3.13. The molecule binds in a deep hole between subunits with the phosphate groups at the bottom and the adenine base at the surface. Almost all the hydrogen bonds are to the phosphate moieties and are shared between the two subunits at the interface. The contributions from Lys 213′, Lys 214′ and His 215′ from the loop between α8 to βH are

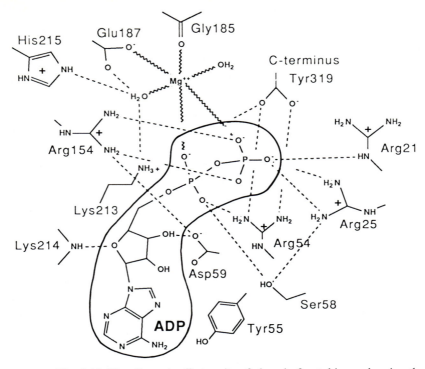

Fig. 3.13. The allosteric effector site of phosphofructokinase showing the interactions with the activatory ADP. Residues Arg 154, Glu 187, Lys 213, Lys 214, His 219 and Tyr 319 come from an adjacent subunit related by the p axis dyad. (From Shirakihara & Evans, 1988.)

significant because this part of the chain moves on the T to R transition. The importance of some of these side chains has been demonstrated in site-directed mutagenesis experiments (Lau & Fersht, 1987). The p axis interface remains unchanged in the T to R transition. The change in affinity for activator or inhibitor at the allosteric site is achieved by changes in tertiary structure that are linked to changes at the r axis interface. Shifts of the loop 211′–216′ are correlated with movements of Arg 162′ and Glu 161′.

In the T state, the r interface includes direct main chain hydrogen bonds between Val 246 and its symmetry counterpart and between the side chain of Thr 245 and the main chain nitrogen of His 249′ so as to make an appropriate extension of the β sheet across the subunit interface (although the twist of the two βI strands is in the wrong sense for a regular sheet) (Fig. 3.11(*a*)). The fructose 6-P site is effectively closed and Arg 162′ is in the wrong position to contribute to the binding of substrate. On conversion to the R state the subunits rotate approximately 8° about the p axis and move apart with intercalation of a layer of water molecules between the ends of the two βI strands at the r interface (Fig. 3.14). Arginine 162' from one subunit moves into the enlarged catalytic site of the other subunit where it interacts with the 6-P of the substrate. Preliminary analysis (P. R. Evans, unpublished observations) shows how movement of Arg 162' (from the βF strand) allows events at the catalytic site to be correlated with events at the allosteric site within the same subunit as Arg 162'. The separation of the catalytic site to this regulatory site is only 20–25 Å. In the T state, Glu 161' occupies a position similar to that of Arg 162' in the R state (thus blocking the phosphate recognition site with an acidic residue) and T state Arg 162' is swung in towards the loop between α8' and βH' of its own subunit. The α8'/βH' loop carries Lys 213' and forms the recognition site for the ribose of ADP bound to the allosteric effector site at the p interface (Fig. 3.11, Fig. 3.13). The proximity of the two basic groups (Arg 162' and Lys 213') in the T state holds the loop in a conformation which cannot bind ADP. This conformation is stabilised by the allosteric inhibitor PEP where the acidic groups help to neutralise the basic groups.

In summary, the essential features of the allosteric mechanism of phosphofructokinase involve a transition in one of the subunit interfaces (the r interface) from a distorted antiparallel β sheet to a situation in which water is intercalated between the strands. This transition appears to be an all-or-none effect as it is difficult to envisage a symmetric arrangement with only some of the waters present. The changes, especially those at the βI and βF strands, allow an arginine, Arg 162', from one subunit to swing into the catalytic site of the adjacent subunit and form the 6-P recognition site for the substrate. The catalytic groups and the ATP (ADP) substrate binding site are little affected by these movements, thus explaining why the enzyme is a K system. The shift of Arg 162 allows the development of the ADP allosteric effector site. Thus homotropic and heterotropic interactions are strongly correlated so that the T to R transition is promoted by binding of either Fru6P at the catalytic site or ADP at the allosteric effector site.

A major unexplained phenomenon is the observation that *E. coli* phosphofructokinase crystallised in the absence of ligands adopts an R state quaternary structure (Table 3.3; Rypniewski & Evans, 1989). It is most surprising that, under conditions where the equilibrium is 10^6 in favour of the T state, the crystallisation should favour the R state. However, although the catalytic site is in a conformation that can bind Fru6P, there are significant changes at the allosteric site compared to the

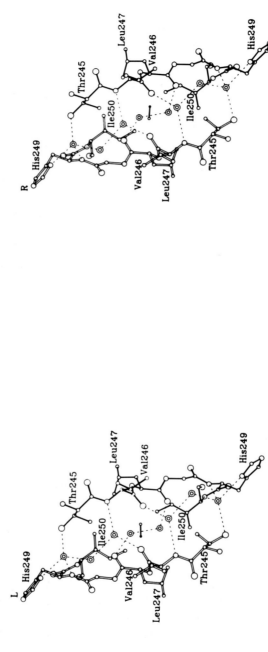

Fig. 3.14. A stereo diagram of part of the phosphofructokinase subunit interface related by the r axis dyad. The double circles are water molecules and the 2 subunits are distinguished by filled and open bonds. The diagram shows the 2 β-sheet strands I around the dyad with the layer of water molecules between the strands in the R state. Residues Leu 247 and Ile 250 from a hydrophobic layer underneath. (From Shirakihara & Evans, 1988.)

Table 3.4. *Crystal structures of glycogen phosphorylase*

State	Enzyme	Condition	Space group and unit cell dimensions	Resolution (and crystallographic R factor)	Reference
T	Phosphorylase *b*	2 mM IMP 10 mM Mg acetate	$P4_32_12$ a = b = 128.5 Å c = 116.3 Å asymmetric unit monomer	1.9 Å (0.18)	Sansom *et al.*, 1985 Johnson *et al.*, 1988 Acharya *et al.*, (unpublished observations)
T	Phosphorylase *a*	50 mM glucose 10 mM Mg acetate	$P4_32_12$ a = b = 128.4 Å c = 115.4 Å asymmetric unit monomer	2.1 Å (0.15)	Sprang & Fletterick, 1979 Sprang *et al.*, (unpublished observations)
R	Phosphorylase *b**	1 M $(NH_4)_2SO_4$	$P2_1$ a = 119.0 Å b = 190.0 Å c = 88.2 Å β = 109.4 Å asymmetric unit tetramer	2.8 Å (0.19)	Barford & Johnson, 1989

* Crystals with very similar unit cell dimensions have been obtained under similar conditions for phosphorylase *b* in the presence of AMP and for phosphorylase *a* (Fasold *et al.*, 1972).

liganded forms of the enzyme. Rypniewski and Evans suggest the structure may represent a form of the enzyme which has been activated by Fru6P but not by the heterotropic activator ADP.

Glycogen phosphorylase

Glycogen phosphorylase (EC 2.4.1.1) catalyses the first step in the intracellular degradation of glycogen.

$$(\alpha\text{-}1,4\text{-glucoside})_n + P_i \rightleftharpoons (\alpha\text{-}1,4\text{-glucoside})_{n-1} + \text{glucose-1-P}$$

Phosphorylase is a classic control protein whose properties are regulated both by allosteric interactions and reversible phosphorylation. Phosphorylase *b* requires AMP for activity and is inhibited by glucose-6-P and ATP. The enzyme exhibits homotropic effects for both substrate and AMP and heterotropic effects which result in a 15-fold increase in affinity for substrate as AMP is increased three-fold to 0.05 mM. The main route for physiological activation is by hormonal or neuronal stimulated conversion to phosphorylase *a* by phosphorylation of a single serine residue, Ser 14. Phosphorylase *a* is active in the absence of AMP although substrate binding remains cooperative. Activity can be augmented ($\approx 10\%$) by AMP and under these conditions cooperativity is abolished. Phosphorylase *a* has a 100-fold increase in affinity for AMP compared with phosphorylase *b*. As with the other enzymes, most of the properties of phosphorylase can be accommodated in the MWC model if the model is extended to include more than two states (see Graves & Wang, 1972; Fletterick & Madsen, 1980; Madsen, 1986; Johnson *et al.*, 1990).

In the absence of effectors, phosphorylase *b* is a dimer composed of two identical subunits molecular weight 97434. Activation either by AMP or by phosphorylation results in an increased tendency of the enzyme to form tetramers. Binding to glycogen dissociates tetramers to yield the more active dimeric form. *In vivo*, a substantial amount of phosphorylase is bound to glycogen particles and the dimer is considered to be the physiologically significant form of the enzyme (Graves & Wang, 1972).

The crystal structures of three different forms of the enzyme have been solved at high resolution (Table 3.4) and numerous binding studies have been carried out. Crystals of phosphorylase *b* were obtained in the presence of IMP, a weak inactivator (Black & Wang, 1968) that confers activity but without the concomitant increase in affinity for substrate that is observed with AMP. These differences in activation properties have led to the suggestion that the AMP-induced activation of phosphorylase *b* consists of two stages: an enhancement of catalytic activity (V system) and an enhancement of affinity for substrate (K system). IMP appears to affect only the first stage. Crystals of phosphorylase a are grown in the presence of glucose, an inhibitor which promotes the T state by binding at the catalytic site and restricting access (Kasvinsky *et al.*, 1978). Solution of the structure of the R state has proved more difficult. The recent R state structure (Table 3.4) was obtained with crystals grown in the presence of ammonium sulphate. Sulphate is an allosteric activator (Engers & Madsen, 1968; Sotiroudis *et al.*, 1979) and in the crystals sulphate mimics phosphate and binds at the catalytic site, the AMP allosteric site and the Ser-P site. The crystals contain the enzyme in the tetrameric state, the state of

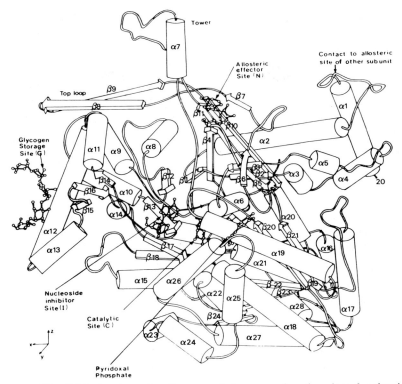

Fig. 3.15. A schematic diagram of glycogen phosphorylase *b* subunit. Helices (numbered α1 to α28) are shown as cylinders and β strands (numbered β1 to β24) as arrows. The catalytic site is at the centre of the subunit and close to the pyridoxal phosphate. The cofactor is behind α19 and linked via a Schiff base to Lys 680 in α21. The allosteric effector site is close to the subunit interface and makes contact to the cap region (loop between α1 and α2) of the other subunit. The ribose and phosphate are situated between the 2 helices α2 and α8. The glycogen storage site and the nucleoside inhibitor site are also shown. The 280s loop, the loop that partially blocks access to the catalytic site, is between α7 and α8.

aggregation associated with R state phosphorylase in the absence of glycogen. This change in aggregation on ligand binding is of relevance for a similar phenomenon observed with some receptors. The interactions involved in the dimer interface are described below. Those interactions that are involved in tetramerisation will be described elsewhere (Barford & Johnson, 1991).

In the T state, the structures of phosphorylase *b* and *a* are very similar (rms difference in Cα positions ⇌ 0.6 Å) (Sprang *et al.*, 1988). The two subunits of the functionally active dimer are related by a crystallographic two-fold symmetry axis. The subunit structure (Fig. 3.15) is composed of two domains, and the catalytic site is located between these domains, and is close to the essential cofactor pyridoxal phosphate. Access to the catalytic site is via a channel some 12 Å in length and severely restricted in diameter in one region. The allosteric activator, AMP, binds to a site some 32 Å away and is located between the two α helices, α2 and α8, and is close to the subunit interface. There is a second nucleoside or nucleotide inhibitor site at the entrance to the catalytic site channel. Binding at this site stabilises the T state and is synergistic with glucose binding (Kasvinsky *et al.*, 1978). In the crystals of T state phosphorylase, oligosaccharides and glycogen analogues do not bind to the catalytic site but bind tightly to a site on the surface of the enzyme, the glycogen storage site, which is about 30 Å and 39 Å from the catalytic site and the allosteric effector site, respectively (Johnson *et al.*, 1988).

The subunit–subunit contacts (Fig. 3.16) are not extensive. In the T state phosphorylase *b* they involve only 7% (2294 Å^2) of the total solvent accessible surface area of the monomer and they increase to 10% (3450 Å^2) in T state phosphorylase *a*. In phosphorylase *b* the dominant interactions involve the cap (residues 36 to 45) and the tower helices α7 (residues 262 to 278), the two structural elements that project out from the subunit surface (Fig. 3.15). The cap interactions provide a link to the C-terminal tail of the other subunit (His 36′ to Asp 838), van der Waals interactions to the α2 helix, hydrogen bonds from the main chain oxygens of Leu 39′ and Val 40′ to Arg 193 (β7 strand) and an ionic link between Lys 41′ and Glu 195. The tower helices run antiparallel to one another with an angle of about 20° between them. There are several hydrogen bonds and van der Waals interactions. Residues from the top of the tower (e.g. Tyr 262) are in van der Waals contact with side chains in the 280s loop of the other subunit. This loop (residues 280 to 287) blocks the access to the catalytic site so that, although the catalytic site itself is not at the subunit interface, there are indirect contacts from the catalytic site to the interface (Fig. 3.16). Finally, the tower is linked to the cap interface by contacts from the tower residues to those in the loop β6′/β7′.

Comparison of the crystal structures of phosphorylase *b* and *a* has shown the conformational changes that represent the first step in the activation of the enzyme by phosphorylation (Sprang *et al.*, 1988). The N-terminal polypeptide segment which carries the Ser 14 phosphorylation site is disordered in phosphorylase *b* but folds into a distorted helix in phosphorylase *a*. The helix is bound to a broad cavity at the subunit interface (Fig. 3.17) through a network of hydrogen bonds, ionic

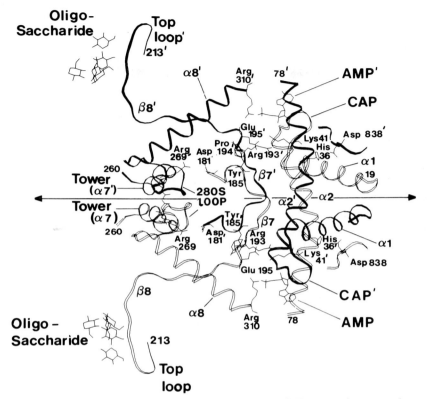

Fig. 3.16. The subunit–subunit contacts of T state glycogen phosphorylase *b*. The view is normal to the two fold dyad axis and down the
crystallographic *z* axis (i.e. the view from the top in Fig. 3.15). One
subunit is shown with open lines and the other with filled lines. Chain
segments shown are 19 to 78 (α1–cap–α2), 181 to 213 (β7–β8–top loop),
260 to 310 (α7 tower helix–280s loop–α8) and 836 to 841 (C terminal tail).
AMP is bound between the helices α2 and α8 and makes contact to the
cap′ of the other subunit. The major polar interactions of the subunit
interface at the cap′ involve (moving from right to left in the diagram)
Asp 838 to His 36′, Lys 41′ to Glu 195, main chain oxygens of 39′ and 40′
to Arg 193. Pro 194 is in van der Waals contact with Tyr 185′ and Asp
181′ is linked to Arg 269 of the tower. The two tower helices are linked by
hydrogen bonds between Asn 270 and Asn 270′. These interactions and
the connection of the 280s loop to the catalytic site are shown more
clearly in Fig. 3.22.

interactions and van der Waals contacts. The phosphate group is bound by
two arginine side chains, one, Arg 69 from its own subunit, and the other
Arg 43′, from the other subunit. It also interacts with the main chain
nitrogen of Val 15 and through a water molecule with Gln 72 (Fig. 3.18).

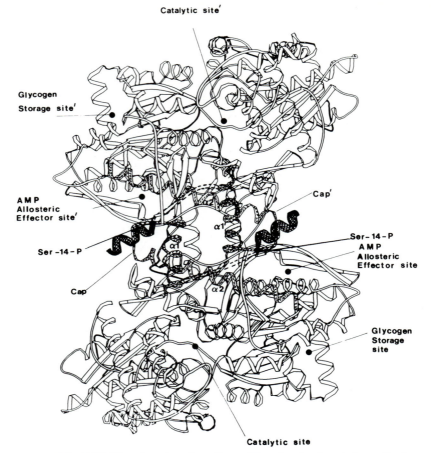

Fig. 3.17. The phosphorylase dimer. The view is almost down the two-fold axis but with a small rotation. The crystallographic *z* axis is vertical as in Fig. 3.15. The α1–cap–α2 regions have been shaded in both subunits. The N-terminal tail containing the Ser-P is shown black in both subunits. The tower helices are on the far side of the molecule. Approximate positions of the AMP allosteric site, the catalytic site and the glycogen storage site are shown. Ribbon drawing produced with software written by J. P. Priestle.

Hydrophobic interactions are important in stabilisation of the peptide/ protein interface. The conserved residues on either side of Ser 14-P, Ile 13 and Val 15, are completely buried in pockets lined with non-polar groups from side chains of the α2 helix. On binding to the subunit interface, the phosphorylated N terminus displaces four C-terminal residues, 838–841, which become disordered in phosphorylase *a*. The ion pair from His 36′ to Asp 838 is disrupted, and movement of the histidine allows Val 15 to bind

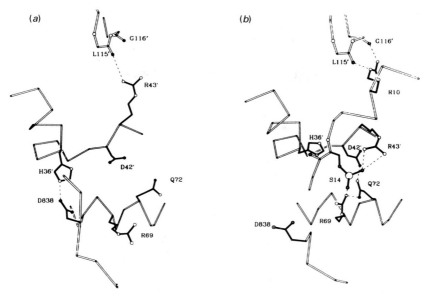

Fig. 3.18. Details of the phosphoserine site in phosphorylase *a* (Fig. 3.18*a*) and *b* (Fig. 3.18*b*). The contacts between Asp 838 and His 36′ and between Arg 43′ and Leu 115′ in phosphorylase *b* are broken in phosphorylase *a* as the Ser-P N terminal tail swings in. The Ser-phosphate contacts Arg 69 (α2 helix) and Arg 43′ (cap′); Gln 72 and Asp 42′ are brought into hydrogen bonding contact and Arg 10 replaces Arg 43′ in contacts to Leu 115′ (main chain oxygen). (From Sprang *et al.*, 1988.)

to its hydrophobic pocket between Val 64 and Ile 68 on the α2 helix. In phosphorylase *b*, the ordered chain begins with residue 19 at the start of the α1 helix, and this part of the chain is turned away from the subunit. The N-terminal tail is highly basic, as was recognised early on from sequence data. Location of the peptide in the position observed for phosphorylase *a*, but without the compensating phosphate group, would result in unfavourable proximity of basic groups. Thus the basic cluster (Lys 9, Arg 10, Lys 11, Arg 16, Arg 43′ and Arg 69) is a feature of the regulatory apparatus that requires the N-terminal tail to be swung away from the subunit in phosphorylase *b* but allows the Ser-P to trigger a functionally active conformation. Clusters of charged residues may play similar roles in other enzymes subject to regulation by cyclic-AMP or Ca^{2+} dependent kinases (Cohen, 1988).

The involvement of at least five of the residues from the N-terminal tail in stabilising the Ser-P 14 at the subunit interface explains why inorganic phosphate is unable to trigger the conformational changes in phosphorylase *b* (Lorek *et al.*, 1984). Since inorganic phosphate is present in the

(a)

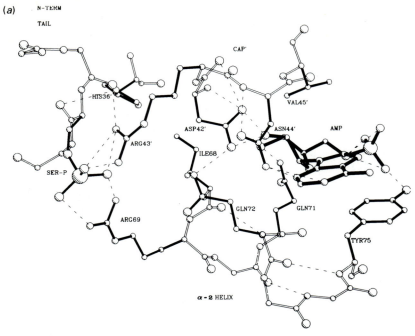

Fig. 3.19(a). For legend see page 75.

cell at a concentration of about 5 mM, this is obviously an important requirement for the regulatory mechanism. Phosphorylation triggers conformational changes remote from the phosphorylation site. In order to make contact with Ser-P, Arg 43′ needs to break its contacts with the C-terminal end of helix α4′. In phosphorylase a, this is encouraged by Arg 10 which replaces the position previously made by Arg 43′ (Fig. 3.18). These changes observed on the binding of the N-terminal peptide tail to the surface of the phosphorylase subunit give some insight into events that might be associated with the binding of a peptide hormone to its receptor site.

The AMP allosteric site is located on the opposite side of the α2 helix to the Ser-P with a separation of about 12 Å. Binding of the Ser-P peptide results in small changes in the cap residues (especially Asp 42′ and Asn 44′ that flank Arg 43′), the α1 helix and the α2 helix that lead to a closed, high affinity AMP site in phosphorylase a. In phosphorylase b, AMP binds with weak van der Waals interactions between the base and Tyr 75 from the α2 helix, a possible hydrogen bond from the 2′-OH of the ribose to Asp 42′, and strong ionic interactions between the phosphate and Arg 309 and Arg 310 from helix α8 (Johnson et al., 1988) (Fig. 3.19). In phosphorylase a, the

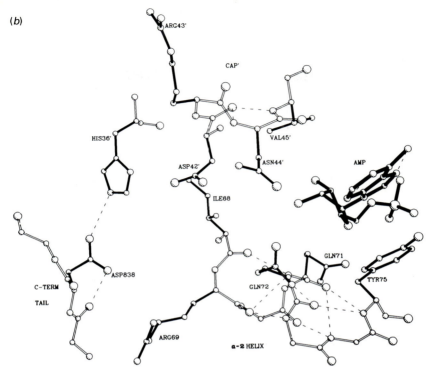

Fig. 3.19(*b*). For legend see page 75.

α2 helix is contracted and forms a more regular helix than in phosphorylase *b* as a result of direct contacts to the Ser-P peptide. This, and the movements of the cap, allow the enzyme to make much closer contacts with AMP, which binds in an alternative orientation (Sprang *et al.*, 1987, 1988) (Fig. 3.19). There are improved contacts to Tyr 75 and hydrogen bonds to Asn 44′, Asp 42′ and Gln 71. The phosphate contacts are similar to those of phosphorylase *b*. The shift of the side chain of Gln 71 explains why glucose 6-P is a good inhibitor of phosphorylase *b* but not of phosphorylase *a*. In phosphorylase *b*, glucose 6-P binds well with its phosphate, making contacts similar to those of AMP but displaced so that it also contacts Arg 242 and the glucose moiety makes several specific hydrogen bonds (Johnson *et al.*, 1988). In phosphorylase *a*, this site is blocked by Gln 71 (Sprang *et al.*, 1988).

Studies on catalysis in the crystal with phosphorylase *b* (McLaughlin *et al.*, 1984; Hajdu *et al.*, 1987) and substrate binding studies with phosphorylase *a* (Madsen *et al.*, 1978) have indicated several important features that are associated with activation at the catalytic site. The natural

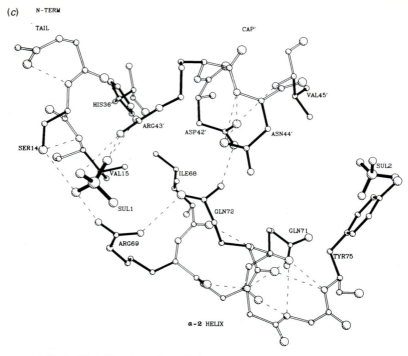

Fig. 3.19(*c*). For legend see facing page.

substrate, glycogen, is too large to diffuse into the crystals and, although oligosaccharides can be diffused in, these bind solely at the glycogen storage site situated on the surface of the enzyme and not at the catalytic site because access to the catalytic site is restricted in the T state crystals. However, the enzyme catalysed reaction of heptenitol and phosphate to heptulose-2-P which mimics the natural reaction (Klein *et al.*, 1986) has proved most informative. Heptulose-2-P is a product inhibitor ($K_i =$ 14 µM) which exhibits properties of a transition state analogue. Heptulose 2-P, formed by catalysis in phosphorylase *b* crystals, binds so that the product phosphate is within hydrogen bonding distance of the 5′-phosphate group of the cofactor pyridoxal phosphate. The constellation of groups provides a satisfactory mechanism for catalysis that involves the cofactor phosphate to first act as a general acid in promoting attack by the substrate phosphate on the heptenitol (Klein *et al.*, 1986; McLaughlin *et al.*, 1984; Johnson *et al.*, 1988). The product phosphate cofactor phosphate contact is stabilised by a number of interactions (Fig. 3.20), the most important of which is that from Arg 569. In the native structure, Arg 569

(d)

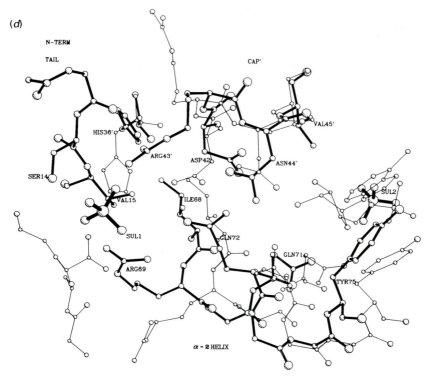

Fig. 3.19. The Ser-P and allosteric effector site for (*a*) T state phosphorylase *a*; (*b*) T state phosphorylase *b*; (*c*) R state phosphorylase *b* and (*d*) R state (thick lines) and T state phosphorylase *b* (thin lines) superimposed. In phosphorylase *a* the cap'/α–2 helix are drawn together so that Tyr 75, Asn 42', Asn 44' and Gln 72 make good interactions with AMP and Gln 71 blocks the binding site for the inhibitor glucose 6-P. In T state phosphorylase *b* AMP makes weak contacts to Tyr 75 and to residues of the cap'. The adjacent site can accommodate the allosteric inhibitor glucose 6-P (not shown) with good contacts. The R state phosphorylase *b* sulphate is bound close to the Ser-P and AMP sites and triggers changes similar to those seen for phosphorylase *a*. (From Barford & Johnson, 1989.)

is turned away from the catalytic site and buried with its charged group in contact with main chain oxygens of Pro 281 and side chain of Asn 133. In the product complex, these contacts are broken and the basic group swings into the catalytic site to make contact with the product phosphate with the simultaneous displacement of an acidic residue, Asp 283, from the catalytic site. This results in shifts of the whole of the loop from 281 to 285, the loop that previously blocked access to the site. These changes also provide an

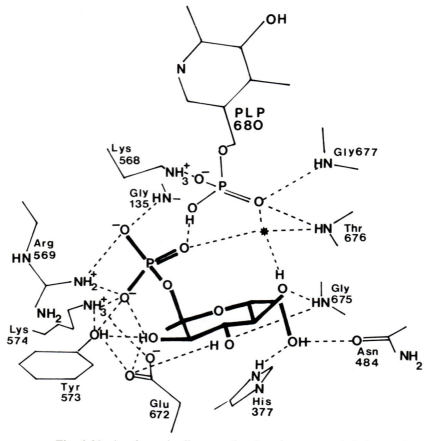

Fig. 3.20. A schematic diagram showing the contacts of the product heptulose-2-P to the catalytic site of phosphorylase *b*. The direct contact between the pyridoxal phosphate 5′-phosphate group and the product phosphate is stabilised by ionic interactions, especially those involving Arg 569 which has shifted significantly from its position in the native structure. In the diagram the positions of Lys 574 and Glu 672 have been shifted slightly in order to show their contacts more clearly. (Johnson *et al.*, 1988).

explanation for the change in state of ionisation of the pyridoxal phosphate as a result of activation (Helmreich & Klein, 1980). In the T state, Asp 283 is located in the catalytic site pocket and linked indirectly through water molecules to the 5′-phosphate of pyridoxal phosphate (Oikonomakos *et al.*, 1987), favouring a monoanionic state of the cofactor phosphate. On activation, the acidic group is replaced by Arg 569 which would favour the dianionic state. The close proximity of the substrate phosphate to that of the cofactor would result in a shift back towards the monoanion in the enzyme substrate complex since the two phosphates must share a proton between them. The essential role of Arg 569 was anticipated from chemical modification studies (Vandenbunder & Buc, 1983).

In the R state structure of phosphorylase *b* (Table 3.4; Barford & Johnson, 1989), sulphate mimics phosphate and binds at the catalytic site, the AMP allosteric site and the Ser-P site. The overall tertiary structural changes between the T and R states are small (rms difference in Cα coordinates 1.3 Å), but sulphate binding induces significant local conformational changes. At the catalytic site, the whole of the 280s loop is displaced and Arg 569 swings down to contact the sulphate anion, adopting a similar conformation to that seen in the heptulose 2-P complex. At the Ser-P site, sulphate induces an ordering of the N terminal peptide from residues 10 to 18, and the conformational changes in this region are almost identical to those that were observed in the comparison of the T state phosphorylase *a* and *b* structures. As a result of changes at the cap′/α2 helix interface, the adenine–ribose recognition part of the AMP allosteric site is in a high affinity state. Sulphate is bound at the AMP phosphate recognition site but this, by itself, appears to cause relatively few changes between the T and R states.

These small changes in tertiary structure are accompanied by large changes in quaternary structure. One subunit rotates 10° with respect to the other subunit about an axis normal to the two-fold axis of the dimer that intersects this axis close to the cap′/α2 interface (Fig. 3.21). As a result, the tower helices pull apart and the cap′ α2 interface is tightened. The relative movements of the two antiparallel tower helices provides a simple and effective mechanism for transmission of information that has relevance for signal transduction through membrane spanning helices of receptor molecules. In the R state structure, the tower helices pull apart so that the total displacement corresponds to two turns of an α helix. The polar contacts between Asn 270 and Asn 274′ (and Asn 274 and Asn 270′) are broken and replaced by non-polar contacts between Val 266 and Leu 267′ (and Leu 267 and Val 266′). The tilt of the two helices relative to one another is greatly increased so that the top of the tower helix no longer contacts the 280s loop of the other subunit (Fig. 3.22). These movements allow critical changes at the catalytic site that enable ionic groups to adopt their correct orientation for binding and catalysis. Thus shifts in residues at the bottom of the tower perturb Ile 15 which, in turn, perturbs Pro 281.

(a)

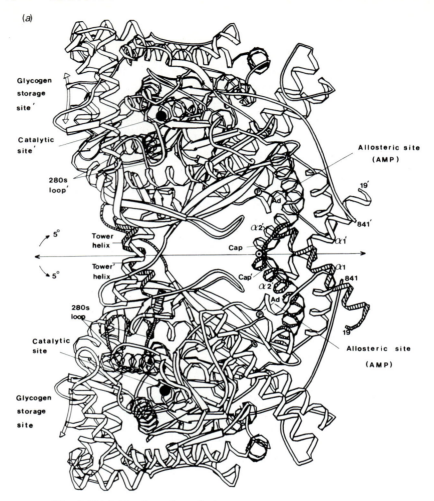

Fig. 3.21(*a*). For legend see facing page.

Arg 569 breaks its contacts to Pro 281, swings into the catalytic site, displaces Asp 283 and allows the 280s loop to become mobile.

Analogously to the tower–tower interface, structural interdependence of quaternary and tertiary changes is observed at the cap′/α-2 interface (Fig. 3.19(*c*) and (*d*)) although the amplitude of both changes is smaller. Movement of the cap′ towards the α-2 helix is facilitated and stabilised by concerted side chain motions of residues on these structural elements which are promoted by the binding of sulphate at the Ser-phosphate site. Residues 10–18 which are disordered in the T-state become ordered. Rotations of the side chains of Arg 43′ and Arg 69 allow the guanidinium

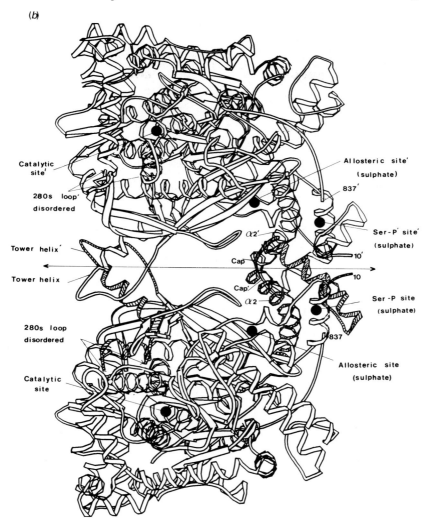

Fig. 3.21. Ribbon representation indicating the structural elements of (*a*) T-state and (*b*) R-state phosphorylase dimer, viewed normal to the molecular dyad axes. Structural elements α1–cap–α2, the tower helix–280s loop and the start of α8 are indicated by shading in the lower subunit. The allosteric effector site and Ser-P site are situated at the cap′/α2 interface. The catalytic site is close to the 280s loop at the termination of the tower helix. The glycogen storage site, to which phosphorylase dimers are attached to glycogen particles *in vivo*, is situated on the 'catalytic face' of the molecule and forms a dimer–dimer contact in the tetramer. On transition to the R-state, each subunit rotates by 5° about an axis positioned close to the cap′/α2 interface (marked by ringed dot in Fig. 3.21*a*). Ribbon diagram by a program by J. B. Priestle.

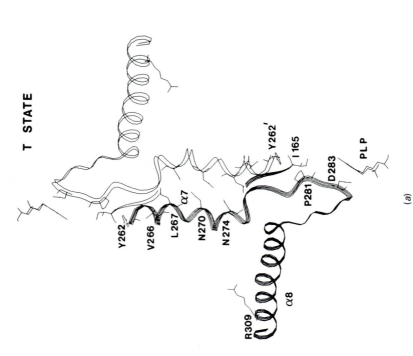

T STATE

T STATE

(a)

(b)

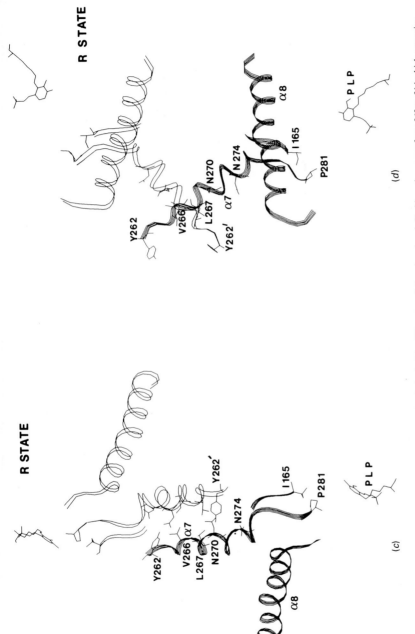

Fig. 3.22. Orthogonal views of the contacts between the subunits in the T and R states of phosphorylase b. Residues shown are from 262 to 314 which comprise the tower (α7) helix–280s loop–α8 helix (that connects to the allosteric site) and residues 162 to 165 that form a short β sheet with residues 276–279. The catalytic site is marked by the pyridoxal phosphate (PLP). The view in (a) and (b) is down the two-fold dyad axis and in (c) perpendicular to the dyad as in Fig. 3.21. In the T state (a and c) the tower helices pack antiparallel with a −20° tilt and there are good hydrogen bonds between Asn 270 and Asn 274′. Pro 281 is in van der Waals contact with Tyr 262′ and Asp 283 is located in the catalytic site. In the R state (b and d) the tower helices pull apart and change their angle of tilt to about −70°. Val 266 and Leu 267′ are now nearly opposite one another and the contacts Pro 281 to Ile 165 and Pro 281 to Tyr 262′ are broken. The loop 282 to 286 becomes disordered and Asp 283 is displaced from the catalytic site. Arg 569 (residue not shown) breaks its contacts with Pro 281 and moves into the catalytic site (Barford & Johnson, 1989).

groups to hydrogen bond with the sulphate. These interactions are only favourable following the closer association of the cap′ and α-2 helix. Coupling of quaternary and tertiary change is observed on the rotation of His 36′ about the Cα–Cβ bond into a previously unoccupied cavity to prevent steric conflict with Ile 68 as the subunits rotate. Val 15 packs into a pocket formed between His 36′ and Ile 68. In the T-state structure, the intercalation of His 36′ between Val 64 and Ile 68 of the α-2 helix, induces the helix to unwind by 1 Å towards its C-terminus. Rotation of His 36′ on transition to the R-state allows reordering and contraction of the helix, bringing Gln 72 into a position to form a H-bond to Asp 42′ of the cap′. Comparison with T-state phosphorylases *b* and *a* suggests that the side chains of Asp 42′, Asn 44′ and Gln 72 are correctly positioned to H-bond to AMP, although this has to be confirmed by direct binding studies. Both AMP and Ser-P or sulphate stabilise the R-state by interacting with residues from both subunits which are only available in the R-state conformation. Communications between Ser-phosphate 14 and the allosteric effector site is mediated through a closer association of the cap′/α-2 helix and reordering of α-2.

Thus, our present understanding of the allosteric mechanism of phosphorylase rests on the intimate connection between tertiary structure and quaternary structure and the conservation of symmetry as the structure goes from one state to the other. Changes at the Ser-P site and the AMP directly involve subunit–subunit contacts so that homotropic effects are communicated via conservation of symmetry. Likewise, although the catalytic site is not in direct contact with residues at the subunit interface (it is about 15 Å from the interface), it is indirectly connected so that changes induced in tertiary structure require a change in quaternary structure through repacking of the tower helices. It is the change in quaternary structure that allows communication between the catalytic sites which are separated by 66 Å.

Conclusions

The allosteric constants for the four proteins reviewed are summarised in Table 3.5. Although, for some of these systems, the MWC model is an over-simplification for a complete interpretation of the kinetic results, it is instructive to compare the same parameters for the different proteins. Phosphofructokinase exhibits the most extreme values of the allosteric parameters. In the absence of ligand, the equilibrium is 10^6 in favour of the

Table 3.5. *Allosteric constants*

Protein	Haemoglobin (a)	(b)	Aspartate carbamoyl-transferase	Phospho-fructokinase	Glycogen phosphorylase b	a
$K_T(M)$	1.3×10^{-4}	1.1×10^{-5}	1.3×10^{-1}	2.5×10^{-2}	3.7×10^{-2}	3.0×10^{-2}
$K_R(M)$	3.3×10^{-7}	2.6×10^{-7}	6.5×10^{-3}	1.3×10^{-5}	2.2×10^{-3}	1.8×10^{-3}
C	2.6×10^{-3}	2.4×10^{-2}	5×10^{-2}	5×10^{-4}	6×10^{-2}	6×10^{-2}
L	3.0×10^{6}	3.7×10^{3}	2.5×10^{2}	4×10^{6}	2.1×10^{3}	3–13

K_T and K_R are dissociation constants for ligand binding.
$C = K_R/K_T$. L is the ratio of the concentration of T state molecules to R state molecules in the absence of ligand ($L = T_o/R_o$).
Haemoglobin; reference: Imai, 1982. (a) Physiological conditions pH 7.4, 0.1 M Cl⁻, 2 mM DPG; (b) pH 7.4, 0.007 M Cl⁻.
Aspartate carbamoyltransferase; reference: Howlett *et al.* (1977). K_R and K_T are for aspartate. In the presence of activator (ATP) L = 70, in the presence of inhibitor (CTP) L = 1250; K_R and K_T are unchanged.
Phosphofructokinase; reference: Blangy *et al.* (1968). K_R and K_T are for Fru6P. K_T and K_R for the inhibitor PEP are 7.5×10^{-4} M and $> 7.5 \times 10^{-1}$ M respectively.
Glycogen phosphorylase; reference: Helmreich *et al.*, 1967; Kastenschmidt *et al.*, 1968; Helmreich, 1969. K_T and K_R for AMP with phosphorylase *b* are 1.5×10^{-2} M and 1.5×10^{-5} M, respectively, and with phosphorylase *a* are 1.2×10^{-5} M and 1×10^{-6} M respectively.

T state, a figure only matched by haemoglobin in the presence of the physiological inhibitors chloride and DPG. The allosteric constants *L* for haemoglobin (in the absence of inhibitors) and for phosphorylase *b* are similar; *L* for aspartate carbamoyltransferase is an order of magnitude smaller. The L constant for phosphorylase *a*, which is already activated by the Ser-P is still greater than 1, and this demonstrates that in spite of activation the enzyme is not completely in the R state, a situation which can only be achieved in the presence of AMP. The change in affinity for substrate on the T to R transition follows the same trend as the allosteric constant *L*. Phosphofructokinase shows a change in affinity of four orders of magnitude, haemoglobin under physiological conditions a change of three orders of magnitude and without inhibitors two orders of magnitude, and aspartate carbamoyltransferase and phosphorylase *b* two to three orders of magnitude. For haemoglobin and aspartate carbamoyltransferase, the R state molecules exhibit similar affinities for their respective ligands as do the free subunits and it is the association of subunits in the T state conformation that leads to a relative lowering of affinity. In each of

Table 3.6. *Summary of structural changes observed on T to R transitions*

Protein	Haemoglobin	Aspartate carbamoyltransferase	Phosphofructokinase	Glycogen/phosphorylase	
				(a) Ser-P $b \rightarrow a$	(b) b T $\rightarrow$ R
Changes in tertiary structure on substrate binding.	Oxygen binding pocket opens to admit ligand. 6-co-ordinated iron moves 0.5 Å into plane of heme.	Catalytic subunit interdomain contacts are strengthened and domains close around catalytic site. Movements of 240s loop and 80s loop allow Arg 229 and Lys 84′ (from another subunit) to contact substrate.	Catalytic site enlarged and one arginine (Arg 162′) from adjacent subunit moves in to contact 6-phosphate group of Fru1,6P.	Ser-P N terminal tail induces changes in two arginines at subunit interface (Arg 69 and Arg 43′) and interactions lead to movement of α2 helix and cap region. These movements create high affinity AMP site some 12 Å away from Ser-P site.	Sulphate mimics phosphate and induces an arginine (Arg 569) to move into catalytic site to create high affinity site. The 280s loop including Asp 283 is displaced. Access to the catalytic site is opened. Sulphate binding at the Ser-P site triggers similar conformational changes to those seen in the $b \rightarrow a$ comparison.

Changes in quaternary structure on substrate binding.	Helix C of α chains slides past FG corner of β chains to admit oxygen. T state stabilised by ionic linkages that are broken in **R** state. Water filled cavity down the 2-fold axis narrows.	Movement of the 240s loop forces the subunits apart by 12 Å. Intersubunit contacts are weakened especially those between C1 and C4, C1 and R4.	Subunit interface contacts weakened by intercalation of water molecules between distorted β sheet. Movement of βI and βF strands allows arginines to move into catalytic site.	Subunit–subunit contacts tightened mostly through disorder to order transition of N-terminal 16 residues and ionic contacts of Ser-P. Little overall change in quaternary structure because in the crystal phosphorylase a is locked in T state by inhibitor glucose binding.	Subunits rotate about axis normal to 2-fold axis of dimer and pull apart by about 2 Å. The anti-parallel tower helices at the subunit interface slide past each other and change their angle of tilt. These changes are communicated to the catalytic site.
Transmission of heterotropic effects.	All effectors (H^+, Cl^-, DPG, CO_2) stabilise T state structure through ionic interactions. All effector sites are at subunit–subunit interfaces.	Regulatory site located within regulatory subunit and 60 Å from catalytic site. Heterotropic effects not fully understood at structural level.	Regulatory site located at different subunit interface from that involved in the catalytic site. Movement of arginine to complete catalytic site promotes changes that create high affinity ADP activator site.		Communication between Ser-P and AMP site at the cap/α-2 helix interface and the catalytic site takes place through linked tertiary and quaternary structural changes. The rotation of the subunits on T → R transition creates both high affinity catalytic and regulatory sites.

these proteins a shift in the equilibrium between the T and R states dramatically affects the amount of ligand bound and leads to the apparent communication between sites whose separation may be greater than 60 Å.

The essential features of the structural response are summarised in Table 3.6. In each of the four proteins, the changes in the constellation of groups required to bring residues into their correct orientation for optimum binding of substrate are linked to changes at the subunit–subunit interface. The conformational response at the ligand-binding sites are small compared with changes elsewhere and, indeed, in each protein a significant portion of the recognition site is already available in the T state. In haemoglobin, a movement of 0.5 Å of the haem iron is required to create a high affinity oxygen binding site; in aspartate carbamoyltransferase, two basic groups swing in to recognise the aspartate; in phosphofructokinase, and in phosphorylase, movement of only one basic group creates a phosphate recognition site. Perhaps the relative smallness of these changes is not surprising. Site-directed mutagenisis experiments have shown that 100-fold changes in K_m can be achieved with a single amino acid replacement (Wilkinson *et al.*, 1984). Indeed, the most significant aspects of the changes required to create the high affinity site were anticipated on the basis of a single structure (either T or R state) for haemoglobin, phosphofructokinase and phosphorylase. In contrast, it was impossible to predict the large changes in quaternary structure required to achieve these tertiary structural changes on the basis of a single structure (T or R). Knowledge of both T and R state structures were essential to elucidate the routes for communication between the sites.

The nature of the subunit–subunit interfaces in the four proteins are diverse and there are no common features among the types of secondary structures involved. In haemoglobin, the interface involves docking of the end of one α helix into the groove of another α helix; in aspartate carbamoyltransferase there is packing of a loop of chain against the corresponding loop of the other subunit; in phosphofructokinase there is a distorted β structure; and in glycogen phosphorylase the packing of two α helices and loop to β strand interactions. Despite this diversity, the structures have in common the feature that each interface appears to be designed to admit only two possible modes of docking. This is achieved in haemoglobin by docking of the FG corner into adjacent grooves on the C α helix, in aspartate carbamoyltransferase by movement of the loop which allows it to dock either side by side or opposed to the corresponding loop on the adjacent subunit, in phosphofructokinase the contacts at the interface are weakened in the R state by intercalation of water between the β sheet interface, and in phosphorylase the helices twist and slide past each other to adopt an alternative mode of packing. In almost all of the examples the subunit interactions are reinforced further by ionic linkages, and loss of even one of these ionic linkages is sufficient to shift the equilibrium in favour of the R state. In general, the subunit–subunit interfaces are more extensive and more highly constrained in the T state that in the R state. Phosphorylase *a* is an exception. Here, the ordering of the N-terminal tail on phosphorylation results in stronger interactions including the important ionic bonds between the phosphate and the two arginine residues. However, the basic character of the N-terminal tail and the Ser-P recognition site create an electrostatic repulsion that leads to the disorder of the N-terminal tail. This repulsion can be viewed as a restraint. Moreover, the shift in the tower helices on the T to R conversion of phosphorylase *b* leads to a weakening of the subunit contacts of the tower helices so that overall there is little net change in the subunit contact areas between T and R state phosphorylase *b*.

In aspartate carbamoyltransferase and phosphofructokinase, the substrate binding sites are at the subunit interface and the ionisable groups that contribute to the R state high affinity substrate binding site come from the adjacent subunit. Thus, the changes in tertiary structure are directly involved in changes at the subunit interface. In phosphorylase the route is indirect and the substrate to interface link is through two amino acids (Arg 569 and Pro 281). In haemoglobin, the route is through two residues on adjacent turns of an α helix (His 92 (F8) to His 97 (FG4)). All the regulator sites in haemoglobin, phosphofructokinase and phosphorylase are at the subunit interface and effector binding is therefore well placed to stabilise either the T or R quaternary structural states. As shown with the effectors of haemoglobin (H^+, Cl^-, DPG) and phosphorylase (Ser-P, AMP, nucleosides), regulatory sites can occur on different regions on the subunit but regulatory effects are still exerted through the subunit–interface contacts. The same regulatory site can function either as an activator or inhibitor site depending on the ligand bound (ATP or CTP for aspartate carbamoyltransferase, ADP or PEP for phosphofructokinase, AMP or glucose-6-P for phosphorylase).

The rotational and translational movements of the subunits with respect to one another as the structure moves from the T to R state are all such as to conserve the symmetry of the oligomeric assembly (222 symmetry for tetrameric haemoglobin and phosphofructokinase, 32 symmetry for hexameric aspartate carbamoyltransferase, and two-fold symmetry for dimeric phosphorylase). This is most often achieved by a rotation about an axis perpendicular to the symmetry axis of the oligomer (Perutz, 1988, 1989). The cyclic symmetry of the oligomeric assembly implies that there must be a protein-free region on the axis of symmetry, for two groups cannot occupy the same position at the same time. In each of the proteins, there is a water-filled hole down the centre of the molecule, often of considerable dimensions (about 10 Å in diameter in phosphorylase). Movements of subunits relative to one another can lead to a change in the dimension of the hole as seen in haemoglobin. Here, the channel between the β subunits forms the DPG binding site in the T state and the channel between the α subunits forms a non-physiological drug binding site in the R state (Perutz *et al.*, 1986). Changes in diameters of these pores achieved by subunit rotations are of relevance to a possible understanding of receptor based ion channels.

Once solved, these structures provide a satisfying explanation for the correlation between tertiary structure and quaternary structure. Interactions at the subunit interface and at the ligand-binding sites are exquisitely tailored to produce a switch between the T and R states and hence to allow events at one binding site to be communicated to a distant binding site. Understanding receptor-based site communication appears a formidable problem at the present time, but the allosteric models provide some clues for the type of structural changes that might be anticipated.

Acknowledgements

I am most grateful to D. Barford, P. R. Evans, W. N. Lipscomb and M. F. Perutz for their help in the preparation of this manuscript and to A. Roper and S. Ley for help with the diagrams.

Note added in proof

Since the time of submission of this chapter in 1989, a number of crystallographic papers have been published that add to the understanding of the allosteric proteins described.

Aspartate carbamoyltransferase

Gouaux, J. E. & Lipscomb, W. N. (1990). Crystal structure of phosphonoacetamide ligated T and phosphonoacetamide and malonate ligated R states of ATPase at 2.8 Å resolution and neutral pH. *Biochemistry*, **29**, 389–402.

Stevens, R. C. Gouaux, J. E. & Lipscomb, W. N. (1990). Structural consequences of effector binding to the T state of aspartate carbamoyltransferase. *Biochemistry*, **29**, 7691–701.

Gouaux, J. E., Stevens, R. C. & Lipscomb, W. N. (1990). Crystal structures of aspartate carbamoyltransferase ligated with phosphonoacetamide, malonate and CTP or ATP at 2.8 Å resolution and neutral pH. *Biochemistry*, **29**, 7702–15.

Phosphofructokinase

Schirmer, T. & Evans, P. R. (1990). Structural basis of the allosteric behaviour of phosphofructokinase. *Nature*, London, **343**, 140–5.

Berger, S. A. & Evans, P. R. (1990). Active site mutants altering the co-operativity of *E. coli* phosphofructokinase. *Nature*, London, **343**, 575–6.

Kundrot, C. E. & Evans, P. R. (1991). Designing an allosterically locked phosphofructokinase. *Biochemistry*, **30**, 1478–84.

Glycogen phosphorylase

Johnson, L. N. & Barford, D. (1990). Glycogen phosphorylase; the structural basis of the allosteric response and comparison with other allosteric proteins. *Journal of Biological Chemistry*, **265**, 2409–12.

Barford, D., Hu, S-H. & Johnson, L. N. (1991). Structural mechanism for glycogen phosphorylase control by phosphorylation and AMP. *Journal of Molecular Biology*, **218**, 233–60.

Fructose-1,6-bisphosphatase Recent structure determination of an additional allosteric protein.

Ke, H., Zhang, Y. & Lipscomb, W. N. (1990). Crystal structure of fructose-1,6-bisphosphatase complexed with fructose 6-phosphate, AMP and magnesium. *Proceedings of the National Academy of Sciences*, USA, **87**, 5243–7.

Ke, H., Liang, J-Y., Zhang, Y. & Lipscomb, W. N. (1991). Conformational transition of fructose-1,6-bisphosphatase: structural comparison between the AMP complex (T form) and the fructose 6-phosphate complex (R form). *Biochemistry*, **30**, 4412–20.

References

Baldwin, J. M. & Chothia, C. (1979). Haemoglobin: the structural changes related to ligand binding and its allosteric mechanism. *Journal of Molecular Biology*, **129**, 183–91.

Barford, D. & Johnson, L. N. (1989). The allosteric mechanism of glycogen phosphorylase *b*. *Nature*, London, **340**, 609–16.

Bennett, W. S. & Huber, R. (1984). Structural and functional aspects of domain motions in proteins. *Critical Reviews in Biochemistry*, **15**, 291–384.

Bethell, M. R., Smith, K. E., White, J. S. & Jones, M. E. (1968). Carbamoyl phosphate: an allosteric substrate for aspartate carbamoyltransferase from *E. coli*. *Proceedings of the National Academy of Sciences*, USA, **60**, 1442–9.

Black, W. & Wang, J. H. (1968). Studies on the allosteric activation of glycogen phosphorylase *b* by nucleotides. *Journal of Biological Chemistry*, **243**, 5892–8.

Blangy, D., Buc, H. & Monod, J. (1968). Kinetics of the allosteric interactions of phosphofructokinase from *E. coli. Journal of Molecular Biology*, **31**, 13–35.

Brzowowski, A., Derewenda, Z., Dodson, E., Dodson, G., Grabowski, M., Liddington, R., Skarzynski, T. & Vallely, D. (1984). Bonding of molecular oxygen to T state hemoglobin. *Nature*, London, **305**, 74–76.

Changeux, J. P., Devillers-Thiery, A. & Chemouille, P. (1984). The acetylcholine receptor is an allosteric protein. *Science*, **225**, 1335–45.

Changeux, J. P., Giraudat, J. & Dennis, M. (1987). The nicotinic acetylcholine receptor: molecular architecture of a ligand regulated ion channel. *Trends in Pharmacological Science*, **8**, 459–65.

Cohen, P. (1988). Protein phosphorylation and hormone action. *Proceedings of the Royal Society B*, **234**, 115–44.

Dickerson, R. E. & Geiss, I. (1983). *Haemoglobin: Structure, Function, Evolution, Pathology.* Benjamin/Cummings Publishing Co., Kenlo Park.

Eklund, H., Samana, J. P., Wallen, L., Branden, C-I., Akeson, A. & Jones, T. A. (1981). Structure of a triclinic ternary complex of horse liver alcohol dehydrogenase at 2.9 Å resolution. *Journal of Molecular Biology*, **146**, 561–87.

Engers, H. D. & Madsen, N. B. (1968). The effect of anions on the activity of phosphorylase b. *Biochemical and Biophysical Communications*, **33**, 49–54.

Evans, P. R., Farrants, G. W. & Hudson, P. J. (1981). Phosphofructokinase: structure and control. *Philosophical Transactions of the Royal Society series B*, **293**, 52–62.

Evans, P. R., Farrants, G. W. & Lawrence, M. C. (1986). Crystallographic structure of allosterically inhibited phosphofructokinase at 7 Å resolution. *Journal of Molecular Biology*, **191**, 713–20.

Evans, P. R. & Hudson, P. J. (1979). Structure and control of phosphofructokinase from *Bacillus stearothermophilus*. *Nature*, London, **279**, 500–4.

Fasold, H., Ortanderl, F., Huber, R., Bartels, K. & Schwager, P. (1972). Crystallisation and crystallographic data for rabbit muscle phosphorylase *a* and *b. FEBS Letters*, **21**, 229–32.

Fermi, G., Perutz, M. F., Shanaan, B. & Fourme, R. (1984). The crystal structure of human deoxyhemoglobin at 1.74 Å resolution. *Journal of Molecular Biology*, **175**, 159–74.

Fletterick, R. J. & Madsen, N. B. (1980). The structure and related functions of phosphorylase *a. Annual Reviews in Biochemistry*, **49**, 31–61.

Foote, J. & Schachman, H. K. (1985). Homotropic effects in aspartate transcarbamoylase: what happens when enzyme binds single molecule of bisubstrate analog *N*-phosphonacetyl-L-aspartate? *Journal of Molecular Biology*, **186**, 175–84.

Gerhart, J. C. & Pardee, A. B. (1962). The enzymology of control by feedback inhibition. *Journal of Biological Chemistry*, **237**, 891–6.

Gerhart, J. C. & Schachman, H. K. (1968). Allosteric interactions in aspartate carbamoyltransferase: evidence for different conformational states of protein in the presence and absence of specific ligands. *Biochemistry*, **7**, 538–52.

Gilman, A. G. (1987). G proteins: transducers of receptor generated signals. *Annual Reviews in Biochemistry*, **56**, 615–49.

Giraudat, J., Dennis, M., Heidman, T., Haumont, P-Y., Lederer, F. & Changeux, J. P. (1987). Structure of the high affinity binding site for non-competitive blockers of the acetylcholine receptor: [³H] chlorpromazine labels homologous residues in the β and δ chains. *Biochemistry*, **26**, 2410–18.

Gouaux, J. E., Krause, K. L. & Lipscomb, W. N. (1987). The catalytic mechanism of *Escherichia coli* aspartate carbamoyltransferase: a molecular modelling study. *Biochemical and Biophysical Research Communications* **142**, 893–7.

Gouaux, J. E. & Lipscomb, W. N. (1988). Three dimensional structure of carbamoyl phosphate and succinate bound to aspartate carbamoyl transferase. *Proceedings of the National Academy of Sciences, USA*, **85**, 4205–8.

Gouaux, J. E. & Lipscomb, W. N. (1989). Structural transitions in native aspartate carbamoyltransferase. *Proceedings of the National Academy of Sciences, USA*, **86**, 845–8.

Graves, D. J. & Wang, J. H. (1972). α-glucan phosphorylases – chemical and physical basis of catalysis and regulation. In: *The Enzymes* (3rd ed. Boyer, P. ed.) vol. 7, pp. 435–82, Academic Press, New York.

Hajdu, J., Acharya, K. R., Stuart, D. I., McLaughlin, P. J., Barford, D., Klein, H. & Johnson, L. N. (1987). Catalysis in the crystal: synchrotron radiation studies with glycogen phosphorylase *b*. *EMBO Journal*, **6**, 539–45.

Hellinga, H. W. & Evans, P. R. (1987). Mutations in the active site of *Escherichia coli* phosphofructokinase. *Nature*, London, **327**, 437–9.

Helmreich, E. (1969). Allosteric control of phosphorylase. *FEBS Symposium*, **19**, 131–48.

Helmreich, E. J. M. & Klein, H. W. (1980). The role of pyridoxal phosphate in the catalysis of glycogen phosphorylase. *Angew. Chem. Int. Ed. Eng.*, **19**, 441–55.

Helmreich, E., Michaelides, M. C. & Cori, C. (1967). Effects of substrates and a substrate analog on the binding of 5′-adenylic acid to muscle phosphorylase *a*. *Biochemistry*, **6**, 3695–710.

Hervé, G. (1988). Aspartate carbamoyltransferase from *Escherichia coli* in *Allosteric Enzymes* (Herve, G., ed.) CRC Press, Boca Raton, Florida.

Honzatko, R. B., Crawford, J. L., Monaco, H. L., Ladner, J. E., Edwards, B. F. P., Evans, D. R., Warren, S. G., Wiley, D. C., Ladner, R. C. & Lipscomb, W. N. (1982). Crystal and molecular structure of native and CTP liganded aspartate carbamoyltransferase from *Escherichia coli*. *Journal of Molecular Biology*, **160**, 219–63.

Howlett, G. J., Blackburn, M. N., Compton, J. G. & Schachman, H. K. (1977). Allosteric regulation of aspartate carbamoyltransferase: analysis of structural and functional behaviour in terms of a two state model. *Biochemistry*, **16**, 5091–9.

Howlett, G. J. & Schachman, H. K. (1977). Allosteric regulation of aspartate carbamoyltransferase. Changes in sedimentation coefficient promoted by the bisubstrate analog *N*-phosphoacetyl-L-aspartate. *Biochemistry*, **16**, 5077–83.

Huber, R. & Bode, W. (1978). Structural basis of the activation and action of trypsin. *Accounts of Chemical Research*, **11**, 114–22.

Imai, K. (1982). *Allosteric Effects in Haemoglobin*. Cambridge University Press, Cambridge, UK.

Janin, J. & Wodak, S. J. (1983). Structural domains in proteins and their role in the dynamics of protein function. *Progress in Biophysical and Molecular Biology*, **42**, 21–78.

Johnson, L. N., Acharya, K. R., Jordan, M. & McLaughlin, P. J. (1990). The refined structure of the phosphorylase *b*-heptulose 2-P complex. *Journal of Molecular Biology*, **211**, 645–61.

Johnson, L. N., Hajdu, J., Acharya, K. R., Stuart, D. I., McLaughlin, P. J., Oikonomakos, N. G. & Barford, D. (1988). Glycogen phosphorylase *b*. In: *Allosteric Proteins* (Herve, G., ed.) pp. 81–127 CRC Press, Boca Raton, Florida.

Kantrowitz, E. R. & Lipscomb, W. N. (1988). *Escherichia coli* aspartate carbamoyltransferase: the relationship between structure and function. *Science*, **241**, 669–74.

Kantrowitz, E. R., Pastra-Landis, S. C. & Lipscomb, W. N. (1980). *Trends in Biochemical Science*, **5**, 124–53.

Kastenschmidt, L. L., Kastenschmidt, J. & Helmreich, E. (1968). The effect of temperature on the allosteric transitions of rabbit muscle phosphorylase *b*. *Biochemistry*, **7**, 4543–56.

Kasvinsky, P. J., Shechosky, S. & Fletterick, R. J. (1978). Synergistic regulation of phosphorylase *a* by glucose and caffeine. *Journal of Biological Chemistry*, **253**, 9102–6.

Ke, H. M., Honzatko, R. B. & Lipscomb, W. N. (1984). Structure of unligated aspartate carbamoyltransferase of *Escherichia coli* at 2.6 Å resolution. *Proceedings of the National Academy of Sciences, USA*, **81**, 4037–40.

Ke, H., Lipscomb, W. N., Cho, Y. & Honzatko, R. B. (1988). Complex of *N*-phosphonacetyl-L-aspartate with aspartate carbamoyltransferase. *Journal of Molecular Biology*, **204**, 725–47.

Kim, K. H., Pan, Z., Honzato, R. B., Ke, H. M. & Lipscomb, W. N. (1987). Structural asymmetry in the CTP-liganded form of aspartate carbamoyltransferase from *Escherichia coli*. *Journal of Molecular Biology*, **196**, 853–75.

Klein, H. W., Im, M. J. & Palm, D. (1986). Mechanism of the phosphorylase reaction: utilisation of D-gluco-hept-1-enitol in the absence of primer. *European Journal of Biochemistry*, **157**, 107–14.

Koshland, D. E. (1958). Application of a theory of enzyme specificity to protein synthesis. *Proceedings of the National Academy of Sciences, USA*, **44**, 98–104.

Koshland, D. E. (1969). Conformational aspects of enzyme regulation. *Current Topics in Cell Regulation*, **1**, 1–27.

Koshland, D. E. & Neet, K. E. (1969). The catalytic and regulatory properties of enzymes. *Annual Reviews in Biochemistry*, **37**, 359–410.

Koshland, D. E., Nemethy, G. & Filmer, D. (1966). Comparison of experimental binding data and theoretical models in proteins containing subunits. *Biochemistry*, **5**, 365–85.

Krause, K. L., Volz, K. W. & Lipscomb, W. N. (1987). 2.5 Å structure of aspartate carbamoyltransferase complexed with the bisubstrate analog *N*-(phosphonacetyl)-L-aspartate. *Journal of Molecular Biology*, **193**, 527–53.

Ladjimi, M. M., Kelleher, K. S. & Kantrowitz, E. R. (1988). The relationship between domain closure and binding, catalysis and regulation in *E. coli* aspartate carbamoyltransferase. *Biochemistry*, **27**, 268–76.

Lau, F. T-K. & Fersht, A. R. (1987). Conversion of allosteric inhibition to activation in phosphofructokinase by protein engineering. *Nature*, London, **326**, 811–12.

Lesk, A. M. & Chothia, C. (1984). Mechanisms of domain closure in proteins. *Journal of Molecular Biology*, **174**, 175–91.

Liddington, R., Derewenda, Z., Dodson, G. & Harris, D. (1988). Structure of liganded T-state of haemoglobin identifies the origin of cooperative oxygen binding. *Nature*, London, **331**, 725–8.

Lorek, A., Wilson, K. S., Sansom, M. S. P., Stuart, D. I., Stura, E. A., Jenkins, J. A., Zanotti, G., Hajdu, J. & Johnson, L. N. (1984). Allosteric interactions of glycogen phosphorylase *b*. *Biochemical Journal*, **218**, 45–60.

McLaughlin, P. J., Stuart, D. I., Klein, H. W., Oikonomakos, N. G. & Johnson, L. N. (1984). Substrate cofactor interactions with glycogen phosphorylase *b*: a binding study in the crystal with heptenitol and heptulose-2-phosphate. *Biochemistry*, **23**, 5862–73.

Madsen, N. B. (1986). Glycogen phosphorylase. In: *The Enzymes*, 3rd edn (Boyer, P. D. & Krebs, E. G., eds) **17**, pp. 366–94. Academic Press, New York.

Madsen, N. B., Kasvinsky, P. J. & Fletterick, R. J. (1978). Allosteric transitions of phosphorylase *a* and the regulation of glycogen metabolism. *Journal of Biological Chemistry*, **253**, 9097–101.

Marmorstein, R. Q. & Sigler, P. B. (1988). Structure and mechanism of the trp repressor/operator system. In: *Nucleic Acids and Molecular Biology* (Eckstein, F., ed.) Springer Verlag, Heidelberg.

Monaco, H. L., Crawford, J. L. & Lipscomb, W. N. (1978). The three dimensional structures of aspartate carbamoyl transferase from *Escherichia coli* and its complex with CTP. *Proceedings of the National Academy of Sciences, USA*, **75**, 5276–80.

Monod, J., Wyman, J. & Changeux, J. P. (1965). On the allosteric transitions: a plausible model. *Journal of Molecular Biology*, **12**, 88–118.

Oikonomakos, N. G., Johnson, L. N., Acharya, K. R., Stuart, D. I., Barford, D., Hajdu, J., Varvill, K. M., Melpidou, A. E., Papageorgiou, T., Graves, D. J. & Palm, D. (1987). Pyridoxal phosphate site in glycogen phosphorylase *b*: structure in native enzyme and in three derivatives with modified cofactors. *Biochemistry*, **26**, 8381–9.

Otwinowski, Z., Schevitz, R. W., Zhang, R. G., Lawson, C. L., Joachimiak, A., Marmorstein, R. Q., Luisi, B. & Sigler, P. B. (1988). The crystal structure of the trp repressor/operator complex at atomic resolution. *Nature*, London, **335**, 321–9.

Perutz, M. F. (1970). Stereochemistry of cooperative effects in hemoglobin. *Nature*, London, **228**, 726–39.

Perutz, M. F. (1987). Molecular anatomy, physiology, and pathology of hemoglobin. In: *Molecular Basis of Blood Diseases* (Stammatoyanopoulos, G., Niehuis, A. W., Leder, P. & Majerus, P. W., eds) pp. 127–78. W. B. Saunders Co., Philadelphia, USA.

Perutz, M. F. (1988). Control by phosphorylation. *Nature*, London (News and Views), **336**, 202–3.

Perutz, M. (1989). Cooperativity and allosteric regulation in proteins. *Quarterly Reviews in Biophysics*, **22**, 139–237.

Perutz, M. F., Fermi, G., Abraham, D. J., Poyart, C. & Bursaux, E. (1986). Hemoglobin as a receptor of drugs and peptides: X-ray studies of the stereochemistry of binding. *Journal of the American Chemical Society*, **108**, 1064–78.

Poorman, R. A., Randolph, A., Kemp, R. G. & Heinrikson, R. (1984). Evolution of phosphofructokinase – gene duplication and creation of new effector sites. *Nature*, London, **309**, 467–9.

Robey, E. A., Wente, S. R., Markby, W., Flint, A., Yang, Y. R. & Schachman, H. K. (1986). Effect of amino acid substitutions on the catalytic and regulatory properties of aspartate carbamoyltransferase. *Proceedings of the National Academy of Science, USA*, **83**, 5934.

Rypniewski, W. R. & Evans, P. R. (1989). The crystal structure of unliganded phosphofructokinase from *Escherichia coli*. *Journal of Molecular Biology* **207**, 805–21.

Sansom, M. S. P., Stuart, D. I., Acharya, K. R., Hajdu, J., Mclaughlin, P. J., Johnson, L. N. (1985). Glycogen phosphorylase *b* – the molecular anatomy of a large regulatory enzyme. *Journal of Molecular Structure*, **123**, 3–25.

Schachman, H. K. (1988). Can a simple model account for the allosteric transition of aspartate transcarbamoylase? *Journal of Biological Chemistry*, **263**, 18583–6.

Schlessinger, J. (1988). The epidermal growth factor as a multifunctional allosteric protein. *Biochemistry*, 27, 3119–23.

Shanaan, B. (1983). Structure of oxyhemoglobin at 2.1 Å resolution. *Journal of Molecular Biology*, **171**, 31–59.

Shirakihara, Y. & Evans, P. R. (1988). Crystal structure of the complex of phosphofructokinase from *Escherichia coli* with its reaction products. *Journal of Molecular Biology*, **204**, 973–94.

Sharzynski, T. & Wonacott, A. J. (1988). Coenzyme induced conformational changes in glyceraldehyde-3-phosphate dehydrogenase from *Bacillus stearothermophilus*. *Journal of Molecular Biology*, **203**, 1097–118.

Sotiroudis, T. G., Oikonomakos, N. G. & Evangelopoulos, A. E. (1979). Effect of sulphated polysaccharides and sulphate anions on the AMP dependent activity of phosphorylase *b*. Biochemical and Biophysical Research Communications, **90**, 234–9.

Sprang, S. R., Acharya, K. R., Goldsmith, E. J., Stuart, D. I., Varvill, K., Fletterick, R. J., Madsen, N. B. & Johnson, L. N. (1988). Structural changes in glycogen phosphorylase induced by phosphorylation. *Nature*, London, **336**, 215–21.

Sprang, S. R. & Fletterick, R. J. (1979). The structure of glycogen phosphorylase *a* at 2.5 Å resolution. *Journal of Molecular Biology*, **131**, 523–51.

Sprang, S. R., Goldsmith, E. J. & Fletterick, R. J. (1987). Structure of the nucleotide activation switch in glycogen phosphorylase *a*. *Science*, **237**, 1012–19.

Steitz, T. A., Shoham, M. & Bennett, W. S. (1981). Structural dynamics of yeast hexokinase during catalysis. *Philosophical Transactions of the Royal Society Ser B*, **298**, 43–52.

Toyoshima, C. & Unwin, N. (1988). The ion channel of acetylcholine receptor reconstituted from images of post-synaptic membranes. *Nature*, London, **336**, 247–50.

Unwin, N., Toyoshima, C. & Kubalek, E. (1988). Arrangement of the acetylcholine receptor subunits in resting and desensitised states, determined by cryoelectron microscopy of crystallised Torpedo post synaptic membranes. *Journal of Cell Biology*, **1107**, 1123–38.

Vandenbunder, B. & Buc, H. (1983). The arginine residues interacting with glucose-1-phosphate in glycogen phosphorylase. *European Journal of Biochemistry*, **133**, 509–13.

Wiegand, G., Remington, S., Deisenhofer, J. & Huber, R. (1984). Crystal structure analysis and molecular model of a complex of citrate synthase with oxaloacetate and S-acetonyl coenzyme A. *Journal of Molecular Biology*, **174**, 205–19.

Wilkinson, A., Fersht, A. R., Blow, D. M., Carter, P. & Winter, G. (1984). A large increase in enzyme–substrate affinity by protein engineering. *Nature*, London, **307**, 187–8.

Zhang, R. G., Joachimiak, A., Lawson, C. L., Schevitz, R. W., Otwinowski, Z. & Sigler, P. B. (1987). The crystal structure of trp repressor at 1.8 Å shows how binding tryptophan enhances DNA affinity. *Nature*, London, **327**, 591–7.

4

Subunits of signal-transducing receptors: types, analysis and reconstitution

ERIC A. BARNARD

Introduction

This chapter will consider the general features of the subunits of membrane-located receptors for neurotransmitters, neuromodulators and circulating hormones. These are the receptor types that are concerned in signal transmission between neurones or between glia and neurones, or between neurones and muscles or other effector organs, as well as in responses of glands and muscles to circulating signals. This will exclude proteins involved primarily in transport processes within cells or gene regulation (e.g. the steroid hormone and cognate receptors, for which see Rosenfeld, this volume), or in the transport of nutrients into cells, and will also exclude the receptors specific to the immune systems, all of which serve entirely different functions from those considered here. Since the receptors of the categories included are all membrane-spanning and signal-transducing proteins, they can be treated together, although their protein structures can differ widely. This group must comprise many hundreds of receptor types and sub-types.

Such knowledge as we initially had of their composition came, of course, by way of their isolation and their analysis in denaturing gel electrophoresis. In the past few years, this knowledge came, instead, mostly from DNA cloning. We may take the watershed in our knowledge of receptors to be in the year 1982. That was the point at which the first recombinant receptor protein was described, i.e. the α-subunit of the nicotinic acetylcholine (ACh) receptor from *Torpedo* electric organ (Sumikawa *et al.*, 1982; Noda *et al.*, 1982). From then on, the field has undergone a revolution, in that: (*a*) entire receptor polypeptide sequences became known from clones, these being totally unknown hitherto due to the limitations of the required technology at the protein level; (*b*) only a handful of receptors could be completely purified, but large numbers were now nevertheless obtained in recombinant form.

Hence, explosive growth has occurred in the number of receptors whose subunit composition is known. It is a chastening thought that, despite many years of work in numerous laboratories on receptor purification and subunit analysis, almost all of the receptor subunit compositions that we know have come to us only in these past 8 years, through cDNA cloning.

Types of subunits in signal-transducing receptors

The deduced amino acid sequences of a large number of receptor subunit types are now known, and within some of those types multiple sub-types or isoforms have yielded further sequences, and species variants thereof in some cases have also been analysed thus. By these means, in all, several hundred receptor subunit primary structures are now available. By considering this data-base, some generalisations can be reached about their structures overall. It is found that these receptors can contain a monomeric polypeptide, or can contain two or more identical subunits, i.e. are homo-oligomeric, or they contain two to four types of non-identical subunits, i.e. are hetero-oligomeric. Being integral membrane proteins, all of their subunits contain at least one hydrophobic, trans-membrane domain (TM) and also extracellular (EC) and intracellular (IC) domains. It is noteworthy that, in all of the structures determined so far for their subunits, the EC domain is a large one and so is the IC domain, the other links between the trans-membrane domains being very short and unlikely to be able to contain any entire ligand-binding site.

The range of different subunit structures which has emerged can also be considered. Prior to 1982, the classification of receptors was necessarily based upon their pharmacologies alone, but the new knowledge of their subunits from molecular biology makes possible a structurally based classification, as shown in Tables 4.1 and 4.2. Individual receptor types must still be defined by their pharmacology, of course, but those types can logically be grouped into classes by the nature of their subunits. It was surely unanticipated that only three classes of receptor structure would be found when their sequences were deciphered, but to date that is in fact the case. It is the nature of the receptor subunit which determines these three classes, as Table 4.1 summarizes. The subunit type, including its hydro-phobicity distribution, determines the arrangement of the assembled receptor in relation to the cell membrane, and it is this that provides the basic principle of this classification and relates it to the corresponding

Table 4.1. *Signal-transducing receptors: structural classification*

Class	Subunit structure	Transduction system	Examples of ligands
1. Transmitter-gated ion channels (A) extra-cellularly activated (B) intra-cellularly activated	Hetero-oligomeric[a] or homo-oligomeric	Unitary receptor/channel	See Table 4.2
2. Seven-hydrophobic-domain polypeptide	Monomers or homo-dimers	Via a G-protein: (A) plus a diffusible messenger (B) acting directly on a channel[b]	(A) Almost all small transmitters and neuropeptides (B) Atrial muscarinic[c]; neuronal α_1-adrenergic[d]
3. Single-hydrophobic-domain polypeptide	Monomers or post-translational hetero-oligomers	The subunit itself is: (A) a tyrosine kinase[e] (B) a guanylate cylase[e] (C) non enzymic	All are polypeptides (A) Many mitogens; insulin (B) Natriuretic peptides[e] (C) Nerve growth factor[f]; growth hormone

[a] Nicotinic, γ-Aminobutyrate-A (GABA$_A$), Glycine, Glutamate receptors. [b] Reviewed by Brown & Birnbaumer (1990). [c] Yatani *et al.* (1987). [d] Lipscombe *et al.* (1989). [e] See Fig. 4.1. [f] Vale *et al.*, this volume.

Table 4.2. *Transmitter-gated ion channels*

Operator ligand	Ion selectivity	Super-family	Trans-membrane domains	References
A. Extra-cellularly activated				
GABA$_A$	Cl⁻	IA	4	Barnard et al. (1987)
Glycine	Cl⁻	IA	4	Langosch et al. (1990)
ACh (nicotinic, muscle type)	Na⁺, K⁺	IA	4	Maelicke, this volume
ACh (nicotinic, neuronal)	Na⁺, K⁺	IA	4	Maelicke, this volume
Glutamate:non-NMDA	Na⁺, K⁺	IB	4	Barnard & Henley (1990)
Glutamate:NMDA	Na⁺, K⁺, Ca²⁺	IB	4	Moriyoshi et al. (1990)
5-HT$_3$	Na⁺, K⁺	IA	4	Maricq et al. (1991)
ATP (P$_{2x}$, channel-opening)	Ca²⁺, Na⁺, Mg²⁺	?		Burnstock & Kennedy (1986); Benham & Tsien (1987)
B. Intra-cellularly activated				
cGMP (photoreceptors)	Na⁺, K⁺	II	(6)	Kaupp (1991)
cAMP (olfactory neurones)	Na⁺, K⁺	II	(6)	Dhallan et al. (1990)
ATP (channel-closing)	K⁺	?		Bernardi et al., (1988), Dunne et al. (1989)
IP$_3$ (organelles and plasma membrane)	Ca²⁺	III	(?8)	Suppatopone et al. (1988); Furiuchi et al. (1989); Mignery et al. (1989); Kuno & Gardner (1987)
? (ryanodine receptor)	Ca²⁺	III	(4)	Takeshima et al. (1989)

This table lists all of the receptors of the nervous system so far known in which an endogenous signalling molecule activates the opening or closing of an ion channel contained in the receptor structure. Where their protein sequences are known, they fall into the super-families of homologous proteins indicated. Super-families 1A and 1B show only a low degree of resemblance. The trans-membrane domains are inferred from a hydropathy plot of the protein sequence as in Fig. 4.2: in some cases the assignment is less clear, indicated by the numbers in parentheses.

Abbreviations: GABA, γ-aminobutyrate; ACh, acetylcholine; NMDA, *N*-methyl-D-aspartate; 5-HT, serotonin; IP$_3$, inositol triphosphate; P$_{2x}$, purinergic receptor, type 2x.

transduction mechanism. It can now be seen that the logical basis of classification of the signal-transducing receptors is a hierarchy.

One Class of subunit organisation → several transduction

mechanisms → many individual receptor pharmacologies

In Class 1, the subunits have structures which lead them to assemble in an oligomer surrounding a pore through the membrane. Transduction here is the opening *or* the closing of a cation *or* an anion channel. In Class 2, most of the subunit mass is packed within the membrane and each subunit carries a G-protein-interactive sequence on its intracellular face. This organisation is adapted in turn to one of several signal transduction mechanisms (using one of a variety of second messengers or a $G\alpha$ subunit directly) and each of these to one of numerous possible agonist recognition sites. The Class 3 subunit structure is designed for a minimum exposure of the subunit to the membrane lipids (one traversal of the bilayer), which facilitates receptor migration and internalisation (Fig. 4.1) The transduction systems of the receptors of Classes 3A and 3C (see the chapters in this volume by E. Shooter and by K. Siddle) are still largely undefined and hence the principles in common which lead various receptor types to have the Class 3 subunit topology are not known. It can be noted, however, that all of the receptors in that Class produce long-term (e.g. nuclear-based) signalling effects, as a part of their function.

It is important to note that it is the secondary structure of the subunit which is found to determine the receptor type, and not primary structure homology. Within each of the three Classes, some of the members may share no sequence homology whatsoever, as will be noted.

Transmitter-gated ion channels

These are the receptors which perform fast signalling and hence are recognised electrophysiologically by their instantaneous response to an applied agonist and by their independence of any intra-cellular or membrane-diffusible factor. The latter feature can best be demonstrated for them in patch-clamping: in excised, sealed membrane patches, freed of diffusible components the channels still respond but only when the agonist is applied inside the pipette. By such criteria, the receptors listed in Table 4.2 contain an intrinsic ion channel.

These features are not entirely exclusive, in that after an initial period such receptors may show a need for some factor (such as ATP), but this is merely for their maintenance in an activatable state over a long period.

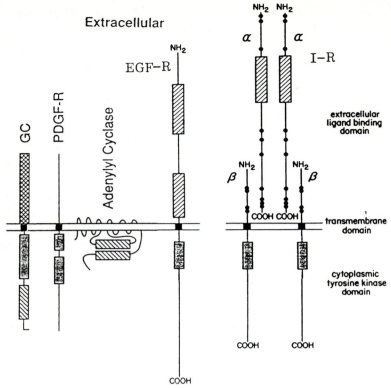

Fig. 4.1. Representative members of Class 3 of the receptor subunits, those which have a single membrane-spanning segment (usually of 23 residues). GC is the guanylate cyclase/natriuretic peptide receptor. PDGF-R is the platelet-derived growth factor receptor. EGF-R is the epidermal growth factor receptor. I-R is the insulin receptor. Corresponding homologous regions have the same shading. Also shown is the deduced organisation of membrane-bound brain adenylate cyclase (Krupinski *et al.*, 1989). Two co-homologous catalytic domains of that cyclase are also homologous to the identified catalytic domain of GC (shaded similarly). A cytoplasmic tyrosine kinase domain exists in the growth factor receptors and the insulin receptor, and a homologous but non-catalytic domain (of unknown significance) exists in GC. The transmitter-binding site is known to be in the extracellular N-terminal domain in all these receptors. Highly cysteine-rich segments there are associated with the binding domain in some of the growth factor receptors (shown by the single-hatched upper boxes). GC has little homology to the others in the extracellular domain. Note that adenylate cyclase, which is not a receptor, has a very different structure, with no extracellular domain. The insulin receptor α and β chains are formed by a specific proteolytic cleavage of a single gene product (prior to glycosylation) of 154 kD. Based upon Ullrich *et al.* (1985) for the human insulin and EGF receptors, on Krupinski *et al.* (1989) for adenylate cyclase and Schulz *et al.* (1989) for GC.

Likewise, such a receptor type might not be employed in fast transmission at *all* of its locations: $GABA_A$ receptor and non-NMDA glutamate receptors, for example, occur on some glial cells as well as on neurones.

The prototypic member of this class is, of course, the nicotinic ACh receptor of the muscle or electric organ type, where the case was first established absolutely that the subunits of the receptor form an oligomer which spans the membrane and encloses a central ion channel. The details of this structure and of the organisation of the subunits in it are described by A. Maelicke in this volume. In that case there are five homologous subunits in the oligomer, only two of them identical ($\alpha_2\beta\gamma\delta$ structure). It is instructive to note that a cation channel of essentially the same properties is formed within the neuronal type of nicotinic receptor by homologous subunits of only two sequece categories, α and β. Here again the structure is pentameric, being $\alpha_2\beta_3$, at least in some cases. There is evidence that the glycine receptor also has the latter structure (Langosch *et al.*, 1990), although forming an anion channel. For the $GABA_A$ receptor, an anion channel can be formed from three types of subunit but the stoichiometry has not been determined to date and appears to vary between its sub-types: the complex possibilities there are reviewed elsewhere in this volume. For the mammalian non-NMDA glutamate receptors, numerous isoforms of a 100 kD subunit have been recognised by cDNA cloning (Hollman *et al.*, 1989; Sommer *et al.*, 1990) and it appears that they make up, as with the $GABA_A$ receptor, a variety of subtypes; the subunit compositions of these are at present unknown. Certainly, an $\alpha_2\beta_3$ or $\alpha_2\beta_2\gamma$ arrangement would not be inconsistent with what is known so far. It appears, therefore, that this type of receptor channel (Classes 1A and 1B of Table 4.1) employs related but non-identical subunits (of a minimum of two types) in an oligomeric structure, probably five in total but the use of four in some cases has not to date been excluded.

The agonist molecule may act to close, not to open, the channel in some receptors in this class. This is known, so far, to be so with the K^+ channel on which ATP acts (Table 4.2). This case indicates the uncertain boundary between transmitter-gated channels and ligand-modulated channels. The ATP-inhibited channel protein in question might conceivably be related structurally to other K^+ channels but be endowed with an extra site which permits ATP to close the channel. For this reason the term 'transmitter-gated ion channel' for this class seems more appropriate than the hitherto-common term 'ligand-gated ion channel', since all types of ion channel contain sites where some ligand or other, especially a toxin, can bind to decrease or increase the ease of opening. The familiar voltage-gated Na^+ channel of axons could, thus, logically fall under the heading 'ligand-gated ion channel', with tetrodotoxin as a ligand. It is also known that some natural ligands can act to modulate strongly some voltage-gated channels, including certain α-subunits of G proteins (Table 4.1). These cases are not otherwise related to the receptors which are classed together in Table 4.2, and it seems more rational to separate the channels which are voltage-activated, albeit with some ligand modulation, from those which depend totally upon the binding of a natural transmitter for their change from the resting state, and on present knowledge this distinction is related to different subunit construction in the two cases. On the other hand, it does not seem logical to lay it down for this class that the transmitter molecule must act only on the outside membrane face of a channel: ATP and the other phosphates listed in the second section of Table 4.2 are acting there as intracellular transmitters to perform direct signalling through a membrane and hence they come within the scope of those receptors considered here. Indeed, the same intracellularly-activated receptor may function in an intracellular membrane and (in another cell type) in the plasma membrane. This is exemplified by the receptor/channel activated by IP_3, which operates between two intracellular compartments between organelles within neurones, but which elsewhere can be activated by cytosolic IP_3 to open a Ca^{2+} channel in the plasma membrane (Table 4.2). Such examples strengthen the proposal that the internally and externally activated receptors of Table 4.2 are members of one general receptor/channel Class.

A borderline case for classification is that where the Ca^{2+} ion acts on a channel. For example, the Ca^{2+}-dependent K^+ channel could from one viewpoint be classed as a transmitter-gated (or ligand-gated) ion channel, but it has not been included here. The 'ryanodine receptor', conversely, is indicated as a transmitter-gated ion channel although its natural transmitter or modulator is uncertain (and might include Ca^{2+}). It has a subunit structure with some similarity to that of the IP_3 receptor and it therefore seems reasonable to classify the two together. For the latter two receptors, the signalling in question is across an intracellular membrane, but in principle this is, as with all the other cases of Table 4.2, operated by a natural transmitter molecule to pass a signal between two compartments.

The neurotransmitter duality principle

The agonists involved as extracellular neurotransmitters in Table 4.2 are also employed thus with other receptor types which are, instead, G-protein-linked. While this dichotomy is familiar for the nicotinic and muscarinic ACh pair and the $GABA_A$ and $GABA_B$ pair, we can note the recent identification of a channel directly activated by 5-HT (Derkach *et al.*, 1989) in addition to the 5-HT receptors using one of several second messengers, and the recent elucidation by cloning of a metabotropic 7-TM receptor for glutamate (Masu *et al.*, 1991). The purinergic receptors (Burnstock & Kennedy, 1986) generally act in slow signalling and show evidence of second messenger dependence, *except* for the P_{2x} type where ATP opens an intrinsic cation channel (Benham & Tsien, 1987).

We should note that no ligand exerts this dual activity between Class 3 and either of the other classes of subunit type. It can be presumed that the duality arises from a need to produce both immediate and slower actions from the same small transmitter released from nerve endings, which would not occur with the large polypeptides involved in the signalling of Class 3.

The subunit structures (where known) are entirely different for the same transmitter in these dual cases, and no similar regions in the linear sequences of any pair can be found to suggest a common binding site for the same ligand. Glycine is, on present knowledge, the one exception to this type of duality: no 7-TM-polypeptide type of glycine receptor is known. However, glycine is also used as an obligatory co-transmitter with glutamate at the NMDA receptor (Johnson & Ascher, 1987) so that it does, again, exert two distinct functions on entirely different receptor structures.

Trans-membrane domains in receptor subunits

From the known subunit sequences of the receptors, it has become recognised that stretches of 19–27 amino acid residues with a high mean hydrophobicity (Fig. 4.1) occur frequently, these being structures which will form an α-helix in a non-polar environment. With about 20 or so residues, this helix will just span the membrane. With significantly more, the helix may be tilted, or an end region may protrude into a vestibule structure above or below the membrane. These deductions have been validated by direct physical analysis in the case of bacteriorhodopsin in the purple bacterial membrane by Henderson and co-workers, where the seven domains of this type correspond directly to seven membrane-spanning helices (Henderson *et al.*, 1990). A further direct validation by crystallographic methods has been made for such domains, including tilted longer helices, in bacterial (Deisenhofer & Michel, 1989) or plant chloroplast (Kühlbrandt & Wang, 1991) membrane light-transducing proteins. The purple membrane analysis is particularly relevant to receptor subunit

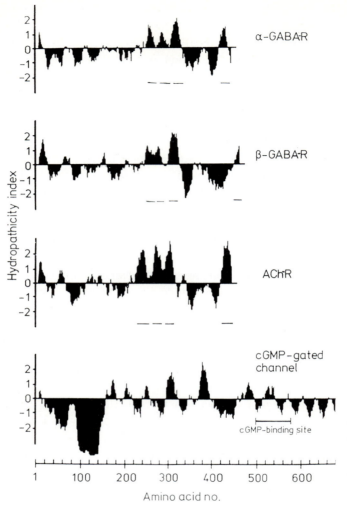

Fig. 4.2. The distribution of hydrophobic domains in some examples of transmitter-gated ion channels. In each case the hydropathy of the side chains is computed, moving along the subunit from the N-terminus on the left. Positive scores are hydrophobic and negative are hydrophilic. α-GABA-R, the α1 subunit, and β-GABA-R, the β1 subunit, are of the bovine GABA$_A$ receptor (Barnard *et al.*, 1987). Note their near-identical profiles, despite a 63 % mean difference in their sequences. AChR, the *Torpedo* nicotinic ACh receptor α-subunit (Noda *et al.*, 1982). Bottom, the cGMP-gated channel from bovine retina. The hydrophobic segment at positions 1–20 is the signal peptide, which is cleaved to leave the mature subunit in the membrane and therefore should be disregarded. There is none in the cGMP receptor, and the topology is assumed to be opposite to the others, with the N- and C-terminal segments intracellular. Note that the distribution of the four TMs in the GABA and ACh receptor

structures, since there is sequence homology between rhodopsins and the G-protein-linked receptor super-family. The seven trans-membrane domains (TMs) of bacteriorhodopsin can, in fact, be related structurally (Henderson *et al.*, 1990) to the seven hydrophobic sequences which are found in the subunits of those receptors. The organisation in the membrane of these receptors of Class 2, at the least, as deduced in outline from their hydrophobicity plots, can therefore be taken as valid. Structural requirements within that outline are discussed in the chapters by B. F. O'Dowd *et al.*, and by M. I. Schimerlik in this volume.

Plots of hydrophobicity along the subunit sequence for some of the receptor types of Class 1 are exemplified in Fig. 4.2. The nicotinic ACh, GABA and glycine receptor subunits have precisely the same distribution of TMs along the chain. This, together with partial sequence homologies observed over almost the whole length of the three subunit types, led to the discovery of a super-family of these transmitter-gated channels (Barnard *et al.*, 1987). This must now be assigned as super-family IA of that Class as a whole (Table 4.1): other subunits there either differ, in sequence greatly (IB, C, glutamate: see Barnard, this volume) or completely (II, III). It is important to note that of the order of 80 different sequences are now known of the super-family IA subunits and that this comprises those on neurones (central and peripheral) or on glia or on muscle cells or electroplaques and ranges phylogenetically from man to lower vertebrates to invertebrates, yet all fit exactly the 4-TM spacing pattern shown for the examples in Fig. 4.2 (upper three plots). In the case of the *Torpedo* ACh receptor subunits, this pattern can be related by chemical and physical evidence to the structure of the assembled receptor in the membrane, as is discussed in depth by A. Maelicke in this volume. Hence, the subunit topology can be regarded as constant throughout that super-family, although it contains both cation and anion channels. The question of whether this extends to the glutamate receptors (type IB) will be considered in Chapter 6.

Other arrangements of the TMs occur in the other subunit types of the transmitter-gated ion channels, where these are known from cDNA cloning. The subunits of the receptors for cGMP and for cAMP exhibit six stretches which may qualify as TMs (Fig. 4.2), but since some of these hydrophobic extents are small and only a single case has been analysed for

subunits (marked by bars) is the same for the two types. The glycine receptor subunits (α and β) have the same distribution of TMs (Langosch *et al.*, 1990). A consensus site for the binding of cGMP is marked, in the C-terminal domain of that receptor. The cAMP-gated channel (Table 4.2) has an essentially identical hydrophobicity profile to that for cGMP.

each ligand type, there is uncertainty about the interpretation (Kaupp, 1991). In the case of the IP_3 receptor, the hydrophobicity plot is complex and even less amenable to interpretation, and only a range from three to seven TMs has been deduced (Furiuchi *et al.*, 1989). Similar uncertainty applies to the ryanodine receptor (Table 4.2). This has good homology with the IP_3 receptor over two TMs, but much less elsewhere. The cGMP (bovine) and cAMP (rat) receptors are 57 % identical in sequence (Dhallan *et al.*, 1990). The latter pair has no homology at all with the IP_3-ryanodine receptor type and neither set has any with the super-families IA or IB.

Hence, a minimum of three entirely distinct super-families is required to cover the subunits of the transmitter-gated channels, and this number could be as large as seven. Single sequences from the types marked as unknown in Table 4.2 would end this present uncertainty. Even at this stage, one can see that there is no universality in Class 1 of the principle of clustering the TMs towards the C-terminus of the subunit, and of having a long intracellular loop between the third and fourth TMs, features of the known non-NMDA glutamate receptor sequences as well as those of the ACh–GABA–GLY super-family. The other structurally elucidated transmitter-gated ion channels do not show those features and have, in contrast to the latter super-family, a considerable hydrophilic domain at their C-terminal end also (Fig. 4.2). This may be related to the fact that in these other receptors the C-terminal domain has been predicted to be intracellular, containing the phosphate ligand binding site.

The subunit stoichiometry in the membrane for the cGMP, cAMP, IP_3 and ryanodine receptors has not been directly established. However, these latter two have exceptionally large subunits (565 kD deduced for the ryanodine receptor: Takeshima *et al.*, 1989) and are likely to be monomers. Several cGMP (80 kD) or cAMP (76 kD) receptor subunits must be arranged to form the channel, and the Hill coefficient of ~ 3 (Kaupp, 1991) suggests at least four cGMP receptor subunits (which have only one binding domain each) are assembled, so the principles of channel construction may be similar throughout Class 1.

Further consideration of the structural differences between the subunits of various anion and cation channel-containing receptors will be given in Chapter 6.

Receptor structure in classes 2 and 3

For the 7-TM subunits of Class 2, the general features of their structure are illustrated in this volume in the accounts of the β and α adrenergic receptors by B. F. O'Dowd *et al.* and the muscarinic types by M. Schimerlik. It can be noted here that while the subunits in this class generally appear to act as monomers, in some cases there is evidence that they exist in the membrane as dimers. Reviewing studies on the purified receptors by hydrodynamic analyses (for those cases where a method is used (Haga *et al.*, 1990) which corrects for the severe perturbation by the bound detergent) and by radiation inactivation, it has been concluded that the $\beta 1$ and $\beta 2$ receptors in heart and in lung are dimers, whereas muscarinic receptors are monomers (Kerlavage *et al.*, 1987; Haga *et al.*, 1990). There is evidence from a strong co-operativity in ligand binding – which, with one ligand site per subunit, provides independent evidence for association – that the opioid μ and δ receptors exist in the membrane as oligomers (Demoliou-Mason & Barnard, 1986). In most other cases, evidence is lacking to establish that 7-TM receptor subunits do exist as their monomers.

The quite homologous structures of the muscarinic and the α and β adrenergic series of receptors are represetative of many others of Class 2 (as illustrated in Fig. 7.1 of O'Dowd *et al.*, this volume). However, there have recently become known several other types of structure in G-protein-linked receptors which, while maintaining the main features of the seven TM regions, diverge greatly elsewhere from that concensus structure. The most divergent is the metabotropic glutamate receptor, which is coupled to the G-Protein-IP_3-Ca^{2+} signalling pathway as are a number of the 7-TM receptors having the consensus-type structure of that Class. But, as Masu *et al.* (1991) found, there is no sequence similarity with the others of that Class, and it has an exceptional C-terminal tail of 370 residues, by far the longest known. The 'C3' cytoplasmic loop, in contrast, is very small. The N-terminal domain, potentially glycosylated and assumed to be extra-cellular, is also exceptionally large, 570 residues. Long extensions there are found in the other 7-TM receptors for the large glycoprotein hormones, but this is unique for a small-molecule transmitter receptor. It is interesting to note that this receptor could not have been cloned by cross-hybridisation from any other known receptor sequence (oocyte expression being feasible for it), and this may apply to some other refractory 7-TM receptors.

The luteinising hormone (LH) receptor (McFarland *et al.*, 1989; Loosfelt *et al.*, 1989) has 26 % identity overall to the $\beta 2$-adrenergic receptor, but has a 341-residue extracellular N-terminus. Similar large initial domains have been found with the other recombinant receptors so far known for glycoprotein hormoes, and are presumed to be necessary for their binding.

In Class 3 (Fig. 4.1), all of the subunits have a very large N-terminal extracellular domain which can be, by excision or by cross-linking, shown to contain the binding site for the polypeptide ligand. In most this site is marked by a high density of cysteines, but in some (e.g. the PDGF or natriuretic receptors) this is much less. In many (Class 3A) there is a protein kinase active site in the long single cytoplasmic extension. In Class 3B a very similar sequence is present but inactive, while a guanylate cyclase active site (homologous to adenylate cyclase) is present (Fig. 4.1). Hence in both 3A and 3B a receptor-enzyme is involved. There is a further group, 3C where both of these sites are absent: see E. Shooter, this volume.

It is interesting that, in a few cases, such as the insulin receptor (Chapter 10), the subunit is cleaved after membrane insertion to form a hetero-oligomer (Fig. 4.1). In general, the transduction systems for Class 3 receptors are still uncertain. In many cases the receptor eventually becomes internalised via clathrin interaction at a coated pit, and a 1-TM configuration seems likely to facilitate that process.

Reconstitution of receptors

The purification of signal-transducing receptors has been covered in depth by many authors in a recent compendium (Hulme, 1990), as well as the subsequent recognition of their subunits in gel electrophoresis. This provides a necessary complement to their DNA cloning, firstly to provide the probes for this in most cases, and secondly to report on the types and ratios of subunits naturally present in a given source. The latter is particularly important when a number of isoforms are found by DNA cloning.

A necessary final stage in that quest is the reconstitution of the receptor in an active form. This is required to identify the subunit(s) as indeed part of the receptor, to report upon its native, undergraded state, and to show that all of the structures for full activity have been recovered. An extension of this task is the expression of isolated sectors of the structure to localise functions within them. These end stages will be briefly reviewed here.

Reconstitution of purified receptor/channels in lipid bilayers

The transmitter-gated ion channel proteins, after purification, can be reconstituted successfully if lipid is maintained at all stages of the process, e.g. with the *Torpedo* ACh receptor (Changeux *et al.*, 1979; Anholt *et al.*, 1980). Both cholesterol and anionic lipids facilitate this process for that receptor (Jones *et al.*, 1988). Function can then be analysed either by forming liposomes and measuring Na^+ or Rb^+ flux, as in the afore-mentioned studies, or by forming the receptor-containing bilayer at the tip of a patch pipette and recording channel opening (Suarez-Isla *et al.*, 1983). The most convenient method is to allow receptor-containing liposomes to incorporate into the bilayer in the pipette. By these means, it was established in those studies that the isolated receptor contained an ACh-activated cation channel of nicotinic pharmacology and that all four types of subunit were present when this was functional. The methods required have been given in detail in a compendium (Miller, 1986).

Similar procedures have been applied with brain receptor channels, namely the purified $GABA_A$ and non-NMDA glutamate receptors. The pure $GABA_A$ receptor from bovine brain can be incorporated into asolectin liposomes, which are purified on a sucrose density gradient and found to have the native high-affinity binding of GABA agonists and of benzo-diazepines as well as of a specific channel ligand (Sigel *et al.*, 1985). This preparation can be rendered more stable by the use, instead, of brain lipids from the region from which the receptor is purified (Bristow & Martin, 1987). These liposomes then show a Cl^- flux when GABA is applied, which is potentiated by benzodiazepines and barbiturates (Fig. 4.3). Hence not only the gated channel but its allosteric interactions are preserved in this type of reconstitution, and it was confirmed that all of the complex GABA receptor functions reside in the proteins purified.

With a purified non-NMDA glutamate receptor in asolectin : cholesterol liposomes, the bilayer patch-clamp technique was recently applied. Single cation channels of 5–10 pS conductance or multiples thereof (as *in situ*) were opened by several non-NMDA ligands, and the predicted phar-macology was found (Barnard *et al.*, 1991). Likewise, a 63 kD polypeptide purified as the cGMP receptor from bovine retina was, via asolectin liposomes, put into a planar bilayer (Hanke *et al.*, 1988). cGMP (EC_{50} 30 μM) activated cation channels of native (26 pS) conductance and Hill coefficient, but three-fold longer mean open time. This agrees with the above-mentioned proposal that the subunit forms a homo-oligomer. Hence, reconstitution of all type of the Class 1 receptors is an accessible tool for validating their purification.

Reconstitution of purified G-protein-linked receptors

In this case, the objectives are usually more restricted, namely to test for coupling to a G-protein. Adrenergic and muscarinic receptors have been reconstituted into liposomes and their coupling to added G-proteins has

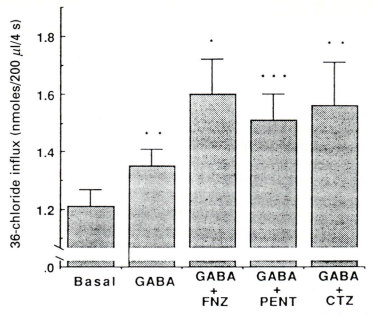

Fig. 4.3. GABA-stimulated ^{36}Cl flux due to reconstituted GABA$_A$ receptors in liposomes. Whole rat brain lipids plus cholesterol hemisuccinate were used, to incorporate the purified rat brain receptor. The basal flux (no GABA) is increased ($p < 0.05$) by 100 μM GABA, and further significantly increased by 1 μM flunitrazepam (FNZ) or 5 μM pentobarbital (PENT) or 1 μM cartazolate (CTZ) (Bristow & Martin, 1987).

been demonstrated by the introduction of a guanine nucleotide-sensitive high affinity binding of agonists. An example is given by Haga *et al.* (1989). In those studies on muscarinic receptors, a requirement for cholesterol was again found. The ability to couple to different G protein types can be tested in these cases. However, reconstitution evidence is necessarily limited in such cases, since a mixture of subtypes will usually be purified, especially from the brain, and each may have its own Gα requirement.

Reconstitution of receptor segments

A new approach has been made recently to the identification of channel-forming elements in a receptor/channel subunit. The isolated TMs can be synthesised and these peptides alone can be reconstituted in bilayers and patch-clamped. Obviously, ligand-gating will be absent but Montal and co-workers have shown that channels can be formed and can open and

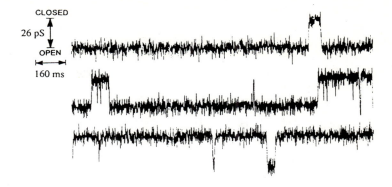

Fig. 4.4. Single-channel recordings from lipid bilayers containing the synthetic tetrameric 'synporin' constructed with the M2 (upper) or M1 (lower) sequence from the δ subunit of the *Torpedo* ACh receptor. Note the discrete spontaneous channel openings with M2 (conductance 26 pS) but none with M1 (from Montal *et al.*, 1990).

close spontaneously if an appropriate TM is used. Initially the M2 segment of the *Torpedo* ACh receptor δ subunit was shown to give cation-selective channels, but this has been refined to tether four 23-mer M2 peptides on a nine-residue backbone, to increase the chance of self-assembly (Montal *et al.*, 1990). The channels then open readily and, if M2 is used, give most of the behaviour of the ACh-gated receptor channel (Fig. 4.4). In contrast, exchange for M1 gave no discrete channels. This provides a direct test of much other evidence that M2 helices line the channel in the nicotinic receptor structure. Similar positive results have recently been obtained (unpublished) with a glutamate receptor subunit TM2 peptide. It can be expected that this 'synporin' approach will become a useful tool in aiding interpretation of the receptor/channel subunit sequences.

Conclusion

In summary, the demonstration of functional activity in a receptor of any class is a necessary requirement for asserting that it has been purified without loss or damage, and receptor reconstitution into liposomes is now an established technique for this purpose.

References

Anholt, R., Lindstrom, J. & Montal, M. (1980). Functional equivalence of monomeric and dimeric forms of purified acetylcholine receptors from *Torpedo californica* in reconstructed lipid vesicles. *European Journal of Biochemistry*, **109**, 481–7.

Bernardi, H., Fosset, M. & Lazdunski, M. (1988). Characterization, purification and affinity labelling of the brain [^{3}H]glibenclamide binding protein, a putative neuronal ATP-regulated K^+ channel. *Proceedings of the National Academy of Science, USA*, **85**, 9816–20.

Barnard, E. A., Darlison, M. G. & Seeburg, P. H. (1987). Molecular biology of the $GABA_A$ receptor: the receptor/channel super-family. *Trends in Neurosciences* **10**, 502–9.

Barnard, E. A. & Henley, J. (1990). The non-NMDA receptors: types, protein structure and molecular biology. *Trends in Pharmacological Science*, **11**, 500–7.

Barnard, E. A., Ambrosini, A., Sudan, H., Prestipino, G., Lu, Q., Rodriquez-Ithurralde, D., Rossier, J., Usherwood, P. N. R. & Henley, J. M. (1991). Identification and properties of a functional unitary non-NMDA receptor from *Xenopus* brain. In *Fidia Research Symposia*, Meldrum B., *et al.*, **5**, 135–44.

Benham, C. D. & Tsien, R. W. (1987). A noval receptor-operated Ca^{2+}-permeable channel activated by ATP in smooth muscle. *Nature*, London, **328**, 275–8.

Bristow, D. R. & Martin, L. L. (1987). Solubilisation of the γ-aminobutyric acid/benzodiazepine receptor from rat cerebellum: optimal preservation of the modulatory responses by natural brain lipids. *Journal of Neurochemistry*, **49**, 1386–93.

Brown, A. M. & Birnbaumer, L. (1990). Ionic channels and their regulation by G protein subunits. *Annual Review of Physiology*, **52**, 218–48.

Burnstock, G. & Kennedy, C. (1986). A dual function for adenosine-5'-triphosphate in the regulation of vascular tone. *Circulation Research*, **58**, 319–30.

Changeux, J. P., Heidmann, T., Popot, J. L. & Sobel, A. (1979) Reconstitution of a functional acetylcholine receptor under defined conditions. *FEBS Letters*, **105**, 181–7.

Deisenhofer, J. & Michel, H. (1989). The photosynthetic reaction center from the purple bacterium *Rhodopseudomonas viridis*. *Science*, **245**, 1463–73.

Demoliou-Mason, C. D. & Barnard, E. A. (1986). Distinctive subtypes of the opioid receptor with allosteric interactions in brain membranes. *Journal of Neurochemistry* **46**, 118–28.

Derkach, V., Suprenant, A. & North, R, A. (1989). 5-HT$_3$receptors are membrane ion channels. *Nature.*, London, **339**, 706–9.

Dhallan, R. S., Yau, K.-W., Schrader, K. A. & Reed, R. R. (1990). Primary structure and functional expression of a cyclic nucleotide-activated channel from olfactory neurons. *Nature*, London, **347**, 184–9.

Dunne, M. J., Bullett, M. J., Li, G., Wollheim, C. B. & Peterson, O. H. (1989). Galanin activates nucleotide-dependent K^+ channels in insulin-secreting cells via a pertussis toxin sensitive G-protein. *EMBO Journal*, **8**, 413–20.

Furiuchi, T., Yoshikawa, S., Miyawaki, A., Wada, K., Maeda, N. & Mikoshiba, K. (1989). Primary structure and functional expression of the inositol 1,4,5 trisphosphate-binding protein P$_{400}$. *Nature, London*, **342**, 32–8.

Haga, T., Haga, K. & Hulme, E. C. (1990). Solubilization, purification and molecular characterization of mucarinic acetylcholine receptors. In: *Receptor Biochemistry, A Practical Approach*, Hulme, E. C., ed. pp. 51–78, IRL Press, Oxford.

Haga, K., Uchiyama, H., Haga, T., Ichiyama, A., Kangawa, K. & Matsuo, H. (1989). Cerebral muscarinic acetylcholine receptors interact with three kinds of GTP-binding proteins in a reconstitution system of purified components. *Molecular Pharmacology*, **35**, 286–94.

Hanke, W., Cook, N. J. & Kaupp, U. B. (1988). cGMP-dependent channel protein from photoreceptor membranes: single channel activity of the purified and reconstituted protein. *Proceedings of the National Academy of Sciences, USA*, **85**, 94–8.

Henderson, R., Baldwin, J. M., Ceska, T. A. Zemlin, S., Beckmann, E., & Downing, K. H. (1990). A model for the structure of bacteriorhodopsin based on high resolution electron cryo-microscopy. *Journal of Molecular Biology*, **213**, 899–929.

Hollmann, M., O'Shea-Greenfield, A., Rogers, S. W. & Heinemann, S. (1989). Cloning by functional expression of a member of the glutamate receptor family. *Nature, London*, **342**, 643–8.

Hulme, E. C. (ed.). 1990. In: *Receptor Biochemistry: A Practical Approach, IRL Press, Oxford.*

Johnson, J. W. & Ascher, P. (1987). Glycine potentiates the NMDA response in cultured mouse neurons. *Nature, London*, **325**, 529–31.

Jones, O. T., Eubanks, J. H., Earnest, J. P. & McNamee, M. G. (1988). A minimum number of lipids are required to support the functional properties of the nicotinic acetylcholine receptor. *Biochemistry*, **27**, 3733–42.

Kaupp, U. B. (1991). The cyclic nucleotide-gated channels of vertebrate photoreceptors and olfactory epithelium. *Trends in Neurosciences*, **14**, 150–7.

Kerlavage, A. R., Fraser, C. M., Chung, F.-Z. & Venter, J. C. (1987). Molecular structure and evolution of adrenergic and cholinergic receptors. *Proteins*, **1**, 287–301.

Kühlbrandt, W. & Wang, D. N. (1991). Three-dimensional structure of plant light-harvesting complex determined by electron crystallography. *Nature, London*, **350**, 130–4.

Kuno, M. & Gardner, P. (1987). Ion channels activated by inositol 1,4,5-triphosphate in plasma membrane of human T-lymphocytes. *Nature, London*, **326**, 301–4.

Krupinski, J., Coussen, F., Bakalyar, H. A., Tang, W.-J., Feinstein, P. G., Orth, K., Slaughter, C., Reed, R. R. & Gilman, A. G. (1989). Adenyl cyclase amino acid sequence: possible channel- or transporter-like structure. *Science*, **244**, 1558–64.

Langosch, D., Becker, C.-M. & Betz, H. (1990). The inhibitory glycine receptor. A ligand-gated chloride channel of the central nervous system. *FEBS Letters*, **194**, 1–8.

Lipscombe, D., Kongsamut, S. & Tsien, R. W. (1989). α-Adrenergic inhibition of sympathetic neurotransmitter release mediated by modulation of N-type calcium-channel gating. *Nature, London*, **340**, 639–42.

Loosfelt, H., Misrahi, M., Atger, M., Salesse, R., Thi, M. T. V. H.-Lu, Jolivet, A., Guiochon-Mantel, A., Sar, S., Jallal, B., Garnier, J. & Milgrom, E. (1989). Cloning and sequence of porcine LH-hCG receptor cDNA: variants lacking transmembrane domain. *Science*, **245**, 525–8.

McFarland, K. C., Sprengel, R., Phillips, H. S., Köhler, M., Rosemblit, N., Nikolics, K., Segaloff, D. L. & Seeburg, P. H. (1989). Lutropin-choriogonadotropin receptor: an unusual member of the G protein-coupled receptor family. *Science*, **245**, 494–9.

Maricq, A. V., Peterson, A. S., Brake, A. J., Myers, R. M. & Julius, D. (1991). Primary structure and functional expression of the 5-HT$_3$ receptor, a serotonin-gated ion channel. *Science*, **254**, 432–7.

Masu, M., Tanabe, Y., Tsuchida, K., Shigemoto, R. & Nakanishi, S. (1991). Sequence and expression of a metabotropic glutamate receptor. *Nature, London*, **349**, 760–5.

Mignery, G. A., Südhof, T. C., Takei, K., & De Camilli, P. (1989). Putative receptor for inositol 1,4,5-triphosphate similar to ryanodine receptor. *Nature, London*, **342**, 192–5.

Miller, C. ed. *Ion Channel Reconstitution*. (1986). pp. 107–21, New York: Plenum.

Montal, M., Montal, M. S. & Tomich, J. M. (1990). Synporins – synthetic proteins that emulate the pore structure of biological ionic channels. *Proceedings of the National Academy of Sciences, USA*, **87**, 6929–33.

Moriyoshi, K., Masu M., Ishii, T., Shigemoto, R., Mizuno, N. & Nakanishi, S. (1991). Molecular cloning and characterization of the rat NMDA receptor. *Nature, London*, **354**, 31–7.

Noda, M., Takahashi, H., Tanabe, T., Toyosato, M., Furutani, Y., Hirose, T., Asai, M., Inayama, S., Miyata, T. & Numa, S. (1982). Primary structure of α-subunit precursor of *Torpedo californica* acetylcholine receptor deduced from cDNA sequence. *Nature, London*, **299**, 793–7.

Schulz, S., Chinkers, M., & Garbers, D. L. (1989). The guanylate cyclase/receptor family of proteins. *FASEB Journal*, **3**, 2026–35.

Sigel, E., Mamalaki, C. & Barnard, E. A. (1985). Reconstitution of the purified γ-aminobutyric acid-benzodiazepine receptor complex from bovine cerebral cortex into phospholipid vesicles. *Neuroscience Letters*, **61**, 165–70.

Sommer, B., Keinanen, K., Verdoorn, T. A., Wisden, W., Burnashev, N., Herb, A., Kohler, M., Takagi, T., Sakmann, B. & Seeburg, P. H. (1990). Flip and Flop: A cell-specific functional switch in glutamate operated channels of the CNS. *Science*, **249**, 1580–5.

Suarez-Isla, B. A., Wan, K., Lindstrom, J. & Montal, M. (1983). Single-channel recordings from purified acetylcholine receptors reconstituted in bilayers formed at the tip of patch pipets. *Biochemistry*, **22**, 2319–23.

Sumikawa, K., Houghton, M., Smith, J. C., Bell, L., Richards, B. M. &

Barnard, E. A. (1982). The molecular cloning and characterisation of cDNA coding for the α subunit of the acetylcholine receptors. *Nucleic Acids Research*, **10**, 5809–22.

Suppattapone, S., Worley, P. F., Baraban, J. M. & Snyder, S. H. (1988). Solubilization, purification, and characterization of an inositol trisphosphate receptor. *Journal of Biological Chemistry*, **263**, 1530–4.

Takeshima, H., Nishimura, S., Matusumoto, T., Ishida, H., Kangawa, K., Minamino, N., Matsuo, H., Ueda, M., Hanaoka, M., Hirose, T. & Numa, S. (1989). Primary structure and expression from complementary DNA of skeletal muscle ryanodine receptor. *Nature, London*, **339**, 439–45.

Ullrich, A., Bell, J. R., Chen, E. Y., Herrera, R., Petruzzelli, L. M., Dull, T. J., Gray, A., Coussens, L., Liao, Y.-C., Tsubokawa, M., Mason, A., Seeburg, P. H., Grunfeld, C., Rosen, O. M. & Ramachandran, J. (1985). Human insulin receptor and its relationship to the tyrosine kinase family of oncogenes. *Nature, London*, **313**, 756–61.

Yatani, A., Codina, J., Brown, A. M. & Birnbaumer, L. (1987). Direct activation of mammalian atrial muscarinic potassium channels by GTP regulatory protein G_k. *Science*, **235**, 207–11.

5

The nicotinic acetylcholine receptor: towards the structure–function relationship

ALFRED MAELICKE

The nicotinic acetylcholine receptor (nAChR) is the archetype of a superfamily of neuroreceptors that act as ligand-gated ion channels. The members of this superfamily display common structural features, the most distinct ones being four putative membrane-spanning regions (M1–M4) and a disulphide bridge consensus sequence in each of their polypeptides. It is likely that all ligand-gated ion channels have a quasi-symmetric, probably pentameric arrangement of subunits around a central pore (ion channel) lined by transmembrane regions M2. Each of the five neuro-receptors cloned so far, the nAChR (this chapter), the GABA receptor (Barnard, Darlison & Seeburg, 1987), the glycine receptor (Betz, 1987), the kainate type glutamate receptor (Hollmann et al., 1989; Boulter et al. 1990), and the ryanodine receptor (Takeshima et al., 1989) exist in iso-forms and hence form their own families of molecules (Maelicke, 1988a, b; Maelicke, 1987a, b) within the superfamily of ligand-gated ion channels. Differences in distribution, electrophysiological, pharmacological, anti-genic and biochemical properties of neuroreceptor isoforms from various species and tissues are well established. A central motif of present receptor research therefore relates to the question of how these functional differences correlate to neuroreceptor structure.

To date, the nicotinic acetylcholine receptor is still the most thoroughly studied of all neuroreceptors. This is due mainly to the availability of (i) relatively simple cellular preparations for electrophysiological studies, (ii) a very large body of ligands of well-established pharmacology, and (iii) a variety of receptor preparations able to provide the large amounts of nAChR required for quantitative biochemical studies. This fortunate situation has provided a wealth of information on practically all levels of analysis unmatched by any other membrane protein. As the only major deficit, a high resolution X-ray structure is still missing (Stroud, McCarthy

& Shuster, 1990). In spite of this fact, the nicotinic acetylcholine receptor may well become the first neuroreceptor the structure–function relationship of which is understood on a truly molecular level.

In this chapter, only those studies thought to be essential for correlating structure and function of the nAChR have been assembled. This chapter therefore is, by necessity, subjective and selective. For accounts of more general scope and of specialized topics, the reader is referred to the following reviews of recent years: general aspects (Changeux, Giraudat & Dennis, 1987; Changeux *et al.*, 1988; Maelicke, 1988*a*, *b*, 1990; Claudio *et al.*, 1989; Stroud *et al.*, 1990; Changeux *et al.*, 1987*a*, *b*; Karlin, Kao & DiPaolo, 1986; McCarthy *et al.*, 1986), electrophysiology (Colquhoun, 1986; Adams, 1987; Colquhoun & Sakmann, 1985), molecular genetics and biosynthesis (Claudio, 1986, 1989; Anderson, 1987; Heinemann *et al.*, 1986; Merlie & Smith, 1986; Merlie & Kornhauser, 1989), developmental regulation (Bloch & Pumplin, 1988; Shuetze & Role, 1987; Salpeter, 1987; Brehm & Henderson, 1988), immunology (Engel, 1987; Heilbronn, 1985; Lindstrom *et al.*, 1986), and learning and memory (Changeux *et al.*, 1987*a*, *b*). If not otherwise, the reported data including numbers of sequence positions refer to *Torpedo* α-subunit.

Basic structural data

Primary structures

The application of recombinant DNA techniques has provided the primary structures of nAChR polypeptides from many species and tissues, including fish electric organ, mammalian muscle and brain, frog and snake muscle, chicken muscle and brain, worm and insect nervous tissue (Changeux, 1988; Claudio, 1989; Hermsen *et al.*, unpublished observations; Heinemann *et al.*, 1989). The aligned sequences exhibit considerable homology in sequence (Fig. 5.1) and hydrophobicity profile (Fig. 5.2) (Noda *et al.*, 1983; Claudio *et al.*, 1983; Popot & Changeux, 1984). They typically consist of a large amino-terminal hydrophilic region of 210–225 amino acids followed by a hydrophobic core of approximately 70 amino acids, another hydrophilic region of 100–150 amino acids and a hydrophobic carboxy-terminal region of approximately 20 amino acids.

Secondary structures

In view of the high level of primary structure homology among nAChRs, their secondary structures are expected to be very similar (Fig. 5.2). There is general agreement that (i) the N-terminal region with its *N*-glyosylation

```
                                      50
SEAEERLV-YLF↓EDYNK-IRPV-N-S--V-VQFGLSLAQLI-VDEVNQIMTTNVWLKQEW-DYKLKWKP-DY-GVKKIRIPSEKIWRPDᵛ
       100                              150                                                 ᴵ
VLYNNADGDFAV----TKAL-YᵀG-Iᵀ WTPPAIᶠKSSCKIDVTHFPFDQQNCTMKFGSᵂTYDKAKIDLVLISS--DL-DFWESGEWVIV-A
                    D V    Y
        200                                              250
PG-KᴴE-KY-CC↓Y-DITY-FIIRRLPLFYTINLIIPCLLISFLTVLVFYLPSDCGEKᴹTLCISᵛLLSLTVFLLᴸITEIIPSTSLVᵛPLIG
    N             300    └→M1-region            V└────→M2-region  V                    ᴵ
EᵧLLFTMᴵFVTLSIVITVFVLNᵥHHRSPTTHTMP-ᵂVR-VFLD--PRLᴸFMKRP-----------------350------------↓↓E
K└→M3 ᵛ region          400                        M        └→variable cytoplasmic region
Aᴵ EGVKYIAᴰHMKᴬED-D-SV-EDᵂKYVAMVIDRIFLWVFILVCLLGTVGLFLQPL-A---↓
 V         E  S                   └────→M4-region
```

Fig. 5.1. Generalized primary structures of nicotinic acetylcholine receptors from different tissues and species. The sequence shown is a composite of amino acid sequences of nAChRs from vertebrate muscle and brain. The one-letter code was used. Bold-face letters refer to conserved amino acids, light-printed letters refer to residues conserved in at least half of the known sequences. The figure indicates a high level of sequence homology between nAChR subunits.

site is oriented toward the synaptic cleft, (ii) the hydrophobic core region consists of three transmembrane α-helical domains, (iii) the smaller hydrophilic region is oriented toward the cytosol, and (iv) the C-terminal region also contains an α-helical domain (Figs. 5.2, 5.3). Beyond this level, there is still considerable disagreement as to the transmembrane topology of nAChR polypeptides (Maelicke *et al.*, 1989, Maelicke, 1990).

There is solid evidence that the N-termini of subunits are on the extracellular side of the nAChR (Anderson *et al.*, 1982) although, in the native receptor, they are inaccessible to antibodies (Ratnam & Lindstrom, 1984). Likewise the single *N*-glycosylation site (in *Torpedo* α-subunit at Asn-141) and, based on affinity labelling studies, the two adjacent cysteines in positions 192 and 193 (Kao & Karlin, 1986), Tyr 93, Trp 149, Tyr 190 (Trp 86, Tyr 151, Tyr 198) (Dennis *et al.*, 1988) are on the extracellular side. Binding site and epitope mapping studies using synthetic peptides established that major parts, if not all of the sequence regions $\alpha55$–85 (Conti-Tronconi *et al.*, 1990; Barkas *et al.*, 1987; Das & Lindstrom, 1989; Saedi *et al.*, 1990), $\alpha127$–153 (Conti-Tronconi *et al.*, 1990), and $\alpha181$–205 (Wilson *et al.*, 1984; Mosckovitz & Gershoni, 1988; Radding *et al.*, 1988; Ralston *et al.*, 1987; Barkas *et al.*, 1988; Conti-Tronconi *et al.*, 1990) are extracellular. From the positions of phosphorylation sites and from immunochemical and proteolytic fragmentation studies, the region between residues 333 and 377 probably is on the cytosolic side of the nAChR (Huganir & Miles, 1989; LaRochelle *et al.*, 1985). Similarly, the region previously suggested to contain an amphipathic transmembrane domain ($\alpha425$–455) (Finer-Moore & Stroud, 1984; Guy, 1984) has been shown to be freely accessible for antibody from the cytosolic side (Ratnam *et al.*, 1986a, b; Maelicke *et al.*, 1989). Since dimers of *Torpedo* nAChR can be dissociated by the cleavage of a disulphide bridge formed between the penultimate cysteines of their δ-subunits (Dunn *et al.*, 1986; McCrea *et al.*,

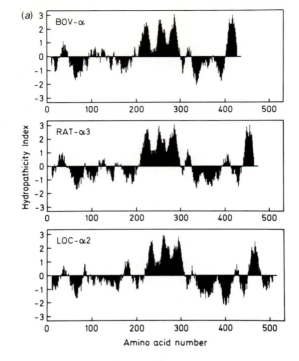

Fig. 5.2. (*a*) Aligned hydropathy profiles of α-subunit sequences from bovine muscle, rat brain (α₃) and *Locusta migratoria* (α₂). Amino acid sequence data from Noda *et al.* (1983), Boulter *et al.* (1986) and Hermsen *et al.* (unpublished observations). Hydropathy was calculated according to Kyle and Doolittle (1982) employing a program developed by Thomas Karsh, Lehrstuhl für Allgemeine Botanik, Ruhruniversität, D-4630 Bochum, Germany. (*b*) Generalized domain structure of nAChR subunits. M1–M4 refer to the putative transmembrane domain obtained from hydropathy profiles. In addition, the position of cysteines and of the simple *N*-glycosylation site are indicated.

1987; DiPaola, Czajkowski & Karlin, 1989), it appears likely that the δ-subunit C-terminus is at the extracellular side (Popot & Changeux, 1984). Epitope mapping studies suggest, however, that the C-termini of nAChR polypeptides are either cytosolic or otherwise inaccessible (similar to the N-termini) in the native conformation of nAChR (Ratnam *et al.*, 1986*a*, *b*; Young *et al.*, 1985). As anti-peptide antibodies often do not have the sequence specificity expected (Maelicke *et al.*, 1989), there is no need at

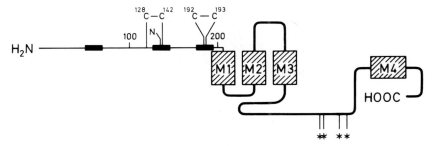

Fig. 5.3. Model of the transmembrane topography of the α-subunit from *Torpedo* nAChR. Areas shown to contain elements of the neuro-transmitter binding site (Conti-Tronconi *et al.*, 1990) are indicated by boxes. * indicate phosphorylation sites. The sideness of the C-terminus has been disputed.

present to diverge from the originally proposed models (based on hydropathy profiles) with four transmembrane α-helical domains (Claudio *et al.*, 1983; Noda *et al.*, 1983; Devillers-Thiery *et al.*, 1983), even though there exists contradicting evidence (Ratnam *et al.*, 1986*a, b*; Young *et al.*, 1985; Pedersen *et al.*, 1990).

Tertiary structures

At the present time, the key information on the three-dimensional structure of the nAChR stems from electron microscopy and low resolution (approximately 20 Å) X-ray diffraction studies (Mitra, McCarthy & Stroud, 1989; Toyoshima & Unwin, 1990). In particular, tubular vesicles with two-dimensional quasicrystalline arrays of receptor dimers were used for image analysis. The extracellular part of the nAChR monomer is of cylindrical shape extending approximately 54 Å above the membrane plane. The outer diameter of the vestibule averages 77 Å, with a protein wall of 25 Å inner diameter surrounding a central pit. Suggested by electrophysiological data (Sanchez *et al.*, 1986), but not resolved by EM or X-ray studies, the diameter of the pit (the ion channel) probably narrows to approximately 7 Å at the entry of the lipid bilayer level. The receptor protrudes approximately 15 Å from the cytosolic side of the membrane and, in intact cells, is attached via anchoring proteins to the cytoskeleton (Brisson & Unwin, 1985; Tsui, Cohen & Fischbach, 1990; Toyoshima & Unwin, 1988). The predominantly five-fold symmetry of its cylindrical structure suggests a regular arrangement of the five subunits of *Torpedo* nAChR around a central pore which forms the integral cation channel (Fig. 5.4).

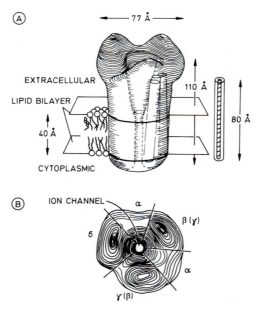

Fig. 5.4. Model of the three-dimensional structure of *Torpedo* nAChR. Side view A and top view B of membrane-bound nAChR according to electromicroscopic data in combination with biochemical data. The receptor is a pentamer of five glycoproteins forming the lining of the integral cation channel. (Reproduced, with kind permission, from Kistler *et al.*, 1982.)

Quaternary structures

Torpedo nAChR is a heteropentamer composed of four polypeptide chains with the stoichiometry $\alpha_2\beta\gamma\delta$. This stoichiometry is also observed in muscle nAChR from various species (Turnbull, Harrison & Lunt, 1985; Nathanson & Hall, 1979; Froehner, Reiness & Hall, 1977; Barnard *et al.*, 1986), and is suggested for one type of neuronal nAChR (Barnard *et al.*, 1986; Conti-Tronconi *et al.*, 1985; Schneider *et al.*, 1985). In the nervous systems of vertebrates, there exists another type of nAChR composed of only two polypeptide chains (Whiting & Lindstrom, 1986, 1987, 1988; Baron, 1989; Boulter *et al.*, 1990; Deneris *et al.*, 1989; Wada *et al.*, 1990). Prompted by their similar ion conducting properties as compared to muscle nAChR, the fully functional receptors of this type are assumed to also be heteropentamers. In insects, only neuronal nAChR exists (LaRochelle *et al.*, 1985). In addition to heteropentamers composed of two types of subunits, there may also exist in insects homopentameric nAChR (Breer *et al.*, 1986, 1989; Maelicke, 1987*a*, *b*; Marshall *et al.*, 1990). Based on sequence homology considerations, and chromosomal linkages observed for nAChR genes, the receptor probably has evolved from a primordial homooligomer by two rounds of gene duplication (Kubo *et al.*, 1986), with the α-polypeptide having evolved more slowly than the others.

The topographical arrangement of subunits is still being debated even for *Torpedo* nAChR. As a minimal conclusion of related studies (Karlin, 1987; Hamilton, Pratt & Eaton, 1985), the two α-subunits are not adjacent to each other.

Molecular mass

The calculated molecular weights of *Torpedo* nAChR polypeptides (50 116; 53 681; 56 279; 57 567) differ considerably from those estimated by SDS gel electrophoresis (40 000; 50 000; 59 000; 67 000). These discrepancies cannot be accounted for by the well-established shortcomings of SDS gel electrophoresis of glycosylated and membrane proteins. Considering that all nAChR polypeptides are glycosylated, phosphorylated and the α and β chains probably also contain covalently bound phospholipid (Olson, Glaser & Merlie, 1984), these discrepancies point to conformational differences between the subunits, being particularly pronounced for the α-subunit.

From the peptide sequences of *Torpedo* nAChR subunits, the molecular mass calculates at 268 kD. Due to post-translational modifications, the molecular mass of the mature nAChR is approximately 290 kD (Doster *et al.*, 1980). Native nAChR from *Torpedo*, *Narcine* and *Electrophorus* can form dimers which are stabilized by disulphide bridges between δ-subunits (DiPaola *et al.*, 1989). They can be dissociated by disulphide-reducing agents and by detergent-forced disulphide rearrangement (Rüchel, Watters & Maelicke, 1982); their functional significance is debatable (Fels, Wolff & Maelicke, 1982; Boheim *et al.*, 1981; Schindler, Spillecke & Neumann, 1984; Hamilton, McLaughlin & Karlin, 1979).

Conformational states

Electrophysiological and biochemical evidence suggests that the nAChR can exist in a multiplicity of conformational states. These states develop and decay in the subsecond to minute time range and are closely related to receptor function. The structure–function relationship of the nAChR therefore cannot be unravelled in full without an intimate understanding of the conformational states including their equilibria and kinetics of conversion.

Basic functional data

The nAChR is a cation-selective ion channel (Adams, Dwyer & Hille, 1980; Lewis & Stevens, 1983) with a cut-off internal diameter of 6.5 Å (Sanchez *et al.*, 1986). From single channel studies the conductance of the channel is independent of the activating ligand and only weakly depends on the membrane potential (Gardner, Ogen & Colquhound, 1984; Sakmann & Neher, 1984). Binding of acetylcholine and/or its agonists induces transient openings of the nAChR cation channel (Fig. 5.5).

Concentration dependence of response

From dose–response studies of ligand-induced ion flux into *Torpedo* membrane vesicles, the concentrations of half-maximal response (apparent equilibrium dissociation constants) are 30 μM and 400 μM for acetylcholine and carbamoylcholine, respectively (Forman & Miller, 1988, Forman *et al.*, 1989). These data agree with single channel studies of *Torpedo* receptor expressed in *Xenopus* oocytes yielding for acetylcholine apparent dissociation constants of 5 μM and 500 μM for the two classes of sites (Sine & Steinbach, 1987). The dose–response curves yield a Hill coefficient of approximately 2 indicating that two molecules of ACh cooperatively interact to activate the channel. In other systems, Hill coefficients between 1.5 and 2.7 were found (Adams, 1975; Dionne, Steinbach & Stevens, 1978; Dreyer, Peper & Sterz, 1978). There is an inherent discrepancy, however, between some apparent equilibrium dissociation constants and Hill coefficients determined. In several reports, the dissociation constants determined for monoliganded and diliganded nAChR, respectively, point to *anti*cooperative behaviour (corresponding to Hill coefficients of less than 1) while the same studies report Hill coefficients corresponding to *positively* cooperative behaviour (Sine, Claudio & Sigworth, 1990; Jackson, 1988). Other single current studies have reported modest positive cooperativity of acetylcholine binding sites (Auerbach & Lingle, 1987; Sine & Steinbach, 1987; Colquhoun & Ogden, 1988).

Rates of response

Recordings of single channel events at low concentrations of acetylcholine show that the channel opens within a few microseconds (Colquhoun & Sakmann, 1985; Sine & Steinbach, 1986*b*, 1987, Sine *et al.*, 1990). As inferred from the concentration dependence of single channel events, association of acetylcholine with nAChR occurs at a nearly diffusion-controlled rate (Auerbach & Lingle, 1987; Sine & Steinbach, 1987, Sine *et*

$$R \xrightleftharpoons[k_{-1}]{k_1} AR \xrightleftharpoons[k_{-2}]{k_2} ARA \xrightleftharpoons[\alpha]{\beta} ATA \rightleftharpoons ADA$$

Fig. 5.5. Simplified reaction scheme for the interaction of agonists with the nAChR. R refers to closed channel states of the nAChR having low affinity for agonist binding. T is the open channel state ('transmitting'), D is a closed channel state ('desensitized') having high affinity for agonist binding. At low concentrations of agonists (and with active acetylcholine esterase), occupation of the D-state is of little significance. The R-state exists in three substates (free, monoliganded and diliganded nAChR) of which ARA is also termed 'activatable'.

al., 1990; Colquhoun & Ogden, 1988). The mean open time of the channel is of the order of milliseconds. It is determined by the rate of channel closing which, in most systems, is of the same order of magnitude as that of acetylcholine dissociation from the low affinity state (but see below). Thus, the simplest models suggested that the channel is opened only once during nAChR occupation by two agonist molecules.

It is well established though that a single activation of the nAChR channel often produces multiple openings in quick succession. This 'nachschlag' phenomenon depends on the relative values of the channel opening rate constant and of the acetylcholine dissociation rate constant. Due to considerable differences in the latter, the probability that a channel reopens within one cycle of activation is rather low for *Torpedo* nAChR (Sine *et al.*, 1990) but much higher (in increasing order) for nAChR from frog muscle (Colquhoun & Sakmann, 1985), *Xenopus* myocytes (Auerbach & Lingle, 1987) and BC3H-1 cells (Sine & Steinbach, 1986a, b). A high rate constant of acetylcholine dissociation is a key prerequisite for fast termination of synaptic response (Colquhoun & Sakmann, 1985).

Additional evidence against a simple correlation between acetylcholine dissociation and channel closing comes from studies of nAChR with a covalently attached agonist (Chabala & Lester, 1986; Siemen *et al.*, 1986). Under these conditions, the channel lifetime is lengthened but the channel does not remain open persistently, suggesting that channel closing and ligand dissociation are independent phenomena (Siemen *et al.*, 1986).

At elevated concentrations of acetylcholine, the patterns of single channel events are complicated by the effects of desensitization and direct channel block. Accordingly, in the concentration range typical for cholinergic signal transduction at neuromuscular endplates, the rates of the essential transitions (from closed to open channel state and from open to closed channel state), and the transmitter dissociation rates are not (yet) accessible to direct determination. Using the rate constants obtained at very low concentrations of acetylcholine, the binding step and the channel opening step are similarly fast (neither one is rate-limiting), the efficacy of channel opening is close to 1, and the channel lifetime is very short, which is all consistent with the kinetics of macroscopic response (Colquhoun, 1986; Colquhoun & Ogden, 1988). As a general conclusion, the rates of channel opening and closing vary by orders of magnitude between synaptic and junctional receptors, and, to a more limited extent, also between nAChR species (Colquhoun & Sakmann, 1985; Sine & Steinbach, 1986*a*, *b*, 1987; Sine *et al.*, 1990). This also applies to the rates of desensitization (Siara *et al.*, 1990).

Pharmacology of responses

The affinities of acetylcholine binding to the two sites per nAChR monomer and the level of cooperativity between sites, as inferred from the dose dependence of single channel events, vary considerably for the species of nAChR employed (Auerbach & Lingle, 1987; Sine & Steinbach, 1987; Colquhoun & Ogden, 1988). The same applies to binding of agonists of acetylcholine, and the ion channel block induced by them at elevated concentrations. For a given degree of activation, acetylcholine appears to produce less direct block than any other agonists (Colquhoun & Ogden, 1986), with most partial agonists (such as decamethonium) producing very strong channel block (Adams & Sakmann, 1978). As a consequence, it is still not clear whether the putative 'partial agonists' do have a low probability of channel opening (low β/α), as is inferred by their definition, or whether their strong self-blocking action merely *mimics* such a property (Colquhoun, Ogden & Mathie, 1987).

Based on the results of ion flux studies, it has been suggested that acetylcholine produces a particularly strong blockade of the *Torpedo* nAChR channel by binding to some additional (peripheral) sites ('isosteric inhibition') (Udgaonkar & Hess, 1987*a*, *b*). Single channel studies of *Torpedo* nAChR expressed in mouse fibroblasts did not reveal such properties (Sine *et al.*, 1990).

Relatively few physiological data exist as to the action of antagonists on the channel. The ideal non-depolarizing antagonists are supposed to act predominantly by competition with acetylcholine for receptor binding, but few, if any, low molecular weight ligands exhibit only these properties (Maelicke, 1988*a*, *b*). As an example, tubocurarine and gallamine are capable of voltage-dependent channel blockade (Colquhoun, 1986; Strecker & Jackson, 1989). Similarly, 'direct channel blockers' such as the local anaesthetics of the organic amine type often do not merely reduce the number of open channels but also change the lifetimes of open and closed states (Neher, 1983; Colquhoun & Hawkes, 1983). It is not even clear whether they indeed require an open channel to gain access to their site(s). These deviations from simple open channel block have not yet been considered in the reaction schemes used in the analysis of single channel currents.

It is increasingly recognized that additional parameters may affect the properties of nAChR channels. To these belong the integrity of the membrane environment (Covarrubias & Steinbach, 1990), the level of glycosylation (Covarrubias, Kopta, & Steinbach, 1989; Buller & White, 1990) and other (as yet unidentified) post-translational and post-insertional modifications (Rohrbough & Kidokoro, 1990; Leonard *et al.*, 1988*a*, *b*; Li *et al.*, 1990).

Neuronal nicotinic receptors have only recently become accessible to in-depth biochemical and electrophysiological analysis. Based on their pharmacology, they can be separated into at least two classes. Thus, there exists in mammalian brain, in the peripheral ganglia of mammals and chick, and in chromaffin and PC12 cells a nAChR that is not blocked by α-bungarotoxin but instead by κ-bungarotoxin and κ-flavitoxin (Loring & Zigmond, 1988; Freeman, Schmidt & Oswald, 1980; De La Garza *et al.*, 1987; Wada *et al.*, 1988; Oortgiesen & Vijverberg, 1989). Hexamethonium, decamethonium and tubocurarine appear to act as channel blockers rather than as true antagonists (Ascher, Large & Rang, 1979) at this neuronal nAChR while the block produced by trimetaphan is likely to be competitive (similar to κ-bungarotoxin) (Loring *et al.*, 1989*a*, *b*). The high potency (and use-dependency) of hexamethonium has been suggested to result from its action as 'trapped' channel blocker (Gurney & Rang, 1984). The single channel conductance in ganglia is generally much smaller than in muscle of the same species (Mathie, Colquhoun & Cull-Candy, 1990; Derkach *et al.*, 1987; Lipscombe & Rang, 1988) and there exist considerable differences in the current patterns.

In insects, nAChRs are confined to the central nervous system in which neurons with distinct function have been identified and distinct cholinergic pathways have been located (Breer & Sattelle, 1987; Sattelle *et al.*, 1983). As in the mammalian central nervous system, there apparently exist a class of nAChR that can be blocked by α-bungarotoxin (David & Sattelle, 1984, 1990; Benson, 1988), and another one insensitive to this neurotoxin (Goodman & Spitzer, 1980; Lane, Sattelle & Hafnagel, 1982). A third class of nAChR has overlapping pharmacology with respect to α-bungarotoxin and κ-bungarotoxin, but also with respect to strychnine and bicuculline (Marshall *et al.*, 1990; Benson, 1988; Chiappinelli *et al.*, 1989; Bucking-ham, Sattelle & Hue, 1990), which are ligands typical for glycine receptors

(Betz, 1987). It has been suggested that some insect nAChRs may have properties of an ancestral receptor (Breer, Kleene & Hinz, 1985; Breer *et al.*, 1989; Maelicke, 1987*a*, *b*). In support of this hypothesis, a nAChR from *Locusta migratoria*, consisting of apparently only one type of polypeptide chain (Breer *et al.*, 1985), has been purified and reconstituted (Hanke & Breer, 1986), and expression in *Xenopus* oocytes of a single subunit RNA coding for nAChR from *Schistocerca gregaria* yielded a functional channel (Marshall *et al.*, 1990).

Non-classical responses

An increasing number of natural and synthetic compounds previously not thought to be ligands of nAChRs have been shown to bind and act at the receptor. Anatoxin-a and some analogues are potent cholinergic agonists (Koskinen & Rapoport, 1985; Swanson *et al.*, 1986, 1989; Kofuji *et al.*, 1990), and the slowly reversible acetylcholine esterase (AChE) inhibitor, physostigmine and related compounds act as agonists and channel blockers (Shaw *et al.*, 1985; Bradley, Sterz & Peper, 1986). ATP activates nAChR channels in cultured muscle cells (Igusa, 1988; Kolb & Wankelam, 1983). Forskolin, cAMP and chloramphenicol have direct channel blocking effects (McHugh & McGee, 1986; Eriksson *et al.*, 1986; Henderson *et al.*, 1986; Deana & Scuka, 1990) as have calcium channel effectors of the dihydropyridine type (Adam & Henderson, 1990). Philanthotoxin acts as a non-competitive inhibitor at nAChRs from vertebrates and insects (Rozental *et al.*, 1989), and some opiates modulate nAChR response by direct binding (Costa *et al.*, 1990; Oswald, Michel & Bigelow, 1986). In contrast, the effects of calcitonin gene-related peptide (CGRP) and of substance P on nAChR response (Eusebi *et al.*, 1989; Fontaine *et al.*, 1986; Mulle *et al.*, 1988; New & Mudge, 1986) appear to be by indirect mechanisms prompted by stimulation of the cAMP second messenger system (Caratsch & Eusebi, 1990; Simmons, Schuetze & Role, 1990). Except for physostigmine and its derivatives, the molecular basis of these effects remains to be established. It is likely though, that they all play a role in perfecting cholinergic synaptic transmission.

Multiplicity of nAChRs and functional significance

As discussed in the preceding sections, there is ample structural and functional evidence in favour of a family of isomeric forms of the nAChR within the same species. How can we define subfamilies and what are their functional roles?

Multiplicity of nAChR genes

As a particularly instructive example, 12 cDNA clones coding for nAChR polypeptides have so far been identified in the rat (Boulter *et al.*, 1986, 1990; Deneris *et al.*, 1989; Wada *et al.*, 1988, 1990; Gardner, 1990; Isenberg & Meyer, 1989). Based on sequence homology considerations, they subdivide into 5 α, 3 β, 1 γ, 1 δ and 1 ε clone. The $\alpha1,\beta1,\gamma,\delta$ and ε genes encode subunits which assemble to form the embryonic $\alpha1_2\beta1\gamma\delta$ and adult $\alpha1_2\beta1\varepsilon\delta$ forms of nAChR present in the post-synaptic membrane of the neuromuscular junction. The $\alpha2$–$\alpha5$, and $\beta2$ and $\beta3$ subunit genes are expressed in the peripheral and central nervous system. To add to the diversity, at the least the $\alpha4$ gene produces two gene products by alternative splicing. After co-injection of $\beta2$ RNA and either $\alpha2$, $\alpha3$ or $\alpha4$ RNA, functional nAChRs with distinct pharmacological properties are expressed (Papke *et al.*, 1989). Together with data from *in situ* hybridization histochemical studies (Deneris *et al.*, 1989; Wada *et al.*, 1988, 1989), including for $\alpha5$ (Wada *et al.*, 1990), it seems clear that there exist at least four different types of functional neuronal nAChR in the rat. They are expressed in distinct, although overlapping sets of structures of the rat CNS suggesting that they also serve distinct functional purposes. From electrophysiological studies, it is also clear that some cells express more than one type of nAChR channel (Margiotta, 1988; Simmons *et al.*, 1987; Takai *et al.*, 1984).

While the overall patterns of distribution of rat neuronal nAChR transcripts resemble those of acetylcholine binding sites in the brain (Clarke *et al.*, 1985; Clarke, 1986) (and those of some monoclonal antibodies, Deutch *et al.*, 1987; Swanson *et al.*, 1987), they clearly differ from the distribution of binding sites for α-bungarotoxin (Patrick & Shallup, 1977). From indirect evidence, using synthetic peptides matching in sequence elements of the major subsite for binding of α-bungarotoxin, $\alpha1$ and $\alpha5$ but not $\alpha2$, $\alpha3$ and $\alpha4$ subunits have been identified to bind this neurotoxin (McLane, Wu & Conti-Tronconi, 1990). In the chick, two α-bungarotoxin binding polypeptides have been identified, the cDNA clones of which are members of the superfamily of ligand-gated ion channels but are distinct from the gene families of nAChRs from muscle and brain (Schöpfer *et al.*, 1990). Together with the description of yet another novel type of nAChR in the rat central nervous system (CNS) (Mulle & Changeux, 1990), the total number of rat nAChRs may well be 15 or more.

A similar heterogeneity, including alternative splicing of genes (Beeson *et al.*, 1990), has been reported for nAChR from other species (for reviews,

(a)

```
Species              255      259          266    270              280    284
                      -       - +           *      *                -      -

Torpedo α      P T D S G  E K M T L S I  S V L L  S L T V F L L V I V  E L I P  S T S S A V P
bovine α       P T D S G  E K M T L S I  S V L L  S L T V F L L V I V  E L I P  S T S S A V P
mouse α        P T D S G  E K M T L S I  S V L L  S L T V F L L V I V  E L I P  S T S S A V P
rat neur. α2   P S E C G  E K I T L C I  S I L L  S L T V F L L L I T  E I I P  S T S L V I P
Dros. ALS      P S D S G  E K I S L C I  S I L L  S L T V F F L L L A  E I I P  P T S L T V P
Locusta α2     P S D S G  E K V T L C I  S I L L  S L T V F F L L L A  E I I P  P T S L A V P
Torpedo β      P P D A G  E K M S L S I  S A L L  A V T V F L L L L A  D K V P  E T S L S V P
mouse β        P Q D A G  E K M G L S I  F A L L  T L T V F L L L L A  I K V P  E T S L S V P
Dros. ARD      P A E A G  E K V T L G I  S I L L  S L V V F L L L V S  K I L P  P T S L
Torpedo γ      P A Q A G G  Q K C T L S I  S V L L  A Q T I F L F L I A  Q K V P  E T S L N V P
bovine γ       P A K A G G  Q K C T V A I  N V L L  A Q T V F L F L V A  K K V P  E T S Q A V P
bovine ε       P A Q A G G  Q K C T V S I  N V L L  A Q T V F L F L I A  Q K T P  E T S L S V P
Torpedo δ      P A E S G  E K M S T A I  S V L L  A Q A V F L L L T S  Q R L P  E T A L A V P
bovine δ       P A D C G  E K T S M A I  S V L L  A Q S V F L L L I S  K R L P  A T S M A I P
mouse δ        P G D C G  E K T S V A I  S V L L  A Q S V F L L L I S  K R L P  A T S M A I P
```

(b)

$$\text{W—P——I—C————V——FPFD—QNC}^{\text{T}}_{\text{S}}$$

(c)

Species and Tissue	Sequence[a]

```
Torpedo electroplax     Y R G W K H W V Y Y T C C P D T P Y L D
Xenopus laevis muscle   Y R C W K H W V Y Y T C C P D K P Y L D
Chicken muscle          Y R G W K H W V Y Y A C C P D T P Y L D
Bovine muscle           S R G W K H W V F Y A C C P S T P Y L D
Human muscle            S R G W K H S V T Y S C C P D T P Y L D
Mouse muscle            A R G W K H W V F Y S C C P T T P Y L D
Cobra muscle            Y R G F W H S V N Y S C C L D T P Y L D
Rat neurons: α2         A T G T Y N S K K Y D C C A E - I Y P D
Rat neurons: α3         A P G Y K H E I K Y N C C E E - I Y Q D
Rat neurons: α4         A V G T Y N T R K Y E C C A E - I Y P D
Locusta: α1             V P A V R N E K F Y T C C D E - P Y L D
Locusta: α2             V P A T R N E E Y Y P C C V E - P Y S D
Locusta: β                  P Y L N I Y E G N H P - T E T D
```

[a] Conserved and conservatively substituted residues (as compared to the *Torpedo* sequence) are shown in boxes.

Fig. 5.6. Regions of homology within the primary structures of nAChR from different species. (*a*) Sequences of M2 regions of nAChR subunits. Mutations of residues at 255, 259, 266, 270, 280 and 284 affect channel conductance (Imoto *et al.*, 1988; Leonard *et al.*, 1988*a*, *b*). (*b*) Aligned sequence of the disulphide loop between C-128 and C-142 containing the *N*-glycosylation site at position 141 conserved in all species of muscle nAChR (but not in neuronal nAChR). (*c*) Aligned sequences of the region around neighbouring cysteines in positions 192, 193.

see Claudio, 1989; Steinbach & Ifune, 1989). Even in the nervous system of the locust *Locusta migratoria*, generally assumed to be much simpler than that of higher vertebrates, at least four isoforms of the α polypeptide and several non-α polypeptides exist (Hermsen *et al.*, unpublished observations). It is not known, at present, how many different nAChRs are formed from these polypeptides, and how they differ in quaternary structure and in pharmacology. A comparable heterogeneity of nAChRs may also exist in other insect species (Marshall *et al.*, 1990; Baumann, Jonas & Gundelfinger, 1990; Jonas *et al.*, 1990; Sawruk *et al.*, 1990*a*, *b*).

At present, the following subfamilies can be distinguished: (1) nAChR from vertebrate muscle and electric tissue (m-nAChR), consisting of four types of subunits with the stoichometry 2:1:1:1; at least one embryonic and several adult forms bind α-bungarotoxin. (2) N1-family of nAChR (n1-nAChR), in vertebrate CNS; consisting of two types of subunits with unknown stoichiometry; many isoforms of distinct pharmacology and tissue distribution; most members do not bind α-bungarotoxin but bind κ-bungarotoxin. (3) N2-family of nAChR (n2-nAChR), in vertebrate CNS; the two α-subunits so far identified are co-translated and co-expressed within the same molecule; stoichiometry unknown; bind α-bungarotoxin. (4) N3-family of nAChR (n3-nAChR), in insect nervous system; several isoforms; pharmacology overlapping with n1 and n2 receptors but also with other ligand-gated ion channels; stoichiometry unknown, with the possibility of homo-oligomeric and hetero-oligomeric forms.

This classification is supported by distinct differences in the primary structures (Fig. 5.1, 5.6). As a general rule, the sequences of all nAChRs are similar within the putative transmembrane domains (Fig. 5.6(*a*)), in an area of the N-terminal extracellular domain framed by two cysteines and containing an N-glycosylation site (Fig. 5.6(*b*)), and (in the case of α-subunits) around the two adjacent cysteines in front of transmembrane domain M1 (Fig. 5.6(*b*), (*c*)). This suggests that these regions serve common functions of the nAChR, i.e. channel formation and ligand recognition. The sequences diverge most extensively in the putative large cytoplasmic domain which contains sites typical for modulatory control of protein function.

Embryonic and adult nAChR

The identification of two isoforms of calf *γ* cDNA (53 % sequence identity with each other as compared with 45 % identity with calf *δ*-subunit) (Takai *et al.*, 1984, 1985) was followed by the discovery that their expression is developmentally controlled (Takai *et al.*, 1985). Following a more extensive electrophysiological analysis after expression of chimeric *Torpedo*-calf nAChR in *Xenopus* oocytes, it became clear that the γ-subunit represented a (low conductance) embryonic form (with similar conductance and gating properties to the related *Torpedo* nAChR) while the isoform termed ε represented a (high conductance) adult (synaptic) form (Mishina *et al.*, 1986). More recently, the time course of γ and ε gene transcription in the course of embryonic development of the rat (Witzemann *et al.*, 1989) and the assembly and channel properties of foetal and adult rat muscle nAChR (Criado, Koenen & Sakmann, 1990) have been studied in full detail. From these data, the ε- and the γ-subunit compete with each other for participation in channel formation, and the adult form of channel (as judged by single channel studies) appears within hours following the rise in ε gene transcription in subsynaptic nuclei (Brenner, Witzemann & Sakmann, 1990). Both nAChR subtypes are co-expressed during a short postnatal period during which γ gene transcription decreases and ε gene transcription increases (Sakmann & Brenner, 1978). In addition to their different channel properties, the two subtypes differ significantly in their metabolic turnover rates (Gu *et al.*, 1990). Functional changes in the course of embryonic development have also been reported for cultured *Xenopus* myocytes (Leonard *et al.*, 1984, 1988*a*, *b*; Clark & Adams, 1991) but they have not yet been related to separate gene transcripts as in the case of embryonic and adult bovine muscle nAChR.

Embryonic and adult forms have so far only been identified for muscle nAChR with their expression controlled by electric activity and/or trophic factors (see below). Similar forms may not be required in the vertebrate CNS as most of its synapses are formed under much more restrictive spatial conditions and they are immediately stabilized. Exceptions exist, however, e.g. the olfactory system which has a strong potential of regeneration. The distribution of transcripts of cloned neuronal nAChR genes changes considerably in the course of CNS development (Wada *et al.*, 1989, 1990; Boyd *et al.*, 1988; Daubas *et al.*, 1990; Cauley, Agranoff & Goldman, 1990) but there is no evidence yet for a stage-specific isoform exchange mechanism as observed at the neuromuscular junction.

Control of expression of nAChR genes

So far, the best studied example of developmental regulation of nAChR polypeptide expression is the postnatal formation of bovine and rat neuromuscular junctions (Witzemann *et al.*, 1989; Mishina *et al.*, 1986). In

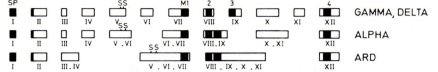

Fig. 5.7. Comparison of exon–intron distributions in the α, γ and δ subunit genes of vertebrate muscle and of *Drosophila ard* gene. Redrawn from Gundelfinger *et al.* (1989). The positions of the signal peptide sequence and of hydrophobic domains M1–M4 are indicated by black boxes. Neuronal nAChR genes (from vertebrate brain and insect nervous system) appear to be simpler in structure than those coding for muscle receptor.

the course of this process, extrajunctional (embryonic) nAChR are removed, and high concentrations of junctional (adult) nAChR cluster in the subsynaptic membrane. This is due in part to the lateral migration of receptors from extrajunctional sites but also to the insertion of newly synthesized nAChR molecules (Fambrough & Devreotes, 1978; Fambrough, 1979; Schuetze & Role, 1987). The regulation of nAChR synthesis could, in principle, occur on either the transcriptional or translation level, or by post-translational level, or by post-translational modifications, or at the level of subunit assembly and membrane insertion. All data presently available point to at least partial control at the transcriptional level (Evans *et al.*, 1987; Buonanno & Merlie, 1986; Buonnano, *et al.*, 1989; Witzemann, 1989; Witzemann *et al.*, 1989; Moss *et al.*, 1989; Klarsfeld & Changeux, 1985; Klarsfeld *et al.*, 1989) (Fig. 5.7). It is likely that the change in the ratio of nAChR isoforms is controlled by nerve-driven muscle activity affecting the transcriptional activity of subsynaptic nuclei (Burszijn, Berman & Gilbert, 1989; Brenner *et al.*, 1990) but conflicting evidence also exists (Burden, 1977; Cohen, 1972). Recent data suggest that the whole subsynaptic sarcoplasm represents a compartment specialized for the transcription, post-translational processing, and stabilization of proteins of the postsynaptic membrane (Jasmin *et al.*, 1989, 1990*a, b*).

Many nAChR genes contain *cis*-acting regulatory sequences that confer neural regulation and tissue-specific transcription (Klarsfeld *et al.*, 1987; Baldwin & Burden, 1988, 1989; Wang *et al.*, 1988; Merlie & Kornhauser, 1989). In chicken and rat clustered nAChR, genes have been identified (Nef *et al.*, 1984; Boulter *et al.*, 1990; Couturier *et al.*, 1990) which may allow transcriptional co-regulation. At present, it is not clear whether, and to what extent, myogenic regulatory factors are involved in transcriptional control of nAChR genes (Baldwin & Burden, 1988, 1989; Piette *et al.*, 1990). There is ample evidence, however, that several neural factors and second messenger systems are involved. To these belong the calcitonin-gene-related peptide (CGRP) (Salpeter & Loring, 1985; Fontaine & Changeux, 1989; Kirilovsky *et al.*, 1989; Österlund *et al.*, 1989), ascorbate (Knaack *et al.*, 1987), the chicken brain polypeptide ARIA (Usdin & Fischbach, 1986; Harris *et al.*, 1988), component(s) of the extracellular matrix (Hartman & Claudio, unpublished observations), agrin (Nitkin & Rothschild, 1990), dystrophin (Jasmin *et al.*, 1990*a*, *b*), cAMP (Laufer & Changeux, 1987; Garcia *et al.*, 1990; Green, Ross & Claudio, 1991) and phosphoinositide (Laufer & Changeux, 1989). Activation of nAChR genes after denervation requires protein synthesis suggesting that there is *de novo* synthesis of transcriptional activator(s) (Tsay *et al.*, 1990), while cAMP stimulation of nAChR transcription is mediated by post-translational mechanisms (Claudio *et al.*, 1989; Paulson & Claudio, 1990; Green *et al.*, 1991), correlating with phosphorylation of nAChR subunits (Green & Claudio, 1988; Green, W. N. *et al.*, unpublished observations).

Post-translational modifications (or non-covalent conformational changes) are also involved in the 'maturation' of subunits (Claudio *et al.*, 1988; Merlie & Smith, 1986; Conroy, Saedi & Lindstrom, 1990), as determined by the ability to bind α-bungarotoxin and to assemble with other subunits. Maturation may involve changes in glycosylation, fatty acid acylation, and phosphorylation (Merlie *et al.*, 1984), with the participation of 'chaperonins' (Pelham, 1988; Ellis & Hemmingsen, 1989). This process probably takes place in intracellular compartments prior to assembly and insertion (Conroy *et al.*, 1990), with the disulphide bridge between cysteines 192 and 193 already formed. Another form of regulatory control may be rapid degradation of immature subunits (Blount & Merlie, 1990).

Much less is known about the regulatory control of neuronal nAChR expression (Berg *et al.*, 1989). It is unclear, for example, whether there exist extrasynaptic nAChR of distinct properties in neurons, and whether nAChR gene regulation also takes place prior to, or in the absence of, innervation. The involvement of presynaptic nerve activity, however, is very likely (Role, 1988; Marshall, 1985). This is further supported by the effects of denervation which, in contrast to the results obtained with denervated muscle, leads to a substantial decline of nAChR transcription and expression (Boyd *et al.*, 1988; Jacob & Berg, 1987). Several second messenger systems have been identified as being involved in neuronal nAChR activity regulation (Downing & Role, 1987; Margiotta, Berg & Dionne, 1987; Margiotta & Gurantz, 1989). Surprisingly, the observed cAMP-dependent increase in response to acetylcholine is not due to changes in gene expression as it does not require protein biosynthesis, and it is not accompanied by an increase in ACh binding sites (Margiotta *et al.*, 1987, Margiotta & Gurantz, 1989; Halvorsen & Berg, 1987; Higgins & Berg, 1988*a*, *b*). In the case of bovine adrenal chromaffin cells in culture, the effect of cAMP depends on nAChR age (Higgins & Berg, 1988*a*, *b*) suggesting that it is part of a maturation process. As a general conclusion, there seem to exist considerable differences in the mechanisms by which synaptic contacts are stabilized and synaptic function is modulated in neuron–neuron synapses as compared to the neuromuscular junction.

Structure of the nAChR ion channel

Single channel recordings from planar lipid bilayers composed of synthetic phospholipids and nAChR provided the first unequivocal proof that the nAChR controlled cation channel is part of the nAChR molecule itself (Boheim *et al.*, 1981; Labarca, Lindstrom & Montal, 1984). This finding was confirmed (Mishina *et al.*, 1984) by the expression of fully functional nAChR channels after injection into *Xenopus* oocytes of mRNA coding for the four types of subunits of *Torpedo* receptor (Sumikawa *et al.*, 1981; Barnard, Miledi & Sumikawa, 1982). Affinity labelling, and electrophysiological studies of mutated nAChR, meanwhile have identified essential elements of the channel structure. Thus, photolabelling of the membrane-bound *Torpedo* nAChR with the local anaesthetic [^{3}H]-chlorpromazine, in the presence of carbamoylcholine (and in the absence and presence of competing local anaesthetics), identified δ-ser 262 (Giraudat *et al.*, 1986), β-ser 254 and β-leu 257 (Giraudat *et al.*, 1987), α-ser 248 (Giraudat *et al.*, 1989) and γ-ser 257, γ-thr 253 and γ-leu 260 (Revah *et al.*, 1990) which all are located in homologous positions within the putative transmembrane region M2 (see Fig. 5.6). Using the same strategy with the channel-blocking cation triphenyl-methylphosphonium (TPMP^{+}), Oberthuer *et al.* identified δ-ser 262 and provided suggestive evidence for the labelling of homologous residues of the α- and β-subunit of *Torpedo* nAChR (Oberthuer *et al.*, 1986*a*, *b*; Oberthuer & Hucho, 1988; Hucho *et al.*, 1986).

Studies with chimeras constructed from M2 (and parts of the sequence connecting M2 and M3) from bovine δ-subunit and the remaining part of δ-subunit, and the other subunits from *Torpedo* showed the lower conductance of the bovine channel while chimeras constructed with *Torpedo* δ-M2 showed the higher conductance of the *Torpedo* channel (Imoto *et al.*, 1986). In subsequent studies, particular amino acid residues controlling channel coductance were identified by site-directed mutagenesis (Imoto *et al.*, 1988; Leonard *et al.*, 1988*a, b*, 1990). Thus, three rings of negatively charged residues neighbouring M2 were found, one of which is located at the extracellular side, the other two at the cytosolic side of the nAChR. The ring of serine residues previously identified by chemical modification studies (see above) was demonstrated to control channel conductance and the residence time of the direct channel blocker QX222 (Leonard *et al.*, 1988*a, b*, 1990; Charnet *et al.*, 1990), suggesting that the serine residues of M2 indeed are located within the high affinity local anaesthetic binding site.

Based on these data, models for the receptor-integral ion channel have been proposed (Hucho *et al.*, 1986*a, b*; Hucho & Hilgenfeld, 1989; Giraudat *et al.*, 1987; Revah *et al.*, 1990; Reuhl, Amador & Dani, 1990). They assume the channel to lie in the axis of quasi-symmetry of the nAChR oligomer, the inner lining of the channel to be formed by the M2 transmembrane domains of all subunits, with its central core region being uncharged. Previous suggestions of a charged channel formed by amphipathic transmembrane domains, initially supported also by site-directed mutagenesis studies (Mishina *et al.*, 1985), have been disproved by the results reported above and by those of other experiments (Dennis *et al.*, 1988; Ratnam *et al.*, 1986; Maelicke *et al.*, 1989). To permit the local anaesthetics to reach their binding site inside the channel from the extracellular side, the M2 helices should be tilted by about 7° with respect to the central axis (Revah *et al.*, 1990; Hucho & Hilgenfeld, 1989; Furois-Corbin & Pullman, 1989*a, b*). This would lead to gaps in the upper part of the channel wall which may be filled by M1 domains (Dani, 1989*a, b*; Leonard *et al.*, 1988*a, b*; DiPaola *et al.*, 1990). As a consequence of such an arrangement, neighbouring helices, in particular M1 should be bent, at least in part. Interestingly, M1 helices contain conserved prolines which may produce the kinks required (Furois-Corbin & Pullman, 1989*a*).

The narrow part at the end of the truncated cone should then represent the entrance to the channel (6.5 Å diameter, Hille, 1984; 3–7 Å in length, Dani, 1989*a, b*) and the location of the local anaesthetics binding site. The serine and threonine rings in this area may contributed to dehydration of permeating ions by substituting some of the water molecules of the hydration shell (Hille, 1984; Changeux *et al.*, 1988).

By determining permeability ratios for selected permeant ions, Dani (1989*a, b*) showed that there is one main cationic binding site inside the permeation pathway. Is this the ring of serine residues (Changeux *et al.*, 1988)? A synthetic peptide matching the amino acid sequence of *Torpedo* δ-subunit, and uncharged synthetic peptides containing serine residues, forms discrete ion channels in phospholipid bilayers (Oiki, Anko & Mortal, 1988, 1990; Boheim *et al.*, 1989; Lear, Wasserman & DeGrado, 1988; Reuhl *et al.*, 1990). Also less flexible model compounds (Carmichael *et al.*, 1989), which therefore are better models for rigid membrane channels (Boheim *et al.*, 1982), produce authentic channel properties. It is noteworthy in this respect that the M2 ring of serine residues is conserved in all nAChRs sequenced so far, and is conservatively replaced in GABA and glycine receptors.

Several models have been proposed to account for an allosteric transition assumed to induce the gating of the nAChR channel (Unwin & Zampighi, 1988; Furois-Corbin & Pullman, 1989*a*; Guy, 1984; Changeux *et al.*, 1984, 1988). Although gross conformational changes in the course of ligand binding are well established (see Maelicke *et al.*, 1977*a*, *b*; Unwin & Zampighi, 1988; Prinz & Maelicke, unpublished observations, 1991), they may relate to desensitization rather than to channel activation (Covarrubias *et al.*, 1986). The rapid opening and closing transitions of nAChR channels suggest that channel activation involves the movement of only a few residues.

Structure of agonist binding sites at the nAChR

Ligand binding is the property of a receptor most central to its definition (Langley, 1905, 1906, 1907). In the case of the nAChR, ligand binding has been shown to be a complex recognition process resulting in ligand-specific electrical responses (Boheim *et al.*, 1981; Labarca *et al.*, 1984; Mishina *et al.*, 1984). The binding sites for cholinergic ligands therefore cannot be simple structures such as single amino acids, or pairs of them (Low, 1980). In contrast, ligand binding is likely to involve multipoint attachment (Maelicke *et al.*, 1977*a*, *b*, 1989; Conti-Tronconi *et al.*, 1991). Indirect biochemical evidence for such properties are among others (i) the sensitivity of binding and response parameters to the presence of lipids (Fong & McNamee, 1986), detergents (Brisson, Devaux & Changeux, 1975; Briley, Devaux & Chageux, 1978; Prinz & Maelicke, unpublished observations), alcohols (Forman, Righi & Miller, 1989; Gage *et al.*, 1975; Bradley *et al.*, 1980; Prinze & Maelicke, unpublished observations, 1991) and other 'denaturing' compounds (Prinz & Maelicke, unpublished observations), (ii) the accelerated dissociation of toxin–nAChR complexes in the presence of cholinergic ligands (Maelicke *et al.*, 1977*a*, *b*, Maelicke, 1987*a*, *b*; Kang & Maelicke, 1980), (iii) the identification of a rather large region of α-cobratoxin involved in nAChR binding (Martin, Chibber & Maelicke, 1983*a*, *b*; Johnson, Cushman & Malekzadeh, 1990), and (iv) the existence of ligand-specific conformational states of the nAChR (Covarrubias *et al.*, 1986).

Several approaches have been used to identify elements of the receptor's primary structure involved in cholinergic ligand binding. Covalent affinity labelling of the membrane-bound nAChR after selective cleavage of the disulphide bridge between Cys-192 and Cys-193 of the α-subunit has suggested that these cysteines are located 'in the vicinity' of the binding site (Kao *et al.*, 1984). Similarly, photoaffinity labelling with tubocurarine (Pedersen, Dreyer & Cohen, 1986), binding of snake α-toxins to α-subunit proteolytic fragments (Wilson *et al.*, 1984; Oblas, Singer & Boyd, 1986), or synthetic peptides (Neumann *et al.*, 1985; Radding *et al.*, 1988; Mulac-Jericevic & Atassi, 1987; Ralston *et al.*, 1987; Conti-Tronconi *et al.*, 1990), or α-subunit fragments expressed in *E. coli* transformants (Gershoni, 1987), or site-directed mutagenesis of cysteines (Mishina *et al.*, 1985) all have pointed to the region surrounding cysteines 192 and 193 as a potential site of ligand interaction.

Using a photoaffinity label acting in the dark as a competitive antagonist (Langenbuch-Cachat *et al.*, 1988), in addition to the two adjacent cysteines Tyr 93, Trp 149 and Tyr 190 (possibly also Trp 86, Tyr 151 and Tyr 19) were identified as in close proximity to the ACh binding site (Dennis *et al.*, 1986, 1988; Galzi *et al.*, 1990). Tyr 190 was also identified in labelling studies with the coral toxin lophotoxin (Abramson *et al.*, 1989).

All labelling studies using highly reactive intermediates are limited by the fact that these intermediates covalently attach to the closest residue towards which their chemical reactivity is directed. Hence, the residues identified in this way may either only lie 'in the vicinity' of the ACh binding site, or one or more of them could represent true attachment points for reversible binding of cholinergic ligands (see below). Avoiding this ambiguity, the cholinergic binding site has recently been mapped by means of high affinity reversible ligands and synthetic peptides serving as structural models of local regions of the nAChR (Conti-Tronconi *et al.*, 1990). Three sequence regions were identified which were recognized by all the four high affinity ligands (α-bungarotoxin and three competitive antibodies) used; α55–74, α134–153 and α181–200 (Fig. 5.8, see also Fig. 5.3). Acetylcholine and other low molecular weight ligands competed with the high affinity ligands in binding to each of these 'subsites', suggesting that all three of them contain elements of the ACh binding site (Conti-Tronconi *et al.*, 1990; Maelicke *et al.* unpublished observations). Although

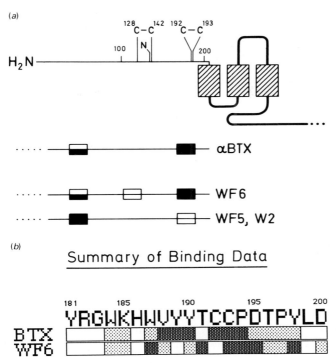

Fig. 5.8. Location of elements of the cholinergic binding site within *Torpedo* α-subunits. (*a*) Location of matching sequences of synthetic peptides recognized by α-BGT and monoclonal antibodies WF6, W2 and WF5. The level of ligand binding to each subsite is from strong ■, through intermediate ▨, to weak binding □. (Data from Conti-Tronconi *et al.*, 1990.) (*b*) Schematic representation of the amino acid residues within peptide Ta181-200$_{unmod}$ which were found to be 'essential' or 'influential' for the binding of αBTX and WF6. Black squares refer to 'essential residues', dotted squares refer to 'influential' residues. (From Conti-Tronconi *et al.*, 1991.)

there exists controversy as to whether other sequence regions of the α-subunit are also involved in toxin binding (Ruan *et al.*, 1990), the results of all studies using this approach consistently point to the existence of several subsites (several sequence regions discontinuously distributed over the N-terminal extracellular region) of the nAChR. A binding region composed of several loops of the nAChR α-polypeptide, possibly stabilized by β-structure (Conti-Tronconi *et al.*, 1991), would represent a complementary structure to the β-stranded binding region of α-cobratoxin previously identified (Martin *et al.*, 1983a, b).

Using a panel of 20 synthetic peptides each differing from *Torpedo*

α181–200 by the exchange of a single residue along the sequence, the residues essential for the interaction with α-bungarotoxin and the competitive monoclonal antibody WF6 were determined (Conti-Tronconi *et al.*, 1991). It was shown that the patterns of essential amino acids were different for the two ligands used. The conclusion of ligand-specific attachment point patterns is further strengthened by the fact that the amino acids serving as essential attachment points for α-bungarotoxin are preserved in all α-subunits from vertebrate muscle and electric tissue sequences to date while some exchanges occur in those α-subunits of neuronal nAChRs that bind κ-bungarotoxin but not α-bungarotoxin (Conti-Tronconi *et al.*, 1991).

Given the large body of binding site mapping studies already available, it seems surprising that the attachment point at the nAChR α-subunit for the trimethyl ammonium head group of acetylcholine has not yet been identified. Affinity labelling studies using an acetylcholine mustard have not been successful (Weiss & Maelicke, unpublished observations), suggesting that the negatively charged acceptor group at the nAChR is not an aspartic acid or glutamic acid residue as was long assumed (Smart *et al.*, 1984; Luyton, 1986; Karlin *et al.*, 1986) and was indirectly supported by X-ray crystallographic analysis of an antibody–phosphorylcholine complex (Padlan *et al.*, 1976). Model studies using methyl-substituted ammonium compounds showed that their binding is improved when the counter structures (macrocyclic compounds) contain ionized aromatic rings rather than carboxylic groups (while unsubstituted ammonium ions prefer carboxylates as counter structures) (Schneider *et al.*, 1986; Behr, Lehn & Vierling, 1976, 1982; Kieffer *et al.*, 1986; Lehn, 1985). The amino acids labelled with [^{3}H]-DDF (Galzi *et al.*, 1990; Changeux *et al.*, 1988) are obvious candidates for such residues. Interestingly, these residues are conserved in all muscle and neuronal nAChR α-subunits sequenced to date. They are absent in the corresponding positions of the other subunits of *Torpedo* nAChR but not always in non-α subunits from other species.

Equilibrium binding studies have established that there exist two cooperatively interacting binding sites for ACh at the membrane-bound nAChR (Neubig & Cohen, 1979; Fels *et al.*, 1982). The Hill coefficients deduced from dose–response curves point in the same direction (Trautmann, 1983; Jackson, 1988). The two binding sites are non-equivalent (Weber & Changeux, 1974*a*, *b*; Maelicke *et al.*, 1976, 1977*a*, *b*; Kang & Maelicke, 1980; Neubig & Cohen, 1979, 1980; Watters & Maelicke, 1983; Prinz & Maelicke, 1983, unpublished observations, 1991; Covarrubias *et al.*, 1986) although the α-subunits of native nAChRs are encoded by a single gene (Merlie *et al.*, 1983; Klarsfeld *et al.*, 1984) and thus are probably identical in primary structure. Studying the binding properties of pairs of α and non-α subunits of mouse muscle nAChR stably expressed in fibroblasts, it was found that different pairs of subunits have different binding properties, suggesting that the non-equivalence of binding sites may result from the different environment of each of the α-subunits in native nAChR (Blount & Merlie, 1989). Photoaffinity labelling studies with tubocurarine suggest that the ligand binding sites may actually be located in part within the interfaces between subunits (Pedersen *et al.*, 1990).

The non-classical responses reported suggest that ligand-binding sites in addition to the cholinergic sites and the sites for direct channel blockers may exist. So far, an additional class of separate binding sites at the α-subunit has only been established for the plant alkaloid physostigmine (eserine) (Okonyo *et al.*, unpublished observations; Kuhlmann *et al.*, 1991; Schrattenholz, unpublished observations). Eserine binds to (and activates) *Torpedo* nAChR even in the presence of saturating concentrations of α-neurotoxins, competitive antibodies, and low molecular weight antagonists of acetylcholine while its binding to the receptor is competed for by an antibody non-competitive with ACh, and by the cholinergic antagonist benzoquinonium (Okonjo *et al.* unpublished observations).

In summary, the ligand binding region at the nAChR α-subunit represents a sophisticated three-dimensional structure, composed of several separate sequence regions (loops), with individual areas serving as subsites for different classes of ligands, and with individual (ligand-specific) attachment point patterns within these subsites (Conti-Tronconi *et al.*, 1990, 1991; Okonjo *et al.*, unpublished observations).

Agonist binding, channel activation and closing, and desensitization

The allosteric model

It is practical to discuss the functional properties of the nAChR in terms of a four-state concerted model of allosteric interactions (Changeux & Podleski, 1968; Changeux, 1981; Changeux *et al.*, 1988) assuming one *resting*, one *active* (open channel) and two *desensitized* states (Fig. 5.9). The equilibrium between these states, which pre-exists in the absence of ligands (Monod, Wyman & Changeux, 1965; Jackson, 1984; Steinbach *et al.*, 1986; Neubig, Boyd & Cohen, 1982; Sine & Taylor, 1982), is shifted upon

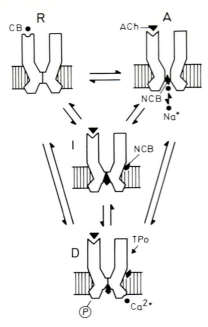

Fig. 5.9. Four-state allosteric model of the nAChR. Minimal four-state model for the allosteric transitions of the acetylcholine receptor. ACh, acetylcholine; CB, competitive blocker; NCB, noncompetitive blocker; TPo, thymopoietine; P, covalently bound phosphate; R, resting; A, active; I, intermediate or quickly desensitized; D, equilibrium or slowly desensitized. (Modified from Changeux *et al.*, 1984.)

ligand binding, giving rise to the various levels of activity detected by electrophysiological and ion flux measurements (Colquhoun & Ogden, 1986; Steinbach, 1989; Sine *et al.*, 1990; Auerbach & Sachs, 1983; Moore & Raftery, 1980; Covarrubias *et al.*, 1984). The existence of different conformational states of the nAChR is well established (Heidmann & Changeux, 1979, 1980; Prinz & Maelicke, 1983*a*, *b*: Covarrubias *et al.*, 1986; McCarthy & Stroud, 1989), as are ligand-induced transitions between such states (Dunn, Blanchard & Raftery, 1980; Dunn, Shelman & Agey, 1989; Prinze & Maelicke, 1983*b*; Covarrubias *et al.*, 1984; Grünhagen & Changeux, 1976). Ligand binding is an entropy-driven process (Maelicke *et al.*, 1977*a*, *b*) involving conformational adjustments of receptor and ligand (Dunn *et al.*, 1980; Kang & Maelicke, 1980; Pearce & Hawrot, 1990; Conti-Tronconi *et al.*, 1991) and long-range allosteric coupling between functional domains (Neubig & Cohen, 1979; Fels *et al.*, 1982, 1986; McCarthy & Stroud, 1989). Taking into account the large body of available electrophysiological and biochemical data, it is also clear, however, that the four-state model is an oversimplification, representing only the upper hierarchical level of conformational states (Maelicke, 1984). As a further limitation, the *ligand-specific* conformational states of the nAChR (Colquhoun, 1979; Boheim *et al.*, 1981; Covarrubias *et al.*, 1986) probably only develop in the presence of ligand (Prinz & Maelicke, unpublished observations see below) but do not pre-exist.

Within the frame of the four-state allosteric model, and based on time-resolved ligand binding and ion flux studies (Weber & Changeux, 1974*a, b*; Sine & Taylor, 1979; Dunn *et al.*, 1980; Blanchard *et al.*, 1989; Heidmann & Changeux, 1979; Boyd & Cohen, 1980; Prinz & Maelicke, 1983*b*; unpublished observations), the majority of nAChR molecules in native membranes pre-exist in a state which binds cholinergic agonists with low affinity (resting state). Prolonged exposure of the receptor to agonist leads to a conformational transition to states of high affinity of agonist binding while the affinity of binding for antagonists is changed only insignificantly, if at all. Two rates of affinity increase are resolved (Neubig *et al.*, 1982; Heidmann *et al.*, 1983; Karpen *et al.*, 1983) and correlate well with the time courses of electrophysiological desensitization (Katz & Thesleff, 1957; Feltz & Trautmann, 1982), suggesting that the two phenomena are related to each other. This is further supported by the effects of local anaesthetics which increase the rate of the affinity increase for agonists (Weiland & Taylor, 1979) and the rate of desensitization (Ogden, Siegelbaum & Colquhoun, 1981). Recently, it was shown, however, that the state of high agonist affinity can be induced not only by agonists but also by antagonists, direct channel blockers and a variety of membrane structure-disturbing compounds (Prinz & Maelicke, unpublished observations, 1991) suggesting that this state is a general low energy conformation of the nAChR forming the final 'sink' for all conformational transitions of this protein. Thus, the exact correlation between electrophysiologically defined desensitized states and biochemically defined states of high agonist affinity still remains to be elucidated.

Similarly, a proper biochemical characterization of the active (open channel) state is still missing. Recent kinetic studies using the fluorescent agonist NBD-5-acylcholine (Jürss, Prinz & Maelicke, 1979; Meyers *et al.*, 1983) and nAChR-rich *Torpedo* membrane vesicles have identified a transient receptor state bearing close similarities with the electrophysiological 'activatable' state in terms of rates of formation and decay (Prinz & Maelicke, unpublished observations). This state is particularly stable (slow rate of conversion into the state of high agonist affinity) when the nAChR is fully saturated with agonist. Its protection from rapid decay (Prinz & Maelicke, unpublished observations) may be the molecular basis for a fully saturated receptor having a far higher probability of channel opening than only partially saturated or free receptor (Jackson, 1986; Neubig *et al.*, 1982).

Because identical models were used for data analysis, the fluorescence kinetic data of agonist binding to *Torpedo* membrane vesicles (Prinz & Maelicke, unpublished observations) and the single channel currents recorded from *Torpedo* nAChR stably expressed in mouse fibroblasts (Sine *et al.*, 1990) can be compared. The rate constants of formation and decay, and the equilibrium dissociation constants of monoliganded and bi-liganded nAChR obtained by the two approaches are very similar indeed, demonstrating for the first time the basic equivalence of the biochemical and electrophysiological methods. Since the single channel current data (Sine *et al.*, 1990) do not extend to conditions of desensitization, the above noted inconsistencies with respect to its correlation with a state of high agonist affinity are not resolved by this comparison.

A 'third' binding site for agonist?

Several studies have provided circumstantial evidence for the existence of additional low affinity binding sites at the nAChR (Ziskind & Dennis, 1978; Dunn *et al.*, 1983; Pasquale *et al.*, 1983; Kang & Maelicke, 1980; Prinz & Maelicke, unpublished observation; Feltz & Trautmann, 1982). From our recent study (Prinz & Maelicke, unpublished observations), such additional sites cannot be detected at equilibrium but they may exist transiently in the course of an activation–inactivation cycle of the nAChR. As one possibility, the conformational transition from the state of low affinity agonist binding to one of high affinity may involve a rearrangement of sites in the course of which additional sites may be transiently exposed (Prinz & Maelicke, unpublished observations). A similar rearrangement of sites has been suggested for the mechanisms of action of acetylcholine esterase (Rosenberry, 1975); and models of multiple ligand binding sites at the nAChR have been proposed on the basis of sequence homologies (Smart *et al.*, 1984; Kosower, 1983; Cockcroft *et al.*, 1990).

Independent of these considerations, a 'third' low affinity agonist site is not necessarily required for nAChR function, as was previously suggested (Dunn, Conti-Tronconi & Raftery, 1983; Takeyasu, Udgaonkar & Hess, 1983). As shown by the quantitative analysis of fluorescence kinetic and single channel current data (Prinz & Maelicke, unpublished observations;

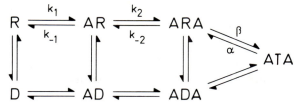

Fig. 5.10. Model of ligand-induced conformational states of the nAChR. R refers to closed channel states of low agonist affinity. T is the open channel (transmitting) state. D refers to closed channel states with high affinity for agonist binding. An equilibrium between free receptor in the R or D state may (e.g. *Torpedo* nAChR) or may not (e.g. *Electrophorus* nAChR) exist in the absence of ligands (Prinz & Maelicke, 1983*b*; unpublished observations). There also exist transitions from free and monoliganded receptor (R and AR) to open channel states, albeit at very low probability (Jackson, 1986; Neubig *et al.*, 1982). From the kinetics of agonist binding, formation of AR proceeds with a rate which is close to diffusion controlled while formation of ARA proceeds one order of magnitude more slowly. AR and ARA have short half-lives as they rapidly decay by ligand dissociation and by isomerization (Prinz & Maelicke, 1983*b*; unpublished observations).

Sine *et al.*, 1990), the active state is satisfactorily described by a transient short-lived low affinity state of biliganded receptor which is terminated by one of two alternative processes, ligand-dissociation or desensitization.

The model of ligand-induced conformational states

The specific allosteric model (Monod *et al.*, 1965; Karlin, 1967; Changeux *et al.*, 1967) is based on a pre-existing equilibrium of conformational states which is shifted in the presence of ligand (see above). This shift occurs because the conformational states have different affinities of binding for the ligand under consideration. As was recently shown, however (Prinz & Maelicke, unpublished observations), *Torpedo* nAChR can be induced to undergo large conformational transitions (resulting in an increase in agonist affinity typical for transition to a desensitized state) that are not accompanied by an affinity change for the inducing ligand (hence, the different states do not differ in their affinity for the ligand under consideration). This observation therefore is inconsistent with basic requirements of the allosteric model. It can be reconciled, however, if ligand binding is assumed to consist of a set of reaction steps (conformational changes) as is typical for classical transition kinetics (Fig. 5.10). Continuing this line of argument, an initial state (free nAChR) and a final state (fully saturated nAChR, high affinity for agonist), and a set of intermediates between these states are suggested. The intermediate states are different for different ligands because they have ligand-specific conformations (Covarrubias *et al.*, 1986; Kuhlmann, Okonjo & Maelicke, 1991) and pharmacology (Boheim *et al.*, 1981). For each individual ligand, the formation of the final state proceeds with individual kinetics along ligand-specific pathways of intermediates. One of the intermediates passed is the activatable state, with its (ligand-specific) kinetics of formation and decay defining the activity pattern of the channel, including desensitization (Prinz & Maelicke, unpublished observations).

Basically, this model agrees with the definition of the nAChR as an allosteric protein (Changeux *et al.*, 1967; Changeux, 1981) but removes the limitations of the specific allosteric model (Monod *et al.*, 1965). These limitations are the consideration of ligand-*independent* states (which is at odds with the pharmacological specificity of ligands) and the requirement of a pre-equilibrium of states with different ligand affinity.

Conclusion

Elucidating the structure and function of the nicotinic acetylcholine receptor has been one of the major research efforts of the last two decades in the life sciences. The wealth of information assembled in this way is impressive, in some areas even overwhelming. It is particularly satisfying that the different approaches of electrophysiology, pharmacology, biochemistry and molecular genetics increasingly produce converging results suggesting that we are approaching a basic understanding on the molecular level of this prototype of an integral signal transducer. Nevertheless, an intimate understanding is still a long way ahead with central elements of the puzzle still missing, e.g. X-ray structure, molecular basis of the pharmacological classification of ligands, molecular mechanisms of ligand-induced channel gating, individual pharmacology of isoforms, physiological significance of post-translational modifications, mechanism of membrane insertion and turnover, activity-controlled gene regulation of neuronal receptors, information transfer between postsynaptic membrane (and extracellular matrix) and subsynaptic nuclei, among other, in some cases not yet identified, problems.

Acknowledgements

Many fruitful discussions with my past and present co-workers and with colleagues from both sides of the Atlantic have contributed to this manuscript. My secretary Veronika Wölfe has protected me from many phone calls and visitors, and my wife Sigrid Reinhardt-Maelicke has helped in collecting the cited literature, in the layout of the manuscript and by critical reading and discussion. Work from my laboratory is supported by grants from the Deutsche Forschungsgemeinschaft, the Fonds der Chemischen Industrie, the Bayer AG and the Naturwissenschaftlich-Medizinisches Forschungszentrum of the Johannes-Gutenberg University.

References

Abramson, S. N., Li, Y., Culver, P. & Taylor, P. (1989). *Journal of Biological Chemistry*, **264**, 12666–72.

Adam, L. P. & Henderson, E. G. (1990). *Pflugers Arch.* **416**, 586–93.

Adams, D. J., Dwyer, T. M. & Hille, B. (1980). *Journal of General Physiology*, **75**, 493–510.

Adams, P. R. (1975). *Pflugers Arch.* **360**, 145–53.

Adams, P. R. (1987). *Neurology and Neurobiology*, **23**, 317–59.

Adams, P. R. & Sakmann, B. (1978). *Proceedings of the National Academy of Sciences, USA*, **75**, 2994–8.

Anderson, D. J. (1987). *Neurology and Neurobiology*, **23**, 285–315.

Anderson, D. J., Walter, P. & Blobel, G. (1982). *Journal of Cell Biology*, **93**, 501–6.

Ascher, P., Large, W. A. & Rang, H. P. (1979). *Journal of Physiology*, **295**, 139–70.

Auerbach, A. & Lingle, C. J. (1987). *Journal of Physiology*, **393**, 437–66.

Auerbach, A. & Sachs, F. (1983). *Biophysics Journal*, **42**, 1–10.

Baldwin, T. J. & Burden, S. J. (1988). *J. Cell Biol.*, **107**, 2271–9.

Baldwin, T. J. & Burden, S. J. (1989). In: *Molecular Biology of Neuroreceptors and Ion Channels.* (Maelicke, A., ed.) NATO ASI Series H: Cell Biology, vol. 32, pp. 519–40. Springer Press, Berlin.

Barkas, T., Gabriel, J. M., Mauron, A., Hughes, G. J., Roth, B., Alliod, C., Tzartos, S. J. & Ballivet, M. (1988). *Journal of Biophysical Chemistry*, **263**, 5916–20.

Barkas, T., Mauron, A., Roth, B., Alliod, C., Tzartos, S. J. & Ballivet, M. (1987). *Science*, **235**, 77–80.

Barnard, E. A., Beeson, D. W. M., Cockroft, V. B., Darlison, M. G., Hicks, A. A., Lai, F. A., Moss, S. J. & Squire, M. D. (1986). In: *Structure and Function of the Nicotinic Acetylcholine Receptor.* (Maelicke, A. ed.), NATO ASI Series H: Cell Biology, vol. 3, pp. 389–415. Springer Press Berlin.

Barnard, E. A., Darlison, M. G. & Seeburg, P. H. (1987). *Trends in Neurosciences*, **10**, 502–9.

Barnard, E. A., Miledi, R. & Sumikawa, K. (1982). *Proceedings of the Royal Society of London*, **215**, 241–6.

Baron, M. (1989). In *Molecular Biology of Neuroreceptors and Ion Channels.* NATO ASI Series, Series H: Cell Biology, vol. 32, pp. 645–9. (Maelicke, A., ed.), Springer Press, Berlin.

Baumann, A., Jonas, P. & Gundelfinger, E. D. (1990). *Nucleic Acids Research*, **18**, 3640.

Beeson, D., Morris, A., Vincent, A. & Newsom-Davis, J. (1990). *EMBO Journal*, **9**, 2101–6.

Behr, J. P., Lehn, J. M. & Vierling, P. (1976). *JCS Chemical Communications*.1976, 621–3.

Behr, J. P., Lehn, J. M. & Vierling, P. (1982). *Helvetica Chimica Acta* **65**, 1853–67.

Benson, J. A. (1988). *Nicotinic Acetylcholine Receptors*, (Clementi, F., Gotti, C. & Sher, E., eds), NATO ASI Series., pp. 227–40.

Berg, D. K., Boyd, R. T., Halvosen, S. W., Higgins, L. S., Jacob, M. H. & Margiotta, J. F. (1989). *Trends in Neurosciences*, **12**, 16–21.

Betz, H. (1987). *Trends in Neurosciences*, **10**, 113–17.

Blanchard, S. G., Quast, U., Reed, K., Lee, T., Schimerlik, M. I., Vandlen, R., Claudio, T., Strader, C. D., Moore, H. P. & Raftery, M. A. (1989). *Biochemistry*, **18**, 1875–83.

Bloch, R. J. & Pumplin, D. W. (1988). *American Journal of Physiology*, **254**, 345–64.

Blount, P. & Merlie, J. P. (1989). *Neuron*, **3**, 349–57.

Blount, P. & Merlie, J. P. (1990). *Journal of Cell Biology*, **111**, 2613–22.

Boheim, G., Hanke, W., Barantes, F. J., Eibl, H., Sakmann, B., Fels, G. & Maelicke, A. (1981). *Proceedings of the National Academy of Sciences, USA*, **78**, 3586–90.

Boheim, G., Hanke, W., Methfessel, C., Eibl, H., Kaupp, U. B., Maelicke, A. & Schultz, J. E. (1982). In: *Transport in Biomembranes: Model Systems and Reconstitution*. (Antolini, R. *et al.*, eds), pp. 87–97. Raven Press, New York.

Boheim, G., Helbig, I., Meder, S., Franz, B. & Jung, G. (1989). In: *Molecular Biology of Neuroreceptors and Ion Channels*. NATO ASI Series, Series H; Cell Biology, vol. 32, pp. 401–12. (Maelicke, A., ed.), Springer Press, Berlin.

Boulter, J., Evans, K., Goldman, D., Martin, G., Treco, D., Heinemann, S. & Patrick, J. (1986). *Nature, London*, **319**, 368–74.

Boulter, J., O'Shea-Greenfield, A., Duvoisin, R. M., Connolly, J. G., Wada, E., Jensen, A., Gardner, P. D., Ballivet, M., Deneris, E. S., McKinnon, D., Heinemann, S. & Patrick, J. (1990). *J. Biol. Chem.*, **265**, 4472–82.

Boyd, N. D. & Cohen, J. B. (1980). *Biochemistry*, **19**, 5353–8.

Boyd, R. T., Jacob, M. H., Couturier, S., Ballivet, M. & Berg, D. K. (1988). *Neuron* **1**, 495–502.

Bradley, R. J., Peper, K. & Sterz, R. (1980). *Nature, London*, **284**, 60–2.

Bradley, R. J., Sterz, R. & Peper, K. (1986). *Brain Research*, **376**, 199–203.

Breer, H., Hanke, W., Benke, D., Tareilus, E. & Krieger, J. (1989). In: *Molecular Biology of Neuroreceptors and Ion Channels*. NATO ASI Series H: Cell Biology, vol. 32, pp. 55–68. (Maelicke, A., ed.), Springer Press Berlin.

Breer, H., Hinz, G., Mädler, U. & Hanke, W. (1986). In: *Structure and Function of the Nicotinic Acetylcholine Receptor*. NATO ASI Series H: Cell Biology, vol. 3, pp. 319–32. (Maelicke, A., ed.), Springer Press Berlin.

Breer, H., Kleene, R. & Hinz, G. (1985). *Journal of Neuroscience*, **5**, 3386–92.

Breer, H. & Sattelle, D. B. (1987). *Journal of Insect Physiology* **33**, 771–90.

Brehm, P. & Henderson, L. (1988). *Developments in Biology*, **129**, 1–11.

Brenner, H. R., Witzemann, V. & Sakmann, B. (1990). *Nature, London*, **344**, 544–7.

Briley, M. S., Devaux, P. F. & Changeux, J.-P. (1978). *European Journal of Biochemistry*, **84**, 429–39.

Brisson, A., Devaux, P. F. & Changeux, J.-P. (1975). *Compte Rendu Seances Academie Sciences* (Paris), **280**, 2153–6.

Brisson, A. & Unwin, P. N. (1985). *Nature, London*, **315**, 474–7.
Buckingham, S. D., Sattelle, D. B. & Hue, B. (1990). *Journal of Experimental Biology*, in press.
Buller, A. L. & White, M. M. (1990). *Journal of Membrane Biology*, **115**, 179–89.
Buonanno, A., Casabo, L., Kornhauser, J., Crowder, C. M. & Merlie, J. P. (1989). In: *Molecular Biology of Neuroreceptors and Ion Channels*. NATO ASI Series H: Cell Biology, vol. 32, pp. 541–52 (Maelicke, A., ed.), Springer Press, Berlin.
Buonanno, A. & Merlie, J. P. (1986). *Journal of Biological Chemistry*, **261**, 11452–5.
Burden, S. J. (1977). *Developments in Biology*, **57**, 317–29.
Burszijn, S., Berman, S. A. & Gilbert, W. (1989). *Proceedings of the National Academy of Sciences, USA*, **86**, 2928–32.
Caratsch, C. G. & Eusebi, F. (1990). *Neuroscience Letters*, **111**, 344–50.
Carmichael, V. E., Dutton, P. J., Fyles, T. M., James, T. D., Swan, J. A. & Zojaji, M. (1989). *Journal of the American Chemical Society*, **111**, 767–9.
Cauley, K., Agranoff, B. W. & Goldman, D. (1990). *Journal of Neuroscience*, **10**, 670–83.
Chabala, L. D. & Lester, H. A. (1986). *Journal of Physiology*, **379**, 83–108.
Changeux, J.-P. (1981). pp. 85–254. New York: Academic Press.
Changeux, J.-P. (1988). *FIDIA Research Foundation Neuroscience Award Lectures*, **4**, 21–168.
Changeux, J.-P., Devillers-Thiéry, A. & Chemouilli, P. (1984). *Science*, **225**, 1335–45.
Changeux, J.-P., Devillers-Thiéry, A., Giraudat, J., Dennis, M., Heidmann, T., Revah, F., Mulle, C., Heidmann, O., Klarsfeld, A., Fontaine, B., Laufer, R., Nghiêm, H. O., Kordeli, E. & Cartaud (1987). In: *Journal of Neuroscience*, Hayaishi, O., ed. pp. 29–76. BV Utrecht: Japan Scientific Societies Press Tokyo VNU Science Press.
Changeux, J.-P., Giraudat, J. & Dennis, M. (1987*a*). *Trends in Pharmacological Science*, **8**, 459–65.
Changeux, J.-P., Klarsfeld, A. & Heidmann, T. (1987*b*). In: *The Neural and Molecular Basis of Learning* (Changeux, J.-P. & Koniski, M., eds), pp. 31–84.
Changeux, J.-P., Llinas, R. R., Purves, D. & Bloom, F. E. (1988). In: *Functional Architecture and Dynamics of the Nicotinic Acetylcholine Receptor: An Allosteric Ligand-gated Ion Channel*, pp. 21–168, Fidia Research Foundation, Neuroscience Award Lectures, New York.
Changeux, J.-P. & Podleski, T. (1968). *Proceedings of the National Academy of Sciences*, **59**, 944–50.
Changeux, J.-P., Thiéry, J., Tung, Y. & Kittel, C. (1967). *Proceedings of the National Academy of Sciences*, **57**, 335–41.
Charnet, P., Labarca, C., Leonard, R. J., Vogelaar, N. J., Czyzyk, L., Gouin, A., Davidson, N. & Lester, H. A. (1990). *Neuron*, **4**, 87–95.
Chiappinelli, V. A., Hue, B., Mony, L. & Sattelle, D. B. (1989). *Journal of Experimental Biology*, **141**, 61–71.
Clark, R. B. & Adams, P. R. (1991). *Society for Neuroscience*, **7**, 838.
Clarke, P. B. S. (1986). In: *Structure and Function of the Nicotinic Acetylcholine Receptor*. NATO ASI Series H: Cell Biology, vol. 3 (Maelicke, A., ed.), pp. 345–57, Springer Press, Berlin.

Clarke, P. B. S., Schwartz, R. D., Poaul, S. M., Pert, C. B. & Pert, A. (1985). *Journal of Neuroscience*, **5**, 1307–15.

Claudio, T. (1986). *Trends in Neuroscience*, **7**, 308–12.

Claudio, T. (1989) Molecular genetics of acetylcholine receptor-channels. In: *Frontiers in Molecular Biology: Molecular Neurobiology* (Glover, D. M. & Hames, B. D. eds), pp. 63–142. IRL Press Oxford.

Claudio, T., Ballivet, M., Patrick, J. & Heinemann, S. (1983). *Proceedings of the Academy of Sciences, USA*, **80**, 1111–15.

Claudio, T., Paulson, H. L., Green, W. N., Ross, A. F., Hartman, D. S. & Hayden, D. (1989). *Journal of Cell Biology*, **108**, 2277–90.

Claudio, T., Paulson, H. L., Hartman, D., Sine, S. & Sigworth, F. J. (1988). *Current Topics in Membranes and Transport*, **33**, 219–47.

Cockcroft, V. B., Osguthorpe, E. A. B. & Lunt, G. G. (1990). *Proteins*, **8**, 386–97.

Cohen, M. W. (1972). *Brain Research*, **41**, 457–63.

Colquhoun, D. (1979). In: *The Receptors: A Comprehensive Treatise*. (O'Brian, R. D., ed.), pp. 93–142, Plenum Press, New York.

Colquhoun, D. (1986). In: *Handbook of Experimental Pharmacology*, vol. 79, (Kharkevich, D. A., ed.), pp. 59–113. Springer Verlag, Berlin-Heidelberg.

Colquhoun, D. & Hawkes, A. G. (1983). In: *Single Channel Recording*, (Sakman, B. & Neher, E., eds), pp. 135–75. Plenum Press, New York.

Colquhoun, D. & Ogden, D. C. (1986). In: *Structure and Function of the Nicotinic Acetylcholine Receptor*, NATO ASI Series H: Cell Biology, vol. 3, (Maelicke, A., ed.), pp. 197–218, Springer Press, Berlin.

Colquhoun, D. & Ogden, D. (1988). *Journal of Physiology*, **395**, 131–59.

Colquhoun, D., Ogden, D. C. & Mathie, A. (1987). *Trends in Pharmacological Science*, **8**, 465–72.

Colquhoun, D. & Sakmann, B. (1985). *Journal of Physiology*, **369**, 501–57.

Conroy, W. G., Saedi, M. S. & Lindstrom, J. (1990). *Journal of Biological Chemistry*, **265**, 21642–51.

Conti-Tronconi, B. M., Diethelm, B. M., Wu, X., Tang, F., Bertazzon, R., Schröder, B., Reinhardt-Maelicke, S., Maelicke, A. (1991). In press.

Conti-Tronconi, B. M., Dunn, S. M., Barnard, E. A., Dolly, J. O., Lai, F. A., Ray, N. & Raftery, M. A. (1985). *Proceedings of the National Academy of Sciences*, **82**, 5208–12.

Conti-Tronconi, B. M., Tang, F., Diethelm, B. M., Spencer, S. R., Milius, R., Reinhardt-Maelicke, S. & Maelicke, A. (1990). *Biochemistry*, **29**, 6221–30.

Costa, A. C. S., Swanson, K. L., Aracava, Y., Aronstam, R. S. & Albuquerque, E. X. (1990). *Journal of Pharmacological Exp. Therapy*, **252**, 507–16.

Couturier, S., Erkman, L., Valera, S., Rungger, D., Bertrand, S., Boulter, J., Ballivet, M. & Bertrand, D. (1990). *Journal of Biological Chemistry*, **265**, 17560–7.

Covarrubias, M., Kopta, C. & Steinbach, J. H. (1989). *Journal of General Physiology*, **93**, 765–83.

Covarrubias, M., Prinz, H. & Maelicke, A. (1984). *FEBS Letters*, **169**, 229–33.

Covarrubias, M., Prinz, H., Meyers, H.-W. & Maelicke, A. (1986). *Journal of Biological Chemistry*, **261**, 14955–61.

Covarrubias, M. & Steinbach, J. H. (1990). *Pflugers Arch.*, **416**, 385–92.

Criado, M., Koenen, M. & Sakmann, B. (1990). *FEBS Letters*, **270**, 95–9.

Dani, J. A. (1989*a*). *Current Opinion in Cell Biology* **1**, 753–64.

Dani, J. A. (1989*b*). *Journal of Neuroscience*, **9**, 884–92.

Das, M. K. & Lindstrom, J. (1989). *Biochemical and Biophysical Research Communications*, **165**, 865–71.

Daubas, P., Devillers-Thiéry, A., Geoffroy, B., Martinez, S., Bessis, A. & Changeux, J. P. (1990). *Neuron*, **5**, 49–60.

David, J. A. & Sattelle, D. B. (1984). *Journal of Experimental Biology*, **108**, 119–36.

David, J. A. & Sattelle, D. B. (1990). *Journal of Experimental Biology*, **151**, 21–39.

De La Garza, R., McGuire, T. J., Freeman, R. & Hoffer, B. J. (1987). *Neuroscience*, **23**, 887–91.

Deana, A. & Scuka, M. (1990). *Neuroscience Letters*, **114**, 272–6.

Deneris, E. S., Boulter, J., Connolly, J., Wada, E., Wada, E., Wada, K., Goldman, D., Swanson, L. W., Patrick, J. & Heinemann, S. (1989). *Clinical Chemistry*, **35**, 731–7.

Deneris, E. S., Boulter, J., Swanson, L. W., Patrick, J. & Heinemann, S. (1989). *Journal of Biological Chemistry*, **264**, 6268–72.

Dennis, M., Giraudat, J., Kotzyba-Hibert, F., Goeldner, M., Hirth, C., Chang, J.-Y. & Changeux, J.-P. (1986). *FEBS Letters*, **207**, 2443–9.

Dennis, M., Giraudat, J., Kotzyba-Hibert, F., Goeldner, M., Hirth, C., Chang, J.-Y., Lazure, C., Chretien, M. & Changeux, J.-P. (1988). *Biochemistry*, **27**, 2346–57.

Derkach, V. A., North, R. A., Selyanko, A. A. & Skok, V. I. (1987). *Journal of Physiology*, **388**, 141–51.

Deutch, A. Y., Holliday, J., Roth, H. R., Chun, L. L. Y. & Hawrot, E. (1987). *Proceedings of the National Academy of Sciences, USA*, **84**, 8697–701.

Devillers-Thiéry, A., Changeux, J. P., Paroutaud, P. & Strosberg, A. D. (1983). *Proceedings of the National Academy of Sciences, USA*, **80**, 2067–071.

Dionne, V. E., Steinbach, J. H. & Stevens, C. F. (1978). *Journal of Physiology*, **218**, 421–44.

DiPaola, M., Czajkowski, C. & Karlin, A. (1989). *Journal of Biological Chemistry*, **264**, 15457–63.

DiPaola, M., Kao, P. N. & Karlin, A. (1990). *Journal of Biological Chemistry*, **265**, 11017–29.

Doster, W., Hess, B., Watters, D. & Maelicke, A. (1980). *FEBS Letters*, **113**, 312–14.

Downing, J. E. G. & Role, L. W. (1987). *Proceedings of the National Academy of Sciences, USA*, **84**, 7739–43.

Dreyer, F., Peper, K. & Sterz, R. (1978). *Journal of Physiology*, **281**, 395–419.

Dunn, S. M. J., Blanchard, S. G. & Raftery, M. A. (1980). *Biochemistry*, **19**, 5645–52.

Dunn, S. M. J., Conti-Tronconi, B. M. & Raftery, M. A. (1983). *Biochemistry*, **22**, 2512–18.

Dunn, S. M. J., Conti-Tronconi, B. M. & Raftery, M. A. (1986). *Biochemical and Biophysical Research Communications*, **139**, 830–7.

Dunn, S. M. J., Shelman, R. A. & Agey, M. W. (1989). *Biochemistry*, **28**, 2551–7.

Ellis, R. J. & Hemmingsen, S. M. (1989). *Trends in Biochemical Science*, **14**, 339–42.

Engel, A. G. (1987). *The Vertebrate Neuromuscular Junction*. pp. 361–424.

Eriksson, H., Salmonsson, R., Liljeqvist, G. & Helbronn, E. (1986). In: *Dynamics of Cholinergic Function* (Hamin, I., ed.), pp. 933–937. Plenum Press, New York.

Eusebi, F., Farini, D., Grassi, F. & Santoni, A. (1989). *Pflugers Arch.* **415**, 150–5.

Evans, S., Goldman, D., Heinemann, S. & Patrick, J. (1987). *Journal of Biological Chemistry*, **262**, 4911–16.

Fambrough, D. M. (1979). *Physiological Reviews*, **59**, 165–227.

Fambrough, D. M. & Devreotes, P. N. (1978). *Journal of Cell Biology*, **76**, 237–44.

Fels, G., Pluemer-Wilk, R., Schreiber, M. & Maelicke, A. (1986). *Journal of Biological Chemistry*, **261**, 15746–54.

Fels, G., Wolff, E. K. & Maelicke, A. (1982). *European Journal of Biochemistry*, **127**, 31–8.

Feltz, A. & Trautmann, A. (1982). *Journal of Physiology*, **322**, 257–72.

Finer-Moore, J. & Stroud, R. M. (1984). *Proceedings of the National Academy of Sciences, USA*, **81**, 155–9.

Fong, T. M. & McNamee, M. G. (1986). *Biochemistry*, **25**, 830–40.

Fontaine, B. & Changeux, J.-P. (1989). *Journal of Cell Biology*, **108**, 1025–37.

Fontaine, B., Klarsfeld, A., Hökfelt, T. & Changeux, J. P. (1986). *Neuroscience Letters*, **71**, 59–65.

Forman, S. A. & Miller, K. W. (1988). *Biophysical Journal*, **54**, 149–58.

Forman, S. A., Righi, D. L. & Miller, K. W. (1989). *Biochimica et Biophysica Acta*, **987**, 95–103.

Freeman, J. A., Schmidt, J. T. & Oswald, R. E. (1980). *Neuroscience*, **5**, 929–42.

Froehner, S. C., Reiness, C. G. & Hall, Z. W. (1977). *Journal of Biological Chemistry*, **252**, 8589–96.

Furois-Corbin, S. & Pullman, A. (1989*a*). *Biochimica et Biophysica Acta*, **984**, 339–50.

Furois-Corbin, S. & Pullman, A. (1989*b*). *FEBS Letters*, **252**, 63–8.

Gage, P. W., McBurney, R. N. & Schneider, G. T. (1975). *Journal of Physiology*, **244**, 409–29.

Galzi, J.-L., Revah, F., Black, D., Goeldner, M., Hirth, C. & Changeux, J.-P. (1990). *Journal of Biological Chemistry*, **265**, 10430–7.

Garcia, L., Picon-Raymond, G. R., Changeux, J.-P., Lazdunski, M. & Rieger, F. (1990). *FEBS Letters*, **263**, 147–152.

Gardner, P. D. (1990). *Nucleic Acids Research*, **18**, 6714.

Gardner, P. D., Ogen, D. C. & Colquhoun, D. (1984). *Nature, London*, **309**, 160–2.

Gershoni, M. (1987). *Proceedings of the National Academy of Sciences, USA*, **84**, 4318–21.

Giraudat, J., Dennis, M., Heidmann, T., Chang, J.-Y. & Changeux, J.-P. (1986). *Proceedings of the National Academy of Sciences, USA*, **83**, 2719–23.

Giraudat, J., Dennis, M., Heidmann, T., Haumont, P. T., Lederer, F. & Changeux, J.-P. (1987). *Biochemistry*, **26**, 2410–18.

Giraudat, J., Galzi, J. L., Revah, F., Changeux, J.-P., Haumont, P. Y. & Lederer, F. (1989). *FEBS Letters*, **253**, 190–8.

Goodman, C. S. & Spitzer, N. C. (1980). In: *Receptors for Trasmitters, Hormones and Pheromones in Insects*, pp. 195–207. Elsevier, Amsterdam.

Green, W. N. & Claudio, T. (1988). *Society of Neuroscience Abstracts*, **14**, 1045.

Green, W. N., Ross, A. F. & Claudio, T. (1991). *Science*, in press.

Grünhagen, H. H. & Changeux, J.-P. (1976). *Journal of Molecular Biology*, **106**, 497–516.

Gu, Y., Franco, A., Jr., Gardner, P. D., Lansman, J. B., Forsayeth, J. R. & Hall, Z. W. (1990). *Neuron*, **5**, 147–157.

Gundelfinger, E. D., Hermans-Borgmeyer, I., Schloss, P., Sawruk, E., Udri, C., Vingron, M., Betz, H. & Schmitt, B. (1989). In *Molecular Biology of*

Neuroreceptor and Ion Channels. NATO ASI Series H: Cell Biology, vol. 32, pp. 69–81, Maelicke, A., ed. Springer Press, Berlin.

Gurney, A. M. & Rang, H. P. (1984). *British Journal of Pharmacology*, **82**, 225–42.

Guy, H. R. (1984). *Biophysics Journal*, **45**, 249–61.

Halvorsen, S. W. & Berg, D. K. (1987). *Journal of Neuroscience*, **7**, 2547–55.

Hamilton, S. L., McLaughlin, M. & Karlin, A. (1979). *Biochemistry*, **18**, 155–63.

Hamilton, S. L., Pratt, D. R. & Eaton, D. C. (1985). *Biochemistry*, **24**, 2210–19.

Hanke, W. & Breer, H. (1986). *Nature London*, **321**, 171–4.

Harris, D. A., Falss, D. L., Dill-Devor, R. M. & Fischbach, G. D. (1988). *Proceedings of the National Academy of Sciences, USA*, **85**, 1983–7.

Heidmann, T., Bernhardt, J., Neumann, E. & Changeux, J.-P. (1983). *Biochemistry*, **22**, 5452–9.

Heidmann, T. & Changeux, J.-P. (1979). *European Journal of Biochemistry*, **94**, 255–79.

Heidmann, T. & Changeux, J.-P. (1980). *Biochemical and Biophysical Research Communications*, **97**, 889–96.

Heilbronn, E. (1985). In: *Handbook of Neurochemistry* (Lajtha, A., ed.), pp. 241–284. Plenum Press, New York.

Heinemann, S., Boulter, J., Connolly, J., Goldman, D., Evans, K., Treco, D., Ballivet, M. & Patrick, J. (1986). In: *Structure and Function of the Nicotinic Acetylcholine Receptor*. NATO ASI Series H: Cell Biology, vol. 3, pp. 359–387 (Maelicke, A., ed.), Springer Press, Berlin.

Heinemann, S., Boulter, J., Deneris, E., Connolly, J., Gardner, P., Wada, E., Wada, K., Duvoisin, R., Ballivet, M., Swanson, L. & Patrick, J. (1989). In: *Molecular Biology of Neuroreceptors and Ion Channels*. NATO ASI Series H: Cell Biology (Maelicke, A., ed.), vol. 32, pp. 13–30, Springer Press, Berlin.

Henderson, F., Prior, C., Dempster, J. & Marshall, I. G. (1986). *Molecular Pharmacology*, **29**, 52–64.

Higgins, L. S. & Berg, D. K. (1988a). *Journal of Cell Biology*, **107**, 1147–56.

Higgins, L. S. & Berg, D. K. (1988b). *Journal of Cell Biology*, **107**, 1157–65.

Hille, B. (1984). *Ionic Channel of Excitable Membranes*, Sinauer, Sunderland, MA.

Hollmann, M., O'Shea-Greenfield, A., Rogers, S. W. & Heinemann, S. (1989). *Nature, London*, **342**, 643–8.

Hucho, F. & Hilgenfeld, R. (1989). *FEBS Letters*, **257**, 17–23.

Hucho, F., Oberthuer, W. & Lottspeich, F. (1986a). *FEBS Letters*, **205**, 137–42.

Hucho, F., Oberthuer, W., Lottspeich, F. & Wittmann-Liebold, B. (1986b). In: *Structure and Function of the Nicotinic Acetylcholine Receptor*. NATO ASI Series H: Cell Biology, (Maelicke, A., ed.), vol. 3, pp. 115–27. Springer Press, Berlin.

Huganir, R. L. & Miles, K. (1989). *CRC Critical Reviews in Biochemistry and Molecular Biology*, **24**, 183–215.

Igusa, Y. (1988). *Journal of Physiology London*, **405**, 169–85.

Imoto, K., Busch, C., Sakmann, B., Mishina, M., Konno, T., Nakai, J., Bujo, H., Mori, Y., Fukuda, K. & Numa, S. (1988). *Nature, London*, **335**, 645–8.

Imoto, K., Methfessel, C., Sakmann, B., Mishina, M., Mori, Y., Konno, T., Fukuda, K., Kurasaki, M., Bujo, H., Fujita, Y. & Numa, S. (1986). *Nature, London*, **324**, 670–4.

Isenberg, K. E. & Meyer, G. E. (1989). *Journal of Neurochemistry*, **52**, 988–91.

Jackson, M. B. (1984). *Proceedings of the National Academy of Sciences, USA*, B1, 3901–4.

Jackson, M. B. (1986). *Biophysics Journal*, **49**, 663–72.

Jackson, M. B. (1988). *Journal of Physiology*, **397**, 555–83.

Jacob, M. H. & Berg, D. K. (1987). *Journal of Cell Biology*, **105**, 1847–54.

Jasmin, B. J., Cartaud, J., Bornens, M. & Changeux, J.-P. (1989). *Proceedings of the National Academy of Sciences, USA*, **86**, 7218–22.

Jasmin, B. J., Cartaud, A., Ludosky, M. A., Changeux, J.-P. & Cartaud, J. (1990*a*). *Proceedings of the National Academy of Sciences, USA*, **87**, 3938–41.

Jasmin, B. J., Changeux, J.-P. & Cartaud, J. (1990*b*). *Nature, London*, **344**, 673–5.

Johnson, D. A., Cushman, R. & Malekzadeh, R. (1990). *J. Biol. Chem.*, **265**, 7360–8.

Jonas, P., Baumann, A., Merz, B. & Gundelfinger, E. D. (1990). *FEBS Letters*, **269**, 264–8.

Jürss, R., Prinz, H. & Maelicke, A. (1979). *Proceedings of the National Academy of Sciences, USA*, **76**, 1064–8.

Kang, S. & Maelicke, A. (1980). *Journal of Biological Chemistry*, **255**, 7326–32.

Kao, P. N., Dwork, A. J., Kaldany, R. J., Silver, M. L., Wideman, J., Stein, S. & Karlin, A. (1984). *Journal of Biological Chemistry*, **259**, 11662–5.

Kao, P. N. & Karlin, A. (1986). *Journal of Biological Chemistry*, **261**, 8085–8.

Karlin, A. (1967). *Journal of Theoretical Biology*, **16**, 306–20.

Karlin, A. (1987). *Nature, London*, **329**, 286–7.

Karlin, A., Kao, P. N. & DiPaola, M. (1986). *Trends in Pharmacological Sciences*, **308**, 304–8.

Karpen, J. W., Sachs, A. B., Cash, D. J., Pasquale, E. B. & Hess, G. P. (1983). *Analytical Biochemistry*, **135**, 83–94.

Katz, B. & Thesleff, S. (1957). *Journal of Physiology*, **138**, 63–80.

Kieffer, B., Goeldner, M., Hirth, C., Aebersold, R. & Chang, J. Y. (1986). *FEBS Letters*, **201**, 91–6.

Kirilovsky, J., Ducclert, A., Fontaine, B., Devillers-Thiéry, A., Österlund, M. & Changeux, J.-P. (1989). *Neuroscience*, **32**, 289–96.

Kistler, J., Stroud, R. M., Klymkowsky, M. W., Lalancette, R. A. & Fairclough, K. H. (1982). *Biophysics Journal*, **37**, 371–88.

Klarsfeld, A. & Changeux, J.-P. (1985). *Proceedings of the National Academy of Sciences, USA*, **82**, 4558–62.

Klarsfeld, A., Daubas, P., Bourrachot, B. & Changeux, J.-P. (1987). *Molecular and Cell Biology*, **7**, 951–5.

Klarsfeld, A., Devillers-Thiéry, A., Giraudat, J. & Changeux, J.-P. (1984). *EMBO Journal*, **3**, 35–41.

Klarsfeld, A., Laufer, R., Fontaine, B., Devillers-Thiéry, A., Dubreuil, C. & Changeux, J.-P. (1989). *Neuron*, **2**, 1229–36.

Knaack, D., Podleski, T. R. & Salpeter, M. M. (1987). *Annals of the New York Academy of Science*, **498**, 77–89.

Kofuji, P., Aracava, Y., Swanson, K. L., Aronstam, R. S., Rapoport, H. & Albuquerque, E. X. (1990). *Journal of Pharmacology and Experimental Therapy* **252**, 517–25.

Kolb, H. A. & Wankelam, M. J. O. (1983). *Nature, London*, **303**, 621–3.

Koskinen, A. M. & Rapoport, H. (1985). *Journal of Medical Chemistry*, **28**, 13101–9.

Kosower, E. M. (1983). *Biochemical and Biophysical Research Communications*, **116**, 17–22.

Kubo, T., Fukuda, K., Mikami, A., Maeda, A., Takahashi, H., Mishina, M., Haga, T., Haga, K., Ichiyama, A., Kangawa, K., Kojima, M., Matsuo, H., Hirose, T. & Numa, S. (1986). *Nature, London*, **323**, 411–16.

Kuhlmann, J., Okonjo, K. & Maelicke, A. (1991). *FEBS Letters*, in press.

Kyle, J. & Dolittle, R. J. (1982). *Journal of Molecular Biology*, **157**, 105–32.

Labarca, P., Lindstrom, J. & Montal, M. (1984). *Journal of General Physiology*, **83**, 473–96.

Lane, N. J., Sattelle, D. B. & Hafnagel, L. A. (1982). *Tissue and Cell*, **14**, 489–500.

Langenbuch-Cachat, J., Bon, C., Mulle, C., Goeldner, M., Hirth, C. & Changeux, J.-P. (1988). *Biochemistry*, **27**, 2337–45.

Langley, J. N. (1905). *Journal of Physiology*, **33**, 374–413.

Langley, J. N. (1906). *Proceedings of the Royal Society of London*, **78**, 170–94.

Langley, J. N. (1907). *Journal of Physiology*, **36**, 347–84.

LaRochelle, W. J., Wray, B. E., Sealock, R. & Froehner, S. C. (1985). *Journal of Cell Biology*, **100**, 684–91.

Laufer, R. & Changeux, J.-P. (1987). *EMBO Journal*, **6**, 901–6.

Laufer, R. & Changeux, J.-P. (1989). *Journal of Biological Chemistry*, **264**, 2683–9.

Lear, J. D., Wasserman, Z. R. & DeGrado, W. F. (1988). *Science*, **240**, 1177–81.

Lehn, J. M. (1985). *Science*, **227**, 849–56.

Leonard, R. J., Charnet, P., Labarca, C., Vogelaar, N. J., Czyzyk, L., Gouin, A., Davidson, N. & Lester, H. (1990). *Biophysics Journal*, **57**, A209.

Leonard, R. J., Labarca, C. G., Charnet, P., Davidson, N. & Lester, H. A. (1988*a*). *Science*, **242**, 1578–81.

Leonard, R. J., Nakajima, S., Nakajima, Y. & Carlson, C. G. (1988*b*). *Journal of Neuroscience*, **8**, 4038–48.

Leonard, R. J., Nakajima, S., Nakajima, Y. & Takahshi, T. (1984). *Science*, **226**, 55–7.

Lewis, C. A. & Stevens, C. F. (1983). *Proceedings of the National Academy of Sciences, USA*, **80**, 6110–13.

Li, L., Schuchard, M., Palma, A., Pradier, L. & McNamee, M. G. (1990). *Biochemistry*, **29**, 5428–36.

Lindstrom, J., Criado, M., Lam, H., Le Nguyen, D. L., Luther, M., Ralston, S., Rivier, J., Swanson, L., Whiting, P., Berg, D., Jacob, M., Smith, M., Stollberg, J., Sargent, P. & Sarin, V. (1986). In: *Structure and Function of the Nicotinic Acetylcholine Receptor*. NATO ASI Series H: Cell Biology, (Maelicke, A., ed.), vol. 3, pp. 19–33. Springer Press Berlin.

Lipscombe, D. & Rang, H. P. (1988). *Journal of Neuroscience*, **8**, 3258–65.

Loring, R. H., Aizenman, E., Lipton, S. A., & Zigmond, R. E. (1989*a*). *Journal of Neuroscience*, **9**, 2423–31.

Loring, R. H., Schulz, D. W. & Zigmond, R. E. (1989*b*). *Brain Research*, **79**, 109–16.

Loring, R. H. & Zigmond, R. E. (1988). *Trends in Neuroscience*, **11**, 73–8.

Low, B. W. (1980). In: *Handbook of Experimental Pharmacology*, vol. 52 (Lee, C. Y., ed.), pp. 213–57, Springer Press, Berlin.

Luyton, W. (1986). *Journal of Neuroscience Research*, **16**, 51–73.

Maelicke, A. (1984). *Angew. Chem. Int. Edit.* **23**, 195–221.

Maelicke, A. (1987*a*). In: *Molecular Biology of the Nervous System*, de Belleroche, J., ed., pp. 108–12. *Biochemical Society Transactions*.

Maelicke, A. (1987*b*). *Trends in Neuroscience*, **3**, 107.

Maelicke, A. (1988*a*). The cholinergic synapse. In: *Handbook of Experimental*

Pharmacology (Born, G. V. R., Farah, A., Herken, H. & Welch, A. D., eds), vol. 86, pp. 267–313, Springer Press, Berlin.

Maelicke, A. (1988*b*). *Trends in Pharmacological Science*, **13**, 199–202.

Maelicke, A. (1990). In: *Monographs in Anaesthesiology, Muscle Relaxants* (Agoston, S. & Bowman, W. C., eds), vol. 19, pp. 19–58, Elsevier, Amsterdam.

Maelicke, A. & Conti-Tronconi, B. M. (1989). *Journal of Protein Chemistry*, **8**, 326–7.

Maelicke, A., Fulpius, B. W., Klett, R. P. & Reich, E. (1977*a*). *Journal of Biological Chemistry*, **252**, 4811–30.

Maelicke, A., Fulpius, B. W. & Reich, E. (1977*b*). *Handbook of Physiology – The Nervous System* I, pp. 493–519. Bethesda: American Physiological Society.

Maelicke, A., Plümer-Wilk, R., Fels, G., Spencer, S. R., Engelhard, M., Veltel, D. & Conti-Tronconi, B. M. (1989). *Biochemistry*, **28**, 1396–405.

Maelicke, A. & Reich, E. (1976). *Cold Spring Harbor Symposia in Quantitative Biology*, XL, 231–5. [Abstract].

Margiotta, J. F. (1988). *Biophysical Journal* **53**, 358a.

Margiotta, J. F., Berg, D. K. & Dionne, V. E. (1987). *Journal of Neuroscience*, **7**, 3612–22.

Margiotta, J. F. & Gurantz, D. (1989). *Developmental Biology*, **135**, 326–39.

Marshall, J., Buckingham, S. D., Lunt, G. G., Shingai, R., Goosey, M. W., Darlison, M. G., Sattelle, D. B. & Barnard, E. A. (1990). *EMBO Journal*, **9**, 4391–8.

Marshall, L. M. (1985). *Nature, London*, **317**, 621–3.

Martin, B. M., Chibber, B. A. & Maelicke, A. (1983*a*). *Toxicon*, **3**, 273–6.

Martin, B. M., Chibber, B. A. & Maelicke, A. (1983*b*). *Journal of Biological Chemistry*, **258**, 8714–22.

Mathie, A., Colquhoun, D. & Cull-Candy, S. G. (1990). *Journal of Physiology* (*Lond*), **427**, 625–55.

McCarthy, M. P., Earnest, J. P., Young, E., Choe, S. & Stroud, R. M. (1986). *Annual Reviews in Neurosciences*, **9**, 383–413.

McCarthy, M. P. & Stroud, R. M. (1989). *Biochemistry*, **28**, 40–8.

McCrea, P. D., Popot, J.-L. & Engleman, D. M. (1987). *EMBO Journal*, **6**, 3619–26.

McHugh, E. M. & McGee, R. (1986). *Journal of Biological Chemistry*, **261**, 3103–6.

McLane, K. E., Wu, X. & Conti-Tronconi, B. M. (1990). *Journal of Biological Chemistry*, **265**, 9816–24.

Merlie, J. P. & Kornhauser, J. M. (1989). *Neuron*, **2**, 1295–300.

Merlie, J. P., Sebbane, R., Gardner, S. & Lindstrom, J. (1983). *Proceedings of the National Academy of Sciences, USA*, **80**, 3845–9.

Merlie, J. P., Sebbane, R., Gardner, S., Olson, E. & Lindstrom, J. (1984). *Cold Spring Harbor Symposia Quantitative Biology*, 48 Pt1, 135–46.

Merlie, J. P. & Smith, M. M. (1986). *Journal of Membrane Biology*, **91**, 1–10.

Meyers, H.-W., Jürss, R., Brenner, H. R., Fels, G., Prinz, H., Watzke, H. & Maelicke, A. (1983). *European Journal of Biochemistry*, **137**, 399–404.

Mishina, M., Kurosaki, T., Tobimatsu, T., Morimoto, Y., Noda, M., Yamamoto, T., Terao, M., Lindstrom, J., Takahashi, T., Kuno, M. & Numa, S. (1984). *Nature, London*, **307**, 604–8.

Mishina, M., Takai, T., Imoto, K., Noda, M., Takahashi, T., Numa, S., Methfessel, C. & Sakmann, B. (1986). *Nature, London*, **321**, 406–11.

Mishina, M., Tobimatsu, T., Imoto, K., Tanaka, K., Fujita, Y., Fukuda, K., Kurasaki, M., Takahashi, H., Morimoto, Y., Hirose, T., Inayama, S., Takahashi, T., Kuno, M. & Numa, S. (1985). *Nature, London*, **313**, 364–9.

Mitra, A. K., McCarthy, M. P. & Stroud, R. M. (1989). *Journal of Cell Biology*, **109**, 755–74.

Monod, J., Wyman, J. & Changeux, J.-P. (1965). *Journal of Molecular Biology*, **12**, 88–118.

Moore, H.-P. H. & Raftery, M. A. (1980). *Proceedings of the National Academy of Sciences, USA*, **77**, 4509–13.

Mosckovitz, R. & Gershoni, J. M. (1988). *Journal of Biological Chemistry*, **263**, 1017–22.

Moss, S. J., Darlison, M. G., Beeson, D. M. W. & Barnard, E. A. (1989). *Journal of Biological Chemistry*, **264**, 20199–205.

Mulac-Jericevic, B. & Atassi, M. Z. (1987). *Biochemical Journal*, **248**, 847–52.

Mulle, C., Benoit, P., Pinset, C., Roa, M. & Changeux, J.-P. (1988). *Proceedings of the National Academy of Sciences, USA*, **85**, 5728–32.

Mulle, C. & Changeux, J.-P. (1990). *Journal of Neuroscience*, **10**, 169–75.

Nathanson, N. M. & Hall, Z. W. (1979). *Biochemistry*, **18**, 3392–401.

Nef, P., Mauron, A., Stalder, R., Alliod, C. & Ballivet, M. (1984). *Proceedings of the National Academy of Sciences, USA*, **81**, 7975–9.

Neher, E. (1983). *Journal of Physiology*, **339**, 663–78.

Neubig, R. R., Boyd, N. D. & Cohen, J. B. (1982). *Biochemistry*, **21**, 3460–7.

Neubig, R. R. & Cohen, J. B. (1979). *Biochemistry*, **18**, 5464–75.

Neubig, R. R. & Cohen, J. B. (1980). *Biochemistry*, **19**, 2770–9.

Neumann, D., Gershoni, J. M., Fridkin, M. & Fuchs, S. (1985). *Proceedings of the National Academy of Sciences, USA*, **82**, 3490–3.

New, H. V. & Mudge, A. W. (1986). *Nature, London*, **323**, 809–11.

Nitkin, R. M. & Rothschild, T. C. (1990). *Journal of Cell Biology*, **111**, 1161–70.

Noda, M., Takahashi, H., Tanabe, T., Toyosato, M., Kikyotani, S., Furutani, Y., Hirose, T., Takashima, H., Inayama, S., Miyata, T. & Numa, S. (1983). *Nature, London*, **302**, 528–32.

Oberthuer, W. & Hucho, F. (1988). *Journal of Protein Chemistry*, **7**, 141–50.

Oberthuer, W., Muhn, P., Baumann, H., Lottspeich, F., Wittmann-Liebold, B. & Hucho, F. (1986). *EMBO Journal*, **5**, 1815–19.

Oblas, B., Singer, R. H. & Boyd, N. D. (1986). *Molecular Pharmacology*, **29**, 649–56.

Ogden, D. C., Siegelbaum, S. A. & Colquhoun, D. (1981). *Nature, London*, **289**, 596–8.

Oiki, S., Anko, W. & Mortal, M. (1988). *Proceedings of the National Academy of Sciences, USA*, **85**, 8703–7.

Oiki, S., Madison, V. & Montal, M. (1990). *Proteins*, **8**, 226–36.

Olson, E. N., Glaser, L. & Merlie, J. P. (1984). *Journal of Biological Chemistry*, **259**, 5364–7.

Oortgiesen, M. & Vijverberg, H. P. M. (1989). *Neuroscience*, **31**, 169–79.

Österlund, M., Fontaine, B., Devillers-Thiéry, A., Geoffroy, B. & Changeux, J.-P. (1989). *Neuroscience*, **32**, 279–87.

Oswald, R. E., Michel, L. & Bigelow, J. (1986). *Molecular Pharmacology*, **29**, 179–87.

Padlan, E. A., Davis, D. R., Rudikoff, S. & Potter, M. (1976). *Immunochemistry*, **13**, 945–9.

Papke, R. L., Boulter, J., Patrick, J. & Heinemann, S. (1989). *Neuron*, **3**, 589–96.

Pasquale, E. B., Takeyasu, K., Udgaonkar, J. B., Cash, D. J., Severski, M. C. & Hess, G. P. (1983). *Biochemistry*, **22**, 5967–73.

Patrick, J. & Shallup, W. (1977). *Journal of Biological Chemistry*, **252**, 8629–36.

Paulson, H. L. & Claudio, T. (1990). *Journal of Cell Biology*, **110**, 1705–17.

Pearce, S. F. & Hawrot, E. (1990). *Biochemistry*, **29**, 10649–59.

Pedersen, S. E., Bridgman, P. C., Sharp, S. D. & Cohen, J. B. (1990). *Journal of Biological Chemistry*, **265**, 569–81.

Pedersen, S. E. & Cohen, J. B. (1990). *Proceedings of the National Academy of Sciences, USA*, **87**, 2785–9.

Pedersen, S. E., Dreyer, E. B. & Cohen, J. B. (1986). *Journal of Biological Chemistry*, **261**, 13735–43.

Pelham, H. (1988). *Nature, London*, **332**, 776–7.

Piette, J., Bessereau, J.-L., Huchet, M. & Changeux, J.-P. (1990). *Nature, London*, **345**, 353–5.

Popot, J. L. & Changeux, J.-P. (1984). *Physiological Review*, **64**, 1162–239.

Prinz, H. & Maelicke, A. (1983a). *Journal of Biological Chemistry*, **258**, 10273–82.

Prinz, H. & Maelicke, A. (1983b). *Journal of Biological Chemistry*, **258**, 10263–82.

Radding, W., Corfield, P. W. R., Levinson, L. S., Hashim, G. A. & Low, B. W. (1988). *FEBS Letters*, **231**, 212–16.

Ralston, S., Sarin, V., Thanh, H. L., Rivier, J., Fox, J. & Lindstrom, J. (1987). *Biochemistry*, **26**, 3261–6.

Ratnam, M., Le Nguyen, D. L., Sargent, P. B. & Lindstrom, J. (1986a). *Biochemistry*, **25**, 2633–43.

Ratnam, M. & Lindstrom, J. (1984). *Biochemical Biophysical Research Communications*, **122**, 1225–33.

Ratnam, M., Sargent, P. B., Sarin, V., Fox, J. L., Le Nguyen, D. L., Rivier, J., Criado, M. & Lindstrom, J. (1986b). *Biochemistry*, **25**, 2621–32.

Reuhl, T. O. K., Amador, M. & Dani, J. A. (1990). *Brain Research Bulletin*, **25**, 433–5.

Revah, F., Galzi, J.-L., Giraudat, J., Haumont, P.-Y., Lederer, F. & Changeux, J.-P. (1990). *Proceedings of the National Academy of Sciences, USA*, **87**, 4675–9.

Rohrbough, J. & Kidokoro, Y. (1990). *Journal of Physiology, London*, **425**, 245–69.

Role, L. W. (1988). *Proceedings of the National Academy of Sciences, USA*, **85**, 2825–9.

Rosenberry, T. L. (1975). *Advances in Enzymology*, **43**, 103–218.

Rozental, R., Scoble, G. T., Albuquerque, E. X., Idriss, M., Sherby, S., Sattelle, D. B., Nakanishi, K., Konno, K., Eldefrawi, T. & Eldefrawi, M. E. (1989). *Journal of Pharmacology and Experimental Therapy*, **249**, 123–30.

Ruan, K.-H., Spurlino, J., Quiocho, F. A. & Atassi, M. Z. (1990). *Proceedings of the National Academy of Sciences, USA*, **87**, 6156–60.

Rüchel, R., Watters, D. & Maelicke, A. (1982). In: *Neuroreceptors*, (Hucho, F., ed.), pp. 263–73. Walter de Gruyter & Co., Berlin, New York.

Sackmann, B. & Neher, E. (1984). *Annual Reviews in Physiology*, **46**, 455–72.

Saedi, M. S., Anand, R., Conroy, W. G. & Lindstrom, J. (1990). *FEBS Letters*, **267**, 55–59.

Sakmann, B. & Brenner, H. R. (1978). *Nature, London*, **276**, 401–2.

Sakmann, B. & Neher, E. (1984). *Annual Reviews in Physiology*, **46**, 55–72.

Salpeter, M. M. (1987). *Neurology and Neurobiology*, **23**, 55–115.

Salpeter, M. & Loring, R. H. (1985). *Progress in Neurobiology*, **25**, 297–325.

Sanchez, J. A., Dani, J. A., Siemen, D. & Hille, B. (1986). *Journal of General Physiology*, **87**, 985–1001.

Sattelle, D. B., Harrow, I. D., Pelhate, M., Gepner, J. I. & Hall, L. M. (1983). *Journal of Experimental Biology*, **107**, 473–89.

Sawruk, E., Schloss, P., Betz, H. & Schmitt, B. (1990*a*). *EMBO Journal*, **9**, 2671–7.

Sawruk, E., Udri, C., Betz, H. & Schmitt, B. (1990*b*). *FEBS Letters* **273**, 177–81.

Schindler, H., Spillecke, F. & Neumann, E. (1984). *Proceedings of the National Academy of Sciences, USA*, **81**, 6222–6.

Schneider, H. J., Güttes, D. & Schneider, U. (1986). *Angew. Chem. Int. Ed. Engl.* **25**, 647–9.

Schneider, M., Adee, C., Betz, H. & Schmidt, J. (1985). *Journal of Biological Chemistry*, **260**, 14505–12.

Schöpfer, R., Conroy, W. G., Whiting, P., Gore, M. & Lindstrom, J. (1990). *Neuron*, **5**, 35–48.

Schuetze, S. M. & Role, L. W. (1987). *Annual Reviews in Neuroscience*, **10**, 403–57.

Shaw, K. P., Aracava, Y., Akaike, A., Daly, J. W., Rickett, D. L. & Albuquerque, E. X. (1985). *Molecular Pharmacology*, **28**, 527–38.

Siara, J., Ruppersberg, J. P. & Rüdel, R. (1990). *Pflugers Arch.* **415**, 701–6.

Siemen, D., Hellmann, S. & Maelicke, A. (1986). In: *Structure and Function of the Nicotinic Acetylcholine Receptor*. NATO ASI Series H: Cell Biology, vol. 3, pp. 233–41 (Maelicke, A., ed.), Springer Press, Berlin.

Simmons, L. K., Schuetze, S. M. & Role, L. W. (1987). *Society of Neuroscience Abstracts*, **13**, 704.

Simmons, L. K., Schuetze, S. M. & Role, L. W. (1990). *Neuron*, **4**, 393–403.

Sine, S. M., Claudio, T. & Sigworth, F. J. (1990). *Journal of General Physiology*, **96**, 395–437.

Sine, S. M. & Steinbach, J. H. (1986*a*). *Journal of Physiology*, **370**, 357–79.

Sine, S. M. & Steinbach, J. H. (1986*b*). *Journal of Physiology*, **373**, 129–62.

Sine, S. M. & Steinbach, J. H. (1987). *Journal of Physiology*, **385**, 325–59.

Sine, S. M. & Taylor, P. (1979). *Journal of Biological Chemistry*, **254**, 3315–25.

Sine, S. M. & Taylor, P. (1982). *Journal of Biological Chemistry*, **257**, 8106–14.

Smart, L., Meyers, H. W., Hilgenfeld, R., Saenger, W. & Maelicke, A. (1984). *FEBS Letters*, **178**, 64–68.

Steinbach, J. H. (1989). *Annual Reviews in Physiology*, **51**, 353–65.

Steinbach, J. H., Covarrubias, M., Sine, S. M. & Steele, J. (1986). In: *Structure and Function of the Nicotinic Acetylcholine Receptor*. NATO ASI Series H: Cell Biology (Maelicke, A., ed.), vol. 3, pp. 219–32, Springer Press, Berlin.

Steinbach, J. H. & Ifune, C. (1989). *Trends in Neurosciences*, **12**, 3–6.

Strecker, G. J. & Jackson, M. B. (1989). *Biophysical Journal*, **56**, 795–806.

Stroud, R. M., McCarthy, M. P. & Shuster, M. (1990). *Biochemistry* **29**, 11009–23.

Sumikawa, K., Houghton, M., Emtage, J. S., Richards, B. M. & Barnard, E. A. (1981). *Nature, London*, **292**, 862–4.

Swanson, K. L., Allen, C. N., Aronstam, R. S., Rapoport, H. & Albuquerque, E. X. (1986). *Molecular Pharmacology*, **29**, 250–7.

Swanson, K. L., Aracava, Y., Sardina, F. J., Rapoport, H., Aronstam, R. S. & Albuquerque, E. X. (1989). *Molecular Pharmacology*, **35**, 223–31.

Swanson, L. W., Simmons, D. M., Whiting, P. J. & Lindstrom, J. (1987). *Journal of Neuroscience*, **7**, 3334–42.

Takai, T., Noda, M., Furutani, Y., Takahashi, H., Notake, M., Shimizu, S.,

Kayano, T., Tanabe, T., Tanaka, K., Hirose, T., Inayama, S. & Numa, S. (1984). *Journal of Biochemistry*, **143**, 109–15.

Takai, T., Noda, M., Mishina, M., Shimizu, S., Furutani, Y., Kayano, T., Ikeda, T., Kubo, T., Takahashi, H., Takahashi, T., Kuno, M. & Numa, S. (1985). *Nature, London*, **315**, 761–4.

Takeshima, H., Nishimura, S., Matsumoto, T., Ishida, H., Kangawa, K., Minamino, N., Hisayuki, M., Ueda, M., Hanaoka, M., Hirose, T. & Numa, S. (1989). *Nature, London* **339**, 439–45.

Takeyasu, K., Udgaonkar, J. B. & Hess, G. P. (1983). *Biochemistry*, **22**, 5973–8.

Toyoshima, C. & Unwin, N. (1988). *Nature, London*, **336**, 247–50.

Toyoshima, C. & Unwin, N. (1990). *Journal of Cell Biology*, **111**, 2623–35.

Trautmann, A. (1983). *Proceedings of the Royal Society of London (Biol.)* **218**, 214–51.

Tsay, H.-J., Neville, C. M. & Schmidt, J. (1990). *FEBS Letters*, **274**, 69–72.

Tsui, H.-C. T., Cohen, J. B. & Fischbach, G. D. (1990). *Developments in Biology*, **140**, 437–46.

Turnbull, G. M., Harrison, R. & Lunt, G. G. (1985). *International Journal of Developmental Neuroscience* **3**, 123–34.

Udgaonkar, J. B. & Hess, G. P. (1987a). *Trends in Pharmacological Science*, **8**, 190–2.

Udgaonkar, J. B. & Hess, G. P. (1987b). *Trends in Pharmacological Science*, **8**, 294–5.

Unwin, P. N. T. & Zampighi, G. (1988). *Journal of Cell Biology*, **107**, 1123–38.

Usdin, T. B. & Fischbach, G. D. (1986). *Journal of Cell Biology*, **103**, 493–507.

Wada, E., McKinnon, D., Heinemann, S., Patrick, J. & Swanson, L. W. (1990). *Brain Research*, **526**, 45–53.

Wada, E., Wada, K., Boulter, J., Deneris, E., Heinemann, S., Patrick, J. & Swanson, L. W. (1989). *Journal of Comparative Neurology*, **284**, 314–35.

Wada, K., Ballivet, M., Boulter, J., Connolly, J., Wada, E., Deneris, E. S., Swanson, L. W., Heinemann, S. & Patrick, J. (1988). *Science*, **240**, 330–4.

Wang, Y., Xu, H.-P., Wang, X.-M., Ballivet, M. & Schmidt, J. (1988). *Neuron*, **1**, 527–43.

Watters, D. & Maelicke, A. (1983). *Biochemistry*, **22**, 1811–19.

Weber, M. & Changeux, J.-P. (1974a). *Molecular Pharmacology*, **10**, 1–13.

Weber, M. & Changeux, J.-P. (1974b). *Molecular Pharmacology*, **10**, 14–34.

Weiland, G. & Taylor, P. (1979). *Molecular Pharmacology*, **15**, 197–212.

Whiting, P. J. & Lindstrom, J. M. (1986). *Biochemistry*, **25**, 2082–93.

Whiting, P. J. & Lindstrom, J. M. (1987). *Proceedings of the National Academy of Sciences, USA*, **84**, 595–9.

Whiting, P. J. & Lindstrom, J. M. (1988). *Journal of Neuroscience*, **8**, 3395–404.

Wilson, P. T., Gershoni, J. M., Hawrot, E. & Lentz, T. L. (1984). *Proceedings of the National Academy of Sciences, USA*, **81**, 2553–7.

Witzemann, V. (1989). In: *Molecular Biology of Neuroreceptors and Ion Channels*. NATO ASI Series H; Cell Biology, vol. 32, pp. 509–17 (Maelicke, A., ed.), Springer Press, Berlin.

Witzemann, V., Barg, B., Criado, M., Stein, E. & Sakmann, B. (1989). *FEBS Letters*, **242**, 419–24.

Young, E. F., Ralston, E., Blake, J., Ramachandran, J., Hall, Z. W. & Stroud, R. M. (1985). *Proceedings of the National Academy of Sciences, USA*, **82**, 626–30.

Ziskind, L. & Dennis, M. J. (1978). *Nature, London*, **276**, 622–623.

6

Subunits of GABA$_A$, glycine, and glutamate receptors

ERIC A. BARNARD

Introduction

This chapter deals with other transmitter-gated ion channels, providing a fuller account of the subunits of the GABA$_A$ (γ-aminobutyrate) and glycine receptors. The structural information available on subunits of the various glutamate receptors is reviewed; since this information is so far much more limited than that on the GABA$_A$ and glycine receptors, the treatment is brief. A comparison is made with the nicotinic receptor.

The GABA$_A$ receptors

The GABA$_A$ receptor protein

GABA is the major inhibitory neurotransmitter of the vertebrate brain. For fast signalling the GABA$_A$ receptor is employed in inhibition on the great majority of its neurones, as well as occurring on many glia. This receptor is, therefore, ubiquitous in the nervous system and of major physiological and pharmaceutical importance.

Pharmacological and membrane-binding studies (for review see Olsen and Venter (1986) and Barnard (1988)) have identified at least six types of binding site in the GABA$_A$ receptor: (i) the GABA agonist/antagonist site; (ii) the benzodiazepine site, where binding of these tranquillizer drugs allosterically causes an increase in the frequency of GABA-induced channel opening; (iii) a channel gating site, where agents such as picrotoxin block the channel; (iv) the depressant site, recognising the CNS-depressant barbiturates and similar drugs, which prolong the lifetime of the GABA-activated channel; (v) a site for certain neurosteroids (brain metabolites of progesterone, deoxycorticosterone and pregnenolone) which act similarly to depressants in some but not all ways (Peters *et al.*, 1988); (vi) sites binding the channel-permeating anions (but not other ions). Each of these types of ligand site can interact allosterically with one or more of the other types. From the network of interactions found, it can be deduced that several of these sites can be occupied by their respective ligands simultaneously and that each of the six types of site must be physically distinct. The $GABA_A$ receptor is, therefore, a highly regulated structure. It is presumed that natural modulators – neurosteroids, peptides, etc. act *in vivo* on it at the sites where these experimentally employed ligands bind.

Despite its ubiquity, the absolute receptor abundance is low. Purification of the receptor was eventually accomplished on a benzodiazepine affinity column with a cationic spacer arm, followed by an ion-exchange step (Sigel *et al.*, 1982). The single protein obtained binds ligands for all of the sites listed above (Sigel *et al.*, 1983; Sigel & Barnard, 1984). (The neurosteroid site was not identified until later). The rank order of drug potencies in these classes is preserved from the *in vivo* state to the membrane, and to the purified protein state. The purified protein also shows the characteristic allosteric interactions between the sites, and when reconstituted in liposomes forms the GABA-activatable anion channel with its allosteric modulations (see the section on Reconstitution of Receptors in Chapter 4).

In the purified receptor from bovine or chick brain, two subunit classes were detectable in denaturing gel electrophoresis, α (M_r 53000) and β (M_r 57000). These were recognised in broad bands possibly containing multiple forms. Using irreversible labelling with the benzodiazepine flunitrazepam (which behaves as a photoaffinity reagent) the α subunit was found to carry the high-affinity benzodiazepine site (Sigel *et al.*, 1983). However, it was known by performing this labelling on membranes from different brain regions and animal ages (Sieghart & Drexler, 1983) that the reacted subunit could exist in several forms: these have been identified as different isoforms of the α subunit (Fuchs, Adamiker & Sieghart, 1990). In the mixture of isolated pure receptors, the β subunits are the site of photoaffinity-labelling by the GABA agonist muscimol (Casalotti, Stephenson & Banard, 1986; Deng, Ransom & Olsen, 1986), and hence contain the high-affinity GABA site. The micro-heterogeneity noted persists after full deglycosylation (Mamalaki, Stephenson & Barnard, 1987), when the sizes seen are in reasonable agreement with those of the α and β subunits identified by cDNA cloning from bovine brain.

The molecular weight of the receptor glycoproteins was determined (by hydrodynamic methods which are not perturbed by the bound detergent) to be in range of 240 kD to 290 kD (Mamalaki, Barnard & Stephenson, 1989). This size is in agreement with the target size in radiation inactivation; molecular weights of 50 kD–100 kD were estimated thus by several laboratories (for review see Stephenson, 1988), but the observed value depends upon the conditions used and the full receptor was found (Chang & Barnard, 1982) to have a target size of 230 kD, with the muscimol and benzodiazepine sites being inactivated together by the radiation. These two approaches both lead to a size for a unitary GABA/benzodiazepine receptor corresponding to five subunits in the molecule. Overall, the retention on the benzodiazepine column and elution by a benzodiazepine of the muscimol and the other binding activities, the common target size, the co-migration of the GABA, benzodiazepine and channel-blocker sites which was shown in the physical separations applied, and the reconstitution evidence for the presence of the channel in the same protein, clearly demonstrated that all of the sites known for this system reside on one protein, the GABA/benzodiazepine receptor.

cDNAs encoding the GABA receptor subunits

Purified GABA$_A$ receptor proteins from bovine cerebral cortex were originally used for cDNA cloning. Peptide sequences obtained from these were used to generate a set of oligonucleotide probes for cDNA library screening. Full-length cDNAs encoding the α1 subunit and the β1 subunit were obtained (Schofield *et al.*, 1987). For expression, translation in *Xenopus* oocytes was employed, from α1 and β1 RNAs synthesised in the SP6 *in vitro* transcription system. When α1 and β1 RNAs were co-injected, strong responses to GABA developed which were due to a GABA-receptor type of chloride channel (Schofield *et al.*, 1987; Levitan *et al.*, 1988*a*). The binary receptor expressed in heterologous cells shows the receptor drug responses of the types (1) to (6) above, with the exception of the normal benzodiazepine potentiation of the GABA-evoked current.

There is a similar distribution of four trans-membrane regions (TMs) in these recombinant subunits, deduced from their hydrophobicity plots, as illustrated in Fig. 6.1 (right-hand diagram). The mature α1 and β1 polypeptides are also clearly homologous in amino acid sequence, with 35% of the amino acids identical. A striking feature is their unmistakeable relationship to the acetylcholine receptor subunits, in sequence and in deduced topology (Fig. 6.1). Compared to the α subunit of the latter from muscle of the same species, this identity level is 19% and identical plus

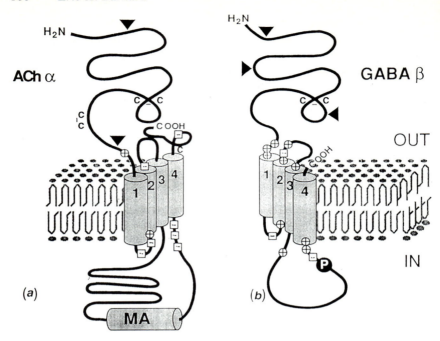

Fig. 6.1. Comparison of models for the topology in the membrane of (a) a nicotinic acetylcholine receptor (muscle or neuronal) α subunit and (b) a GABA$_A$ receptor β-subunit. Four membrane-spanning domains in each subunit are shown as cylinders (1–4). The structure in the extracellular domain is drawn in an arbitrary manner; the presumed β-loop formed by the disulphide bond at cysteins 128 and 142 (*Torpedo* nicotinic receptor α-subunit numbering) is shown. Potential extracellular sites for N-glycosylation are indicated by triangles and a site for cAMP-dependent serine phosphorylation, present in the GABA$_A$ receptor β-subunits, is denoted by an encircled P. The latter is the only feature shown which differs between the α, β and γ subunits of the GABA receptor. Those charged residues that are located close to the ends of the membrane-spanning domains are shown as circles with positive charge marked, or as open squares for negative charges. Note the large excess of positive charge which will be at the mouths of the channel when four or five of the GABA$_A$ subunits are assembled to form the receptor. Note also that there will be a small excess of negative charge at the channel mouths in the corresponding structure formed from the nicotinic receptor subunits. Constant cysteine residues (C) are also shown: the vicinal pair in the nicotinic α-subunit (only), and the pair 15 residues apart in the 'Cys–Cys loop' in both receptor types (Table 6.1). The latter pair are shown as disulphide-bridged, which has been established to be the case in the nicotinic receptor. MA, the predicted amphipathic helix peculiar to the nicotinic receptor, is shown in (a) as part of the long cytoplasmic loop between M3 and M4; that loop occurs in all the subunits of this super-family. MA may be involved in folding compactly the tertiary structure of the loop in the nicotinic receptors.

conservatively substituted positions are 38%. As discussed in Chapter 2, the subunits of the GABA$_A$and nicotinic receptor types, plus those of the glycine receptor, form a super-family of related subunits.

The general picture obtained of the GABA$_A$ receptor has been confirmed and enhanced by the analysis of other subunits, recognised by cross-hybridisation screening. Thus, isoforms of the α subunit (Levitan *et al.*, 1988*b*) were found, as well as a third category, the γ subunit (Pritchett *et al.*, 1989), which confers the typical benzodiazepine sensitivity. Further isoforms of each type have been found, so that, to date, the sequences of 7 α, 3 β and 3 γ subunits have been reported from several mammalian species Olsen & Tobin, 1990; Sigel *et al.*, 1990; Lüddens *et al.*, 1990; Kato, 1990; Ymer *et al.*, 1990; Cutting *et al.*, 1991; Kofuji *et al.*, 1991). From a more distant vertebrate species, the chicken, various α, β and γ recombinant subunits have also been described (Bateson *et al.*, 1990, 1991*a*; Glencorse *et al.*, 1990*a*) as well as a new β type (β4) (Bateson *et al.*, 1991*b*). The percentage identities between any two of these categories in any one species lie typically, e.g. in the rat, between 32% and 39%. Hence, the α, β and γ subunits are always distinct; each exists in a set of isoforms of 70–80% identity.

This list includes with the α isoforms a somewhat different type of subunit, *rho*, which is primarily expressed in the retina, although low levels of its transcript are also seen in Northern blots of some other brain regions (Cuting *et al.*, 1991). It has 30–38% amino acid sequence identity to the other GABA$_A$ receptor subunits, but here it is provisionally included as a specialised form of α because it lacks the potential phosphorylation sites in the deduced TM3–TM4 intracellular loop which are found there in β and γ subunits. Its four TMs are very homologous to those of all other GABA$_A$ receptor subunits. It can form a homo-oligomeric GABA-activated channel (Cutting *et al.*, 1991). Its sequence differences and distribution suggest that a specialised form of GABA$_A$ receptor is located at some retinal locations.

Another category of subunit, δ, has been described, which is further away in sequence (29% identity) from any α but closer to the γ2 subunit (39%), and of which only one form has so far been found (Shivers *et al.*, 1989). The δ subunit confers no observed pharmacological properties of its own when expressed in combinations with α and β, and hence its significance as a GABA receptor subunit is as yet uncertain.

It is interesting that all of these isoforms (α1, α2, β1, β2, etc.) are products of separate genes. This is clear from their sequences, and where analysed from genomic DNA cloning or from their human chromosomal localisations, which are scattered (Buckle *et al.*, 1989; Sommer *et al.*, 1990*b*). Additionally, in the $\gamma2$ and $\beta4$ cases, alternative splicing of the mRNA precursor has recently been found to generate a further isoform in each case (Glencorse *et al.*, 1990*a, b*; Whiting, McKernan & Iverson, 1990; Bateson *et al.*, 1991*b*; Kofuji *et al.*, 1991).

Sites present on the α, β and γ subunit types

These three principal subunit types appear to perform different functions in the receptor. Present evidence indicates that the benzodiazepine site is located on the α subunit but the stabilisation or completion of that site in the assembled structure requires also the γ subunit. Weak benzodiazepine effects on channel activation by GABA can in some cases be obtained with expressed $\alpha\beta$ combinations (Malherbe *et al.*, 1990), but the behaviour seen then with ligands for the benzodiazepine site is not of the normal type. The photo-affinity labelling of the benzodiazepine site has been confirmed, using isoform-specific antibodies, to be on an α and not a γ subunit (Stephenson, Duggan & Pollard, 1990). While the α and γ subunits are needed for this site, a ternary combination including also a β subunit appears necessary to produce the native type of benzodiazepine pharmacology (Pritchett *et al.*, 1989) and to generate native amplitudes of the GABA response and of its benzodiazepine potentiation (Sigel *et al.*, 1990; Verdoorn *et al.*, 1990).

Each of these three subunit types, however, shows evidence for possession of a site which can bind GABA, at the low apparent affinity ($EC_{50} >$ 5 μM) required for its activation of native $GABA_A$ receptors. An α subunit ($\alpha1, \alpha2, \alpha3, \alpha4$) or a β subunit alone was shown to produce GABA-activated chloride channels on expression in the oocyte (Blair *et al.*, 1988; Olsen & Tobin, 1990). The same is true in mammalian cells transfected to express an α or β or γ subunit alone (Shivers *et al.*, 1989). It is presumed that these assemble with low efficiency into single-subunit (homo-oligomeric) receptors, which would not be seen on neurones *in situ* expressing all three types. Perhaps the high-affinity GABA site seen by photo-affinity labelling on β subunits only (see above) is involved in receptor desensitisation. It is that subunit, and not α, which possesses a site for serine phosphorylation (Fig. 1), which may regulate desensitisation.

The single-subtype complexes noted above also exhibited the native type of modulation by picrotoxin or a barbiturate, so that the two latter classes of site, also, exist on each subunit type. The anion selectivity and the multiple conductance levels characteristic of native $GABA_A$ receptor channels were likewise shown to be encoded in each of the subunit types (Blair *et al.*, 1988). The potentiation by endogenous neurosteroids still occurs strongly in the absence of a γ subunit (Shingai, Sutherland & Barnard, 1991). However, it is further influenced (but not increased) by the addition of γ and it depends strongly on the α subtype.

In general, therefore, some of the six specific binding sites listed earlier occur on all three of these subunit types, whereas others of them do not. Probably the barbiturate, picrotoxin and neurosteroid sites are mainly or entirely in the transmembrane domain, which is very similar in all of the subunits.

The multiplicity of $GABA_A$ receptor types

It is presumed that the native benzodiazepine-sensitive $GABA_A$ receptors are usually formed from the assembly of α, β and γ subunits, but the stoichiometry is as yet unknown. $GABA_A$ receptors not coupled to benzodiazepine sites have been recognised electrophysiologically or auto-radiographically at some locations (Unnerstall *et al.*, 1981; Alger & Nicoll, 1982; McCabe & Wamsley, 1986; de Blas, Victorica & Friedrich, 1988; Osmanovic & Shefner, 1990). Therefore, a significant population of $GABA_A$ receptors *in situ* may lack a γ subunit: this might be replaced by δ (Shivers *et al.*, 1989) or by an extra α or β in the complex. Further, in the granule cells of the mammalian cerebellum the presence of the $\alpha6$ subunit (unique to that location) produces a receptor which has no affinity for benzodiazepine agonists even if a γ subunit is present (Kato, 1990; Lüddens *et al.*, 1990), so that the presence of a γ subunit, while a necessary condition for normal benzodiazepine sensitivity of the receptor, does not guarantee it. Hence, there are potentially many subunit combinations which could produce benzodiazepine-insensitive $GABA_A$ receptors. For the benzodiazepine-sensitive majority, there could be up to 56 possibilities if only one isoform of α, of β and of γ can be present (e.g. $\alpha_2\beta_2\gamma$). Sigel *et al.* (1990) tested various combinations of three, four or five different recombinant isoforms, and found, for example, that $\alpha1\alpha3\beta2\gamma2$ gave higher benzodiazepine potentiation and n_H than with only one α present with $\beta2$ and $\gamma2$, so that many more combinations, e.g. of four isoforms, could in theory be possible. In all, these various possibilities could allow a theoretical total of several hundred $GABA_A$ receptor subtypes.

It is not believed that all of these combinations actually exist *in vivo*, even if they can be found to assemble in the artifical situation of co-expression from recombinant RNAs in a heterologous cell. Attemps to identify the natural combinations which exist could be via several approaches, employing:

(i) *In situ* hybridisation of the various RNAs, using isoform-specific oligonucleotides as probes. Patterns of co-distribution, both in brain regions and in single neurones, can be sought, e.g. to see if a particular α is always expressed with a particular β.

(ii) Isoform-specific anti-peptide antibodies, for similar co-distribution studies, which will be feasible in both the light and electron microscopes.

(iii) Use of such antibodies in immuno-precipitation of extracted native receptors, e.g. to see if two different αs can be co-precipitated.

(iv) Identification of electrophysiological or pharmacological features that are diagnostic for a particular combination or sub-set, expressed *in vitro*, and matching of this with receptors at particular locations *in vivo*. An example is the recognition that the combinations $\alpha 6\beta\gamma$ have lost the usual γ-dependent benzodiazepine sensitivity but have gained a unique sensitivity to the atypical bezodiazepine Ro 15–4513 and that this has only so far been found on cerebellar granule cells (Lüddens *et al.*, 1990).

So far as such studies have yet gone, a considerable multiplicity of combinations can be inferred to occur, varying with the region of the brain (or spinal cord or peripheral nervous system) and between cell types within one structure. Complex distribution patterns of various α isoforms were found by *in situ* isoform-specific RNA studies in bovine brain (Wisden *et al.*, 1988, 1989*a*, *b*) and of γ and δ RNAs in rat brain (Shivers *et al.*, 1989). This has been extended to other isoforms (for review see Olsen & Tobin, 1990). At single-cell resolution, adjacent neurones can be found in some cases to have different patterns (Wisden *et al.*, 1989*b*, 1991). Overall, a different pattern of distribution is found with every isoform which is localised, so that a considerable number of different combinations must indeed occur naturlly. The most common form of the α subunit mRNA is that of $\alpha 1$, which generally corresponds (Wisden *et al.*, 1989*b*; Verdoorn *et al.*, 1990) to the location of type I, the major form of GABA$_A$ receptor which was originally recognised pharmacologically, although $\alpha 1$ in certain combinations does not completely correspond to type I (Bateson *et al.*, 1991*a*). The other form previously recognised pharmacologically, type II, is actually seen to be due, from its regional distribution, to arise from averaging many non-$\alpha 1$-containing combinations. So far, in a very few cases a cell type has been found where expression of a single α isoform has been established, e.g. $\alpha 2$ in the Bergmann glial cells of the bovine cerebellum (Wisden *et al.*, 1989*a*).

Method (iii) has begun to be applied, e.g. in proving that $\alpha 1$ and $\gamma 2$ can occur in the same protein (Stephenson *et al.*, 1990*a*). Method (ii) is now commencing to be available (Benke *et al.*, 1991). It could be applied to three subunits in the same cell, using the multiple-colour fluorescent labels now available. However, its greatest future prospect will be to detect triple co-localisations of the type $\alpha\beta\gamma$ at the electron microscope level, which is in principle possible using three types of gold particle labelling of the antibodies. This could resolve the ambiguity in co-localisations on and around a particular cell, inherent in the mRNA autoradiography. Thus, single synapses could be resolved and if two isoforms of α, say, are found, it could be determined by the multiple gold-label method if they are in fact on one cell and at one type of site on it.

What is the reason for the great multiplicity of combinations of these subunits *in vivo*? It is presumed that each natural combination represents one subtype of the GABA$_A$ receptors, which differs from other subtypes in the detailed configuration of one or more of the functional sites of the receptor. Significant functional differences have been detected when the isoform of α in a given combination is changed, as in the EC$_{50}$ for GABA (Levitan *et al.*, 1988) or for benzodiazepine site ligands (Ymer *et al.*, 1990) or for a neurosteroid (Shingai *et al.*, 1990). Also, the programming of neuronal development to provide the switching-on of two different isoform genes in different pathways, or at different stages, may permit their differential regulation even if the two receptors are very similar. A high multiplicity of the receptors may, for both of these reasons, allow GABA to be employed in constructing a variety of complex neuronal circuits, even though it is a ubiquitous, single inhibitory transmitter in the nervous system.

Isolated receptor subtypes

If each permitted combination of isoforms indeed represents a different subtype of GABA$_A$ receptor, the pharmacology of this receptor must be far more complex than any which can be analysed by conventional methods. The different pharmacologies can only be defined by expressing different mixtures of recombinant isoforms. This should preferably be done in permanent cell lines, for constant reference and general availability. Since a comparable level of stable expression of several DNAs is required together, this operation is less straightforward to perform successfully than for single-subunit receptors. Multiple GABA$_A$-receptor subunits can be expressed when an inducible promoter system is used (Moss *et al.*, 1990). Such cells are suitable for patch-clamping analysis of channel properties, as well as for pharmacological distinctions both by ligand binding measurements and by recording effects on channel parameters.

Glycine receptor subunits

Glycine is another inhibitory transmitter in the spinal cord and (to a lesser extent than GABA) in the brain of vertebrates. Its receptor has been purified and contains α (48 kD) and β (58 kD) subunits, whose sequences have been determined by cDNA cloning (Grenningloh *et al.*, 1987, 1990). The protein as isolated contained an additional polypeptide of 93 kD, but this is a peripheral membrane protein which is not glycosylated and which by immuno-electron microscopy it is seen to be cytoplasmic and separate from the receptor/channel structure, although it might associate with it. The α subunit was shown by photo-labelling to contain the strychnine-binding site (Graham, Pfeiffer & Betz, 1983). The β amino acid sequence is 47 % identical to that of α (in the rat). The β subunit sequence differs mainly in the presumed intracellular domain between TM3 and TM4 (see Fig. 6.1), and in particular introduces potential phosphorylation sites there for protein kinases, as in the β and γ subunits of the GABA$_A$ receptor.

Either the α or the β subunits alone, when expressed in oocytes, produce glycine-activated chloride currents (Grenningloh *et al.*, 1990). The β-homo-oligomer responds only weakly to glycine, with a Hill coefficient (n_H) of 1, and is strychnine-insensitive. In contrast, the α-homo-oligomer is highly strychnine-sensitive, has an EC_{50} for glycine 100 times lower than that of α and has n_H of 3.0. The expressed $\alpha\beta$ combination shows no difference in function from α alone and co-assembly of the α and β subunits in heterologous cells has not been demonstrated (Grenningloh *et al.*, 1990). Nevertheless, by cross-linking and reaction of the extracted native receptor with anti-α and anti-β subunit antibodies, evidence for a structure of $\alpha_3\beta_2$ for the glycine receptor has been reported (Langosch *et al.*, 1990).

Several isoforms of the α subunit have been found by cross-hybrid-isation: $\alpha1$, $\alpha2$, $\alpha3$ and $\alpha4$ forms have been recognised so far (Langosch *et al.*, 1990). Moreover, alternative mRNA splicing can create an 8-amino-acid insertion in the loop between TM3 and TM4 (Langosch, Becker & Betz, 1990), i.e. in the equivalent region to that where such an exonic addition can occur in the GABA$_A$ receptor β and γ subunits (see above). The $\alpha2$ isoform has been characterised: it is 79 % identical to $\alpha1$ and is expressed selectively in embryonic and neonatal spinal cord. This produces a second, neonatal subtype of the glycine receptor which is of low sensitivity to strychnine (Becker *et al.*, 1988). The β subunit transcript is more widely distributed in the nervous system and during development than the mRNA for any of the α isoforms (Grenningloh *et al.*, 1990), suggesting that a constant β is usually combined with one α out of several.

Comparison of the subunits of the receptor super-family IA

This super-family, as defined in Chapter 4 on the basis of homology, comprises the GABA$_A$, glycine, muscle nicotinic and neuronal nicotinic acetylcholine receptor types. The structural features of its subunits are shown in Fig. 6.1. This model shows an extracellular domain which extends over the entire N-terminal half of the subunit. There is a high concentration and excess of positively charged amino acid side chains in the immediate vicinity of the ends of the proposed transmembrane domains of all of the GABA (and glycine) receptor subunits. In contrast, most of these positions are occupied by neutral or negatively charged residues in the nicotinic ACh receptors. Clusters of arginines and lysines at the channel mouth, therefore, are presumed to act in the anion channel. A common feature of interest in this super-family is the long cytoplasmic loop, found in all cases between the TM3 and TM4 domains. In the nicotinic receptor this contains also a marked amphipathic structure (MA), which could form an α-helix with one face polar and the other hydrophobic. This is present in all of the acetylcholine receptor subunits known, both from muscle and brain and including invertebrate subunits (Marshall *et al.*, 1990). It is always absent in 5HT$_3$, GABA$_A$ and glycine receptors, so it is not required for the channel structure in general.

Taking all of the sequenced isoforms of the subunits of the GABA$_A$, glycine, muscle nicotinic and neuronal nicotinic receptors, from all species investigated, a set of more than 80 sequences of this super-family is now available, covering a wide range of species, down to invertebrates. This data-base allows us to discern the sequence features which are invariant, and hence likely to be important (Table 6.1). Some of these are common to the cation and the anion channel subunits and some are specific to each of those, as noted. The consensus sequence noted of the Cys–Cys potential loop (which has been shown to be disulphide-bridged in the nicotinic case) is suitable for molecular modelling, when a folded, amphipathic structure, with a fixed position for an aspartic side-chain, is predicted to be invariant in tertiary structure in all of the subunit types of these IA receptors (Cockcroft *et al.*, 1990).

TM2 is always the most hydrophilic of the four TMs, but it is distinctly

Table 6.1. *Features of the GABA$_A$, glycine, and nicotinic acetylcholine receptor subunits[a]*

Feature	GABA and glycine receptors	Comparison with ACh and 5HT$_3$ receptors[b]
(i) TM2	Contains a common[c] hydroxy-rich sequence Thr-Thr-Val-Leu-Thr-Met-Thr(Ser)- and with a total of 8 Ser or Thr in each TM2	This sequence is completely different. TM2 has two additional acidic groups
(ii) TM1	Invariant Pro at eleventh position	The same, in all cases. Consensus sequence there: -Pro-Cys-Ser/Thr-X-Leu/Met-, in all three receptor types
(iii) TM4	Pro at fourth position	Absent
(iv) Charges (per subunit) within eight residues of the ends of the TMs, in EC and IC	Always a high positive charge density: up to 13 total there, and always an excess (up to 9) of positive over negative	A much lower density of positive charges. A small excess of negative charges is always present
(v) β-loop could form between Cys-139 and Cys-153	8/15 positions identical or very conservatively substituted, in all subunits	Consensus sequence there, for all known subunits of the four receptor types: -Cys-X-Hy-X-Hy-X-X-Hy-Pro-Hy-Asp-X-(Gln/His)-X-Cys-X-Hy-X-Hy

[a] Abbreviations: TM, trans-membrane α-helix, assuming the structure shown in Fig. 6.1; EC, extracellular domain; IC, intracellular domain. Hy is a strongly hydrophobic residue (Ile, Met, Leu, Val, Tyr, Phe, or Trp); X is any amino acid. The numbering is for the GABA receptor α1 subunit.
[b] Nicotinic acetylcholine receptor subunits from electric organs, muscles and neurones (see Chapter 5).
[c] The glycine receptor β subunit is distinct from all the others: in its TM2 there are only five Ser or Thr residues, and in it the first, sixth and seventh Thr residues of the common sequence of all the others are changed to hydrophobic residues.

different in sequence between the anion and cation channels, having many hydroxy side chains in the former and more negative charges in the latter (Barnard *et al.*, 1987). These features are consistent with the proposal that TM2 provides the aqueous channel interaction site for ions in both cases.

Genes and evolution of subunits of the receptor super-family 1A

For the subunits of $GABA_A$ receptors, cDNA cloning has now covered a wide range of animal species. A highly exceptional constancy of amino acid sequence in phylogeny has been revealed. For example, the $\alpha 1$ subunit in the chicken is 98% identical to the corresponding form in the cow or in man (Bateson *et al.*, 1991*a*). Other subunit types examined so far show almost as much conservation between the avian and the human isoforms (Bateson *et al.*, 1990, 1991*b*; Glencorse *et al.*, 1990*a*). Within the mammals, only differences between 0 and 3% have been found in the sequences of any isoform. For the neuronal nicotinic receptor subunits, the variation found in the vertebrates of a given isoform is greater; for example, the identity score is 72% for chicken and rat $\beta 2$ subunits (Nef *et al.*, 1988). The striking evolutionary stability of the GABA receptor subunits is much greater than when isoenzymes are compared across the vertebrates, and supports the concept that each subunit isoform of the $GABA_A$ receptor has a separate and significant functional role.

Further, an invertebrate $GABA_A$ receptor cDNA has recently been cloned (Harvey *et al.*, 1991), from the mollusc *Lymnaea stagnalis*. This encodes an expressing β-type subunit of the same structural pattern as the vertebrate subunits. When compared to mammalian β, this shows $\sim 50\%$ identity. In the TMs 1 to 3 (the most conserved region) this rises to 76%, but in TM4 it is at a maximum of only 41% identity to any mammalian isoform. Although these values are still high considering the great phylogenetic distance involved, it may be that conservation is also lowered because the invertebrate subunit type does not correspond to any identified mammalian subunit, or it may be due to a partial change in the receptor function in vertebrate phylogeny. Its pharmacology shows some differences from the vertebrate (Harvey *et al.*, 1991). Construction of the most parsimonious phylogenetic tree for this super-family suggests that the divergence of the $GABA_A$ and acetylcholine receptors arose early in metazoan evolution (Cockcroft *et al.*, 1991).

The structure of the genes for GABA receptor isoforms has been determined for representative cases from man and from the chicken (Barnard *et al.*, 1991*b*; Lasham *et al.*, 1991) and from the mouse (Sommer *et al.*, 1990*b*). There are nine exons in all cases (plus a tenth, not always used, found so far only in $\gamma 2$: see legend to Fig. 6.2). The positions of the exons are constant or extremely close in all the cases where they have been determined, to give the fixed pattern shown in Fig. 6.2. There (top two lines) the exons (between the arrows) are shown under the corresponding

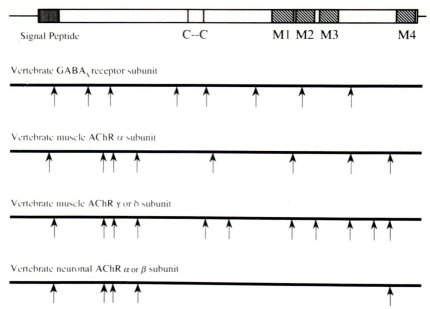

Fig. 6.2. Schematic representation of the genomic organisation for the subunits of the GABA$_A$ and nicotinic acetylcholine receptors. At the top, the encoded polypeptide is shown, with features common to all of the subunits of both receptor types: the positions of the signal peptide, the cysteine-loop region (C–C) and the four membrane-spanning domains (M1 to M4) are shown. For the genes as designated below this, the arrows indicate the positions of the introns relative to the coding regions above. (The subunit lengths differ and are not drawn to scale). The first of these shows the exons found in the subunit genes of the GABA$_A$ receptor of vertebrates. The locations shown are from complete or partial mappings of the α, β, γ (Barnard *et al.*, 1991*b*; Lasham *et al.*, 1991) and δ (Sommer *et al.*, 1990*a, b*) subunit genes from either human, mouse or chicken. At the last intron shown (between M3 and M4) the alternative splicing which is referred to in the text occurs in the primary gene transcript. This has been found in the β4 and γ2 subunits, so that in the latter case an extra exon (of only 24 nucleotides, not shown) is present. The lower 3 lines show the exons in the nicotinic receptor from: muscle α subunit; muscle γ and δ subunits; the brain α and β subunits. The subunits of the nicotinic receptors from man, mouse, rat and chicken have, where determined identical genomic patterns in each of these cases.

points in the subunit polypeptide. In the latter there is always a hydrophobic N-terminal signal peptide of 20–30 residues (of extremely variable sequence) which is removed intracellularly to form the mature subunit. Two exons cover TM1–TM3, with TM4 always separated. The Cys–Cys loop in the extracellular region (see Table 6.1) is always in a separate small exon. The constancy of the intron positions in vertebrate evolution and in the different subunit types is remarkable.

Further, analysis of the gene for the molluscan β-subunit (Harvey *et al.*, 1991) shows – as far as has been determined – that the positions of these boundaries are maintained with high precision. This invariant genomic organisation of the subunits is, again, highly exceptional and reinforces the argument given above for the very high selection pressure on the subunit structure in this series.

The genomic organisation has also been determined for the nicotinic acetylcholine receptors, for various subunits from muscle receptor from man (Noda *et al.*, 1983; Shibahara *et al.*, 1985), mouse (Buonanno *et al.*, 1989) and chicken (Nef *et al.*, 1984). This can be compared with that for the neuronal nicotinic receptor subunits from chicken (Nef *et al.*, 1988) and rat (Boulter *et al.*, 1990*b*). The pattern is constant for all the neuronal subunits analysed, whereas the muscle gene pattern varies somewhat with the subunit type (Fig. 6.2). The neuronal nicotinic receptors have a simpler genomic organisation: the first four exons have the same locations as in the muscle genes, but the fifth exon covers four or seven exons of the latter.

It can be seen (Fig. 6.2) that few features of the pattern are common to $GABA_A$ and nicotinic receptor genes. The Cys–Cys loop is usually within a single, relatively small exon. However, the preservation of the organisation of the $GABA_A$ receptor genes through all subunit types, isoforms and evolution stands out as a feature. The intronic sequences where examined vary, of course, greatly between species, whereas the exons and their boundaries are maintained almost constant for each isoform. The combination of a high multiplicity of the genes and an invariancy of each of them may be required, therefore, as discussed above, for the subunits of such a receptor which is employed in the construction of most neuronal pathways.

Excitatory amino acid receptors

The vertebrate receptors activated by glutamate have been classified into three general groups: (i) kainate and AMPA (α-amino-3-hydroxy-5-methylisoxazole-4-propionic acid) receptors, which are transmitter-gated (ionotropic) ion channels; (ii) the NMDA (*N*-methyl-D-aspartate) receptor, another ionotropic receptor type, which has multiple and specific allosteric modulatory sites; (ii) the metabotropic glutamate receptor, which can be activated by glutamate and by quisqualate to stimulate inositol phospholipid metabolism (Monaghan *et al.*, 1989). The ionotropic channels are permeated by Na^+ and K^+ ions, plus Ca^{2+} ions in the case of the NMDA receptor.

The metabotropic receptor is linked to a G protein. The recent cloning of its cDNA by Masu and co-workers (1991) has shown that, as predicted, this receptor is built from a single polypeptide chain which includes seven hydrophobic domains (Masu *et al.*, 1991). It has no structural relationship to known ionotropic glutamate receptors. For the NMDA receptor, a recombinant 105-kD subunit which can form a functional receptor has been obtained by Moriyoshi *et al.* (1991). Much progress has also been made in the isolation and characterisation of some of the non-NMDA receptors and of cDNAs encoding them.

Hampson and Wenthold (1988) isolated a kainate binding protein from the CNS of the frog *Rana pipiens* by affinity chromatography on a column of domoate (a kainate-like agonist). This contained a single polypeptide of 48 kD, from which a 26-residue peptide sequence was obtained. Oligo-nucleotide probes were constructed and used to derive DNA clones that encoded the protein (Wada *et al.* , 1989). The final cDNA could be expressed in transfected cells, where kainate binding sites but not functional receptors were detected.

Another type of kainate-binding protein was recognised by Teichberg and co-workers as exceptionally abundant in the chicken cerebellum. From this source it was purified by sequential fractionation procedures without recourse to affinity chromatography (Gregor *et al.*, 1989). A single polypeptide (M_r 49000) was again obtained. Again a cDNA was cloned, which encoded the isolated polypeptide and which hybridised to chicken cerebellar mRNA. However, no functional translation product could be obtained in oocytes.

While those two proteins bind kainate but not AMPA, a receptor with high affinity for both types of ligand has been recognised and affinity-purified from the CNS of *Xenopus laevis* (Henley *et al.*, 1989; Barnard *et al.*, 1991a). On reconstitution in liposomes this protein produces Na^+, K^+-channels, which are specifically activated by low concentrations of either kainate or AMPA (Ambrosini, Barnard & Prestipino, 1991) and in binding studies those ligands are mutually competitive, so that this is a prototypic unitary non-NMDA receptor. Several isoforms of a subunit of apparent $M_r \sim 42000$ are present, and cDNA cloning has shown these *Xenopus* subunits to be $\sim 65\%$ homologous to the *Rana* subunits (Barnard *et al.*, 1991a).

From rat brain, oocyte-expression cloning techniques were used to isolate a different form of non-NMDA receptor cDNA encoding a subunit

(GluR1) of $M_r \sim 100\,000$ (Hollman *et al.*, 1989) and by cross-hybridisation a series of related recombinant receptor subunits have been obtained, GluR 2, 3 and 4. These show approximately 70% sequence identity overall. Each yields a functional AMPA and kainate-gated ion channel and high-affinity [^{3}H]-AMPA binding sites. Cells transfected with a combination of GluR-1 and -2 subunits respond well to both kainate and AMPA, with the characteristics of some native non-NMDA receptors (Boulter *et al.*, 1990*a*; Keinanen *et al.*, 1990). A further isoform, GluR5, is rather more distant, with only 40% amino acid identity to the others and responding only to glutamate on expression (Bettler *et al.*, 1990). Two forms can be derived by alternative splicing from some of the RNAs, and these show distinct expression patterns in rat brain and differential functional characteristics (Sommer *et al.*, 1990*a*). In these subunits, the membrane-spanning regions include sequences that are nearly identical (98% if conservative substitutions are counted as identities) in the region of the putative cytoplasmic loop between TM3 and TM4. Surprisingly, this occurs in the region which shows the greatest sequence variation between the subunits in the other transmitter-gated ion channels.

It appears, therefore, that polypeptides encoded by these cloned cDNAs represent subunits of one set of mammalian non-NMDA receptors, with probably at least two subunit types in each oligomer. It is not possible, from the *in vitro* expression results, to predict and true stoichiometry. These are indications that there is a much greater heterogeneity of the subunits of this major excitatory receptor class, than with the major inhibitory (GABA$_A$) receptor.

Subunit structures and the super-family IB

The proteins derived from the glutamate receptor clones so far isolated have all been predicted (from their hydrophobicity plots) to contain four transmembrane domains (TM1–TM4). The overall topology, therefore, fits a model (shown in Fig. 6.3) like that for the nicotinic acetylcholine receptor subunits (Fig. 6.1). Sites for the binding of lectins are shown on the extracellular domain (e d), due to the strong effect of the binding of concavalin A or wheatgerm lectin in suppressing the rapid desensitisation to AMPA responses (for review, see Barnard & Henley, 1990). Also suggested in the model is an associated regulatory protein, the affinity of the AMPA sites has been deduced by Honoré, Drejer & Nielson (1986), from radiation inactivation data, to be depressed by a large additional polypeptide.

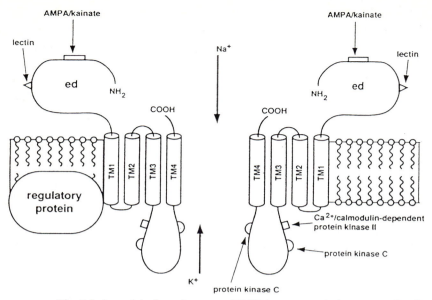

Fig. 6.3. A model of a unitary non-NMDA receptor. At least two subunit types, which are closely related but not identical, are assumed to be present. By analogy with the muscle nicotinic acetylcholine receptor a total of five subunits seems likely, but there is as yet no information on the subunit stoichiometry. Four trans-membrane segments per subunit would then give 20 per receptor, arranged to form a channel and its wall. ed, the large N-terminal domain is presumed to be extracellular; this is about twice as large in the rat sequence as in the others known (from Barnard & Henley, 1990).

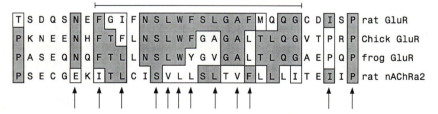

Fig. 6.4. Alignment with a rat brain nicotinic acetylcholine receptor ($\alpha2$ subunit) sequence (lowest line) of the TM2 region of the rat, chick and frog GluR-type proteins. Shading indicates amino acids which are identical and arrows indicate positions at which highly conservative substitutions cover all the forms. The bar shows the 19-residue hydrophobic segment proposed as TM-2. (The region assigned here as TM2 would correspond to a TM3 region as defined by Hollmann *et al.*, 1989.)

Table 6.2. *Percentage sequence identities*

	Rat GluR	Chick KBP	Frog KBP
Rat GluR1	–	38	39
Chick kainate-binding protein (KBP)	–	–	56
Nicotinic (mouse muscle α)	22	15	16
Nicotinic (rat brain α2)	26	20	20

For references to the sequences, see text and Chapter 5. Inserted stretches in pairwise alignments are not included in the calculation of the percentages shown of identical amino acids at the aligned positions.

Homology, albeit rather low, can be discerned between the putative TM2 regions here and the corresponding TM2 in neuronal nicotinic acetyl-choline receptors (Fig. 6.4). Within the set of non-NMDA proteins the homology is good, especially in the transmembrane regions (15 identical or strongly homologous residues out of 26 in or near TM2). This suggests that there are common features in the channel structures between these three non-NMDA types, despite their low overall sequence similarity (Table 6.2).

Conclusions

The sequences so far known for the non-NMDA receptor subunits suggest that the rat protein is of a different subtype to that of the chicken or frog proteins (Table 6.2). The subunits of rat GluRs and NMDA-R are more than twice the length of the chicken, frog or *Xenopus* subunits and this extra length is all present as a major extension of the deduced N-terminal extracellular domain (Fig. 6.3). All of them have some features in common, however, with the GABA/glycine/nicotinic receptor super-family. They have four putative transmembrane domains at about the same spacing, and also have the characteristic long and very variable putative in-tracellular loop between TM3 and TM4, with potential sites there for regulation by phosphorylation, as indicated in Fig. 6.3. There are prolines and negatively charged amino acids flanking TM2 (Fig. 6.4) and also TM1, in each of the three types of sequence, as in the nicotinic acetylcholine sequences. There is no evidence yet that TM2 is the channel-lining segment in non-NMDA receptors, as there is for the nicotinic acetylcholine receptors.

All of the other structural features of the super-family I subunits, listed in Table 6.1, are missing in these glutamate receptor subunits, with the exception of the small excess of negative charges predicted to be near the mouths (Fig. 6.3) of the channel, as with the nicotinic receptor (Fig. 6.1). In view of this, and since the homologies outside TM2 are so low between these and the $GABA_A$/glycine/nicotinic subunits, it does not seem justified to include the non-NMDA glutamate receptor subunits in the same super-family as the latter. It therefore seems better, on the examples available so far, to classify the glutamate receptors in a different but related super-family, designated IB.

References

Alger, B. E. & Nicholl, R. A. (1982). Pharmacological evidence for two kinds of GABA receptor on rat hippocampal pyramidal cells studied *in vitro*. *Journal of Physiology*, **328**, 125–41.

Ambrosini, A., Barnard, E. A. & Prestipino, G. (1991). AMPA and kainate operated channels reconstituted in artificial bilayers. *FEBS Letters* **281**, 27–9.

Barnard, E. A., Darlison, M. G. & Seeburg, P. H. (1987). Molecular biology of the $GABA_A$ receptor: the receptor/channel super-family. *Trends in Neurosciences* **10**, 502–9.

Barnard, E. A. (1988). The structure of the GABA/benzodiazepine receptor complex with its gated ion channel. In: *GABA and Benzodiazepine Receptors* (Squires, R. H., ed.), vol. II, pp. 104–27, CRC Press, Boca Raton.

Barnard, E. A. & Henley, J. M. (1990). The non-NMDA receptors: types, protein structure and molecular biology. *Trends in Pharmacological Sciences*, **11**, 500–7.

Barnard, E. A., Ambrosini, A., Henley, J. M., Ishimaru, H., Lu, Q, Rodriguez-Ithurralde, D., Sudan, H. & Usherwood, P. N. (1991*a*). A unitary non-NMDA receptor and an NMDA receptor purified from *Xenopus* central nervous system. In: *Transmitter Amino Acid Receptors: Structures, Transduction and Models for Drug Development* (Barnard, E. A. and Costa, E., eds), Fidia Research Foundation Symposium Series, vol. 7, Raven Press, New York, pp. 357–65.

Barnard, E. A., Bateson, A. N., Darlison, M. G., Harvey, R. J., Hicks, A. A., Lasham, A. & Shingai, R. (1991*b*). Structures of genes for brain GABA receptors and their expression. In: *GABAergic Synaptic Transmission*. (Biggio, G., ed.), Advances in Biochemical Pharmacology, Raven Press, New York, in press.

Bateson, A. N., Harvey, R. J., Bloks, C. C. M. & Darlison, M. G. (1990). Sequence of the chicken $GABA_A$ receptor β3-subunit cDNA. *Nucleic Acids Research*, **18**, 5557.

Bateson, A. N., Harvey, R. J., Wisden, W., Glencorse, T. A., Hicks, A. A., Hunt, S. P., Barnard, E. A. & Darlison, M. G. (1991*a*). The chicken $GABA_A$ receptor α1 subunit: cDNA sequence and localization of the corresponding mRNA. *Molecular Brain Research*, **9**, 333–9.

Bateson, A. N., Lasham, A. & Darlison, M. G. (1991*b*). γ-Aminobutyric acid$_A$ receptor heterogeneity is increased by alternative splicing of a novel β-subunit gene transcript. *Journal of Neurochemistry*, **56**, 1437–40.

Becker, C. M., Hoch, W. & Betz, H. (1988). Glycine receptor heterogeneity in rat spinal cord during postnatal development. *EMBO Journal*, **7**, 3717–26.

Benke, D., Mertens, S., Trceziak, A., Gillessen, D. & Möhler, H. (1991). Identification of the γ2-subunit in native GABA$_A$-receptors in brain. *European Journal of Pharmacology*, **189**, 337–40.

Benke, D., Mertens, S., Trceziak, A., Gillessen, D. & Möhler, H. (1991). GABA$_A$- receptors: association of the γ2-subunit with the α1-subunit and the β2/3-subunits. *Journal of Biological Chemistry* (in press).

Bettler, B., Boulter, J., Hermans-Borgmeyer, I., O'Shea-Greenfield, A., Deneris, E. S., Moll, C., Borgmeyer, U., Hollmann, M. & Heinemann, S. (1990). Cloning of a novel glutamate receptor subunit, GluR5: expression in the nervous system during development. *Neuron*, **5**, 583–95.

Blair, L. A. C., Levitan, E. S., Dionne, V. E. & Barnard, E. A. (1988). Single subunits of the GABA$_A$ receptor form ion channels with properties characteristics of the native receptor. *Science*, **242**, 577–9.

Boulter, J., Hollmann, M., O'Shea-Greenfield, A., Hartley, M., Deneris, E., Maron, C. & Heinemann, S. (1990*a*). Molecular cloning and functional expression of glutamate receptor subunit genes. *Sicece*, **249**, 1033–7.

Boulter, J., O'Shea-Greenfield, A., Duvosin, R. M., Connolly, J. G., Wada, E., Jensen, A., Gardner, P. D., Ballivet, M., Deneris, E. S., McKinnon, D., Heinemann, S. & Patrick, J. (1990*b*). α3, α5, and β4: three members of the rat neuronal ncotinic acetylcholine receptor-related gene family form a gene cluster. *Journal of Biological Chemistry*, **265**, 4472–82.

Buckle, V. J., Fujita, N., Ryder-Cook, A. S., Derry, J. M. J., Barnard, P. J., Lebo, R. V., Schofield, P. R., Seeburg, P. H., Bateson, A. N., Darlison, M. G. & Barnard, E. A. (1989). Chromosomal localization of GABA$_A$ receptor subunit genes: relationship to human genetic disease. *Neuron*, **3**, 647–54.

Buonanno, A., Mudd, J. & Merlie, J. P. (1989). Isolation and characterization of the β ad ε subunit genes of mouse muscle acetylcholine receptor. *Journal of Biological Chemistry*, **264**, 7611–16.

Casalotti, S. O., Stephenson, F. A. & Barnard, E. A. (1986). Separate subunits for agonist and benzodiazepine binding in the γ-aminobutyric acid$_A$ receptor oligomer. *Journal of Biological Chemistry*, **261**, 15013–16.

Chang, L.-R. & Barnard, E. A. (1982). The benzodiazepine/GABA receptor complex: molecular size in brain synaptic membranes and in solution. *Journal of Neurochemistry*, **39**, 1507–18.

Cockcroft, V. B., Osguthorpe, D. J., Barnard, E. A. & Lunt, G. G. (1990). Modeling of agonist binding to the ligand-gated ion channel superfamily of receptors. *Proteins*, **8**, 386–97.

Cockcroft, V. B., Osguthorpe, D. J., Barnard, E. A., Friday, A. F., Wonnacott, S. & Lunt, G. G. (1991). Ligand-gated ion-channels: homology and diversity. *Molecular Neurobiology* (in press).

Cutting, G. R., Bruce, L. L., O'Hara, F., Kasch, L. M., Montrose-Rafizadeh, C., Donovan, D. M., Shimada, S., Antonarakis, S. E., Guggino, W. B., Uhl, G. R. & Kazazian, H. H. Jr. (1991). Cloning of the γ-aminobutyric acid (GABA) ρ_1 cDNA: a GABA receptor subunit highly expressed in the retina. *Proceedings of the National Academy of Sciences, USA*, **88**, 2673–7.

de Blas, A. L., Vitorica, J. & Friedrich, P. (1988). Localization of the GABA$_A$ receptor in the rat brain with a monoclonal antibody to the 57000 M_r peptide of the GABA$_A$ receptor/benzodiazepine receptor/Cl$^-$ channel complex. *Journal of Neuroscience*, **8**, 602–14.

Deng, L., Ransom, R. W. & Olsen, R. W. (1986). [^{3}H]Muscimol photolabels the

GABA receptor site on a peptide subunit distinct from that labeled with benzodiazepine. *Biochemical and Biophysical Research Communications*, **138**, 1308–1314.

Fuchs, K., Adamiker, D. & Sieghart, W. (1990). Identification of α_2- and α_3-subunits of the $GABA_A$-benzodiazepine receptor complex purified from the brains of young rats. *FEBS Letters*, **261**, 52–4.

Glencorse, T. A., Bateson, A. N. & Darlison, M. G. (1990*a*). Sequence of the chicken $GABA_A$ receptor $\gamma2$-subunit cDNA. *Nucleic Acids Research*, **18**, 7157.

Glencorse, T. A., Bateson, A. N. & Darlison, M. G. (1990*b*). The chicken $GABA_A$ receptor: heterogeneity of gamma subunits. *Biochemical Society Transactions*, **19**, 4.

Graham, D., Pfeiffer, F. & Betz, H. (1983). Photoaffinity-labeling of the glycine receptor of rat spinal cord. *European Journal of Biochemistry*, **131**, 519–25.

Gregor, P., Mano, I., Maoz, I., McKeown, M. & Teichberg, V. I. (1989). Molecular structure of the chick cerebellar kainate-binding subunit of a putative glutamate receptor. *Nature, London*, **342**, 689–92.

Grenningloh, G., Rienitz, A., Schmitt, B., Methfessel, C., Zensen, M., Beyreuther, K., Gundelfinger, E. D. & Betz, H. (1987). The strychnine-binding subunit of the glycine receptor shows homology with nicotinic acetylcholine receptors. *Nature, London*, **328**, 215–20.

Grenningloh, G., Pribilla, I., Prior, P., Multhaup, G., Beyreuther, K., Taleb, O. & Betz, H. (1990). Cloning and expression of the 58 kD β subunit of the inhibitory glycine receptor. *Neuron*, **4**, 963–70.

Hampson, D. R. & Wenthold, R. J. (1988). A kainic acid receptor from frog brain purified using domoic acid affinity chromatography. *Journal of Biological Chemistry*, **263**, 2500–5.

Harvey, R. J., Vreugdenhil, E., Zaman, S. H., Bhandal, N. S., Usherwood, P. N. R., Barnard, E. A. & Darlison, M. G. (1991). Cloning of a cDNA encoding a functional invertebrate $GABA_A$ receptor subunit. *EMBO Journal*, **10**, 3239–45.

Heley, J. M., Ambrosini, A., Krogsgaard-Larsen, P. & Barnard, E. A. (1989). Evidence for a single glutamate receptor of the ionotropic kainate/quisqualate type. *The New Biologist*, **1**, 153–8.

Hollmann, M., O'Shea-Greenfield, A., Rogers, S. W. & Heinemann, S. (1989). Cloning by functional expression of a member of the glutamate receptor family. *Nature, London*, **342**, 643–8.

Honoré, T., Drejer, J. & Nielsen, M. (1986). Calcium discriminates two [^{3}H]kainate binding sites with different molecular target sizes in rat cortex. *Neuroscience Letters*, **65**, 47–52.

Kato, K. (1990). Novel $GABA_A$ receptor α subunit is expressed only in cerebellar granule cells. *Journal of Molecular Biology*, **214**, 619–24.

Keinanen, K., Wisden, W., Sommer, B., Werner, P., Herb, A., Verdoorn, A., Sakmann, B. & Seeburg, P. H. (1990). A family of AMPA-selective glutamate receptors. *Science*, **249**, 556–60.

Kofuji, P., Wang, J. B., Moss, S. J., Huganir, R. L. & Burt, D. R. (1991). Generation of two forms of the γ-aminobutyric acid$_A$ receptor γ_2-subunit in mice by alternative splicing. *Journal of Neurochemistry*, **56**, 713–5.

Langosch, D., Becker, C.-M. & Betz, H. (1990). The inhibitory glycine receptor: A ligand-gated chloride channel of the central nervous system. *FEBS Letters*, **194**, 1–8.

Lasham, A., Vreugdenhil, E., Bateson, A. N., Barnard, E. A. & Darlison, M. G.

(1991). Conserved organization of γ-aminobutyric acid$_A$ receptor genes: cloning and analysis of the chicken β4-subunit gene. *Journal of Neurochemistry*, **57**, 352–5.

Levitan, E. S., Blair, L. A. C., Dionne, V. E. & Barnard, E. A. (1988*a*). Biophysical and pharmacological properties of cloned GABA$_A$ receptor subunits expressed in *Xenopus* oocytes. *Neuron*, **1**, 773–81.

Levitan, E. S., Schofield, P. R., Burt, D. R., Rhee, L. M., Wisden, W., Köhler, M., Fujita, N., Rodriguez, H., Stephenson, F. A., Darlison, M. G., Barnard, E. A. & Seeburg, P. H. (1988*b*). Structural and functional basis for GABA$_A$ receptor heterogeneity. *Nature, London*, **335**, 76–9.

Lüddens, H., Pritchett, D. B., Köhler, M., Killisch, I., Keinänen, K., Monyer, H., Sprengel, R. & Seeburg, P. H. (1990). Cerebellar GABA$_A$ receptor selective for a behavioural alcohol antagonist. *Nature, London*, **346**, 648–51.

McCabe, R. T. & Wamsley, J. K. (1986). Autoradiographic localization of subcomponents of the macromolecular GABA receptor complex. *Life Science*, **39**, 1937–46.

Malherbe, P., Draguhn, A., Multhaup, G., Beyreuther, D. & Möhler, H. (1990). GABA$_A$ receptor expressed from rat brain α- and β-subunit cDNAs displays potentiation by benzodiazepine receptor ligands. *Molecular Brain Research*, **8**, 199–208.

Mamalaki, C., Stephenson, F. A. & Barnard, E. A. (1987). The GABA$_A$/benzodiazepine receptor is a heterotetramer of homologous subunits. *EMBO Journal*, **6**, 561–5.

Mamalaki, C., Barnard, E. A. & Stephenson, F. A. (1989). Molecular size of the γ-aminobutyric acid$_A$ receptor purified from mammalian cerebral cortex. *Journal of Neurochemistry*, **52**, 124–34.

Marshall, J., Buckingham, S. D., Shingai, R., Lunt, G. G., Goosey, M. W., Darlison, M. G., Satelle, D. B. & Barnard, E. A. (1990). Sequence and functional expression of a single α subunit of an insect nicotinic acetylcholine receptor. *EMBO Journal*, **9**, 4391–8.

Masu, M., Tanabe, Y., Tsuchida, K., Shigemoto, R. & Nakanishi, S. (1991). Sequence and expression of a metabotropic glutamate receptor. *Nature, London*, **349**, 760–5.

Monaghan, D. T., Bridges, R. J. & Cotman, C. W. (1989). The excitatory amino acid receptors: their classes, pharmacology and distinct properties in the function of the central nervous system. *Annual Reviews in Pharmacology, and Toxicology*, **29**, 365–402.

Moriyoshi, K., Masu, M., Ishii, T., Shigemoto, R., Mizuno, N. & Nakanishi, S. (1991). Molecular cloning and characterization of the rat NMDA receptor. *Nature, London*, **354**, 31–7.

Moss, S. J., Smart, T. G., Porter, N. M., Nayeem, N., Devine, J., Stephenson, F. A., Macdonald, R. L. & Barnard, E. A. (1990). Cloned GABA receptors are maintained in a stable cell line: allosteric and channel properties. *European Journal of Pharmacology*, **189**, 77–88.

Nef, P., Mauron, A., Stalder, R., Alliod, C. & Ballivet, M. (1984). Structure, linkage and sequence of the two genes encoding the δ and γ subunits of the nicotinic acetylcholine receptor. *Proceedings of the National Academy of Sciences, USA*, **81**, 7975–9.

Nef, P., Oneyser, C., Alliod, C., Couturier, S. & Ballivet, M. (1988). Genes expressed in the brain define three distinct nicotinic acetylcholine receptors. *EMBO Journal*, **7**, 595–601.

Noda, M., Furutani, Y., Takahashi, H., Toyosato, M., Tanable, T., Shimizu, S.,

Kikyotani, S., Kayano, T., Hirose, T., Inayama, S. & Numa, S. (1983).
Cloning and sequence analysis of calf cDNA and human genomic DNA
encoding α-subunit precursor of muscle acetylcholine receptor. *Nature,
London*, **305**, 818–23.

Olsen, R. W. & Venter, J. C. (eds). (1986). Benzodiazepine/GABA receptors
and chloride channels: structural and functional properties. *Receptor
Biochemistry and Methodology*, vol. 5, Alan R. Liss, New York.

Olsen, R. W. & Tobin, A. J. (1990). Molecular biology of GABA$_A$ receptors.
FASEB Journal, **4**, 1469–80.

Osmanovic, S. S. & Shefner, S. A. (1990). γ-Aminobutyric acid responses in rat
locus coeruleus neurones *in vitro*: a current-clamp and voltage-clamp study.
Journal of Physiology, **421**, 151–70.

Peters, J. A., Kirkness, E. F., Callachan, H., Lambert, J. J. & Turner, A. J.
(1988). Modulation of the GABA$_A$ receptor by depressant barbiturates and
pregnane steroids. *British Journal of Pharmacology*, **94**, 1257–69.

Pritchett, D., Sontheimer, H., Shivers, B. D., Ymer, S., Kettemann, H.,
Schofield, P. R. & Seeburg, P. (1989). Importance of a novel GABA$_A$
receptor subunit for benzodiazepine pharmacology. *Nature, London*, **338**,
582–5.

Schofield, P. R., Darlison, M. G., Fujita, N., Burt, D. R., Stephenson, F. A.,
Rodriquez, H., Rhee, L. M., Ramachandran, J., Reale, V., Glencorse,
T. A., Seeburg, P. H. & Barnard E. A. (1987). Sequence and functional
expression of the GABA-A receptor shows a ligand-gated receptor
superfamily. *Nature, London*, **328**, 221–7.

Shibahara, S., Kubo, T., Perski, P., Takahashi, H., Noda, M. & Numa, S.
(1985). Cloning and sequence analysis of human genomic DNA encoding γ
subunit precursor of muscle acetylcholine receptor. *European Journal of
Biochemistry*, **146**, 15–22.

Shingai, R., Sutherland, M. L. & Barnard, E. A. (1991). Effects of subunit types
of the cloned GABA$_A$ receptor on the response to a neurosteroid. *European
Journal of Pharmacology*, **206**, 77–80.

Shivers, B. D., Killisch, I., Sprengel, R., Sontheimer, H., Köhler, M., Schofield,
P. R. & Seeburg, P. H. (1989). Two novel GABA$_A$ receptor subunits exist in
distinct neuronal subpopulations. *Neuron*, **3**, 327–37.

Sieghart, W. & Drexler, G. (1983). Irreversible binding of [³H]flunitrazepam to
different proteins in various brain regions. *Journal of Neurochemistry*, **41**,
47–55.

Sigel, E. & Barnard, E. A. (1984). A γ-aminobutyric acid/benzodiazepine
receptor complex from bovine cerebral cortex. Improved purification with
preservation of regulatory sites and their interactions. *Journal of Biological
Chemistry*, **259**, 7219–23.

Sigel, E., Mamalaki, C. & Barnard, E. A. (1982). Isolation of a GABA receptor
from bovine brain using a benzodiazepine affinity column. *FEBS Letters*,
147, 45–8.

Sigel, E., Stephenson, F. A., Mamalaki, C. & Barnard, E. A. (1983). A γ-
aminobutyric acid/benzodiazepine receptor complex of bovine cerebral
cortex: Purification and partial characterization. *Journal of Biological
Chemistry*, **258**, 6965–71.

Sigel, E., Baur, R., Trube, G., Möhler, H. & Mallherbe, P. (1990). The effect of
subunit composition of rat brain GABA$_A$ receptors on channel function.
Neuron, **5**, 703–11.

Sommer, B., Keinanen, K., Verdoorn, T. A., Wisden, W., Burnashev, N., Herb,

A., Kohler, M., Takagi, T., Sakmann, B. & Seeburg, P. H. (1990*a*). Flip and flop: A cell-specific functional switch in glutamate operated channels of the CNS. *Science*, **249**, 1580–5.

Sommer, B., Poustka, A., Spurr, N. K. & Seeburg, P. H. (1990*b*). The murine GABA$_A$ receptor δ-subunit gene: structure and assignment to human chromosome 1. *DNA and Cell Biology*, **9**, 561–8.

Stephenson, F. A. (1988). Understanding the GABA$_A$ receptor: a chemically-gated ion channel. *Biochemical Journal*, **249**, 21–32.

Stephenson, F. A., Duggan, M. J. & Pollard, S. (1990). The γ2 subunit is an integral component of the γ-aminobutyric acid$_A$ receptor but the α1 polypeptide is the principal site of the agonist benzodiazepine photoaffinity labeling reaction. *Journal of Biological Chemistry*, **265**, 21160–5.

Unnerstall, J. R., Kuhar, M. J., Niehoff, D. L. & Palacios, J. M. (1981). Benzodiazepine receptors are coupled to a subpopulation of GABA receptors; evidence from a quantitative autoradiographic study. *Journal of Pharmacology and Experimental Therapeutics*, **218**, 797–804.

Verdoorn, T. A., Draguhn, A., Ymer, S., Seeburg, P. H. & Sakmann, B. (1990). Functional properties of recombinant rat GABA$_A$ receptors depend upon subunit composition. *Neuron*, **4**, 919–28.

Wada, K., Dechesne, C. J., Shimasaki, S., King, R. G., Kusano, K., Buonanno, A., Hampson, D. R., Banner, C., Wenthold, R. J. & Nakatani, Y. (1989). Sequence and expression of a frog brain complementary DNA encoding a kainate-binding protein. *Nature, London*, **342**, 684–9.

Whiting, P., McKernan, R. M. & Iversen, L. L. (1990). Another mechanism for creating diversity in γ-aminobutyrate type A receptors: RNA splicing directs expression of two forms of γ2 subunit, one of which contains a protein kinase C phosphorylation site. *Proceedings of the National Academy of Sciences, USA*, **87**, 9966–70.

Wisden, W., Morris, B. J., Darlison, M. G., Hunt, S. P. & Barnard, E. A. (1988). Distinct GABA$_A$ receptor alpha-subunit mRNAs show differential patterns of expression in bovine brain. *Neuron*, **1**, 937–47.

Wisden, W., McNaughton, L., Hunt, S. P., Darlison, M. G. & Barnard, E. A. (1989*a*). Differential distribution of GABA$_A$ receptor subunit mRNA in bovine cerebellum; localization of α2 in Bergman glial cells. *Neuroscience Letters*, **106**, 7–12.

Wisden, W., Morris, B. J., Darlison, M. G., Hunt, S. P. & Barnard, E. A. (1989*b*). Localization of GABA$_A$ receptor α-subunit mRNAs in relation to receptor subtypes. *Molecular Brain Research*, **5**, 305–10.

Wisden, W., Morris, B. J., Seeburg, P., Barnard, E. A. & Hunt, S. H. (1991). Distribution of GABA receptor subunit mRNAs in rat lumbar spinal cord. *Molecular Brain Research*, **10**, 179–83.

Ymer, S., Draguhn, A., Wisden, W., Werner, P., Keinänen, K., Schofield, P. R., Sprengel, R., Pritchett, D. B. & Seeburg, P. H. (1990). Structural and functional characterization of the γ1 subunit of GABA$_A$/benzodiazepine receptors. *EMBO Journal*, **9**, 3261–7.

7

The family of adrenergic and structurally related G protein-coupled receptors

BRIAN F. O'DOWD, MARK HNATOWICH,
MARC G. CARON AND
ROBERT J. LEFKOWITZ*

Introduction

Numerous biologically active ligands, including hormones, neurotransmitters, drugs and even sensory stimuli such as photons of light, signal to the interior of cells via interaction with receptors which are coupled to signal transducing guanine nucleotide regulatory or 'G' proteins. Examples of such G protein-coupled receptors are the adrenergic receptors which mediate the physiological effects of epinephrine and norepinephrine. The family of adrenergic receptors (α_1, α_2, β_1, β_2) contains the most intensely studied members of this large group. This is because of their ubiquity, their coupling to several different effectors and the importance of numerous drugs which interact with them. The recent purification of these receptors, and successful cloning of their genes has led to an explosion of information concerning their structure, and the structural basis of their function. Some of this information is reviewed here.

Structural similarities of the adrenergic and related receptors

The deduced amino acid sequences of the genes encoding the adrenergic, muscarinic, serotonergic and dopaminergic receptors, aligned in Fig. 7.1, exhibit many structural similarities, indicating that they are variations on a common functional theme. Each of these receptors also shares sequence homology with rhodopsin (O'Dowd et al., 1989a,b), and the structural homologies likely parallel similarities in function and regulation. The single polypeptide chain of each receptor has seven hydrophobic segments that probably represent transmembrane-spanning (TMS) regions, and each receptor is coupled to at least one of a group of structurally conserved

* To whom correspondence should be addressed.

(a)

TM-I

TM-II

TM-III

TM-IV

TM-V

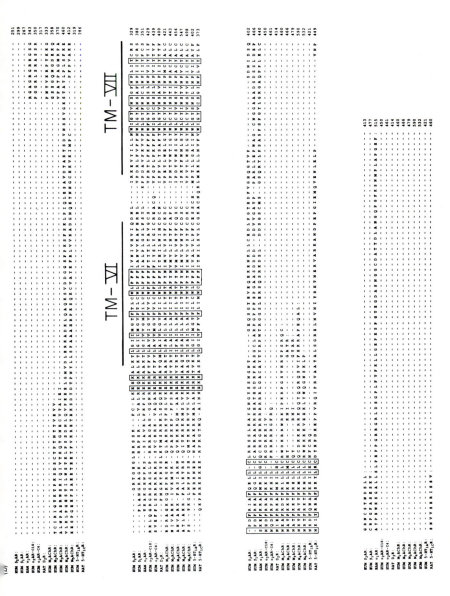

Fig. 7.1. Amino acid sequence homology among the G protein-coupled receptors: human β2 and β1 adrenergic receptors (AR); hamster α1AR; human α2AR (on chromosome 10) and α2AR (on chromosome 4); rat D2-dopaminergic receptor; human muscarinic cholinergic receptor 1,2,3,4, and 5; human serotonin-1A receptor (5-HT1A) and rat serotonin-1C receptor (5-HT1C). Letters enclosed in boxes represent amino acid residues that are identical in all receptors or differ by one residue. The seven putative transmembrane-spanning segments (TM) are indicated by heavy lines labelled with Roman numerals.

guanyl nucleotide-binding regulatory proteins (G proteins). Comparing the coding sequences of the 13 receptors aligned in Fig. 7.1, indicates that the sequence similarities are not distributed uniformly but are concentrated in the seven putative TMS regions. Only 32 amino acids are conserved in all of the receptors and 29 of these residues are found in or near the TMS segments. Furthermore, the majority of conserved residues in the TMS regions are located on the side of the membrane closest to the cytoplasm, perhaps because of the requirement to interact with common cytoplasmic transducing molecules (e.g. the structurally similar G proteins) (O'Dowd *et al.*, 1989*a,b*). However, close inspection of amino acid sequences also reveals that receptors which bind one class of ligands (e.g. either adrenergic or muscarinic cholinergic) are more alike irrespective of which G protein they couple to (i.e. G_s, G_i, G_p, etc.), whereas within a single ligand binding class, receptors coupling to the same G protein are the most similar. When compared with β_2AR, the β_1-, α_2- and α_1ARs share overall homology ranging between 70% and 40%, whereas the muscarinic, serotonergic and D_2-dopaminergic receptors share $\sim 30\%$, $\sim 25\%$ and 25% homology with β_2AR, respectively.

Determination of the three-dimensional structure of a member of this family of G protein-linked receptors awaits preparation of crystallized forms suitable for X-ray diffraction studies. However, based on the combined results of physical and biochemical studies of rhodopsin (Applebury & Hargrave, 1986) and proteolysis studies of β_2AR (Dohlman *et al.*, 1987), models of the topography of β_2AR have been proposed (Fig. 7.2). The hydropathicity profile of each of the receptors aligned in Fig. 7.1 is very similar to the profiles of rhodopsin and bacteriorhodopsin, proteins known to have seven TMS regions (Applebury & Hargrave, 1986). To directly test the topographical model of β_2AR, Dohlman *et al.* (1987) reconstituted purified hamster β_2AR into lipid vesicles and used proteolysis (trypsin and carboxypeptidase Y), as well as treatment with endoglyco-sidase F, to identify salient structural characteristics of the receptor in the plasma membrane. The various features delineated include a carboxyl tail of $\sim 7kD$ that is sensitive to carboxypeptidase Y and contains sites of phosphorylation, a glycosylated amino terminus, and an extracellular trypsin-sensitive site. These features are entirely consistent with a rhodop-sin-like structure for the β_2AR with multiple membrane-spanning regions.

The requirement for multiple TMS regions in the structural organization of each of the G protein-coupled receptors, coupled with the general lack of introns in their coding sequences (at present, only the rat/D_2R and the hamster α_1AR are known to contain introns within the coding block; S.

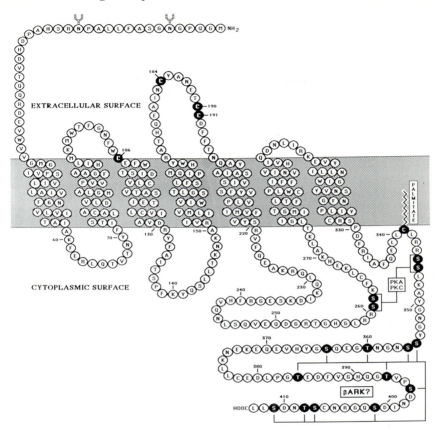

Fig. 7.2. Representation of the possible topography of human β_2AR in the plasma membrane. Putative sites of phosphorylation by protein kinase A (PKA), protein kinases C (PKC), and β_2AR receptor kinase (βARK) are indicated. The site of palmitoylation of Cys 341 and the position of extracellular cysteines are also indicated. The palmitoylation of Cys 341 may anchor the N-terminal portion of the carboxyl tail in the plasma membrane creating a fourth intracellular loop.

Cotecchia, personal communication), has undoubtedly contributed to the conservation of amino acid sequences among these receptors. The only structural features that vary appreciably among the receptors aligned in Fig. 7.1 are the sizes of their third cytoplasmic loops and carboxyl termini. However, a roughly inverse relationship exists between the lengths of these two cytoplasmic domains. Those receptors with 'long' carboxyl tails (e.g. β_2AR) tend to have 'short' third cytoplasmic loops, and vice versa (e.g. M_4AChR). Whereas the significance of this relationship is not presently

Split Receptors

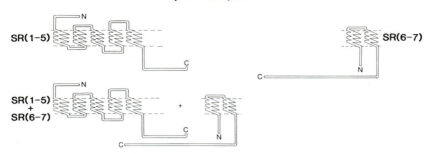

Fig. 7.3. Construction of 'split' β_2ARs. SR (1–5) represents a truncation of the human β_2AR after amino acid 262, while SR (6–7) represents the β_2AR in which amino acids 3–261 have been deleted.

apparent, it may suggest that certain (perhaps common) structural or functional components of these receptors may be found in *either* their third cytoplasmic loops or carboxyl tails. An evolutionary mechanism accounting for the length of the third cytoplasmic loop seems clear in the case of the human platelet α_2AR (α_2AR-C10) which is approximately three times longer than the equivalent loop in β_2AR (Kobilka *et al.*, 1987). The increase in size appears to correlate with a gain of several internal duplications (O'Dowd *et al.*, 1989*a,b*). These internal repeats in the α_2AR gene may have allowed this receptor to uniquely develop in function from other, related receptors.

Post-translational modifications of the receptors

The single polypeptide chains of each of the G protein-linked receptors aligned in Fig. 7.1 are predicted to span the plasma membrane seven times, resulting in four extracellular hydrophilic domains and four cytoplasmic hydrophilic domains (Fig. 7.2). Using truncated forms of the structurally related receptor bovine rhodopsin, Friedlander and Blobel (1985) demonstrated that integration of a polytopic integral membrane protein with multiple TMS domains may require multiple signal sequences. These investigators localized two of these sequences in bovine rhodopsin, one in the first TMS segment and one in the amino terminal portion of the sixth TMS region. The presence of at least two signal sequences in human β_2AR was confirmed by Kobilka *et al.* (1988) in a remarkable way. They constructed a human β_2AR mutant (SR(1-5); Fig. 7.3) by inserting a termination codon after Ser 262 in the third cytoplasmic loop, and a β_2AR mutant (SR(6-7)) by deleting the region between Gly 2 and Ser 262 (Fig.

7.3) (Kobilka *et al.*, 1988). It was possible to express SR(1-5) and SR(6-7), each functionally inactive, together in *Xenopus laevis* oocytes and obtain a functional receptor capable of ligand binding and of mediating activation of adenylyl cyclase. Thus, each portion of this 'split' receptor must contain sequences independently capable of directing polypeptide integration into the plasma membrane.

Glycosylation

Each of the adrenergic receptors is glycosylated and preliminary structural determinations indicate the presence of N-linked complex oligosaccharides (Terman & Insel, 1988; Convents *et al.*, 1988; Benovic *et al.*, 1987; Stiles *et al.*, 1984) and high mannose and complex oligosaccharides on the hamster β_2AR (Benovic *et al.*, 1987). Several lines of evidence indicate that of the four putative glycosylation sites in the hamster β_2AR (two in the N-terminus and two in the carboxyl tail; Dixon *et al.*, 1986), only those in the amino terminus (Asn 6 and Asn 15) are utilized. First, carboxypeptidase Y treatment of hamster β_2AR reconstituted in phospholipid vesicles removes the two sites in the carboxyl tail but the truncated product remains sensitive to Endo F (Dohlman *et al.*, 1987). Second, protein sequencing of hamster β_2AR residues 2–17 reveals two blanks in the amino acid sequence corresponding to the two Asn residues, presumably due to N-linked glycosylation (Dixon *et al.*, 1986). Finally, removal of Asn 6 and Asn 15 in the hamster β_2AR (by deleting residues 6–15), followed by expression in COS-7 cells, results in a deglycosylated form of the receptor (Dixon *et al.*, 1987).

The function(s) of N-linked glycosylation in the adrenergic receptors is presently unclear. Reports by Benovic *et al.* (1987) and George *et al.* (1986) indicate that normal biological features of the β_2AR (e.g. ligand binding and mediation of agonist-stimulated adenylyl cyclase activity) are still observed when the receptors are deglycosylated enzymatically. More recently, however, functional tests of deglycosylated β_2AR from a human epidermoid carcinoma cell line (A431) in a reconstituted system show these receptors to have lost their ability to couple normally to their signal transducing molecules (i.e. G proteins) (Boege *et al.*, 1988). The observation that the extent of glycosylation on the extracellular surface of the receptor may mediate the degree of agonist-induced adenylyl cyclase stimulation is surprising in light of the fact that receptors:G protein coupling occurs on the cytoplasmic and/or TMS aspects of the receptor. However, given that receptor:G protein interactions can be dependent upon translational diffusion rate(s) (TD) of either or both proteins (Wier & Edidin, 1988), increasing the TD of β_2AR by deglycosylation could affect the efficiency of interaction with G proteins and, hence, adenylyl cyclase activity.

Terman and Insel (1988) have recently demonstrated a role for carbohydrates in the normal functioning of ARs. These investigators observed that, not only is glycosylation of α_1ARs important for function, but that the type of attached oligosaccharide also plays a role in the functioning and proper cellular localization of the receptors. By interfering with the conversion of high mannose oligosaccharides to complex oligosaccharides on the α_1AR, they observed a 40% reduction in agonist-stimulated phosphatidylinositol turnover when compared with the normally glycosylated receptor. In addition, the inhibition of glycoprotein processing both increased the number of intracellular sequestered receptors and decreased the number of functional cell surface receptors.

Disulphide bond formation in β_2AR

Dohlman *et al.* (1991) and Moxham *et al.* (1988) have demonstrated a reduction-dependent change in the mobility of the hamster β_2AR on SDS-polyacrylamide gels. Such a mobility shift provides indirect evidence for the existence of intramolecular disulphide bonding. Moreover, Dohlman *et al.* (1991) have also demonstrated that treatment of denatured hamster β_2AR with dithiothreitol leads to a marked ($\gtrsim$ 15-fold) increase in [^{14}C]-iodoacetamide alkylation of the receptor compared with alkylation performed in the absence of reducing agent. The deduced amino acid sequence of hamster β_2AR reveals the presence of 15 cysteine residues. Thus, these investigators concluded that as many as seven disulphide bonds may exist in the mature hamster β_2AR (Dohlman *et al.*, 1991).

Palmitoylation of the human β_2AR

Alignment of amino acid sequences in the N-terminal portions of the carboxyl tails of the G protein-linked receptors (Fig. 7.1) reveals the presence of a conserved cysteine residue (Cys 341 or β_2AR). In a recent study by Ovchinnikov *et al.* (1988), the palmitoylation of vicinal cysteines (Cys 322, 323) at the equivalent position in bovine rhodopsin was reported. In order to determine whether the human β_2AR is also post-translationally modified by fatty acylation, transfected mammalian cells expressing wild-type human β_2AR and a glycine-substituted mutant β_2AR (S341) were incubated with [³H]-palmitic acid. Following receptor solubilization and purification, fluorography of SDS-polyacrylamide gels revealed that wild-type, but not S341, receptor was [³H]-labelled (Fig. 7.3). Subsequent identification of the [³H]-label bound to the β_2AR as palmitic acid was deduced by TLC analysis of the [³H]-radioactivity released upon exposure to KOH (O'Dowd *et al.*, 1989*a,b*). In light of the findings of Dohlman *et al.*, (1991) that as many as seven disulphide bonds are formed within the β_2AR, accounting for 14 of the 15 cysteine residues present in the receptor, it appears likely that the remaining residue (i.e. Cys 341) is thioesterified with palmitic acid.

In the topographical model of human β_2AR, the carboxyl tail is normally shown as being exposed to the cytoplasm. The presence of palmitic acid linked to Cys 341 could, as proposed by Ovchinnikov *et al.* for bovine rhodopsin (1988), promote the association of the N-terminal portion of the carboxyl tail with the plasma membrane, thus forming a fourth intracellular loop (Fig. 7.2). The fact that a cysteine residue equivalent in position to Cys 341 in human β_2AR is conserved in 12 of the 13 receptors aligned in Fig. 7.1, suggests that palmitoylation may represent a general feature of post-translational processing in the family of G protein-linked receptors. Although the human kidney α_2AR (α_2AR-C4) subtype does not contain an equivalent cysteine residue at this position, it does possess one in its third cytoplasmic loop. It will be of interest to determine if palmitoylation of α_2AR-C4 occurs, or if this form of covalent modification is nonessential in some members of the G protein-linked family of receptors.

Arrangement of the seven TMS segments in the plasma membrane

Kobilka *et al.* (1988) have used a series of chimeric β_2AR : α_2ARs to predict the structural arrangement of the seven TMS domains in the plasma membrane. Their model is based on data pertaining to the functional ability of four chimeric receptors (CR2, CR6, CR7 and CR8) to bind α- and β-adrenergic ligands, and mediate agonist-stimulated adenylyl cyclase activity (Fig. 7.4). CR8 was shown to be more efficient at mediating agonist-activation of adenylyl cyclase than was CR2 or CR6. However, CR7 was non-functional even though, like CR8, it contains elements of β_2AR necessary to activate G_s (Kobilka *et al.*, 1988). The model proposes that TMS-7 lies adjacent to TMS-3 and TMS-4. The most stable of the chimeric receptors and, therefore, the receptors most likely to mediate

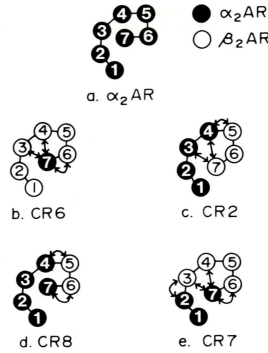

Fig. 7.4. Model of the seven TMS for α_2AR and various chimeric receptors. The TMS are shown in black circles with white numbers while those from β_2AR are shown as white circles with black numbers. Possible destabilizing forces between α-helices are indicated by arrows. The model attempts to explain the experimental observation that CR8 d, functions better than CR6 b, CR2 c, or CR7 e, as assessed by ligand binding and activation of adenylyl cyclase (Kobilka *et al.*, 1988).

adenylyl cyclase stimulation, have TMS-7 of α_2AR with TMS-3 and TMS-4 of α_2AR. Chimeric receptors such as CR6, CR2 and CR7 are apparently destabilized due to unfavourable interactions involving TMS-3, TMS-4 and TMS-7.

Phosphorylation

Perhaps the best characterized form of post-translational modification among the G protein-coupled receptors is phosphorylation. *In vitro* phosphorylation of β_1AR (Stadel *et al.*, 1983), β_2AR (Benovic *et al.*, 1985), α_1AR (Bouvier *et al.*, 1987), α_2AR (Benovic *et al.*, 1987), M_2AChR (Kwatra *et al.*, 1989) and rhodopsin (Palczewski, McDowell & Hargrave, 1988) and *in vivo* phosphorylation of α_1AR (Leeb-Lundberg *et al.*, 1987),

β_1AR (Sibley *et al.*, 1984), β_2AR (Bouvier *et al.*, 1988), M_2AChR (Kwatra *et al.*, 1987) and rhodopsin (Kuhn, 1974) has been demonstrated. In most cases, a strong correlation between receptor phosphorylation and receptor desensitization has been observed, leading to the hypothesis that a causal relationship exists between these two phenomena (Benovic *et al.*, 1988).

To date, at least four kinases have been shown to phosphorylate G protein-coupled receptors: cAMP-dependent protein kinase (PKA) (Bouvier *et al.*, 1987; Benovic *et al.*, 1985), Ca^{2+}/phospholipid-dependent protein kinase (PKC) (Bouvier *et al.*, 1987), β-adrenergic receptor kinase (βARK) (Benovic *et al.*, 1986) and rhodopsin kinase (Palczewski, McDowell & Hargrave, 1988). Strict consensus sequences are difficult to assign for each kinase, especially for PKC and βARK. Moreover, even a 'preferred substrate' need not be utilized under all physiological conditions if the recognition site resides within an unfavourable chemical or structural environment. However, for the purpose of discussion, sequences such as R/K-R/K-X-S/T or R/K-R/K-X-X-S/T are considered suitable for recognition by PKA (Feramisco, Glass & Krebs, 1980); R/K-R/K-S/T, R/K-S/T-R/K or S/T-R/K-R/K by PKC (House, Wettenhall & Kemp, 1987); and S/T residues in the vicinity of acidic amino acids are treated as being susceptible to phosphorylation by βARK (Benovic *et al.*, 1989).

Inspection of the primary structures of the receptors aligned in Fig. 7.1 reveals that putative PKC/PKA recognition sites are common to each. These sites are generally restricted to the C-terminal portion of their second cytoplasmic loops (joining transmembrane-spanning regions (TM) III and IV), and the N- and C-terminal sections of intracellular loop III (joining TM-V and TM-VI). These three areas are characterized by having a predominance of basic amino acid residues. Putative βARK sites, however, appear to be confined to either the middle portion of intracellular loop III or the carboxyl tails in any given receptor, but not both. These latter two regions characteristically gain more acidic character with increased distance from the plasma membrane and generally share a similar chemical composition. This raises the possibility that a single genetic element encoding amino acid sequence for a 'βARK-phosphorylation domain' may have arisen in one of two possible loci in the coding blocks of various G protein-linked receptors, forming *either* a large middle section of intracellular loop III or a long carboxyl tail. Independent of its location in a particular receptor, however, this domain appears to be large enough to spatially extend across the entire cytoplasmic surface of the

β₂-ADRENERGIC RECEPTOR-LIKE   **α₂-ADRENERGIC RECEPTOR-LIKE**

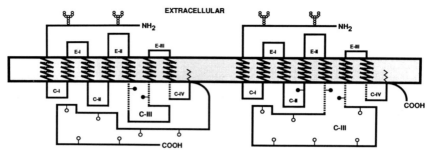

Fig. 7.5. Comparison of structural features of β_2AR-like and α_2AR-like receptors. β_2AR-like molecules are characterized by long carboxyl tails, relatively short third cytoplasmic loops (C-III) and the absence of putative PKA/PKC phosphorylation sites in their C-IIs. α_2AR-like receptors have short carboxyl tails, large C-IIIs and putative PKA/PKC sites in C-II. Putative phosphorylation sites are indicated by ball and stick symbols: filled = PKA/PKC; open = βARK. Dashed lines at the N- and the C-terminal extremes of C-III and the N-terminal segment of C-IV represent regions thought to be involved in receptor:G protein coupling. The jagged line at the C-terminal end of C-IV represents the putative point of attachment of palmitate (or some other fatty acid) to G protein-coupled receptors.

molecule (Fig. 7.5) and may allow for identical or analogous interactions with various other functionally important intracellular domains.

As indicated above, putative PKC/PKA phosphorylation sites are found either at the N- or C-terminal extremes of intracellular loop III, or both. These regions are those which, by site-directed mutagenesis studies, have been directly implicated in receptor:G-protein coupling (O'Dowd *et al.*, 1988; Kobilka *et al.*, 1988; Strader *et al.*, 1987*a,b*). The presence of phosphate groups within or near the β_2AR:G$_s$ binding site could reduce the affinity and/or productivity of this interaction, and suggests a mechanism by which certain forms of receptor desensitization can arise.

Based upon the above observations, the functional domains of G protein-linked receptors may be organized as shown in Fig. 7.5. The N- and C-terminal extremes of the third cytoplasmic loop appear to comprise the main binding/coupling domain for G proteins (see below). Embedded within the principal G protein binding region are sites for regulatory phosphorylation by PKC/PKA that serve to disrupt normal coupling between receptor and G protein. In the case of receptors with a 'βARK-phosphorylation domain' in their third cytoplasmic loops, the C-termini of their second intracellular loops bear putative sites for PKA/PKC regulatory phosphorylation. Conversely, the latter are absent in those receptors whose βARK domain constitutes the carboxyl tail.

Receptor: G protein coupling

Since the successful cloning and sequencing of the main subtypes of ARs has been accomplished, a major goal of subsequent research has been to delineate the structural domain(s) that directly determine G protein recognition and coupling ability. Most interest to date has focused on the third intracellular loop of $β_2$AR as comprising the region principally responsible for interaction with G_s.

An $α_2$AR:$β_2$AR chimeric receptor study by Kobilka *et al.* (1988) showed that insertion of the entire third cytoplasmic loop and some flanking sequence of the fifth and sixth TMS of $β_2$AR into the human platelet $α_2$AR allowed the chimeric receptor to couple to G_s and consequently mediate stimulation of adenylyl cyclase. Although the level of stimulation achieved did not match that of native $β_2$AR, the results clearly pointed to the important involvement of amino acid sequences in this portion of the receptor for productive receptor:G_s coupling.

Recently, mutagenesis studies by O'Dowd *et al.* (1988) and Hnatowich *et al.* (unpublished observations) have pinpointed individual amino acid residues at the N- and C-terminal boundaries of the third cytoplasmic loop of human $β_2$AR as likely being those responsible for the impairment of receptor:G_s coupling associated with larger modifications of the molecule (Strader *et al.*, 1987*a*). We have also demonstrated that palmitoylation of Cys 341 in the N-terminal portion of the cytoplasmic tail of $β_2$AR is important in maintaining a functional G_s binding site (O'Dowd *et al.*, 1989*a,b*).

A series of mutant $β_2$ARs incorporating substitutions with platelet $α_2$AR sequence in the N-terminal third of cytoplasmic loop III were impaired to various extents but were still capable of productive coupling with G_s (O'Dowd, *et al.*, 1988; O'Dowd *et al.*, 1991) (Fig. 7.6). The conservation of basic amino acids in this region in all receptors sequences to date (Fig. 7.1) suggests that the distribution of charged residues may play an important role in coupling. Substitution of two of these basic residues in the $β_2$AR sequence 226–228 (Ala–Lys–Arg) with unrelated sequence (Gly–Ala–Gly), resulted in virtually complete abolition of coupling ability. Thus, we presently believe that the role of the N-terminus of the third intracellular loop is to bind common G protein elements and not to determine the specificity of which G protein may bind.

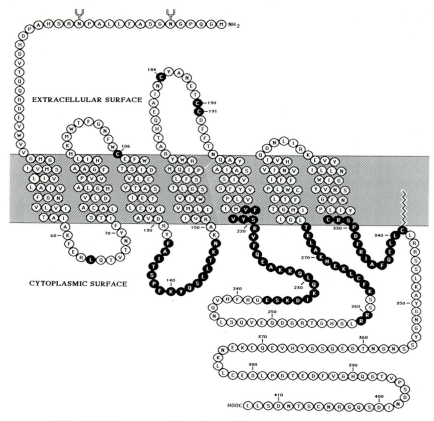

Fig. 7.6. Summary of regions of β_2AR examined by site-directed mutagenesis. Human β_2AR in the membrane. Black circles indicate amino acids either substituted with the corresponding sequences found in the human platelet α_2AR or deleted (see text).

Conservation of amino acid residues at the C-terminus of the third cytoplasmic loop of these receptors suggested to us and others (Strader *et al.*, 1987*a*) that this region may also be involved in receptor:G protein coupling. In our mutagenesis experiments, substitution of amino acid residues in β_2AR, 263–274, 267–274 and 271–274 with sequence from platelet α_2AR all led to profound impairment of the mutant receptor's ability to couple to G_s (O'Dowd *et al.*, 1991). Thus, some of the amino acid sequence that confers G protein binding specificity on β_2AR may reside in this region and specifically within residues 271–274.

Charge distribution in the C-terminus of third cytoplasmic loops is well conserved among G protein-coupled receptors (Fig. 7.1) and may be

β₂AR SPECIES	COUPLING IMPAIRMENT	C-TERMINUS OF INTRACELLULAR LOOP III	TMS-VI
		252 267	
WILD TYPE	0%	- - G R T G H G L R R S S K F C L K E │H K│A L│	K T - -
D263-273	~15%	- - - - - - - - - - - - - - G R T G H G L │R R│S │	S T - -
D263-273 + S259-260	~100%	- - - - - - - - - - - - - G R T G H G L G A S │	S T - -
D267-273	~50%	- - - - - - - - G R T G H G L │R R│S S K F C │	L T - -
D258-270	~85%	- - - - - - - - - - - - - - G R T G H G A L │	K T - -
D238-272	~85%	- - - - - - - - - - - - - - - G R T G H G │	A T - -
S263-274	~90%	- - G R T G H G L R R S S G G Q N R E │K R│F T │	F V - -
S267-274	~90%	- - G R T G H G L R R S S K F C L R E │K R│F T │	F V - -
S271-274	~90%	- - G R T G H G L R R S S K F C L K E │H K│F T │	F V - -

Fig. 7.7. Realignment of amino acid residues at the C-terminus of intracellular loop III of β₂AR following various substitution (S) and deletion (D) mutations. Percent coupling impairment refers to the decrease in the maximal level of agonist-induced adenylyl cyclase activity. Boxed letters indicate residues critical for productive βAR:Gₛ interactions. Substitutions are with α₂AR sequence at the equivalent position of β₂AR. Results with D258–270 and D238–272 are from Strader *et al.* (1987*a*).

important for determining binding to common G protein elements. For instance, disruption of the charge distribution in the equivalent region of bovine rhodopsin completely inactivated the rhodopsin:transducin coupling process (Franke *et al.*, 1988). Surprisingly, an 11-amino acid deletion of residues 263–273 of β₂AR (which included three of the conserved basic residues) produced only a modest effect on the receptor's ability to couple with Gₛ. A smaller deletion involving seven residues, 267–273, is associated with a much greater impairement (O'Dowd *et al.*, 1988). We hypothesized, as illustrated in Fig. 7.7, that realignment of Arg 259–Arg 260 in the 11-amino acid deletion mutant reconstructed the appropriate charge distribution at the C-terminus of intracellular loop III to approximately the correct extent to allow essentially normal Gₛ coupling to occur. The seven-amino acid deletion was not predicted to be consistent with such a favourable charge reorientation and, thus, was associated with marked functional impairment (O'Dowd *et al.*, 1988). The combination of an Arg 259–Arg 260 substitution to Gly–Ala, in itself innocuous (Strader *et al.*, 1987*b*), and our 11-amino acid deletion (263–273) (Fig. 7.7), resulted in complete impairment of coupling function (O'Dowd *et al.*, 1991).

We also have demonstrated involvement of the N-terminus of the carboxyl tail of β₂AR in receptor:Gₛ coupling. In human β₂AR, sub-

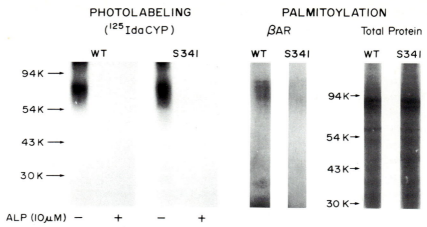

Fig. 7.8. [^{125}I]-photoaffinity labelling and [^{3}H]-palmitoylation of purified β_2AR from wild-type (WT) and S341 (Cys 341 → Gly) cell-derived membranes, and whole cell [^{3}H]-palmitoylation of ST and S341 cells. Photoaffinity labelling of membranes prepared from WT (lanes 1 and 2) and S341 (lanes 3 and 4) cells using [^{125}I]-iodocyanopindolol diaziridine [^{125}Ida Cyp] in the absence (lanes 1 and 3) and presence (lanes 2 and 4) of 10 μM alprenolol (ALP). Fluorographs of [^{3}H]-WT (lane 5) and [^{3}H]-S341 (lane 6) β_2AR following [^{3}H]-palmitate labelling, receptor purification and electrophoresis. Homogenates of total protein from WT (lane 7) and S341 (lane 8) cells following whole cell [^{3}H]-palmitoylation.

stitution of Cys 341 resulted in a non-palmitoylated receptor that fails to couple normally to G_s (O'Dowd *et al.*, 1989*a,b*). Substitution of amino acid sequence between Cys 341 and the plasma membrane with α_2AR sequence (Fig. 7.8) resulted in only modest impairment of the receptor's ability to couple to G_s. Thus, this domain which exhibits considerable homology among receptors may be principally involved in binding common G protein elements or in helping position other regions of the receptor that do so directly.

Ligand binding

In addition to localizing the functional domains responsible for coupling to G proteins and for regulation by covalent modification, efforts from several laboratories have been directed towards defining the ligand binding domain of G protein-linked receptors. To date, both biochemical and molecular biological techniques have been employed with some measure of success to identify specific amino acid residues that directly participate in the ligand binding event. However, as with attempts at delineating the G protein binding/coupling site, distinguishing between those amino acids which directly interact with ligand from those which play supportive or structural roles in forming the ligand binding pocket (or are merely in the vicinity of the latter) has been difficult.

Dixon *et al.* (1987) provided evidence that, like the retinal binding site of rhodopsin, the ligand binding pocket of the hamster β_2AR resides within the hydrophobic core of the protein near the extracellular face of the plasma membrane. Subsequent work by Strader and colleagues showed that residues Asp 113 in TM-III (Strader *et al.*, 1988) and Asn 318 in TM-VII (Strader *et al.*, 1987c) are critical determinants of ligand binding: Asp 113 appears to be the counterion for adrenergic amines, whereas Asn 318 contributes towards agonist binding specificity. Other studies have demonstrated the importance of amino acid residues in TM-II of β_2AR (Dohlman *et al.*, 1988), TM-IV of α_2AR (Matsui *et al.*, 1989), TM-IV, TM-VI and TM-VII of chimeric $\beta_1:\beta_2$AR (Frielle *et al.*, 1988), TM-VII of chimeric $\alpha_2:\beta_2$AR (Kobilka *et al.*, 1988) and TM-VII and another, undetermined site of avian β_1AR (Wong *et al.*, 1988). At present, TM-III, TM-IV and TM-VII appear to be the receptor regions most consistently implicated as direct contributors in ligand binding processes. However, the nature of participation of other membrane domains in these events remains to be elucidated.

References

Applebury, M. L. & Hargrave, P. A. (1986). Molecular biology of the visual pigments. *Vision Research*, **26**, 1881–95.

Benovic, J. L., Pike, L. J., Cerione, R. A., Staniszewski, C., Yoshimasa, T., Codina, J., Caron, M. G. & Lefkowitz, R. J. (1985). Phosphorylation of the mammalian β-adrenergic receptor by cyclic AMP-dependent protein kinase. *Journal of Biological Chemistry*, **260**, 7094–101.

Benovic, J. L., Mayor, F., Jr., Somers, R. L., Caron, M. G. & Lefkowitz, R. J. (1986). Light-dependent phosphorylation of rhodopsin by β-adrenergic receptor kinase. *Nature, London*, **322**, 869–72.

Benovic, J. L., Regan, J. W., Matsui, H., Mayor, F., Jr., Cotecchia, S., Leeb-Lundberg, L. M. F., Caron, M. G. & Lefkowitz, R. J. (1987). Agonist-dependent phosphorylation of the α_2-adrenergic receptor of the β-adrenergic receptor kinase. *Journal of Biological Chemistry*, **262**, 17251–3.

Benovic, J. L., Bouvier, M., Caron, M. G. & Lefkowitz, R. J. (1988). Regulation of adenylyl cyclase-coupled β-adrenergic receptors. *Annual Review of Cell Biology* **4**, 405–28.

Benovic, J. L., Stone, W. C., Caron, M. G. & Lefkowitz, R. J. (1989). Inhibition of the β-adrenergic receptor kinase by polyanions. *Journal of Biological Chemistry*, **264**, 6707–10.

Boege, F., Ward, M., Jurss, R., Hekman, M. & Helmreich, E. (1988). Role of glycosylation for β_2-adrenoceptor function in A431 cells. *Journal of Biological Chemistry*, **263**, 9040–9.

Bouvier, M., Leeb-Lundberg, L. M. F., Benovic, J. L., Caron, M. G. & Lefkowitz, R. J. (1987). Regulation of adrenergic receptor function by phosphorylation. *Journal of Biological Chemistry*, **262**, 3106–13.

Bouvier, M., Hausdorff, W. P., De Blasi, A., O'Dowd, B. F., Kobilka, B. K., Caron, M. G. & Lefkowitz, R. J. (1988). *Nature, London*, **333**, 370–3.

Bunzow, J. R. *et al.* (1988). Cloning and expression of a rat D_2 dopamine receptor cDNA. *Nature, London*, **336**, 783–7.

Convents, A., DeBacker, J.-P., Van Driessche, E., Convents, D., Beeckmans, S. & Vauguelin, G. (1988). Glycoprotein nature of α_2-adrenergic receptors labeled with p-azido [^{3}H]clonidine in calf retina membranes. *FEBS Letters*, **234**, 480–4.

Dixon, R. A., Kobilka, B. K., Strader, D. J., Benovic, J. L., Dohlman, H. G., Frielle, T., Bolanowski, M. A., Bennett, C. D., Rands, E., Diehl, R. E., Mumford, R. A., Slater, E. E., Sigal, I. S., Caron, M. G., Lefkowitz, R. J. & Strader, C. D. (1986). Cloning of the gene and cDNA for mammalian β-adrenergic receptor and homology with rhodopsin. *Nature, London*, **321**, 75–9.

Dixon, R. A. F., Sigal, I. S., Candelore, M. R., Register, R. B., Scattergood, W., Rands, E. & Strader, C. D. (1987). Structural features required for ligand binding to the β-adrenergic receptor. *EMBO Journal*, **6**, 3269–75.

Dohlman, H. G., Bouvier, M., Benovic, J. L., Caron, M. G. & Lefkowitz, R. J. (1987). The multiple membrane spanning topography of the β_2-adrenergic receptor. *Journal of Biological Chemistry*, **262**, 14282–8.

Dohlman, H. G., Caron, M. G., Strader, C. D., Amlaiky, N. & Lefkowitz, R. J. (1988). Identification and sequence of a binding site peptide of the β_2-adrenergic receptor. *Biochemistry*, **27**, 1813–17.

Dohlman, H. G., Caron, M. G., De Blasi, A., Frielle, T., & Lefkowitz, R, J. (1991). Role of extracellular disulphide-bonded cysteines in the ligand binding function of the β_2-adrenergic receptor. *Biochemistry*, **29**, 2335–42.

Feramisco, J. R., Glass, D. B. & Krebs, E. G. (1980). Optimal spatial requirements for the location of basic residues in peptide substrates for the cyclic AMP-dependent protein kinase. *Journal of Biological Chemistry*, **255**, 4240–5.

Franke, R. R., Sakmar, T. P., Oprian, D. D. & Khorana, H. G. (1988). A single amino acid substitution in rhodopsin (Lysine 248 → Leucine) prevents activation of transducin. *Journal of Biological Chemistry*, **263**, 2119–22.

Friedlander, M. & Blobel, G. (1985). Bovine opsin has more than one signal sequence. *Nature, London*, **318**, 338–43.

Frielle, T., Daniel, K. W., Caron, M. G. & Lefkowitz, R. J. (1988). Structural basis of β-adrenergic receptor subtype specificity studied with chimeric β_1/β_2-adrenergic receptors. *Proceedings of National Academy of Sciences, USA*, **85**, 9494–8.

George, S. T., Ruoho, A. F. & Malbon, C. C. (1986). N-glycosylation in expression and function of β-adrenergic receptors. *Journal of Biological Chemistry*, **261**, 16559–64.

House, C., Wettenhall, R. E. H. & Kemp. B. E. (1987). The influence of basic residues on the substrate specificity of protein kinase C. *Journal of Biological Chemistry*, **262**, 772–7.

Kobilka, B. K., Matsui, H., Kobilka, T. S., Yang-Feng, T. L., Francke, Y., Caron, M. G., Lefkowitz, R. J. & Regan, J. W. (1987). Cloning, sequencing, and expression of the gene coding for the human platelet α_2-adrenergic receptor. *Science*, **238**, 650–65.

Kobilka, B. K., Kobilka, T. S., Daniel, K., Regan, J. W., Caron, M. G. & Lefkowitz, R. J. (1988). Chimeric α_2-, β_2-adrenergic receptors: delineation of domains involved in effector coupling and ligand binding specificity. *Science*, **240**, 1310–16.

Kuhn, H. (1974). Light-dependent phosphorylation of rhodopsin in living frogs. *Nature, London*, **250**, 588–90.

Kwatra, M. M., Leung, E., Maan, A. C., McMahon, K. K., Ptasienski, J., Green, R. D. & Hosey, M. M. (1987). Correlation of agonist-induced phosphorylation of chick heart muscarinic receptors with receptor desensitization. *Journal of Biological Chemistry*, **262**, 16314–21.

Kwatra, M. M.. Benovic, J. L., Caron, M. G., Lefkowitz, R. J. & Hosey, M. M. (1989). Phosphorylation of chick heart muscarinic cholinergic receptors by the β-adrenergic receptor kinase. *Biochemistry*, **28**, 4543–7.

Leeb-Lundberg, L. M. F., Cotecchia, S., DeBlasi, A., Caron, M. G. & Lefkowitz, R. J., (1987). Regulation of adrenergic receptor function by phosphorylation. *Journal of Biological Chemistry* **262**, 3098–105.

Matsui, H., Lefkowitz, R. J., Caron, M. G. & Regan, J. W. (1989). Localization of the fourth membrane spanning domain as a ligand binding site in the human platelet α_2-adrenergic receptor. *Biochemistry*, **28**, 4125–30.

Moxham, C. P., Ross, E. M., George, S. & Malbon, C. (1988). β-adrenergic receptors display intramolecular disulphide bridges in *in situ* analysis by immunoblotting and functional reconstitution. *Molecular Pharmacology*, **33**, 486–92.

O'Dowd, B. F., Hnatowich, M., Regan, J. W., Leader, W. M., Caron, M. G. & Lefkowitz, R. J. (1988). Site-directed mutagenesis of the human β_2-adrenergic receptor. *Journal of Biological Chemistry*, **263**, 15985–92.

O'Dowd, B. F., Hnatowich, M., Caron, M. G., Lefkowitz, E. J. & Bouvier, M. (1989*a*). Palmitoylation of the human β_2-adrenergic receptor: Mutation of Cys 341 in the carboxyl tail leads to an uncoupled, non-palmitoylated form of the receptor. *Journal of Biological Chemistry*, **264**, 7564–9.

O'Dowd, B. F., Lefkowitz, R. J. & Caron, M. G. (1989*b*). Structure of the adrenergic and related receptors. *Annual Reviews in Neuroscience*, **12**, 67–83.

O'Dowd, B. F., Hnatowich, M. & Lefkowitz, R. J. (1991). Adrenergic and related G-protein coupled receptors, structure and function. In *Encyclopedia of Human Biology*, vol. 1, pp. 81–92.

Ovchinnikov, Y., Abulaev, N. G. & Bogachuk, A. S. (1988). Two adjacent cysteine residues in the C-terminal cytoplasmic fragment of bovine rhodopsin are palmitoylated. *FEBS Letters*, **230**, 1–5.

Palczewski, K., McDowell, J. H. & Hargrave, P. A. (1988). Rhodopsin kinase: substrate specificity and factors that influence activity. *Biochemistry*, **27**, 2306–13.

Sibley, D. R., Nambi, P., Peters, J. R. & Lefkowitz, R. J. (1984). Phorbol diesters promote β-adrenergic receptor phosphorylation and adenylate cyclase desensitization in duck erythrocytes. *Biochemical and Biophysical Research Communications*, **121**, 973–9.

Stadel, J. M., Nambi, P., Shorr, R. G. L., Sawyer, D. F., Caron, M. G. & Lefkowitz, R. J. (1983). Catecholamine-induced desensitization of turkey erythrocyte adenylate cyclase is associated with phosphorylation of the β-adrenergic receptor. *Proceedings of the National Academy of Sciences, USA*, **80**, 3173–7.

Stiles, G. L., Benovic, J. L., Caron, M. G. & Lefkowitz, R. J. (1984). Mammalian β-adrenergic receptors: distinct glycoprotein populations containing high mannose or complex type carbohydrate chains. *Journal of Biological Chemistry*, **259**, 8655–63.

Strader, C. D., Dixon, R. A. F., Cheung, A. H., Candelore, M. R., Blake, A. D. & Sigal, I. S. (1987a). Mutations that uncouple the β-adrenergic receptor from G_s and increase agonist affinity. *Journal of Biological Chemistry*, **262**, 16439–43.

Strader, C. D., Sigal, I. S., Blake, A. D., Cheung, A. H., Register, R. B., Rands, E., Zemcik, B. A., Candelore, M. R. & Dixon, R. A. F. (1987b). The carboxyl terminus of the hamster β_2-adrenergic receptor expressed in mouse L cells is not required for receptor sequestration. *Cell*, **49**, 855–63.

Strader, C. D., Sigal, I. S., Register, R. B., Candelore, M. R., Rands, E. & Dixon, R. A. F. (1987c). Identification of residues required for ligand binding to the β-adrenergic receptor. *Proceedings of the National Academy of Sciences, USA*, **84**, 4384–8.

Strader, C. D., Sigal, I. S., Candelore, M. R., Rands, E., Hill, W. S. & Dixon, R. A. F. (1988). Conserved aspartic acid residues 79 and 113 of the β-adrenergic receptor have different roles in receptor function. *Journal of Biological Chemistry*, **263**, 10267–71.

Terman, B. I. & Insel, P. A. (1988). Use of 1-deoxymannogirimycin to show that complex oligosaccharides regulate cellular distribution of the α_1-adrenergic receptor glycoprotein in BC_3H_1 muscle cells. *Molecular Pharmacology*, **34**, 8–14.

Wier, M. & Edidin, M. (1988). Constraint of the translational diffusion of a membrane glycoprotein by its external domains. *Science*, **242**, 412–14.

Wong, S. K. -F., Slaughter, C., Ruoho, A. E. & Ross, E. M. (1988). The catecholamine binding site of the β-adrenergic receptor is formed by juxtaposed membrane-spanning domains. *Journal of Biological Chemistry*, **263**, 7925–8.

8

Structure and function of G proteins

R. TYLER MILLER AND
HENRY R. BOURNE

Introduction

Guanine nucleotide regulatory proteins (G proteins) couple receptors communicating with the extracellular environment to effector molecules acting inside cells. Receptors linked to G proteins are triggered by hormones or by sensory signals, including photons and odorant molecules. The effectors include ion channels as well as enzymes that regulate concentrations of second messenger molecules in the cytoplasm. Through cyclic changes in conformation, driven by GTP binding and hydrolysis, G proteins carry information from receptor to effector. This mechanism of signal transduction is ubiquitous in eukaryotes, for several reasons. G proteins mediate unidirectional flow of information, amplify signals, and transfer information with fidelity. The ability of this family of proteins to accomplish these tasks depends upon their structure. To understand G protein-mediated signaling, we will need to know what domains of G proteins contact other proteins and how the GTP-driven conformational change of a G protein alters its structure, as well as the identities, functions, and structures of receptor and effector elements that interact with the G proteins.

In recent years, answers to these questions have begun to flow from biochemical and molecular genetic investigation of the G proteins, as well as biochemical and X-ray crystallographic studies of related GTP binding proteins. Much biochemical investigation has focused on transducin, the retinal G protein that couples photorhodopsin to stimulation of cGMP phosphodiesterase (PDE), because transducin can be purified in large quantities, and on G_s, the G protein that stimulates adenylyl cyclase, because it was the first G protein to be discovered. G_s has also been the target of most genetic studies because the S49 lymphoma cell has provided a useful system for selecting mutants.

In this chapter, the structure and function of G proteins will be discussed, concentrating primarily on their α subunits. What has been learned about the function of different G proteins in cells will also be discussed; this description will be relatively brief because the subject has been thoroughly reviewed (Birnbaumer *et al.*, 1987; Gilman, 1987; Neer & Clapham, 1988; Stryer & Bourne, 1986).

G proteins are heterotrimers composed of α, β and γ polypeptide chains. Different G proteins are distinguished primarily by their α chains, because these exhibit the greatest structural diversity and are responsible for the specificity of many G protein functions, including: GTP binding and hydrolysis; conformational change in response to binding GTP; reversible and specific interactions with receptors, effectors, and $\beta\gamma$ subunits; and service as specific substrates for ADP-ribosylation by cholera and pertussis toxins. To date, 11 distinct α chain cDNAs have been identified (Bray *et al.*, 1986; Gilman, 1987; Fong *et al.*, 1988; Jones & Reed, personal communication). The α chains of G_s and transducin (α_s and α_t, respectively) are well characterized functionally and biochemically; the precise functions of several other G proteins and of their α chains are not as well defined.

The β and γ chains are less well studied than α chains. They are found in a tightly bound subunit complex that does not dissociate under physiological conditions. Two different mammalian β polypeptide chains, the products of separate genes, have been identified (Amatruda *et al.*, 1988). Molecular cloning has revealed the primary structure of one γ chain, that of retinal rod outer segments; in other cells, several other distinct forms appear to exist. $\beta\gamma$ subunits purified in association with different α chains appear to be functionally interchangeable in reconstituted systems (for reviews see Birnbaumer *et al.*, 1987; Gilman, 1987; Neer & Clapham, 1988; Stryer & Bourne, 1986).

$\beta\gamma$ plays a role in anchoring the G protein complex to the cell membrane and facilitates coupling to receptors. Although the α chains of G_s and transducin are responsible for stimulating adenylyl cyclase and cGMP PDE, respectively, recent data suggest that in some cases $\beta\gamma$ may also activate effectors, such as cardiac K^+ channels (Logothetis *et al.*, 1987) and phospholipase A_2 (Jelsema & Axelrod, 1987). Recent studies in *Saccharomyces cerevisiae* suggest that the $\beta\gamma$ subunit of a yeast G protein – and not its α chain – plays a primary role in mediating responses to mating pheromones. Null mutations in genes that encode homologs of mammalian β and γ chains (the STE4 and STE18 genes, respectively) lead to a sterile

Table 8.1. *Taxonomy of G proteins*

Protein	α Chain (kD)	Location	Toxins	Effectors
G_s	45, 52	Everywhere	CT	↑ adenylyl cyclase ↑ Ca^{2+} channels
G_{olf}	45	Olfact. epithelium	CT	↑ adenylyl cyclase
T_r	39	Retinal rods	CT, PT	↑ cGMP PDE
T_c	39	Retinal cones	(CT, PT)	↑ cGMP PDE
G_{i1}	41	Brain	PT	↑ phospholipase C
G_{i2}	40	Everywhere	PT	↑ phospholipase A_2 ↑ K^+ channels
G_{i3}	40	Everywhere	PT	↓ Ca^{2+} channels ↓ adenylyl cyclase
G_o	39	Brain	PT	
G_z	42	Brain	?	?↑ phospholipase C

phenotype; conversely, genetic deletion of the GPA1 gene, which encodes an α chain, constitutively activates the mating pathway, presumably by releasing (and thereby activating) the $\beta\gamma$ subunit (Whiteway *et al.*, 1988).

Functions of G proteins

G proteins are implicated in a signaling process on the basis of one or more of the following characteristics: (1) effects of cholera or pertussis toxins; (2) effects of hydrolysis resistant GDP or GTP analogs; (3) regulation by guanine nucleotides of the receptor's affinity for binding agonist; (4) activation of the process by AlF_4^-; and (5) stimulation of GTP hydrolysis by the agonist. These criteria distinguish G protein-mediated signaling from signal transduction by other mechanisms, but do not necessarily identify the G protein responsible for transducing a specific signal.

Table 8.1 lists the known G proteins, identified by their distinct α chains. Their signaling functions are as follows:

G_s. The activator of hormone-sensitive adenylyl cyclase was the first G protein discovered (for review see Gilman, 1987). G_s is coupled to a large number of receptors for biogenic amines (e.g. β-adrenoceptors, histamine H_2-receptors), peptide hormones (e.g. glucagon, parathyroid hormone, and many others), and lipids (e.g. certain prostaglandins). α_s, like other G protein α chains, was originally thought to activate a single effector, adenylyl cyclase. It is now clear that α_s also can regulate the frequency of opening of voltage-sensitive Ca^{2+} channels in cardiac and skeletal muscle (Brown *et al.*, 1988; Yatani *et al.*, 1987). Four distinct α chain cDNAs, the results of alternative splicing of a single gene, have been identified (Bray *et al.*, 1986; Robishaw, Smigel & Gilman, 1986). These transcripts encode proteins that differ in length by 15 or 16 amino acids; they contain from one to three serine residues, which are potential sites for phosphorylation by cAMP-dependent protein kinase (Bray *et al.*, 1986). On polyacrylamide gels these four α chains are usually resolved into two bands, corresponding to apparent sizes of 52 and 45 kD. $α_s$ is present in virtually all mammalian tissues, although the relative amounts of the two forms varies. Both the 52 and 45 kD forms of $α_s$ activate adenylyl cyclase and regulate Ca^{2+} channels. They differ, however, in their affinities for binding GDP: the 45 kD form has a 2.5-fold greater affinity for GDP (Graziano, Freissmuth, & Gilman, 1988). The physiological significance of this difference is not known.

Transducins These G proteins are expressed only in the retina, where they mediate visual transduction by coupling visual opsins to stimulation of cGMP hydrolysis. Two transducin α chains, $α_{tr}$ and $α_{tc}$, are respectively expressed in rod and cone cells (Lerea *et al.*, 1986). Their amino acid sequences are 78% identical.

G_{olf} Another recently identified α chain, $α_{olf}$, is found only in olfactory epithelium, where it probably couples odorant receptors to stimulation of adenylyl cyclase. $α_{olf}$ is 88% identical in amino acid sequence to the 45 kD form of $α_s$ (Jones & Reed, personal communication).

G_i G_i was originally thought to be a single protein, identified by virtue of its ability to mediate hormonal inhibition of adenylyl cyclase and by the fact that its interactions with receptors could be blocked by pertussis toxin-catalyzed ADP-ribosylation. Molecular cloning has revealed at least three distinct $α_i$-like proteins, ranging in size from 40 to 41 kD. These are products of separate genes, and are expressed in varying ratios in different cell types (Itoh *et al.*, 1988; Jones & Reed, 1987; Neer & Clapham, 1988).

At present, it is not clear to what extent the different $α_i$s have distinct functions. The fact that pertussis toxin blocks G_i-mediated inhibition of adenylyl cyclase suggests that other processes blocked by pertussis toxin, including stimulation of phospholipases C and A_2, as well as activation of K^+ channels may also be mediated by 'G_i' (Table 8.1) (Birnbaumer *et al.*, 1987; Gilman, 1987; Neer & Clapham, 1988). G_i-mediated signals are triggered by a number of receptors, including muscarinic acetylcholine receptors, α_2-adrenoceptors, certain opiate receptors, etc. Thus 'G$_i$' interacts with many receptors and effectors, although it is not clear whether individual α_i chains are selective for specific subsets of receptors or effectors.

Individual hormones can simultaneously regulate multiple 'G$_i$'-mediated processes such as stimulation of phospholipase C, opening of K$^+$ channels, and inhibition of adenylyl cyclase. In theory, a single 'G$_i$' could couple to a single receptor and several different effectors, or multiple 'G$_i$s' could interact with a single receptor class and each activate an individual effector. The problem of specificity of matching receptors and effectors with specific α chains will not be completely resolved by experiments in purified, reconstituted systems. Cells appear to provide organizational information not contained in the receptor-G protein–effector complexes alone. For example, platelet activating factor (PAF) and thrombin stimulate phosphoinositide hydrolysis in whole platelets, but do not affect cAMP synthesis, whereas in membrane preparations – where cytoarchitecture is presumably disrupted – the same agents inhibit adenylyl cyclase *and* stimulate phosphoinositide production (Brass, Woolkalis & Manning, 1988). In a second example, expression of different recombinant muscarinic receptors in Chinese hamster ovary (CHO) cells produce different responses to acetylcholine: inhibition of adenylyl cyclase by M2 and M3 receptor subtypes, and stimulation of phospholipase C by M1 and M4 (Peralta *et al.*, 1988). Because all four receptor subtypes seem to produce their effects by interacting with G proteins, it appears certain that different G proteins mediate inhibition of adenylyl cyclase and stimulation of phospholipase C in CHO cells.

The biochemical mechanism by which hormones inhibit adenylyl cyclase is not clear. In reconstituted adenylyl cyclase systems, $\beta\gamma$ subunits inhibit cAMP synthesis more efficiently than do α_i subunits. This finding led to the notion that subunit dissociation caused by activation of G$_i$ results in an increase in the concentration of 'free' $\beta\gamma$, which in turn slows or reverses activation of α_s (Gilman, 1987; Higashijima *et al.*, 1987c). This idea is not consistent with observations that somatostatin, acting through G$_i$ in S49 *cyc*$^-$ cells (which lack α_s altogether), can inhibit forskolin-stimulated adenylyl cyclase (Katada *et al.*, 1984). ADP-ribosylation of α_s by cholera toxin, which decreases the affinity of α_s for $\beta\gamma$, does not change the relative inhibition of adenylyl cyclase by hormones that act through α_i; this observation also suggests that physiological inhibition of the enzyme may be through a mechanism other than by $\beta\gamma$ binding to α_s (Toro, Montoya & Birnbaumer, 1987).

$G_o\alpha_o$ was first identified as a 39 kD pertussis toxin substrate from brain that copurified with α_i (for review see Gilman, 1987). This protein is the principal pertussis toxin substrate in brain, although it is present at lower concentrations in many other tissues. α_o is distinguished from the three α_i chains by its size, by its inability to inhibit adenylyl cyclase, and by the location of its gene on a different chromosome (Birnbaumer *et al.*, 1987; Neer *et al.*, 1987). Many of the problems in assigning specific functions to the different species of α_i also apply to α_o. Nevertheless, the potential receptors that activate α_o include opiate, muscarinic, and neuropeptide Y receptors (Gilman, 1987; Ewald, Sternweis & Miller, 1988). To date, potential effector functions include activation of phospholipase C and inhibition of neuronal Ca^{2+} channels (Gilman, 1987; Ewald *et al.*, 1988).

$G_z(G_x, G_p)$ G protein-mediated signaling processes that are insensitive to pertussis toxin, including activation of phospholipase C and inhibition of K^+ channels, have been described in several tissues (Table 8.1) (Birnbaumer *et al.*, 1987; Gilman, 1987; Nakajima, Nakajima & Inoue, 1988). The hormones that activate these processes include angiotensin II, α_1-adreno-ceptor agonists, bradykinin, PAF, PGE_2, substance P, thrombin, bombesin, vasopressin, acetylcholine, and thyrotropin releasing hormone (Baukal *et al.*, 1988; Birnbaumer *et al.*, 1987; Fischer & Schonbrunn, 1988; Fong *et al.*, 1988; Negishi *et al.*, 1987).

A candidate for the α chain of a putative pertussis toxin-insensitive G protein was recently identified by cDNA cloning (Fong *et al.*, 1988). Its message is most abundant in nervous tissue, but detectable in peripheral tissues. The new α chain, called α_z, differs from previously described α chains in two respects. α_z lacks consensus sequences for ADP-ribosylation by either pertussis or cholera toxin, and therefore could mediate toxin insensitive processes. In addition, its amino acid sequence diverges from all other α chains in one of the highly conserved guanine nucleotide binding regions. α_z probably represents a new class of G protein α chain. The significance of these structural differences will become clear when the α_z protein is identified and characterized.

The activation cycle of G proteins

G proteins carry information from receptors to effectors by exchanging GTP for bound GDP and assuming a different, 'active' conformation. Fig. 8.1 illustrates the cycle of binding and hydrolysis of GTP by G protein α chains. The essential point of this cycle is that the α chain interacts with

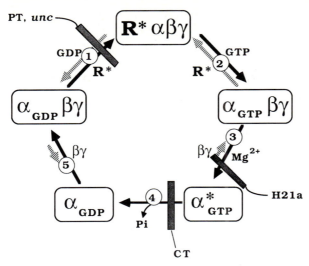

Fig. 8.1. The activation cycle of G proteins, which converts inactive α-GDP to an active form, α*-GTP. See text for details. α and βγ represent the two subunits of the G protein heterotrimer. R* represents the activated form of the receptor. The shaded bars represent points at which mutations (*unc*, H21a), or ADP-ribosylation by pertussis toxin (PT) or cholera toxin (CT) affect the cycle.

different proteins in its GDP- or GTP-bound states. In the GDP-bound conformation, the α subunit binds with high affinity to βγ, but cannot activate effectors. In contrast, the GTP-bound α chain has a low affinity for βγ, but does bind to and activate effectors. Transition of the α chain from its inactive GDP-bound form to the GTP-bound active form depends upon release of GDP, a process that is accelerated by binding to an activated receptor (R*). Once GTP is bound, βγ dissociates from α; the duration of the active state of α (α*) then depends upon the rate at which it hydrolyzes bound GTP. The irreversible hydrolysis of GTP makes the cycle unidirectional, and consequently suitable for transducing information.

In the absence of receptor stimulation, different G proteins exchange GDP for GTP at different rates. The exchange of GDP for GTP at the guanine nucleotide binding site of transducin absolutely requires interaction of transducin with the activated receptor, R* (in this case photorhodopsin). The physiological basis of this strict requirement is clear: in the retina, GDP–GTP exchange in the absence of light-activated rhodopsin would result in apparent detection of a spurious photon. Tight binding of GDP to transducin in the absence of activated rhodopsin ensures that guanine nucleotide exchange does not occur in the dark

(Stryer, 1987). The spontaneous rate of guanine nucleotide exchange is vanishingly small, as indicated by the fact that hydrolysis resistant GTP analogs do not activate transducin in dark adapted rod outer segments. The requirement for R* is not nearly so strict for G_s, because the adenylyl cyclase system can tolerate more 'noise'. In this case, spontaneous GDP–GTP exchange contributes to a basal rate of cAMP synthesis and allows hydrolysis resistant GTP analogs to activate α_s in a time-dependent fashion (Gilman, 1987). Nevertheless, *in vivo*, receptor activation of α_s is required for significant elevations of cAMP.

Let us consider the steps in the α chain activation cycle in more detail (Fig. 8.1). In its inactive state, α-GDP binds tightly to $\beta\gamma$. When $\alpha\beta\gamma$-GDP encounters R* (reaction 1 in the figure), GDP is released; $\alpha\beta\gamma$ then binds R* more tightly. This reaction is reversible; on rebinding GDP the complex will dissociate into its original components. In the absence of guanine nucleotides, however, the $\alpha\beta\gamma$-R* complex is stable and causes R to bind agonists with high affinity (Gilman, 1987). This property can be used to document G protein–receptor interaction (e.g. Bourne et al., 1981) and to copurify receptor G-protein complexes (Couvineau, Amiranoff & Laburthe, 1986). ADP-ribosylation by pertussis toxin (of α_i, α_o, and α_t) (Gilman, 1987; Stryer & Bourne, 1986) and the *unc* mutation (in α_s) (Rall & Harris, 1987; Sullivan et al., 1987) are modifications of the α subunit that block reaction 1 by preventing the $\alpha\beta\gamma$-GDP complex from interacting with R*.

In reaction 2, GTP binds to $\alpha\beta\gamma$-R*, leading to dissociation of $\alpha\beta\gamma$-GTP from the receptor. *In vivo*, binding of GTP (reaction 2) predominates over binding of GDP (reaction 1) because of the excess of GTP over GDP in the cell. R* promotes exchange of GDP for GTP catalytically; one R* can activate many G proteins (Birnbaumer et al., 1987; Gilman, 1987; Stryer & Bourne, 1986). This feature allows amplification of signals.

The conformational change induced by GDP or GTP that causes dissociation of $\alpha\beta\gamma$ from R* is distinct from the GTP-induced conformational change that leads to activation of the effector (reaction 3) (Bourne et al., 1988; Miller et al., 1988). Reaction 3 requires the presence of Mg^{2+}; α-GTP undergoes a conformational change and dissociates from $\beta\gamma$. α*-GTP is then able to activate its effector (adenylyl cyclase, cGMP, PDE, K^+ channels, etc.). This reaction is reversible by the addition of excess $\beta\gamma$ (Gilman, 1987; Higashijima, 1987c).

In reaction 4, the α* chain hydrolyses bound GTP to GDP, reverts to its inactive conformation, and dissociates from the effector. This reaction is irreversible; the α subunit is unable to synthesize GTP from GDP and inorganic phosphate (Stryer, 1987). Irreversibility of this reaction makes G protein-mediated signaling unidirectional. Varying the rate of GTP hydrolysis can amplify signals. Thus, ADP-ribosylation of α_s by cholera toxin decreases the rate of GTP hydrolysis, leading to increased cAMP synthesis (Cassel *et al.*, 1979). Finally, in reaction 5, α-GDP reassociates with $\beta\gamma$, and $\alpha\beta\gamma$-GDP is ready to begin the cycle again.

Structure and function of the guanine nucleotide binding domain

The guanine nucleotide binding domain of the G protein α chain is the key element of the molecular machine that drives the activation cycle. Three lines of evidence indicate that binding of GTP induces a true conformational change in the α chain: (1) Trypsin cleaves the GDP-bound forms of α_t, α_o, and α_s at a characteristic site (corresponding to Arg 204 of α_{tr}), while binding of GTP analogs protects each of these proteins from tryptic cleavage (Bigay *et al.*, 1987; Fung & Nash, 1983; Hurley *et al.*, 1984; Miller *et al.*, 1988). (2) Intrinsic fluorescence of purified α_s, α_i, and α_o changes in response to GTP binding and hydrolysis (Higashijima *et al.*, 1987*a,b*). (3) The ability of sulfhydryl-reactive reagents to label cysteine residues in α_t and α_o depends on which guanine nucleotide, GDP or GTP, is bound (Ho & Fung, 1984; Winslow *et al.*, 1987).

Understanding the details of how G proteins work will also require knowledge of their three dimensional structure in the GDP- and GTP-bound forms, and the structure of complexes with other proteins. Although G protein crystals are not yet available, insights from the solved crystal structures of p21^{ras} and EF-Tu have been valuable (DeVos *et al.*, 1988; Jurnak, 1985; LaCour *et al.*, 1985). The authors' laboratory has reported (Masters, Stroud & Bourne, 1986) a model of G protein α chain structure based upon similarities in stretches of amino acid sequence between parts of the guanine nucleotide binding domains of p21^{ras} and Ef-Tu, on the one hand, and the presumptively cognate domain of G protein α chains.

The crystal structures of p21^{ras} and EF-Tu indicate that, in these proteins, four regions of conserved amino acid sequence occur at turns between β sheets and α helices in the guanine nucleotide binding domain (DeVos *et al.*, 1988; Jurnak, 1985; Kim *et al.*, 1988; LaCour *et al.*, 1985). Although some details of the two guanine nucleotide binding sites differ, the overall structures are quite similar. The G protein α chains share with EF-Tu and p21^{ras} these regions of conserved sequence, which were therefore predicted to form the guanine nucleotide binding site in G protein α chains (Masters *et al.*, 1986). Fig. 8.2 shows the location of these four stretches of sequence in a composite G protein α chain. Masters *et al.* (1986) and Kim *et al.* (1988) predicted that stretches of α_s including amino acids 45–52 (corresponding to the loop made up of residues 10–15 in

Sequence comparison of seven G protein α chains

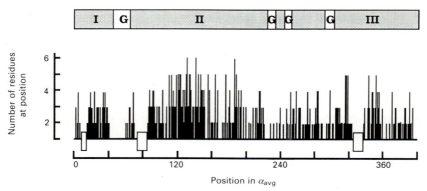

Fig. 8.2. Conservation of primary structure and domains of seven G protein α chains (α_s, α_{i1-3}, α_o, α_{tr}, α_{tc} (Gilman, 1987; Jones & Reed, 1987). The horizontal bar at the top represents a linear sequence of a composite G protein α chain, α_{avg} (Masters *et al.*, 1986). 'G' indicates regions predicted to participate in forming the guanine nucleotide binding site, and Roman numerals I–III indicate the domains into which the remainder of the protein is divided. The graph at the bottom depicts the number of different residues that occupy a given position in seven different α chains; numbers on the bottom line refer to residue positions in α_{avg}. The open spaces on the abscissa mark regions of α_s that do not correspond to parts of the other six α chains. The second open box represents the peptide sequence present in the 52 kD forms of α_s (Bray *et al.*, 1986; Robishaw *et al.*, 1986).

$p21^{ras}$) and 222–228 (residues 57–63 in $p21^{ras}$) should be located near the β and γ phosphoryls of GTP and should also interact with one another. On the basis of the crystal structure of $p21^{ras}$ and biochemical changes produced by mutational substitution of amino acids in the same protein, these regions are thought to be involved in hydrolysis of GTP and in mediating conformational change induced by binding GTP. Functional phenotypes produced by three mutations located in these regions of α_s strongly confirm these predictions and support the notion that all of these proteins are different versions of the same basic molecular machine.

The first of these α_s mutations, in the S49 murine lymphoma cell line called H21a, was selected by virtue of resistance of H21a cells to killing by cholera toxin (Bourne *et al.*, 1981). The α_s gene in H21a carries a point mutation that causes replacement of Gly 226 by alanine (termed G226A, using the single-letter amino acid code) (Miller *et al.*, 1988). This mutation specifically prevents the GTP-induced conformational change (reaction 3, Fig. 8.1) that leads to βγ dissociation and activation of adenylyl cyclase. The mutant G_s couples normally to receptors, as indicated by its ability to

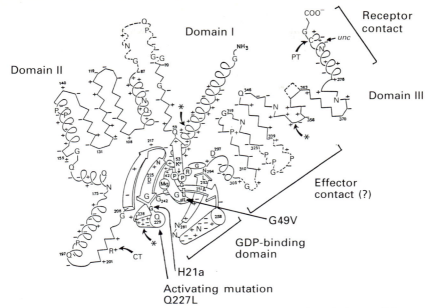

Fig. 8.3. Structural model of a composite G protein α chain, α_{avg} (Masters *et al.*, 1986). In the center is shown the predicted three-dimensional structure of the guanine nucleotide binding domain, based on the solved crystal structure of EF-Tu (Jurnak, 1985). The rest of the protein is shown as predicted secondary structure (α helices as spirals and β strands as zigzags). Dashed lines represent sequences which are unique to α_s. Sites of mutations (*unc*, H21a, G49V, and Q227L [Q227 in α_s corresponds to Q229 in α_{avg}]), covalent modifications (sites ADP-ribosylated by pertussis (PT) or cholera toxins (CT), and predicted functional domains (guanine nucleotide-, effector-, and receptor-binding domains) are marked. Asterisks mark tryptic cleavage sites; cleavage at the site just downstream from residue Q229 is prevented by binding GTP. Residues that are specifically identified are conserved among different α chains.

confer on β-adrenoceptors high affinity for binding agonists, and can bind GTP and GTP analogs, as well as GDP (Bourne *et al.*, 1981; Miller *et al.*, 1988).

In the predicted structure of a composite G protein α chain (see Fig. 8.3), Gly 226 of α_s is presumably located in a conserved region of the GTP binding site, at a turn between a β sheet and an α helix. By introducing a methyl side chain, substitution of alanine for glycine at this position probably prevents normal movement of this portion of α_s in response to GTP binding. The region containing the mutation is unlikely to be directly involved in specific activation of distinct effector molecules by different α chains, because its amino acid sequence is virtually identical in all of the G proteins. Rather, the region probably serves in some fashion to communicate to other parts of the protein the information that GTP has been bound, resulting in $\beta\gamma$ dissociation and ability to activate the effector.

Two other mutations were made in α_s at sites equivalent to codons 12 and 61 in p21^{ras}, and the mutant α chains expressed in cyc^- cells, which are genetically deficient in α_s. Replacement of Gln 227 by leucine (Q227L) produces a constitutively activated α_s (Masters *et al.*, unpublished observations). This mutation was made because the same mutational replacement at the equivalent position (Gln 61) also activates p21^{ras}, leading to increased transforming activity (Barbacid, 1987). In the Q227L α_s mutant, GTP alone stimulates adenylyl cyclase almost as well as does the combination of isoproterenol (a β-adrenoceptor agonist) and GTP in wild-type cells. The mechanisms by which the α_sQ227L and p21^{ras}Q61L mutations activate the two proteins are not clear. The mutation in p21^{ras} reduces its GTPase activity (Barbacid, 1987), and an indirect measure of GTP hydrolysis by the α_sQ227L protein suggests a similar inhibition of its GTPase (Masters *et al.*, unpublished observations). In addition (or alternatively), the amino acid substitution could alter the structure of the guanine nucleotide binding domain to mimic the GTP-bound conformation.

A third mutation in α_s, again based on analogy to p21^{ras}, replaces Gly 49 by valine (G49V) (Masters, unpublished observations). In mammalian G protein α chains (with the single exception of α_z), the amino acid sequence surrounding Gly 49 is highly conserved. In p21^{ras}, the equivalent mutation (G12V) produces a constitutively active protein, at least in part by decreasing GTPase activity; this is the most common p21^{ras} mutation found in human tumors (Barbacid, 1987). Surprisingly, the G49V mutation in α_s does not result in a constitutive elevation of adenylyl cyclase. It does, however, appear to alter the GTP binding properties of the protein, and to decrease GTPase activity. GTP, GTP analogs, and isoproterenol do activate adenylyl cyclase in membranes of cyc^- cells expressing α_sG49V, but not to levels seen in wild type membranes. AlF$_4^-$ ion does not activate adenylyl cyclase in cyc^- membranes containing α_sG49V. Thus the effect of this mutation in α_s differs from its effect in p21^{ras}, although it clearly alters guanine nucleotide binding; these results suggest that this region, too, is involved in binding of guanine nucleotides.

Interactions with $\beta\gamma$

During their activation–deactivation cycle, G protein α chains reversibly associate with $\beta\gamma$. Biochemical studies of the transducin α chain indicate that both its amino and carboxy termini are involved in this interaction

(Hingorani *et al.*, 1988; Navon & Fung, 1987, 1988). In their GDP-bound forms, α subunits have high affinity for binding $\beta\gamma$, while the affinity for $\beta\gamma$ is low in the GTP-bound active conformation (Higashijima, 1987*c*). Only when bound to $\beta\gamma$ can α chains couple to receptors or serve as substrates for ADP-ribosylation by pertussis toxin. Binding of $\beta\gamma$ clearly alters the conformation of α chains, as shown by the abilities of $\beta\gamma$ to reverse GTPγS-induced activation of α_s (for review see Gilman, 1987) and to regulate the affinity of α_o for GDP (Higashijima, 1987*c*). The effect of $\beta\gamma$ on the affinity of α_o for GDP depends on the concentration of Mg^{2+} ion. At low Mg^{2+} concentrations (μM), $\beta\gamma$ increases the affinity of α_o for GDP, while at higher (mM) concentrations – which are probably more physiologically relevant – $\beta\gamma$ facilitates the rate of GDP dissociation, leading to an apparent increase in the rate of GTP hydrolysis.

Receptor binding domain

G proteins transfer information across cell membranes by alternate interactions with receptors and effectors. Understanding this process requires identification of the regions of the G proteins that provide the specificity for these interactions. The α subunit in its GTP-bound form is sufficient to activate effectors, but can only couple to receptors as a complex with $\beta\gamma$, as noted above. The assignment of specific regions of the α chain to the nucleotide binding domain roughly divides the rest of the molecule into three additional domains; these diverge in amino acid sequence (Fig. 8.2), but have similar predicted secondary structures (Fig. 8.3) (Bourne *et al.*, 1988; Masters *et al.*, 1986). These three domains presumably are responsible for binding $\beta\gamma$, receptors and effectors.

Several kinds of evidence indicate that the predicted α helix located at the α chain carboxy terminus (Fig. 8.3) is a contact site for binding receptors. ADP-ribosylation of α_i, α_o, and α_t by pertussis toxin prevents these proteins from interacting with their receptors; the site of this covalent modification in all of these proteins is a conserved cysteine four residues from the carboxy terminus (for reviews see Gilman, 1987; Birnbaumer *et al.*, 1987; Stryer & Bourne, 1986). The *unc* mutation in α_s, which substitutes proline for arginine at a position six residues from the carboxy terminus, prevents interaction with β-adrenoceptors or PGE_1-receptors, while allowing activation of cyclase in response to GTP analogs, AlF_4^-, and cholera toxin (Rall & Harris, 1987; Sullivan *et al.*, 1987). Finally, the amino acid sequence of the carboxy terminus of α_t closely resembles the sequence of an internal region of arrestin, a retinal protein that competes with α_t for binding to photorhodopsin (Wistow *et al.*, 1986).

Despite the fact that the carboxy terminus is critical for receptor coupling, it is not the only portion of the α chain that contributes to the specificity of this interaction. Comparisons of the sequences of the carboxy-terminal twenty residues in α_i, α_o, and α_t, would suggest that receptors that recognize G_i should also recognize transducin, but might couple less well to G_o. This is not the case, however. Indeed, α_2-adrenoceptors couple much better to G_i and G_o than to transducin, indicating that some other part of the α chains must be involved in specifying the interaction with receptors (Cerione *et al.*, 1985*b*; Cerione *et al.*, 1986).

Studies with chimeric G protein α chains indicate that the putative additional receptor contact region in the α chain is probably located somewhere in its carboxy terminal 40 %. The presence of this portion of α_s in two chimeric α chains is sufficient to allow specific coupling to β-adrenoceptors. The amino terminal 60 % of these two chimeras was derived from either α_{i2} or α_{tr}; expression of either chimera in *cyc*$^-$ cells produced cells in which adenylyl cyclase responded to stimulation by β-adrenoceptor agonists (Masters *et al.*, 1988, unpublished observations). In the case of the α_i/α_s chimera, there was no evidence that receptors which normally activate G_i interacted with this protein (Masters *et al.*, 1988).

Activation of G proteins by mastoparans, peptides found in wasp venom, indicates that coupling of receptors to G proteins involves charge interactions. Mastoparans, 14 residue peptides that form amphipathic α helices in lipid bilayers, contain four positive charges exposed to the aqueous phase. The regions of receptors that are thought to contact G proteins are also positively charged. Mastoparan can activate (i.e. promote GDP–GTP exchange in) several G proteins (G_i, G_o, and to a lesser extent transducin and G_s) in much the same way. Activation of G_i and G_o by mastoparan is blocked by pertussis toxin (Higashijima, 1988; Ross *et al.*, 1988), strongly suggesting that the mastoparans work in a fashion similar to that of G protein-coupled receptors.

Effector binding domain

The interactions of G proteins with effectors are less well characterized than are interactions with receptors. α_s is known to bind directly to the catalytic protein (Gilman, 1987), while α_t binds to the γ subunit of retinal cGMP PDE, relieving toxic inhibition of cGMP hydrolysis (Stryer, 1987). Interactions of G proteins with other effector molecules (e.g. K^+ channels, phospholipases) remain to be characterized biochemically.

The portion of α_s that allows specific activation of adenylyl cyclase is known to be in the carboxy-terminal 40 % of the protein, on the basis of studies of the α_{i2}/α_s and α_t/α_s chimeric proteins described above (Masters *et al.*, 1988, unpublished observations). Like normal α_s, the α_i/α_s chimera mediates β-adrenoceptor stimulation of adenylyl cyclase. Because the interaction of α_i with adenylyl cyclase is inhibitory (Katada *et al.*, 1984), this result argues that the effector binding region of α chains is in the carboxy-terminal 40 %. This argument is strengthened by the phenotype of the α_t/α_s chimera, which also causes adrenoceptor stimulation of adenylyl cyclase, although less efficiently than in wild type. Since α_t does not interact with adenylyl cyclase (Cerione *et al.*, 1985a), the portion of the α_t/α_s chimera that activates adenylyl cyclase must be derived from α_s, i.e. somewhere in its carboxy-terminal 40 %.

Conclusion

G proteins represent a widely used, versatile mechanism for transducing hormonal and sensory signals across membranes. Insights into the structure and function of these proteins have come not only from careful biochemical studies of the proteins themselves, but also from exploring similarities of their α chains to other GTP binding proteins whose three-dimensional structures are better characterized, particularly p21[ras] and EF-Tu. Recent studies of mutations in the α chain of G_s and analogies with p21[ras] and EF-Tu suggest close structural similarities among the guanine nucleotide binding domains of these proteins. Further studies, including study of the effects of site-directed mutations and phenotypes of chimeric α chains, will determine how far these analogies can be stretched.

Relatively little is known about the structure or function of G protein $\beta\gamma$ subunits. They may have effector functions of their own, and so add further complexity to G protein-mediated signal transduction mechanisms.

The problems of specificity of information transfer by G proteins, and of matching receptors and effectors with α chains, will not be completely resolved by experiments in purified reconstituted systems. Better understanding of the role of individual α chains in transmembrane signaling will require knowledge of the post-translational processing, topological targeting, and arrangement in membranes of G protein subunits, receptors, effectors, and other associated proteins.

References

Amatruda, T. W. III, Gautam, N., Fong, M. K. W., Northup, J. K. & Simon, M. I. (1988). The 35- and 36-kDa β subunits of GTP-binding regulatory proteins are products of separate genes. *Journal of Biological Chemistry*, **263**, 5008–11.
Barbacid, M. (1987). *ras* genes. *Annual Review of Biochemistry*, **56**, 779–827.

Baukal, A. J., Balla, T., Hunyady, L., Hausdorf, W., Guillemette, G. & Catt, K. J. (1988). Angiotensin II and guanine nucleotides stimulate formation of inositol 1,4,5 triphosphate and its metabolites in permeabilized adrenal glowerulosa cells. *Journal of Biological Chemistry*, **263**, 6087–92.

Bigay, J., Deterre, P., Pfister, C. & Chabre, M. (1987). Fluoride complexes of aluminum or beryllium act on G proteins as reversibly bound analogs of the γ phosphate of GTP. *EMBO Journal*, **6**, 2907–13.

Birnbaumer, L., Codina, J., Mattera, R., Yatani, A., Scherer, N., Toro, M.-J. & Brown, A. M. (1987). Signal transduction by G proteins. *Kidney International 32 Supplement* **23**, **5**, 14–37.

Bourne, H. R., Kaslow, D., Kaslow, H., Salomon, M. & Licko, V. (1981). Hormone sensitive adenylate cyclase mutant phenotype with normally regulated beta-adrenergic receptors uncoupled from catalytic adenylate cyclase. *Molecular Pharmacology*, **20**, 435–41.

Bourne, H. R., Masters, S. B., Miller, R. T., Sullivan, K. A. & Heideman, W. (1988). Mutations probe structure and function of G protein α chains. *Cold Spring Harbor Symposium on Quantitative Biology*, **53**, 221–228.

Brass, L. F., Woolkalis, M. J. & Manning, D. R. (1988). Interactions in platelets between G proteins and the agonists that stimulate phospholipase C and inhibit adenylyl cyclase. *Journal of Biological Chemistry*, **263**, 5348–55.

Bray, P., Carter, A., Simons, C., Guo, V., Puckett, C., Kamholz, J. & Spiegel, A. (1986). Human DNA clones for four species of G_{as} signal transduction protein. *Proceedings of the National Academy of Sciences, USA*, **83**, 8893–7.

Brown, A. M., Yatani, A., Imoto, Y., Kirsch, G., Hamm, H., Codina, J., Mattera, R. & Birnbaumer, L. (1988). Direct coupling of G proteins to ionic channels. *Cold Spring Harbor Symposium on Quantitative Biology*, **53**, 365–73.

Cassel, D., Eckstein, F., Lowe, M. & Selinger, Z. (1979). Determination of the turn-off reaction for hormone-activated adenylate cyclase. *Journal of Biological Chemistry*, **254**, 9835–8.

Cerione, R. A., Codina, J., Kilpatrick, B. F., Staniszewski, C., Gierschik, P., Somers, R. L., Spiegel, A. M., Birnbaumer, L., Caron, M. G. & Lefkowitz, R. L. (1985a). Transducin and the inhibitory nucleotide regulatory protein inhibit the stimulatory nucleotide regulatory protein mediated stimulation of adenylate cyclase in phospholipid vesicle systems. *Biochemistry*, **24**, 4499–503.

Cerione, R. A., Staniszewski, C., Benovic, J., Lefkowitz, R. J., Caron, M. G., Gierschik, P., Somers, R., Spiegel, A. M., Codina, J. & Birnbaumer, L. (1985b). Specificity of the functional interactions of the β adrenergic receptor and rhodopsin with guanine nucleotide regulatory proteins reconstituted into phospholipid vesicles. *Journal of Biological Chemistry*, **260**, 1493–500.

Cerione, R. A., Regan, J. W., Nakata, H., Codina, J., Benovic, J. L., Gierschik, P., Somers, R. L., Spiegel, A. M., Birnbaumer, L., Lefkowitz, R. J. & Caron, M. G. (1986). Functional reconstitution of the α_2 adrenergic receptor with guanine nucleotide regulatory proteins in phospholipid vesicles. *Journal of Biological Chemistry*, **261**, 391–409.

Codina, J., Yatani, A., Grenet, D., Brown, A. M. & Birnbaumer, L. (1987). The α subunit of the GTP binding protein G_k opens atrial potassium channels. *Science*, **236**, 442–5.

Couvineau, A., Amiranoff, B. & Laburthe, M. (1986). Solubilization of the liver vasoactive peptide receptor. *Journal of Biological Chemistry*, **261**, 14482–9.

DeVos, A. M., Tong, L., Milburn, M. V., Matios, P. M., Jancarik, J., Noguchi, S., Nishimura, S., Miura, K., Ohtsuka, E. & Kim, S.-H. (1988). Three dimensional structure of an oncogene protein: catalytic domain of c-H-*ras*p21. *Science*, **239**, 888–93.

Ewald, D. A., Sternweis, P. C. & Miller, R. J. (1988). Guanine nucleotide-binding protein G_o-induced coupling of neuropeptide Y receptors to Ca^{2+} channels in sensory neurons. *Proceedings of the National Academy of Sciences, USA*, **85**, 3633–7.

Fischer, J. B. & Schonbrunn, A. (1988). The bombesin receptor is coupled to a guanine nucleotide-binding protein which is insensitive to pertussis and cholera toxins. *Journal of Biological Chemistry*, **263**, 2808–16.

Fong, H. K. W., Yoshimoto, K. K., Eversole-Cire, P. & Simon, M. I. (1988). Identification of a GTP-binding protein α subunit that lacks an apparent ADP-ribosylation site for pertussis toxin. *Proceedings of the National Academy of Sciences, USA*, **85**, 3066–70.

Fung, B. K.-K. & Nash, C. R. (1983). Characterization of transducin from bovine retinal rod outer segments. *Journal of Biological Chemistry*, **258**, 10503–10.

Gilman, A. G. (1987). G proteins: transducers of receptor-generated signals. *Annual Review of Biochemistry*, **56**, 615–49.

Graziano, M. P., Freissmuth, M. & Gilman, A. G. (1989). Expression of $G_{s\alpha}$ in *Escherichia coli*: purification and properties of two forms of the protein. *Journal of Biological Chemistry*, **264**, 409–18.

Higashijima, T., Ferguson, K. M., Smigel, M. D. & Gilman, A. G. (1987*a*). The effect of GTP and Mg on the GTPase activity and the fluorescence properties of G_o. *Journal of Biological Chemistry*, **262**, 756–61.

Higashijima, T., Ferguson, K. M., Sternweis, P. C., Ross, E. M., Smigel, M. D. & Gilman, A. G. (1987*b*). The effect of activating ligands on intrinsic fluorescence of guanine nucleotide-binding proteins. *Journal of Biological Chemistry*, **262**, 752–6.

Higashijima, T., Ferguson, K. M., Sternweis, P. C., Smigel, M. D. & Gilman, A. G. (1987*c*). Effect of Mg^{2+} and the *βγ* subunit complex on the interactions of guanine nucleotides with G proteins. *Journal of Biological Chemistry*, **262**, 762–6.

Higashijima, T., Uzu, S., Nakajima, T. & Ross, E. M. (1988). Mastoparan, a peptide toxin from wasp venom, mimics receptors by activating GTP-binding regulatory proteins (G proteins). *Journal of Biological Chemistry*, **263**, 6491–4.

Hingorani, V. N., Tobias, D. T., Henderson, J. T. & Ho, Y.-K. (1988). Chemical cross-linking of bovine retinal transducin and cGMP phosphodiesterase. *Journal of Biological Chemistry*, **263**, 6916–26.

Ho, Y.-K. & Fung, B. K.-K. (1984). Characterization of transducin from bovine retinal rod outer segments: the role of sulfhydryl groups. *Journal of Biological Chemistry*, **259**, 6694–9.

Hurley, J. B., Simon, M. I., Teplow, D. B., Robishaw, J. D. & Gilman, A. G. (1984). Homology between signal transducing G proteins and *ras* gene products. *Science*, **226**, 860–2.

Itoh, H., Toyama, R., Kozasa, T., Tsukamoto, T., Matsuoka, M. & Kaziro, Y. (1988). Presence of three distinct molecular species of G_i protein α subunit. *Journal of Biological Chemistry*, **263**, 6656–64.

Jelsema, C. & Axelrod, J. (1987). Stimulation of phospholipase A2 activity in bovine rod outer segments by the *βγ* subunits of transducin and its inhibition by the α subunit. *Proceedings of the National Academy of Sciences, USA*, **84**, 3623–7.

Jones, D. T. & Reed, R. R. (1987). Molecular cloning of five GTP-binding protein cDNA species from rat olfactory neuroepithelium. *Journal of Biological Chemistry*, **262**, 14241–9.

Jurnak, F. (1985). Structure of the GDP domain of EF-Tu and location of the amino acids homologous to *ras* oncogene proteins. *Science*, **230**, 32–6.

Katada, T., Bokoch, G. M., Smigel, M. D., Ui, M. & Gilman, A. G. (1984). The inhibitory guanine nucleotide-binding regulatory component of adenylate cyclase. *Journal of Biological Chemistry*, **259**, 3586–95.

Kim, S.-H., DeVos, A., Tong, L., Milburn, M. V., Matias, P. M. & Jancarik, J. (1988). *Ras* oncogene proteins: three dimensional structures, functional implications, and a model for signal transducer. *Cold Spring Harbor Symposium on Quantitative Biology*, **53**, 273–81.

LaCour, T. F. M., Nyborg, J., Thirup, S. & Clark, B. F. C. (1985). Structural details of the binding of guanosine diphosphate to elongation factor Tu from *E. coli* as studied by X-ray crystallography. *EMBO Journal*, **4**, 2385–8.

Lerea, C. L., Somers, D. E., Hurley, J. B., Klock, I. B. & Bunt-Milam, A. H. (1986). Identification of specific transducin α subunits in retinal rod and cone photoreceptors. *Science*, **234**, 77–80.

Logothetis, D. E., Kurachi, Y., Galper, J., Neer, E. J. & Clapham, D. E. (1987). The β subunits of GTP binding proteins activate the muscarinic K^+ channel in heart. *Nature, London*, **325**, 321–6.

Masters, S. B., Stroud, R. M. & Bourne, H. R. (1986). Family of G protein α chains: amphipathic analysis and predicted structure of functional domains. *Protein Engineering*, **1**, 47–54.

Masters, S. B., Sullivan, K. A., Miller, R. T., Beiderman, B., Lopez, N. G., Ramachandran, J. & Bourne, H. R. (1988). Carboxy terminal domain of $G_{s\alpha}$ specifies coupling of receptors to stimulation of adenylyl cyclase. *Science*, **241**, 448–51.

Miller, R. T., Masters, S. B., Sullivan, K. A., Beiderman, B. & Bourne, H. R. (1988). A mutation that prevents GTP-dependent activation of the α chain of G_s. *Nature, London*, **334**, 712–15.

Nakajima, Y., Nakajima, S. & Inoue, M. (1988). Pertussis toxin-insensitive G protein mediates substance P-induced inhibition of potassium channels in brain neurons. *Proceedings of the National Academy of Sciences, USA*, **85**, 3643–7.

Navon, S. E. & Fung, B. K.-K. (1987). Characterization of transducin from bovine retinal rod outer segments. Participation of the amino terminal region of T_α. *Journal of Biological Chemistry*, **262**, 15746–51.

Navon, S. E. & Fung, B. K.-K. (1988). Characterization of transducin from bovine retinal rod outer segments. *Journal of Biological Chemistry*, **263**, 489–96.

Neer, E. J. & Clapham, D. E. (1988). Roles of G protein subunits in transmembrane signalling. *Nature, London*, **333**, 129–34.

Neer, E. J., Michel, T., Eddy, R., Shaws, T. & Seidman, J. G. (1987). Genes for two homologous G-protein α subunits map to different human chromosomes. *Human Genetics*, **77**, 259–62.

Negishi, M., Ito, S., Tanaka, T., Yokohama, H., Hayashi, H., Katada, T., Ui, M. & Hayaishi, O. (1987). Covalent cross linking of prostaglandin E receptor from bovine adrenal medulla with a pertussis toxin-insensitive guanine nucleotide-binding protein. *Journal of Biological Chemistry*, **262**, 12077–84.

Peralta, E. G., Ashkenazi, A., Winslow, J. W., Ramachandran, J. & Capon, D. J. (1988). Differential regulation of Pi hydrolysis and adenylyl cyclase by muscarinic receptor subtypes. *Nature, London*, **334**, 432–4.

Rall, T. & Harris, B. (1987). Identification of the lesion in the stimulatory GTP-binding protein of the uncoupled S49 lymphoma. *FEBS Letters*, **224**, 365–71.

Robishaw, J. D., Smigel, M. D. & Gilman, A. G. (1986). Molecular basis for two forms of the G protein that stimulates adenylate cyclase. *Journal of Biological Chemistry*, **261**, 9587–90.

Ross, E. M., Wong, S. K.-F., Rubenstein, R. C. & Higishijima, T. (1988). Functional domains in the β adrenergic receptor. *Cold Spring Harbor Symposium on Quantitative Biology*, **53**, 499–506.

Stryer, L. (1987). Visual transduction: Design and recurring motifs. *Chemica Scripta*, **27B**, 161–71.

Stryer, L. & Bourne, H. R. (1986). G proteins: a family of signal transducers. *Annual Review of Cell Biology*, **2**, 391–419.

Sullivan, K. A., Miller, R. T., Masters, S. B., Beiderman, B., Heideman, W. & Bourne, H. R. (1987). Identification of receptor contact site involved in receptor-G protein coupling. *Nature, London*, **330**, 758–60.

Toro, M., Montoya, E. & Birnbaumer, L. (1987). Inhibitory regulation of adenylyl cyclases. Evidence inconsistent with βγ complexes of G_i proteins mediating hormonal effects by interfering with activation of G_s. *Molecular Endocrinology*, **1**, 669–76.

Whiteway, M., Hougan, L., Dignard, D., MacKay, V. & Thomas, D. Y. (1988). Functions of the STE4 and STE18 genes in mating pheromone signal transduction in *Saccharomyces cerevisiae*. *Cold Spring Harbor Symposium on Quantitative Biology*, **53**, 585–90.

Winslow, J. W., Bradley, J. D., Smith, J. A. & Neer, E. J. (1987). Reactive sulfhydryl groups of α_{39}, a guanine nucleotide-binding protein from brain. *Journal of Biological Chemistry*, **262**, 4501–7.

Wistow, G. J., Katiel, A., Craft, C. & Shinohara, T. (1986). Sequence analysis of bovine retinal S-antigen. *FEBS Letters*, **196**, 23–8.

Yatani, A., Codina, J., Imoto, Y., Reeves, J. P., Birnbaumer, L. & Brown, A. M. (1987). A G protein directly regulates mammalian cardiac calcium channels. *Science*, **238**, 1288–91.

9

Muscarinic acetylcholine receptors

MICHAEL SCHIMERLIK

Introduction

Muscarinic acetylcholine receptor (mAcChR)-mediated responses have been demonstrated in the central and peripheral nervous system, smooth muscle, the heart, secretory glands, and several clonal cell lines. Progress in understanding the structure and mechanism of mAcChRs was initially slow due to the low density of mAcChRs in tissues available for biochemical study and the lack of adequate methodology to study coupling to physiological effector systems. A combination of biochemical, molecular, biological, pharmacological, and electrophysiological approaches has resulted in the rapid advancement of our understanding of muscarinic mechanisms at the molecular level. This chapter attempts to summarize results from many laboratories regarding mAcChR structure and signal transduction.

Physiological responses

Experiments using heart and liver preparations showed that addition of acetylcholine (AcCh) caused a decrease in the level of adenosine $3',5'$-cyclic monophosphate (cAMP) (Murad $et\ al.$, 1962) and that guanosine triphosphate (GTP) was necessary to achieve a maximal effect (Jakobs, Aktories & Schultz, 1979). Coupling of mAcChRs to inhibition of adenylyl cyclase (AC) is mediated by guanine nucleotide binding proteins (G protein) (Gilman, 1987). G proteins that regulate AC activity are heterotrimers composed of differing G_α and similar $G_{\beta\gamma}$ subunits. The hormone–receptor complex catalyzes the exchange of GDP on the G_α subunit for GTP. The binding of GTP results in the dissociation of the heterotrimer into $G_\alpha \cdot$ GTP plus $G_{\beta\gamma}$ subunits which then regulate various effector systems in the cell. Hydrolysis of GTP to GDP by the GTPase activity of the G_α subunit terminates the activation of the G protein and the subunits associate to reform the heterotrimer. The $G_\alpha \cdot$GTP subunit of the stimulatory G protein G_s activates AC, while GTP-activated G_i, the inhibitory G protein, or G_o, the 'other' G protein, inhibits the enzyme. The inhibition could arise from the liberation of excess $G_{\beta\gamma}$ subunits which would promote $G_{s\alpha}$ reassociation to give the inactive G_s heterotrimer (Katada *et al.*, 1984). Competition between $G_{i\alpha}$ and $G_{s\alpha}$ for a binding site on AC and direct binding of $G_{\beta\gamma}$ subunits to the enzyme have also been proposed (Katada, Oinuma & Ui, 1986) as potential inhibitory mechanisms. In the case of calmodulin-stimulated AC, it has been suggested that $G_{i\alpha\beta\gamma}$ competes with enzyme for binding of calcium $\cdot$ calmodulin (Katada *et al.*, 1987).

Pertussis toxin-catalyzed ADP-ribosylation of $G_{i\alpha}$ functionally uncouples the mAcChR from inhibition of AC (Kurose & Ui, 1983; Kurose *et al.*, 1983; Brown *et al.*, 1984). In 1321N1 human astrocytoma cells, mAcChR-induced attenuation of cAMP levels was not sensitive to pertussis toxin (Hughes, Martin & Harden, 1984) and occurred as a result of the activation of a calcium $\cdot$ calmodulin-sensitive phosphodiesterase (Tanner *et al.*, 1986). Since adenosine receptors in this cell line are coupled to inhibition of AC in a pertussis toxin-sensitive manner (Hughes & Harden, 1986), these results suggest that the mAcChRs found in 1321N astrocytoma cells did not interact strongly with G_i.

mAcChRs have also been shown to mediate an increase in the concentration of guanosine 3',5'-cyclic monophosphate (cGMP) in the heart (George *et al.*, 1970), autonomic ganglia (Kababian, Steiner & Greengard, 1975), N1E-115 neuroblastoma cells (Richelson, Prendergast & Divinetz-Romero, 1978), and NG108-15 neuroblastoma × rat glioma cells (Kurose & Ui, 1985). In the latter cell line, the cGMP response was attenuated by treatment with pertussis toxin indicating that it may be mediated by G_i. The increase in cGMP levels caused by muscarinic agonists in N1E-115 cells depended on an increase in intracellular calcium levels (Ohako & Deguchi, 1984) and correlated with a mAcChR state that had low affinity for agonists as opposed to high affinity mAcChRs in the same cell line that coupled to inhibition of AC (McKinney & Richelson, 1986*a*). Treatment of N1E-115 cells with phorbol 12-myristate 13-acetate (PMA), an activator of protein kinase C, did not affect ligand binding, but shifted the dose–response curve for the stimulation of cGMP formation by carbachol to higher concentrations and decreased the maximum response about fourfold (Lai & El-Fakahany, 1987). These results suggest that phosphorylation of one or more components of the cGMP effector system by protein kinase C may attenuate the response.

Although the detailed mechanism of cGMP-mediated signal transduction is not completely understood, several potential sites of action have been demonstrated. mAcChR-stimulated cGMP accumulation in N1E-115 cells led to a hyperpolarization of the membrane potential (Wastek, Lopez & Richelson, 1980), and intracellular injection of cGMP induced an outward potassium current in *Xenopus* oocytes that resembled the slow hyperpolarizing potassium current component of the muscarinic response (Dascal, Landau & Lass, 1984). In frog ventricular myocytes, cGMP stimulated a cyclic nucleotide phosphodiesterase that decreased cAMP levels resulting in a reduction of the inward calcium current (Fischmeister & Hartzell, 1987; Simmons & Hartzell, 1988). Muscarinic stimulation of cGMP-dependent protein phosphorylation was demonstrated in vascular smooth muscle cells (Casnellie *et al.*, 1980), but the physiological effects are unknown.

Since the initial observation that mAcChRs mediated an increased incorporation of labeled phosphate into phosphatidic acid and phosphatidyl inositol (PI) (Hokin & Hokin, 1953), it has been shown that many hormone receptors involved in mobilization of intracellular calcium couple to the stimulation of inositol lipid turnover (Fisher & Agranoff, 1987). The mAcChR-initiated stimulation of phospholipase C (PLC) is guanine nucleotide-dependent (Haslam & Davidson, 1984; Cockroft & Gomperts, 1985; Litosch, Wallis & Fain, 1985) and is mediated by a G protein(s) termed G_p or G_q. The exact nature of G_p is obscure; however, partially purified PLC activity from deoxycholate extracts of platelet membranes was stimulated by G_i and G_o (Banno *et al.*, 1987) and mAcChR-mediated stimulation of PLC in *Xenopus* oocytes was inhibited by injection of $G_{\beta\gamma}$ subunits or treatment with pertussis toxin (Moriarty *et al.*, 1988). A complex of PLC plus a 29 kD G protein, identified in platelet supernatants, showed a tenfold increase in the rate of hydrolysis of phosphatidyl inositol-4,5-bisphosphate (PIP_2) in the presence of guanosine 5'-(3-*o*-thio)triphosphate ($GTP\gamma S$) (Baldassare *et al.*, 1988). mAcChR-mediated stimulation of PI metabolism was not sensitive to pertussis toxin treatment in chick heart cells (Masters *et al.*, 1985), 1321N1 astrocytoma cell membranes (Hepler & Harden, 1986), or Flow 9000 embryonic pituitary tumor cells (Lo & Hughes, 1987*a*); however, in the latter system (Lo & Hughes, 1987*b*) pretreatment with cholera toxin uncoupled the mAcChR from G_p. These results and data from other systems in which receptor-stimulated PI metabolism was inhibited by pertussis toxin treatment (Nakamura & Ui,

1985; Okajima, Katada & Ui, 1985) suggest that the G protein that couples calcium-mobilizing receptors to inositol phospholipid metabolism may vary depending on the cell and/or receptor of interest.

Hydrolysis of PIP_2 by PLC results in the formation of inositol 1,4,5-trisphosphate (IP_3) and diacylglycerol. IP_3 can act as a second messenger to release calcium from storage sites in the endoplasmic reticulum (Streb *et al.*, 1983), while diacylglycerol is an activator of protein kinase C (Nishizuka, 1988) which regulates intracellular processes by phosphorylation of effector proteins. Diacylglycerol may be further metabolized by diacylglycerol lipase to release arachidonic acid, a substrate for eicosanoid synthesis. Alternatively, arachidonate can be released by phospholipase A_2 in a process mediated by a pertussis toxin-sensitive G protein (Nakashima *et al.*, 1988). Whether muscarinic responses are mediated by one or both pathways for arachidonate formation is not yet clear. Arachidonic acid release has been observed in many systems including cerebellar cortex slices (Reichman, Nen & Hokin, 1987) and guinea pig parotid acinar cells (Söling *et al.*, 1987) after addition of muscarinic agonists. Inhibitors of lipoxygenase blocked the mAcChR-mediated increase in cGMP formation in N1E-115 cells (McKinney & Richelson, 1986b) implicating an oxidized metabolite of arachidonate in the activation of guanylyl cyclase.

In embryonic chick heart cells (Brown & Brown, 1984) and brain nerve ending preparations (Fisher, Klinger & Agranoff, 1983), the extent of stimulation of PI metabolism by muscarinic agonists depended on agonist structure, AcCh and carbachol being more effective than pilocarpine, arecoline, or oxotremorine. In the chick heart preparation, oxotremorine and carbachol were equally effective in inhibiting AC. These results suggest that different agonists induce mAcChR conformations which differ in their ability to couple to G proteins.

mAcChR-mediated stimulation of PI metabolism in 1321N1 astrocytoma cell membranes (Orellana, Solski & Brown, 1987) and SH-SY5Y neuroblastoma cells (Serra, Smith & Yamamura, 1986) was inhibited by pretreatment with PMA. The inhibition of PI metabolism by PMA in astrocytoma cell membranes occurred only when PI metabolism was stimulated by GTPγS or through a G protein-linked receptor (calcium-stimulated PI metabolism was unaffected) and was mimicked by protein kinase C. These results indicate that persistent activation of protein kinase C results in attenuation of hormone-stimulated PI metabolism by modulating G_p interactions with PLC.

Depending on the system, muscarinic agonists can cause depolarizing or hyperpolarizing responses (Christie & North, 1988). In heart (Breitwieser & Szabo, 1988), bullfrog sympathetic c neurons (Dodd & Horn, 1983), neurons in the rat nucleus parabrachialis (Egan & North, 1986), and in exocrine cells (Peterson & Maruyama, 1984), AcCh increases potassium conductance, resulting in a hyperpolarization of the membrane potential. Hyperpolarization can inhibit or reduce the duration of action potentials or regulate neuronal firing patterns (McCormick & Prince, 1986). Muscarinic agonists reduced a calcium-dependent potassium conductance in olfactory cortex neurons (Constanti & Sims, 1987a) and hippocampal cells (Madison, Lancaster & Nicoll, 1987) which caused the action potential after hyperpolarization. The effects of AcCh in hippocampal cells were mimicked by phorbol esters suggesting that protein kinase C may play a role in the regulation of this channel (Malenka *et al.*, 1986). Muscarinic agonists also inhibit a non-voltage-dependent background potassium current in hippocampal cells (Madison, Lancaster & Nicoll, 1987) and enteric neurons (North, Slack & Surprenant, 1985), giving rise to slow, excitatory postsynaptic potentials. M currents, potassium currents activated at membrane potentials greater than -60 mV, were suppressed by muscarinic agonists in olfactory cortex neurons (Constanti & Sim, 1987b), spinal cord neurons (Nowak & MacDonald, 1983), and gastric smooth muscle cells (Sims, Singer & Walsh, 1985). In the latter system (Sims, Singer & Walsh, 1988), the effect of mAcChR activation was downstream of cAMP formation, perhaps directly on the ion channel.

mAcChRs have also been shown to inhibit an inward chloride current in sympathetic neurons (Brown & Selyanko, 1985; Mochida & Kobayashi, 1986) and to evoke inward chloride currents in *Xenopus* oocytes (Dascal, Gillo & Lass, 1985). The chloride currents observed in *Xenopus* oocytes were induced by intracellular injection of IP_3 (Oron *et al.*, 1985; Parker & Miledi, 1986), indicating that they were the result of calcium release from intracellular stores. Phorbol esters inhibited cholinergic responses in smooth muscle (Baraban *et al.*, 1985), but mimicked AcCh in inhibiting potassium currents evoked by adenosine and cAMP in *Xenopus* oocytes (Dascal *et al.*, 1985), suggesting that, in the former case, a feedback loop involving protein kinase C may exist that attenuates PI-induced calcium release, while in the latter case phosphorylation of protein(s) by protein kinase C may be involved in muscarinic inhibition of purinergic and cAMP response. In rat hippocampus (Worley *et al.*, 1987), the potency of cholinergic agonists in blocking the response to adenosine paralleled their ability to stimulate PI metabolism. Similar effects were caused by addition of a phorbol ester suggesting that activation of protein kinase C may attenuate G protein linked signal transduction. Evidence that the therapeutic action of lithium may be the result of its effect on PI metabolism was obtained in hippocampal slices (Worley *et al.*, 1988) where therapeutic concentrations of lithium reduced mAcChR-mediated attenuation of the inhibitory actions of adenosine.

Of the ion channels activated by mAcChRs, the inward rectifying potassium channel in the heart has been most thoroughly characterized. mAcChR-mediated activation required guanine nucleotides and was blocked by pretreatment with pertussis toxin (Breitwieser & Szabo, 1985; Pfaffinger *et al.*, 1985; Martin, Hunter & Nathanson, 1985). Channel opening was unaffected by phorbol esters or $G_s \cdot GTP\gamma S$ and erythrocyte $G_i \cdot GTP\gamma S$ was one to two orders of magnitude more effective than activated G_o (Yatani *et al.*, 1987). Detailed analysis of the channel activation process (Breitwieser & Szabo, 1988) permitted estimation of the rate of GDP release from unactivated and mAcChR-activated G_i and calculation of the rate constant for GTP hydrolysis (135 min^{-1}) *in vivo*.

Studies with resolved subunits of G proteins remain somewhat controversial. The $G_{i\alpha} \cdot GTP\gamma S$ complex from erythrocytes activated atrial potassium channels in the picomolar concentration range (Codina *et al.*, 1987; Cerbai, Klöckner & Isenberg, 1988) and an antibody against $G_{i\alpha}$ blocked channel activation in inside-out patches of atrial membranes (Yatani *et al.*, 1988*a*), suggesting that this subunit was responsible for mediating the effect of muscarinic agonists. Recombinant $G_{i\alpha 1}$, $G_{i\alpha 2}$, and $G_{i\alpha 3}$ were equally effective in opening atrial potassium channels (Yatani *et al.*, 1988*b*) indicating a lack of specificity among G_i subtypes. $\beta\gamma$ subunits from bovine cerebral cortex ($\beta 35, 36\gamma$) or human placenta ($\beta 35\gamma$), but not transducin ($\beta 36\gamma$) activated the potassium channel in the concentration range of 0.2–10 nM (Logothetis *et al.*, 1988), suggesting that $\beta\gamma$ subunits may also play a role. Although the activated $G_{i\alpha}$ subunits appear to be one or two orders of magnitude more effective than $G_{\beta\gamma}$, interpretation was complicated by variability in reproducing results, differing detergents and potential problems that may arise in application of the hydrophilic $G_{i\alpha}$ subunits compared to the hydrophobic $G_{\beta\gamma}$ subunits in patch clamp experiments.

Ligand binding properties

The introduction of radiolabeled muscarinic antagonists such as [^{3}H]-L-quinuclidinyl benzilate ([^{3}H] L-QNB) (Yamamura & Snyder, 1974) provided highly specific probes for mAcChRs. Binding studies of non-selective antagonists to membranes prepared from rat cortex (Hulme *et al.*, 1978) showed a single class of binding sites while agonist binding data (Birdsall, Burgen & Hulme, 1978) were analyzed in terms of three non-interacting classes of sites having super-high, high, and low affinity for agonists. Multiple classes of agonist binding have been observed in all preparations examined to date, although data are often analyzed in terms of two rather than three agonist affinity states. Agents such as *N*-ethylmaleimide, magnesium and guanine nucleotides which effect interactions between G proteins and G protein linked receptors have been shown (Berrie *et al.*, 1979; Rosenberger, Yamamura & Roeske, 1980) to influence the apparent distribution of agonist binding sites between low and high affinity states. Binding data and the coupling to inhibition of AC (Ehlert, 1985), have generally been interpreted in terms of a model in which the high affinity agonist state was assumed to be coupled to a G protein while the low affinity agonist state represented free receptor (Rodbell, 1980). Modified models in which reciprocal modulation of agonist and antagonist affinities occur (Burgisser, DeLean & Lefkowitz, 1982; Martin, Smith & Harden, 1984) or involve cooperative interactions between multiple binding sites and several levels of modulation by G protein (Mattera *et al.*, 1985) have also been proposed. A mechanism in which several possible affinity states exist for the mAcChR was proposed (Wong, Sole & Wells, 1986) after binding data was found to be incompatible with either a model assuming different non-interconverting classes of sites or the ternary complex models in which agonists bind preferentially to a receptor–G protein complex (Lee, Sole & Wells, 1986). Studies measuring the reciprocal effects of muscarinic ligands and changes in membrane potential mediated by sodium channels (Cohen-Armon & Sokolovsky, 1986; Cohen-Armon, Garty & Sokolovsky, 1988) led to the hypothesis that muscarinic agonist binding and sodium channel activation in some tissues may be coupled through G proteins. The stimulation of PI metabolism may be elicited by both sodium channel-dependent and independent depolarization where the relative contribution of each pathway depends on agonist structure (Gurwitz & Sokolovsky, 1987).

Kinetic analysis of the binding of labeled antagonists such as L-QNB, *N*-methylscopolamine, *N*-methyl-4-piperidinyl benzilate, or pirenzepine (PZ) to the membrane-bound mAcChR (Järv, Hedlund & Bartfai, 1979; Schimerlik & Searles, 1980; Galper *et al.*, 1982; Luthin & Wolfe, 1984; Schreiber, Henis & Sokolovsky, 1985*a*) was consistent with a two-step kinetic mechanism consisting of a rapidly equilibrating step followed by a slower conformational change of the mAcChR·antagonist complex.

Similar results were found for L-QNB binding to detergent-solubilized mAcChRs from porcine atria (Herron *et al.*, 1982).

Binding of AcCh (Gurwitz, Kloog & Sokolovsky, 1985) to rat cerebral cortex membranes showed two kinetic phases while the association of [methyl] oxotremorine binding to rat heart membranes followed a single kinetic phase (Harden, Meeker & Martin, 1983). Kinetic studies of cis-methyldioxolane binding to embryonic chick heart membranes (Galper *et al.*, 1987) supported a model in which the agonist bound to low and high affinity mAcChRs in two parallel reactions and the formation of the hormone · receptor · G protein · GTP complex was formed by random addition of GTP and agonist prior to receptor-G protein uncoupling. Agonist-induced conversion from the low to high affinity mAcChR conformation has not been found, suggesting that the receptor and G protein are pre-coupled or that any association reaction between the two was too rapid to be observed using the available methodology (Schreiber, Henis & Sokolovsky, 1985*b*; Galper *et al.*, 1987). In at least one system (Järv, Hedlund & Bartfai, 1980), kinetic studies of agonist inhibition of labeled antagonist binding were not consistent with a simple competitive model.

The hydrophobicity of the mAcChR binding sites was studied by competition of alkylguanidines with L-QNB (Cremo & Schimerlik, 1983). It was possible to estimate a contribution of -470 cal/mol per methylene group to the total binding energy, suggesting a relatively hydrophobic environment. Quaternary amine antagonists seem to recognize a sub-population of tertiary amine antagonist binding sites (Brown & Goldstein, 1986) and studies in SK-N-SH neuroblastoma cells (Fisher, 1988) supported a model in which only mAcChRs at the cell surface bound positively charged antagonists at 0 °C, but at 37 °C sequestered mAcChRs equilibrated with those at the cell surface, giving rise to equal numbers of sites for tertiary and quaternary antagonists. Agents such as gallamine (Birdsall *et al.*, 1981; Dunlap & Brown, 1983), 4-aminopyridine (Lai, Ramkumar & El-Fakahany, 1985) and the agonist [3-(*m*-chlorophenyl-carbamoyloxy)-2-butynyl trimethylammonium] chloride (McN-A-343) (Birdsall *et al.*, 1983) have also shown anomalous binding behavior which has in some cases been attributed to binding to allosteric sites.

Binding studies utilizing the antagonist PZ (Hammer *et al.*, 1980) permitted the first subclassification of mAcChRs into M1 (high affinity) and M2 (low affinity) subtypes. M2 mAcChRs were further subclassified into cardiac and glandular subtypes with selective antagonists such as (11-[2-[(diethylamino)methyl]-1-piperidinyl]acetyl]-5, 11-dihydro-6H-pyrido-

[2,3-b][1,4]benzodiazepine-6-one) (AF-DX116; Hammer *et al.*, 1986), which binds more tightly to cardiac mAcChRs, and cyclohexyl(4-fluoro-phenyl)(3-piperidinopropylsilanol (p-F-HHSiD; Lambrecht *et al.*, 1988) which has a higher affinity for smooth muscle and glandular M2 mAcChRs. In the central nervous system PZ-differentiated subtypes appeared to correlate well with stimulation of PI metabolism (high affinity for PZ) and inhibition of AC (low affinity) while differing affinities and weaker correlations between binding and response data were found in peripheral tissues (Gil & Wolfe, 1985). In embryonic chick heart (Brown, Goldstein & Masters, 1985), PZ inhibited mAcChR-mediated effects on AC with higher affinity than mAcChR stimulation of PI metabolism. Studies of PZ binding to rat brain mAcChRs solubilized in digitonin have led to differing results. In one case (Roeske & Venter, 1984), high affinity PZ binding decayed after solubilization while L-QNB binding was unaffected, while a second study (Berrie *et al.*, 1985*a*) found that high affinity PZ binding was stable at 4 °C. The membrane-bound heart mAcChR (Birdsall, Hulme & Keen, 1986; Schimerlik *et al.*, 1986) appeared to exist exclusively in the low affinity state for PZ; however, after solubilization, high affinity PZ binding sites appeared. Purified porcine atrial mAcChRs exhibited only low affinity PZ binding (Schimerlik *et al.*, 1986), suggesting that the affinity of the solubilized mAcChR could be modulated by removal from the membrane and/or interaction with other proteins after solubilization. Since it is now known that there are at least five distinct muscarinic receptor subtypes (see below) interpretation of electrophysiological, ligand binding and effector response in terms of two or three pharmacological subtypes may lead to an oversimplification of experimental results.

Purification and properties

mAcChRs which were able to bind ligands have been solubilized in digitonin (Beld & Ariens, 1974), a digitonin–gitonin mixture (Repke, 1987), a digitonin–cholate mixture (Cremo, Herron & Schimerlik, 1981), 3-[(cholamidopropyl)-dimethylammonio]-1-propane sulfonate (CHAPS) (Gavish & Sokolovsky, 1982), and dodecyl β-D-maltoside (Peterson, Rosenbaum & Schimerlik, 1988). mAcChRs solubilized in digitonin (Hurko, 1978; Gorissen *et al.*, 1991), digitonin–cholate (Herron *et al.*, 1982), and dodecyl β-D-maltoside (Peterson, Rosenbaum & Schimerlik, 1988) bound agonists with low affinity while heterogeneous agonist binding was found in CHAPS or digitonin-solubilized preparations from rat brain (Baron, Gavish & Sokolovsky, 1985). Rat myocardial preparations solubilized in digitonin plus magnesium (Berrie *et al.*, 1984*a*) and bovine brain mAcChRs solubilized in CHAPS (Kuno, Shirakawa & Tanaka, 1983) retained sensitivity to guanine nucleotides.

Highly purified preparations of mAcChRs have been obtained from porcine atria (Peterson *et al.*, 1984), porcine cerebellum (Haga & Haga, 1985), rat forebrain (Berrie *et al.*, 1985), and chick heart (Kwatra & Hosey, 1986). These procedures utilize a variety of chromatographic techniques, but all procedures involve affinity chromatography on columns containing the atropine analogue 3-(2'-aminobenzhydryloxy)tropane (Haga & Haga, 1983). Antagonist binding appears to be similar to that observed for the membrane-bound mAcChR and agonists appear to exhibit low affinity binding (Haga & Haga, 1985; Schimerlik *et al.*, 1986), although unexplained high affinity agonist binding has been observed in purified mAcChR preparations (Peterson *et al.*, 1984). Reduction with dithiothreitol affected the mAcChR hydrodynamic properties (Schimerlik *et al.*, 1986) and the binding properties of oxotremorine M and PZ, causing a shift to a lower affinity state (Wheatley *et al.*, 1987).

Estimations of the molecular weight (M_r) of mAcChRs by sodium dodecyl sulfate gel electrophoresis (SDS–PAGE) of preparations labeled with propylbenzilylcholine mustard (PrBCM) (Birdsall, Burgen, & Hulme, 1979) or antagonist photoaffinity analogues (Amitai *et al.*, 1982; Cremo & Schimerlik, 1984) gave values in the range of 70–80 kD in agreement with results from radiation inactivation analysis (Venter, 1983). Hydrodynamic studies using gel chromatography or sucrose gradient centrifugation gave highly variable results depending on the detergent system and tissue, with M_r values ranging from around 65 kD (Dadi & Morris, 1984; Repke & Schmitt, 1987) to as high as 111 kD (Berrie *et al.*, 1984*b*) and a Stokes radius ranging from 4.3 nm in Triton X-405 (Peterson *et al.*, 1986) to 8.5 nm in SDS (Dadi & Morris, 1984). Extensive characterization of the porcine atrial mAcChR by SDS–PAGE, hydrodynamic methods, and compositional analysis gave an estimation of M_r for the protein portion of the molecule of 50–53 kD (Peterson *et al.*, 1986). Electrophoretic studies of PrBCM-labeled mAcChRs from pancreatic acinar cells (Hootman, Picado-Leonard & Burnham, 1985) and 1321N1 astrocytoma cells (Liang, Martin & Harden, 1987) indicated that mAcChRs in these systems have apparent M_r values substantially higher than 70 kD. Studies of bovine brain mAcChRs by radiation inactivation analysis gave values of 91 kD for the L-QNB-sensitive component but 157 kD for the PZ-sensitive species, leading to the hypothesis that mAcChRs with high affinity for PZ may be coupled to other membrane components (Shirakawa & Tanaka, 1985).

The isoelectric point (pI) of rat brain mAcChRs was estimated to be about 4.6 after solubilization in digitonin/gitonin (Repke & Matthies, 1980) and a pI of 5.9 was found for mAcChRs from a variety of sources after extraction of PrBCM-labeled receptors from SDS polyacrylamide gels (Venter *et al.*, 1984).

Studies of the glycoprotein properties of porcine atrial mAcChRs by chromatography on lectin columns indicated that it contained sialic acid and was heterogeneous in its glycosylation (Herron & Schimerlik, 1983). Compositional analysis (Peterson *et al.*, 1986) indicated that the amino sugar content was 26% w/w. Bovine cerebral cortex mAcChRs solubilized in CHAPS (Shirakawa, Kuno & Tanaka, 1983) showed a different lectin binding pattern than those from porcine atria, suggesting different patterns of glycosylation in the two tissues. A 5% decrease in apparent M_r after endoglycosidase treatment was observed for the cerebral cortex receptor and removal of sialic acid residues with neuraminidase resulted in an increase in pI from 4.3 to about 4.5 (Rauh *et al.*, 1986). Although the exact role of carbohydrate groups in the function of the mAcChR has not yet been determined, a recent study (Gies & Landry, 1988) suggested that enzymatic removal of sialic acid residues selectively affected agonist binding properties of the M2 muscarinic subtype.

Reconstitution

Reconstitution of purified G_i and G_o with bovine brain mAcChRs resolved from G proteins showed guanine nucleotide-dependent high affinity agonist binding (Florio & Sternweis, 1985). Both G_α and $G_{\beta\gamma}$ subunits were necessary to promote high affinity agonist binding to the mAcChR and agonist-stimulated guanine nucleotide exchange (Florio & Sternweis, 1989), suggesting that the α subunit has a different conformation in the heterotrimer than it does in the absence of $\beta\gamma$. A twofold stimulation of the GTPase activity of G_i by a muscarinic agonist was demonstrated using purified rat brain G_i and purified porcine cerebellar mAcChRs (Haga *et al.*, 1985) and reconstitution with either G_i or G_o gave guanine nucleotide-sensitive high affinity agonist binding (Haga, Haga & Ichiyama, 1986). Carbachol stimulated GTPγS binding to G_i and increased the maximal GTPase activity of G_i by about 50% (Kurose *et al.*, 1986). Pertussis toxin-catalyzed ADP-ribosylation of G_i prior to reconstitution blocked the agonist-induced stimulation of GTPγS binding and the formation of high affinity agonist binding sites. Reconstitution of purified porcine atrial mAcChRs and purified atrial G_i resulted in a tenfold stimulation of k_{cat} for the GTPase reaction and guanine nucleotide-sensitive high affinity agonist binding (Tota, Kahler & Schimerlik, 1987). The binding of GTP (Tota & Schimerlik, unpublished observations) or $GTP_\gamma S$ to G_i was unaffected by muscarinic agonists; however, the affinity of GDP was decreased by about fifty-fold and the rate constant for GDP dissociation increased by a factor of about 40. Experiments with purified bovine brain mAcChRs showed that the binding of GDP, but not GTP, to purified G_o was inhibited by the agonist oxotremorine and that the effect was most likely attributable to differential effects on the rate constants for both association and dissociation of the two nucleotides (Florio & Sternweis, 1989). In the absence of muscarinic agonists, GDP release appeared to be the rate limiting step in the steady-state mechanism of the GTPase reaction catalyzed by G_i (Tota, Kahler & Schimerlik, 1987). ADP-ribosylation of G_i in the reconstituted system blocked over 90 % of the agonist-stimulated GTPase activity, but affected guanine nucleotide sensitive agonist binding only if done in the presence of low concentrations of GTP and a muscarinic agonist, suggesting that the mAcChR and G_i were pre-coupled and that association with the mAcChR protects G_i from ADP-ribosylation. In the presence of both agonist and guanine nucleotide, the system transiently uncouples and G_i can then be modified by pertussis toxin. Phosphorylation of the atrial mAcChR by cyclic AMP-dependent protein kinase under varying conditions in the reconstituted system (Rosenbaum et al., 1987) suggested that the protein existed in a unique conformation in the presence of both an agonist and G_i such that the apparent phosphorylation stoichiometry was increased. The atrial mAcChR was not phosphorylated by protein kinase C *in vitro* (Rosenbaum et al., 1987; Haga et al., 1988); however, purified porcine cerebral mAcChRs were phosphorylated on both serine and threonine residues (Haga et al., 1988). Phosphorylation by protein kinase C or cAMP-dependent protein kinase *in vitro* did not affect the ligand binding properties of mAcChRs or their ability to associate with G proteins (Rosenbaum et al., 1987; Haga et al., 1988). The G protein specificity of mAcChRs is somewhat unclear at this time since it has been shown that purified G_{i1}, G_{i2} and G_o all appeared to be effective in interacting with mAcChRs (Haga et al., 1988).

Cloning and expression

The complete amino acid sequences of five mAcChR subtypes have been deduced from DNA sequences of clones isolated from human, porcine, and rat genomic and cDNA libraries (Kubo et al., 1986a,b; Peralta et al., 1987a,b; Bonner et al., 1987, 1988; Braun et al., 1987; Gocayne et al.,

1987; Akiba *et al.*, 1988; Stein, Pinkas-Kramarski & Sokolovsky, 1988; Lai *et al.*, 1988). Mapping of M1–M4 subtype mRNA in various brain regions (Bonner *et al.*, 1987; Braun *et al.*, 1987; Brann, Buckley & Boner, 1988) or by Northern analysis (Peralta *et al.*, 1987*b*) suggests that subtype expression is tissue and/or cell type specific. mRNA coding for the M5 subtype was not found in brain (Bonner *et al.*, 1988), or that M5 mAcChRs are not expressed to a significant degree in that tissue. Southern analysis of rat and human genomic DNA suggests that an additional human subtype and four additional rat mAcChR subtypes may exist (Bonner *et al.*, 1987).

Hydropathy analysis (Kyte & Doolittle, 1982) of the amino acid sequence data indicated that all mAcChR subtypes have seven trans-membrane regions and belong to the family of G protein-linked receptors that are structurally similar to rhodopsin (Ovchinnikov, 1982). By analogy with rhodopsin, the amino terminus, which contains potential glyco-sylation sites, is located extracellularly and the carboxyl terminus intra-cellularly. A high degree of amino acid identity exists among all receptors of this class in the seven transmembrane regions and in the first two cytosolic loops, while poor correspondence is found at the amino and carboxyl termini of the molecules. An extremely low amino acid correlation exists in the cytosolic loop connecting transmembrane segments five and six for all receptors of this class and among the five mAcChR subtypes, suggesting that this region and the carboxyl terminus may play a unique role in regulation and/or coupling to different G proteins. All subtypes contain three aspartate residues located just outside or within trans-membrane segments two and three. Mapping studies with PrBCM-labeled mAcChRs (Curtis *et al.*, 1989) suggest that the site of alkylation may be primarily the aspartate well within transmembrane segment three (aspar-tates 105 and 103 for M1 and M2 mAcChR subtypes, respectively) in agreement with site-specific mutagenesis experiments in the β-adrenergic receptor system (Strader *et al.*, 1987). Although mutagenesis experiments have not yet been completed for mAcChRs, it seems probable that the ligand binding site resides within the hydrophobic core of the protein. A comparison of amino acid sequence data for mAcChR subtypes, nicotinic cholinergic receptors and acetylcholinesterase led to the hypothesis that these proteins may have common features that constitute the AcCh binding domain (Peterson, 1989).

The genes coding for the five mAcChR subtypes have been expressed in several different cell lines permitting analysis of the ligand binding and effector coupling properties of each subtype. Expression of the mRNA for M1 and M2 mAcChR subtypes in *Xenopus* oocytes (Kubo *et al.*, 1986*a*; Fukuda *et al.*, 1987) showed high affinity PZ binding for M1 and low

Table 9.1. *HmAcChR subtype coupling to physiological responses*

Subtype	A.A. number	Protein molecular weight	Physiological response
M1	460	51450	Stimulation of PI metabolism
M2	466	51712	Inhibition of adenylyl cyclase
M3	479	53046	Inhibition of adenylyl cyclase
M4	590	66125	Stimulation of PI metabolism
M5	532	60045	Stimulation of PI metabolism

affinity PZ binding for M2 mAcChRs. Both subtypes coupled to activation of an oscillatory increase in chloride conductance which was dependent on an increase in intracellular calcium levels; however, the M2 mAcChRs also activated a second channel in which the current was carried by sodium and potassium ions. Expression of M2 mAcChRs in Chinese hamster ovary cells (Ashkenazi *et al.*, 1987) showed the receptor coupled to both inhibition of AC and stimulation of pI metabolism, but that it was more tightly coupled to inhibition of AC. Both responses were blocked by pertussis toxin treatment; however, inhibition of AC was more sensitive to pertussis toxin than the pI response, suggesting that they were mediated by different G proteins. Analysis of M1–M4 mAcChRs expressed in embryonic kidney cells (Peralta *et al.*, 1988), NG108-15 cells (Fukuda *et al.*, 1988), A9 L cells (Conklin *et al.*, 1988) and M1 mAcChRs expressed in murine B82 cells (Lai *et al.*, 1988) suggest that M1 and M4 preferentially couple to the stimulation of pI metabolism, while M2 and M3 subtypes are coupled more strongly to inhibition of AC. M1 and M4 mAcChR activation also increased levels of arachidonic acid (Conklin *et al.*, 1988) in a PLC-independent manner (as opposed to pI stimulation which was inhibited by phorbol esters) and inhibited mitogenesis. Activation of M1 or M4 mAcChRs activated calcium-dependent potassium currents and inhibited the M current in NG108-15 cells (Fukuda *et al.*, 1988). The M1 mAcChR subtype expressed in A9 L cells (Jones *et al.*, 1988) activated calcium-dependent potassium channels. The M5 subtype also couples preferentially to the stimulation of pI metabolism (Bonner *et al.*, 1988). Thus M1, M4 and M5 subtypes appear to couple preferentially to PI metabolism and the activation of ion channels that are affected by the resulting increase in intracellular calcium, while M2 and M3 subtypes couple preferentially to inhibition of AC (Table 9.1). This result may not be absolute, however, since M1 mAcChRs expressed in RAT-1 cells (Stein, Pinkas-Kramarski & Sokolovsky, 1988) coupled more efficiently to inhibition of AC than to the stimulation of PI metabolism. Pertussis toxin was capable of uncoupling the M1 subtype from inhibition of AC, but not from stimulation of PI metabolism, suggesting that different G proteins mediate the two responses. The ligand binding properties of the five mAcChR subtypes indicated that it was not possible to make a definitive subtype assignment based on the affinity of PZ or AF-DX116 (Peralta *et al.*, 1987*b*; Bonner *et al.*, 1987, 1988).

Regulation

Long-term exposure of mAcChRs to agonists resulted in the attenuation of muscarinic responses and an accompanying decrease in receptor number in cultured chick heart cells (Galper & Smith, 1980; Galper *et al.*, 1982), embryonic chick heart (Halvorsen & Nathanson, 1981) and brain (Meyer, Gainer & Nathanson, 1982), pancreatic acini (Hootman *et al.*, 1986), intact rat brain cells (Lee & El-Fakahany, 1985), and N1E-115 neuroblastoma cells (Cioffi & El-Fakahany, 1986). Treatment of N1E-115 neuroblastoma cells (Liles *et al.*, 1986) or Flow 9000 pituitary cells (Lo & Hughes, 1988) with phorbol esters resulted in a loss of antagonist binding sites from the cell surface. That this phenomenon may be due to receptor internalization was shown by the appearance of 1321N1 astrocytoma mAcChRs in a light vesicle fraction during desensitization (Harden *et al.*, 1985) and the observation that the slow phase of desensitization in cultured heart cells was blocked by inhibitors of microtubule formation (Galper & Smith, 1980). mAcChRs have also been identified in bovine brain coated vesicles (Silva *et al.*, 1986).

The chick heart mAcChR was shown to be phosphorylated *in vivo* (Kwatra & Hosey, 1986) and the phosphorylation appears to correlate with mAcChR desensitization (Kwatra *et al.*, 1987). A loss of mAcChR binding sites in rat brain synaptic membranes under phosphorylating conditions in the presence of calcium · calmodulin (Burgoyne, 1981) and the phosphorylation of purified mAcChRs from brain (Ho *et al.*, 1987) and heart (Ho, Shang & Duffield, 1986; Rosenbaum *et al.*, 1987) in detergent solution by cAMP-dependent protein kinase resulted in a loss of antagonist binding sites that was reversed by treatment with calcineurin. At this time, it is not known whether a specific protein kinase catalyzes the phosphorylation of mAcChRs *in vivo* and whether phosphorylation leads directly to a loss of ligand binding properties or is a signal for mAcChR internalization.

Studies of the ability of mAcChRs to couple to physiological responses in developing chick hearts (Galper, Dziekan & Smith, 1984; Halvorsen & Nathanson, 1984; Liang *et al.*, 1986; Luetje *et al.*, 1987*a*; Liang & Galper, 1987) have shown that responsiveness correlates with the increased levels of G_o and/or G_i that occur upon vagal innervation. After desensitization, newly synthesized mAcChRs in chick embryos showed a lag phase in their ability to couple to physiological responses (Hunter & Nathanson, 1984), indicating that the mAcChRs may have to undergo additional processing steps before being able to effectively couple with G proteins. mAcChRs in newborn rat cerebral cortex appeared to be more tightly coupled to PI metabolism than those in adult rats (Heacock, Fisher & Agranoff, 1987), suggesting either a change in mAcChR structure or a decrease in level of G_p may occur upon ageing. Changes in both apparent molecular weight (Large *et al.*, 1985*a*) and isoelectric point (Large *et al.*, 1985*b*) of mAcChRs have been observed during synaptogenesis.

Immunological studies

Monoclonal antibodies (mAbs) have been raised against purified mAc-ChRs from calf brain (André *et al.*, 1984) and porcine atria (Luetje *et al.*, 1987*b*). One of the mAbs generated against the calf brain mAcChR showed agonist activity (Leiber *et al.*, 1984) and mAbs against the atrial mAcChR showed M2 subtype specificity. Antibodies raised against a peptide corresponding to the carboxyl terminus of the M1 mAcChR immuno-precipitated [^{3}H]-L-QNB binding sites from rat forebrain and [^{3}H] PZ binding activity from brain regions thought to be enriched in M1 mAcChRs (Luthin *et al.*, 1988).

In an immunocytochemical study on human fibroblast cells (Raposo *et al.*, 1987), mAcChR patching and sequestration into uncoated vesicles was observed in the presence of carbachol. This was followed by the formation of multivesicular structures thought to be involved in mAcChR down-regulation. When the antagonist atropine was used, mAcChRs formed clusters on the cell surface but sequestration did not occur, indicating that the agonist-occupied mAcChR was recognized by one or more components of the cell which were responsible for internalization.

Conclusions

The amino acid sequence and physiological responses of five mAcChR subtypes have been determined to date. In general, the expression of individual subtypes and the particular transducing and effector proteins with which they interact appear to be specific for a given cell line or tissue.

The detailed structure of mAcChRs, the mechanism by which the binding of muscarinic agonists alter the conformation of the protein to activate transducing elements such as G proteins, and the identification of other components of the muscarinic signalling system involved in receptor regulation have yet to be resolved.

Acknowledgements

The author would like to acknowledge the typing skills of Sue Conte and the support of USPHS grants HL23632 and ES00210.

References

Akiba, I., Kubo, T., Maeda, A., Bujo, H., Nakai, J., Mishina, M. & Numa, S. (1988). Primary structure of porcine muscarinic acetylcholine receptor III and antagonist binding studies. *Federation of European Biochemical Societies Letters*, **235**, 257–61.

Amitai, G., Avissar, S., Balderman, D. & Sokolovsky, M. (1982). Affinity labeling of muscarinic receptors in rat cerebral cortex with a photolabile antagonist. *Proceedings of the National Academy of Sciences, USA*, **79**, 243–7.

André, C., Guillet, J. G., De Backer, J-P., Vanderheyden, P., Hoebeke, J. & Strosberg, A. D. (1984). Monoclonal antibodies against the native or denatured forms of muscarinic acetylcholine receptors. *EMBO Journal*, **3**, 17–21.

Ashkenazi, A., Winslow, J. W., Peralta, E. G., Peterson, G. L., Schimerlik, M. I., Capon, D. J. & Ramachandran, J. (1987). An M2 muscarinic receptor subtype coupled to both adenylyl cyclase and phosphoinositide turnover. *Science*, **238**, 672–4.

Baldassare, J. J., Knipp, M. A., Henderson, P. A. & Fisher, G. J. (1988). GTP$_\gamma$S-stimulated hydrolysis of phosphatidylinositol-4,5-bisphosphate by soluble phospholipase C from human platelets requires soluble GTP-binding protein. *Biochemical and Biophysical Research Communications*, **154**, 351–7.

Banno, Y., Nagao, S., Katada, T., Nagata, K-I., Ui, M. & Nozawa, Y. (1987). Stimulation by GTP-binding proteins (G_i, G_o) of partially purified phospholipase c activity from human platelet membranes. *Biochemical and Biophysical Research Communications*, **146**, 861–9.

Baraban, J. M., Gould, R. J., Peroutka, S. J. & Snyder, S. H. (1985). Phorbol ester effects on neurotransmission: interaction with neurotransmitters and calcium in smooth muscle. *Proceedings of the National Academy of Sciences, USA*, **82**, 604–7.

Baron, B., Gavish, M. & Sokolovsky, M. (1985). Heterogeneity of solubilized muscarinic cholinergic receptors: binding and hydrodynamic properties. *Archives of Biochemistry and Biophysics*, **240**, 281–96.

Beld, A. J. & Ariens, E. J. (1974). Stereospecific binding as a tool in attempts to localize and isolate muscarinic receptors. *European Journal of Pharmacology*, **25**, 203–9.

Berrie, C. P., Birdsall, N. J. M., Burgen, A. S. V. & Hulme, E. C. (1979).

Guanine nucleotides modulate muscarinic receptor binding in the heart. *Biochemical and Biophysical Research Communications*, **87**, 1000–5.

Berrie, C. P., Birdsall, N. J. M., Hulme, E. C., Keen, M. & Stockton, J. M. (1984*a*). Solubilization and characterization of guanine nucleotide-sensitive muscarinic agonist binding sites from rat myocardium. *British Journal of Pharmacology*, **82**, 853–61.

Berrie, C. P., Birdsall, N. J. M., Haga, K., Haga, T. & Hulme, E. C. (1984*b*). Hydrodynamic properties of muscarinic acetylcholine receptors solubilized from rat forebrain. *British Journal of Pharmacology*, **82**, 839–51.

Berrie, C. P., Birdsall, N. J. M., Hulme, E. C., Keen, M. & Stockton, J. M. (1985*a*). Solubilization and characterization of high and low affinity pirenzepine binding sites from rat cerebral cortex. *British Journal of Pharmacology*, **85**, 697–703.

Berrie, C. P., Birdsall, N. J. M., Dadi, H. K., Hulme, E. C., Morris, R. J., Stockton, J. M. & Wheatley, M. (1985*b*). Purification of the muscarinic acetylcholine receptor from rat forebrain. *Biochemical Society Transactions*, **13**, 1101–3.

Birdsall, N. J. M., Burgen, A. S. V. & Hulme, E. C. (1978). The binding of agonists to brain muscarinic receptors. *Molecular Pharmacology*, **14**, 723–6.

Birdsall, N. J. M., Burgen, A. S. V. & Hulme, E. C. (1979). A study of the muscarinic receptor by gel electrophoresis. *British Journal of Pharmacology*, **66**, 337–42.

Birdsall, N. J. M., Burgen, A. S. V., Hulme, E. C. & Stockton, J. (1981). Gallamine regulates muscarinic receptors in the heart and cerebral cortex. *British Journal of Pharmacology*, **74**, 798P.

Birdsall, N. J. M., Burgen, A. S. V., Hulme, E. C., Stockton, J. M. & Zigmond, M. J. (1983). The effect of McN-A-343 on muscarinic receptors in the cerebral cortex and heart. *British Journal of Pharmacology*, **78**, 257–9.

Birdsall, N. J. M., Hulme, E. C. & Keen, M. (1986). The binding of pirenzepine to digitonin-solubilized muscarinic acetylcholine receptors from the rat myocardium. *British Journal of Pharmacology*, **87**, 307–16.

Bonner, T. I., Buckley, N. J., Young, A. C. & Brann, M. R. (1987). Identification of a family of muscarinic acetylcholine receptor genes. *Science*, **237**, 527–32.

Bonner, T. I., Young, A. C., Brann, M. R. & Buckley, N. J. (1988). Cloning and expression of the human and rat m5 muscarinic acetylcholine receptor genes. *Neuron*, **1**, 403–10.

Brann, M. R., Buckley, N. J. & Bonner, T. I. (1988). The striatum and cerebral cortex express different muscarinic receptor mRNAs. *Federation of European Biochemical Societies Letters*, **230**, 90–4.

Braun, T., Schofield, P. R., Shivers, B. D., Pritchett, D. B. & Seeburg, P. H. (1987). A novel subtype of muscarinic receptor identified by homology screening. *Biochemical and Biophysical Research Communications*, **149**, 125–32.

Breitwieser, G. E. & Szabo, G. (1985). Uncoupling of cardiac muscarinic and β-adrenergic receptors from ion channels by a guanine nucleotide analogue. *Nature*, London, **317**, 538–40.

Breitwieser, G. E. & Szabo, G. (1988). Mechanism of muscarinic receptor-induced K^+ channel activation as revealed by hydrolysis-resistant GTP analogues. *Journal of General Physiology*, **91**, 469–93.

Brown, B. L., Wojcikiewicz, R. J. H., Dobson, P. R. M., Robinson, A. & Irons, L. I. (1984). Pertussis toxin blocks the inhibitory effect of muscarinic cholinergic agonists on cyclic AMP accumulation and prolactin secretion in GH_3 anterior-pituitary tumor cells. *Biochemical Journal*, **223**, 145–9.

Brown, D. A. & Selyanko, A. A. (1985). Two components of muscarine-sensitive membrane current in rat sympathetic neurones. *Journal of Physiology*, **358**, 335–63.

Brown, J. H. & Brown, S. L. (1984). Agonists differentiate muscarinic receptors that inhibit cyclic AMP formation from those that stimulate phosphoinositide metabolism. *Journal of Biological Chemistry*, **259**, 3777–81.

Brown, J. H. & Goldstein, D. (1986). Analysis of cardiac muscarinic receptors recognized selectively by nonquaternary but not by quaternary ligands. *The Journal of Pharmacology and Experimental Therapeutics*, **238**, 580–6.

Brown, J. H., Goldstein, D. & Masters, S. B. (1985). The putative M_1 muscarinic receptor does not regulate phosphoinositide hydrolysis. *Molecular Pharmacology*, **271**, 525–31.

Burgisser, E., De Lean, A. & Lefkowitz, R. J. (1982). Reciprocal modulation of agonist and antagonist binding to muscarinic cholinergic receptor by guanine nucleotide. *Proceedings of the National Academy of Sciences, USA*, **79**, 1732–6.

Burgoyne, R. D. (1981). The loss of muscarinic acetylcholine receptors in synaptic membranes under phosphorylating conditions is dependent on calmodulin. *Federation of European Biochemical Societies Letters*, **127**, 144–8.

Casnellie, J. E., Ives, H. E., Jamieson, J. D. & Greengard, P. (1980). Cyclic GMP-dependent protein phosphorylation in intact medial tissue and isolated cells from vascular smooth muscle. *Journal of Biological Chemistry*, **255**, 3770–6.

Cerbai, E., Klöckner, U. & Isenberg, G. (1988). The α subunit of the GTP binding protein activates muscarinic potassium channels of the atrium. *Science*, **240**, 1782–3.

Christie, M. J. & North, R. A. (1988). Control of ion conductances by muscarinic receptors. *Trends in Pharmacological Sciences*, February Supplement, 30–3.

Cioffi, C. L. & El-Fakahany, E. E. (1986). Short-term desensitization of muscarinic cholinergic receptors in mouse neuroblastoma cells: selective loss of agonist low-affinity and pirenzepine high-affinity binding sites. *Journal of Pharmacology and Experimental Therapeutics*, **238**, 916–23.

Cockroft, S. & Gomperts, B. D. (1985). Role of guanine nucleotide binding protein in the activation of polyphosphoinositide phosphodiesterase. *Nature, London*, **314**, 534–6.

Codina, J., Yatani, A., Grenet, D., Brown, A. M., and Birnbaumer, L. (1987). The α subunit of the GTP binding protein G_K opens atrial potassium channels. *Science*, **236**, 442–5.

Cohen-Armon, M., Garty, H. & Sokolovsky, M. (1988). G-protein mediates voltage regulation of agonist binding to muscarinic receptors: effects on receptor Na^+ channel interaction. *Biochemistry*, **27**, 368–74.

Cohen-Armon, M. & Sokolovsky, M. (1986). Interactions between the muscarinic receptors, sodium channels, and guanine nucleotide-binding protein(s) in rat atria. *Journal of Biological Chemistry*, **261**, 12498–505.

Conklin, B. R., Brann, M. R., Buckley, N. J., Ma, A. L., Bonner, T. I. & Axelrod, J. (1988). Stimulation of arachidonic acid release and inhibition of mitogenesis by cloned genes for muscarinic receptor subtypes stably expressed in A9 L cells. *Proceedings of the National Academy of Sciences, USA*, **85**, 8698–702.

Constanti, A. & Sims, J. A. (1987*a*). Calcium-dependent potassium conductance in guinea pig olfactory cortex neurones *in vitro*. *Journal of Physiology*, **387**, 173–94.

Constanti, A. & Sims, J. A. (1987*b*). Muscarinic receptors mediating suppression of the M-current in guinea pig olfactory cortex neurones may be of the M_2-subtype. *British Journal of Pharmacology*, **90**, 3–5.

Cremo, C. R., Herron, G. S. & Schimerlik, M. I. (1981). Solubilization of the atrial muscarinic acetylcholine receptor: a new detergent system and rapid assays. *Analytical Biochemistry*, **115**, 331–8.

Cremo, C. & Schimerlik, M. I. (1983). Histrionicotoxin and alkylguanidine interactions with the solubilized and membrane-bound muscarinic acetylcholine receptor from porcine atria. *Archives of Biochemistry and Biophysics*, **224**, 506–14.

Cremo, C. & Schimerlik, M. I. (1984). Photoaffinity labeling of the solubilized, partially purified muscarinic receptor from porcine atria by *p*-azidoatropine methyl iodide. *Biochemistry*, **23**, 3494–501.

Curtis, C. A. M., Wheatley, M., Bansal, S., Birdsall, N. J. M., Eveleigh, P., Pedder, E. K., Poyner, D. & Hulme, E. C. (1989). [^{3}H]-Propylbenzilyl-choline mustard labels an acid residue in transmembrane helix 3 of the muscarinic receptor. *Journal of Biological Chemistry*, **264**, 489–95.

Dadi, H. K. & Morris, R. J. (1984). Muscarinic cholinergic receptor of rat brain. Factors influencing migration in electrophoresis and gel filtration in sodium dodecyl sulphate. *European Journal of Biochemistry*, **144**, 617–28.

Dascal, N., Gillo, B. & Lass, Y. (1985). Role of calcium mobilization in mediation of acetylcholine-evoked chloride currents in *Xenopus laevis* oocytes. *Journal of Physiology*, **366**, 299–313.

Dascal, N., Landau, E. M. & Lass, Y. (1984). *Xenopus* oocyte resting potential, muscarinic responses and the role of calcium and guanosine $3',5'$-cyclic monophosphate. *Journal of Physiology*, **352**, 551–74.

Dascal, N., Lotan, I., Gillo, B., Lester, H. A. & Lass, Y. (1985). Acetylcholine and phorbol esters inhibit potassium currents evoked by adenosine and cAMP in *Xenopus* oocytes. *Proceedings of the National Academy of Sciences, USA*, **82**, 6001–5.

Dodd, J. & Horn, J. P. (1983). Muscarinic inhibition of sympathetic C neurones in the bullfrog. *Journal of Physiology*, **334**, 271–9.

Dunlap, J. & Brown, J. H. (1983). Heterogeneity of binding sites on cardiac muscarinic receptors induced by the neuromuscular blocking agents gallamine and pancuronium. *Molecular Pharmacology*, **24**, 15–22.

Egan, T. M. & North, R. A. (1986). Acetylcholine hyperpolarizes central neurones by acting on an M_2 muscarinic receptor. *Nature London*, **319**, 405–7.

Ehlert, F. J. (1985). The relationship between muscarinic receptor occupancy and adenylate cyclase inhibition in the rabbit myocardium. *Molecular Pharmacology*, **28**, 410–21.

Fisher, S. K. (1988). Recognition of muscarinic cholinergic receptors in human SK-N-SH neuroblastoma cells by quaternary and tertiary ligands is dependent upon temperature, cell integrity, and the presence of agonists. *Molecular Pharmacology*, **33**, 414–22.

Fisher, S. K. & Agranoff, B. W. (1987). Receptor activation and inositol lipid hydrolysis in neural tissues. *Journal of Neurochemistry*, **48**, 999–1017.

Fisher, S. K., Klinger, P. D. & Agranoff, B. W. (1983). Muscarinic agonist binding and phospholipid turnover in brain. *Journal of Biological Chemistry*, **258**, 7358–63.

Fischmeister, R. & Hartzell, H. C. (1987). Cyclic guanosine 3′,5′-monophosphate regulates the calcium current in single cells from frog ventrical. *Journal of Physiology*, **387**, 453–72.

Florio, V. & Sternweis, P. (1985). Reconstitution of resolved muscarinic cholinergic receptors with purified GTP-binding proteins. *Journal of Biological Chemistry*, **260**, 3477–83.

Florio, V. & Sternweis, P. (1989). Mechanisms of muscarinic receptor action on G_o in reconstituted phospholipid vesicles. *Journal of Biological Chemistry*, **264**, 3909–15.

Fukuda, K., Higashida, H., Kubo, T., Maeda, A., Akiba, I., Bujo, H., Mishina, M. & Nuna, S. (1988). Selective coupling with K^+ currents of muscarinic acetylcholine receptor subtypes in NG108-15 cells. *Nature*, London, **335**, 355–8.

Fukuda, K., Kubo, T., Akiba, I., Maeda, A., Mishina, M. & Numa, S. (1987). Molecular distinction between muscarinic acetylcholine receptor subtypes. *Nature*, London, **327**, 623–25.

Galper, J. B., Dziekan, L. C., O'Hara, D. S. & Smith, T. W. (1982). The biphasic response of muscarinic cholinergic receptors in cultured heart cells to agonists. *Journal of Biological Chemistry*, **257**, 10344–56.

Galper, J. B., Dziekan, L. C. & Smith, T. C. (1984). The development of physiologic responsiveness to muscarinic agonists in chick embryo heart cell cultures. *Journal of Biological Chemistry*, **259**, 7382–90.

Galper, J. B., Haigh, L. S., Hart, A. C., O'Hara, D. S. & Livingston, D. J. (1987). Muscarinic cholinergic receptors in the embryonic chick heart: interaction of agonist, receptor, and guanine nucleotides studied by an improved assay for direct binding of the muscarinic agonist [^{3}H]*cis*-methyldioxolane. *Molecular Pharmacology*, **32**, 230–40.

Galper, J. B. & Smith, T. W. (1980). Agonist and guanine nucleotide modulation of muscarinic cholinergic receptors in cultured heart cells. *Journal of Biological Chemistry*, **255**, 9571–9.

Gavish, M. & Sokolovsky, M. (1982). Solubilization of muscarinic acetylcholine receptor by zwitterionic detergent from rat brain cortex. *Biochemical and Biophysical Research Communications*, **109**, 819–24.

George, W. J., Polson, J. B., O'Toole, A. G. & Goldberg, N. D. (1970). Elevation of guanosine 3′,5′-cyclic phosphate in rat heart after perfusion with acetylcholine. *Proceedings of the National Academy of Sciences, USA*, **66**, 398–403.

Gies, J-P. & Landry, Y. (1988). Sialic acid is selectively involved in the interaction of agonists with M_2 muscarinic receptors. *Biochemical and Biophysical Research Communications*, **150**, 673–80.

Gil, W. & Wolfe, B. B. (1985). Pirenzepine distinguishes between muscarinic receptor-mediated phosphoinositide breakdown and inhibition of adenylate cyclase. *Journal of Pharmacology and Experimental Therapeutics*, **232**, 608–16.

Gilman, A. G. (1987). G proteins: transducers of receptor-generated signals. *Annual Review of Biochemistry*, **56**, 615–49.

Gocayne, J., Robinson, D. A., Fitzgerald, M. G., Chung, F-Z., Kerlavage, A. R., Lentes, K-U., Lai, J., Wang, C-D., Fraser, C. M. & Venter, J. C. (1987). Primary structure of rat cardiac β-adrenergic and muscarinic cholinergic receptors obtained by automated DNA sequence analysis: further evidence for a multigene family. *Proceedings of the National Academy of Sciences, USA*, **84**, 8296–300.

Gorissen, H., Aerts, G., Ilien, B. & Laduron, P. (1991). Solubilization of muscarinic acetylcholine receptors from mammalian brain: an analytical approach. *Analytical Biochemistry*, **14**, 33–41.

Gurwitz, D., Kloog, Y. & Sokolovsky, M. (1985). High affinity binding of [³H] acetylcholine to muscarinic receptors. Regional distribution and modulation by guanine nucleotides. *Molecular Pharmacology*, **28**, 297–305.

Gurwitz, D. & Sokolovsky, M. (1987). Dual pathways in muscarinic receptor stimulation of phosphoinositide hydrolysis. *Biochemistry*, **26**, 633–8.

Haga, K. & Haga, T. (1983). Affinity chromatography of the muscarinic receptor. *Journal of Biological Chemistry*, **258**, 13575–9.

Haga, K., Haga, T., Ichiyama, A., Katada, T., Kurose, H. & Ui, M. (1985). Functional reconstitution of purified muscarinic receptors and inhibitory guanine nucleotide regulatory protein. *Nature*, London, **316**, 731–3.

Haga, K., Haga, T. & Ichiyama, A. (1986). Reconstitution of the muscarinic acetylcholine receptor. Guanine nucleotide-sensitive high affinity binding of agonists to purified muscarinic receptors reconstituted with GTP-binding proteins (G$_i$ and G$_o$). *Journal of Biological Chemistry*, **261**, 10133–40.

Haga, K. & Haga, T. (1985). Purification of the muscarinic acetylcholine receptor from porcine brain. *Journal of Biological Chemistry*, **260**, 7927–35.

Haga, T., Haga, K., Berstein, G., Nishiyama, T., Uchiyama, H. & Ichiyama, A. (1988). Molecular properties of muscarinic receptors. *Trends in Pharmacological Sciences*, Feb. Suppl., 12–18.

Halvorsen, S. W. & Nathanson, N. M. (1981). *In vivo* regulation of muscarinic acetylcholine receptor number and function in embryonic chick heart. *Journal of Biological Chemistry*, **256**, 7941–8.

Halvorsen, S. W. & Nathanson, N. M. (1984). Ontogenesis of physiological responsiveness and guanine nucleotide sensitivity of cardiac muscarinic receptors during chick embryonic development. *Biochemistry*, **23**, 5813–21.

Hammer, R., Berrie, C. P., Birdsall, N. J. M., Burgen, A. S. V. & Hulme, E. C. (1980). Pirenzepine distinguishes between different subclasses of muscarinic receptors. *Nature*, London, **283**, 90–2.

Hammer, R., Giraldo, E., Schiavi, G. B., Monferini, L. & Ladinsky, H. (1986). Binding profile of a novel cardioselective muscarine receptor antagonist, AF-DX116, to membranes of peripheral tissues and brain in the rat. *Life Sciences*, **38**, 1653–62.

Harden, T. K., Meeker, R. B. & Martin, M. W. (1983). Interaction of a radiolabeled agonist with cardiac muscarinic receptors. *Journal of Pharmacology and Experimental Therapeutics*, **227**, 570–7.

Harden, T. K., Petch, L. A., Traynelis, S. F. & Waldo, G. L. (1985). Agonist-induced alterations in the membrane form of muscarinic cholinergic receptors. *Journal of Biological Chemistry*, **260**, 13060–6.

Haslam, R. J. & Davidson, M. M. L. (1984). Receptor-induced diacylglycerol formation in permeabilized platelets; possible role for a GTP-binding protein. *Journal of Receptor Research*, **4**, 605–29.

Heacock, A. M., Fisher, S. K. & Agranoff, B. W. (1987). Enhanced coupling of neonatal muscarinic receptors in rat brain to phosphoinositide turnover. *Journal of Neurochemistry*, **48**, 1904–11.

Hepler, J. R. & Harden, T. K. (1986). Guanine nucleotide-dependent pertussis-toxin-insensitive stimulation of inositol phosphate formation by carbachol in a membrane preparation from human astrocytoma cells. *Biochemical Journal*, **239**, 141–6.

Herron, G. S., Miller, S., Manley, W-L. & Schimerlik, M. I. (1982). Ligand interactions with the solubilized porcine atrial muscarinic receptor. *Biochemistry* **21**, 515–20.

Herron, G. S. & Schimerlik, M. I. (1983). Glycoprotein properties of the solubilized atrial muscarinic receptor. *Journal of Neurochemistry*, **41**, 1414–20.

Ho, A. K. S., Ling, Q-L., Duffield, R., Lam, P. H. & Wang, J. H. (1987). Phosphorylation of brain muscarinic receptor: evidence of receptor regulation. *Biochemical and Biophysical Research Communications*, **142**, 911–18.

Ho, A. K. S., Shang, K. & Duffield, R. (1986). Calmodulin regulation of the cholinergic receptor in the rat heart during ontogeny and senescence. *Mechanisms of Ageing and Development*, **36**, 143–54.

Hokin, M. R. & Hokin, L. E. (1953). Enzyme secretion and the incorporation of ^{32}P into phospholipids of pancreas slices. *Journal of Biological Chemistry*, **203**, 967–77.

Hootman, S. R., Brown, M. E., Williams, J. A. & Logsdon, C. D. (1986). Regulation of muscarinic acetylcholine receptors in cultured guinea pig pancreatic acini. *American Journal of Physiology*, **251**, G75–83.

Hootman, S. R., Picado-Leonard, T. M. & Burnham, D. B. (1985). Muscarinic acetylcholine receptor structure in acinar cells of mammalian exocrine glands. *Journal of Biological Chemistry*, **260**, 4186–94.

Hughes, A. R. & Harden, T. K. (1986). Adenosine and muscarinic cholinergic receptors attenuate cyclic AMP accumulation by different mechanisms in 1321N1 astrocytoma cells. *Journal of Pharmacology and Experimental Therapeutics*, **237**, 173–8.

Hughes, A. R., Martin, M. W. & Harden, T. K. (1984). Pertussis toxin differentiates between two mechanisms of attenuation of cyclic AMP accumulation by muscarinic cholinergic receptors. *Proceedings of the National Academy of Sciences, USA*, **81**, 5680–4.

Hulme, E. C., Birdsall, N. J. M., Burgen, A. S. V. & Mehta, P. (1978). The binding of antagonists to brain muscarinic receptors. *Molecular Pharmacology*, **14**, 737–50.

Hunter, D. D. & Nathanson, N. M. (1984). Decreased physiological sensitivity mediated by newly synthesized muscarinic acetylcholine receptors in embryonic chicken heart. *Proceedings of the National Academy of Sciences, USA*, **81**, 3582–6.

Hurko, O. (1978). Specific [^{3}H]quinuclidinyl benzilate binding activity in digitonin-solubilized preparations from bovine brain. *Archives of Biochemistry and Biophysics*, **190**, 434–45.

Jakobs, K. H., Aktories, K. & Schultz, G. (1979). GTP-dependent inhibition of cardiac adenylate cyclase by muscarinic cholinergic agonists. *Naunyn-Schmiedeberg's Archives of Pharmacology*, **310**, 113–19.

Järv, J., Hedlund, B. & Bartfai, T. (1979). Isomerization of the muscarinic receptor–antagonist complex. *Journal of Biological Chemistry*, **254**, 5595–8.

Järv, J., Hedlund, B. & Bartfai, T. (1980). Kinetic studies on muscarinic antagonist–agonist competition. *Journal of Biological Chemistry*, **255**, 2649–51.

Jones, S. V. P., Barker, J. L., Bonner, T. I., Buckley, N. J. & Brann, M. R. (1988). Electrophysiological characterization of cloned M1 muscarinic

252   *Michael Schimerlik*

receptors expressed in A9 L cells. *Proceedings of the National Academy of Sciences, USA*, **85**, 4056–60.

Kababian, J. W., Steiner, A. L. & Greengard, P. (1975). Muscarinic cholinergic regulation of cyclic guanosine 3′,5′monophosphate in autonomic ganglia: possible role in synaptic transmission. *Journal of Pharmacology and Experimental Therapeutics*, **193**, 474–88.

Katada, T., Bokoch, G. M., Smigel, M. D., Ui, M. & Gilman, A. G. (1984). The inhibitory guanine nucleotide-binding regulatory component of adenylate cyclase. *Journal of Biological Chemistry*, **259**, 3586–95.

Katada, T., Kusakabe, K., Oinuma, M. & Ui, M. (1987). A novel mechanism for the inhibition of adenylate cyclase via inhibitory GTP-binding proteins. *Journal of Biological Chemistry*, **262**, 11897–900.

Katada, T., Oinuma, M. & Ui, M. (1986). Mechanisms for inhibition of the catalytic activity of adenylate cyclase by the guanine nucleotide-binding protein serving as the substrate of islet-activating protein, pertussis toxin. *Journal of Biological Chemistry*, **261**, 5215–21.

Kubo, T., Fukuda, K., Mikami, A., Maeda, A., Takahashi, H., Mishina, M., Haga, T., Haga, K., Ichiyama, A., Kangawa, K., Kojima, M., Matsuo, H., Hirose, T. & Numa, S. (1986*a*). Cloning, sequencing and expression of complementary DNA encoding the muscarinic acetylcholine receptor. *Nature*, London, 323, 411–16.

Kubo, T., Maeda, A., Sugimoto, K., Akiba, I., Mikami, A., Takahashi, H., Haga, T., Haga, K., Ichiyama, A., Kangawa, K., Matsui, H., Hirose, T. & Numa, S. (1986*b*). Primary structure of porcine cardiac muscarinic acetylcholine receptor deduced from the cDNA sequence. *Federation of European Biochemical Societies Letters*, **209**, 367–72.

Kuno, T., Shirakawa, O. & Tanaka, C. (1983). Regulation of the solubilized bovine cerebral cortex muscarinic receptor by GTP and Na^+. *Biochemical and Biophysical Research Communications*, **112**, 948–53.

Kurose, H., Katada, T., Amano, T. & Ui, M. (1983). Specific uncoupling by islet-activating protein, pertussis toxin, of negative signal transduction via α-adrenergic, cholinergic, and opiate receptors in neuroblastoma × glioma hybrid cells. *Journal of Biological Chemistry*, **258**, 4870–5.

Kurose, H., Katada, T., Haga, T., Haga, K., Ichiyama, A. & Ui, M. (1986). Functional interaction of purified muscarinic receptors with purified inhibitory guanine nucleotide regulatory proteins reconstituted in phospholipid vesicles. *Journal of Biological Chemistry*, **261**, 6423–8.

Kurose, H. & Ui, M. (1983). Functional uncoupling of muscarinic receptors from adenylate cyclase in rat cardiac membranes by the active component of islet-activating protein, pertussis toxin. *Journal of Cyclic Nucleotide and Protein Phosphorylation Research*, **9**, 305–10.

Kurose, H. & Ui, M. (1985). Dual pathways of receptor-mediated cyclic GMP generation in NG108–15 cells as differentiated by susceptibility to islet-activating proteins, pertussis toxin. *Archives of Biochemistry and Biophysics*, **238**, 424–34.

Kwatra, M. M. & Hosey, M. M. (1986). Phosphorylation of the cardiac muscarinic receptor in intact chick heart and its regulation by a muscarinic agonist. *Journal of Biological Chemistry*, **261**, 12429–32.

Kwatra, M. M., Leung, E., Maan, A. C., McMahon, K. K., Ptasienski, J., Green, R. D. & Hosey, M. M. (1987). Correlation of agonist-induced phosphorylation of chick heart muscarinic receptors with receptor desensitization. *Journal of Biological Chemistry*, **262**, 16314–21.

Kyte, J. & Doolittle, R. F. (1982). A simple method for displaying the hydrophobic character of a protein. *Journal of Molecular Biology*, **157**, 105–32.

Lai, J., Mei, L., Roeske, W. R., Chung, F-Z., Yamamura, H. I. & Venter, J. C. (1988). The cloned M_1 muscarinic receptor is associated with the hydrolysis of phosphatidylinositols in transfected murine B82 cells. *Life Sciences*, **42**, 2489–502.

Lai, W. S. & El-Fakahany, E. E. (1987). Phorbol ester-induced inhibition of cyclic GMP formation mediated by muscarinic receptors in murine neuroblastoma cells. *Journal of Pharmacology and Experimental Therapeutics*, **241**, 366–73.

Lai, W. S., Ramkumar, V. & El-Fakahany, E. E. (1985). Possible allosteric interaction of 4-aminopyridine with rat brain muscarinic acetylcholine receptors. *Journal of Neurochemistry*, **44**, 1936–42.

Lambrecht, G., Feifel, R., Forth, B., Strohmann, C., Tacko, R. & Mutschler, E. (1988). p-Fluoro-hexahydrosiladifenidol: the first M2β-selective muscarinic antagonist. *European Journal of Pharmacology*, **152**, 193–4.

Large, T. H., Rauh, J. J., De Mello, F. & Klein, W. L. (1985a). Two molecular weight forms of muscarinic acetylcholine receptors in the avian central nervous system: switch in predominant form during differentiation of synapses. *Proceedings of the National Academy of Sciences, USA*, **82**, 8785–9.

Large, T. H., Cho, N. J., De Mello, F. & Klein, W. L. (1985b). Molecular alteration of a muscarinic acetylcholine receptor system during synaptogenesis. *Journal of Biological Chemistry*, **260**, 8873–81.

Lee, J-H. & El-Fakahany, E. E. (1985). Use of intact rat brain cells as a model to study regulation of muscarinic acetylcholine receptors. *Life Sciences*, **37**, 515–21.

Lee, T. W. T., Sole, M. J. & Wells, J. W. (1986). Assessment of a ternary model for the binding of agonists to neurohumoral receptors. *Biochemistry*, **25**, 7009–20.

Leiber, D., Harbon, S., Guillet, J.-G., André, C. & Strosberg, A. D. (1984). Monoclonal antibodies to purified muscarinic receptor display agonist-like activity. *Proceedings of the National Academy of Sciences, USA*, **81**, 4331–4.

Liang, B. & Galper, J. B. (1987). Reconstitution of muscarinic cholinergic inhibition of adenylate cyclase activity in homogenates of embryonic chick hearts by membranes of adult chick hearts. *Journal of Biological Chemistry*, **262**, 2494–501.

Liang, B. T., Helmich, M. R., Neer, E. J. & Galper, J. B. (1986). Development of muscarinic cholinergic inhibition of adenylate cyclase in embryonic chick heart. *Journal of Biological Chemistry*, **261**, 9011–21.

Liang, M., Martin, M. W. & Harden, T. K. (1987). [³]Propylbenzilyl-choline mustard-labeling of muscarinic cholinergic receptors that selectively couple to phospholipase c or adenylate cyclase in two cultured cell lines. *Molecular Pharmacology*, **32**, 443–9.

Liles, W. C., Hunter, D. D., Meier, K. E. & Nathanson, N. M. (1986). Activation of protein kinase C induces rapid internalization and subsequent degradation of muscarinic acetylcholine receptors in neuroblastoma cells. *Journal of Biological Chemistry*, **261**, 5307–13.

Litosch, I., Wallis, C. & Fain, J. N. (1985). 5-Hydroxytryptamine stimulates inositol phosphate production in a cell-free system from blowfly salivary glands. Evidence for a role of GTP in coupling receptor activation to phosphoinositide breakdown. *Journal of Biological Chemistry*, **260**, 5464–71.

Lo, W. W. Y. & Hughes, J. (1987a). Pertussis toxin distinguishes between muscarinic receptor-mediated inhibition of adenylate cyclase and stimulation of phosphoinositide hydrolysis in Flow 9000 cells. *Federation of European Biochemical Societies Letters*, **220**, 155–8.

Lo, W. W. Y. & Hughes, J. (1987b). A novel cholera toxin-sensitive G-protein (G_c) regulating receptor-mediated phosphoinositide signalling in human pituitary cloned cells. *Federation of European Biochemical Societies Letters*, **220**, 327–31.

Lo, W. W. Y. & Hughes, J. (1988). Differential regulation of cholecystokinin- and muscarinic-receptor-mediated phosphoinositide turnover in Flow 9000 cells. *Biochemical Journal*, **251**, 625–30.

Logothetis, D. E., Kim, D., Northup, J. K., Neer, E. J. & Clapham, D. E. (1988). Specificity of action of guanine nucleotide-binding regulatory protein subunits on the cardiac muscarinic K^+ channel. *Proceedings of the National Academy of Sciences, USA*, **85**, 5814–18.

Luetje, C. W., Gierschik, P., Milligan, G., Unson, C., Spiegel, A. & Nathanson, N. M. (1987a). Tissue-specific regulation of GTP-binding protein and muscarinic acetylcholine receptor levels during cardiac development. *Biochemistry*, **26**, 4876–84.

Luetje, C. W., Brumwell, C., Norman, M. G., Peterson, G. L., Schimerlik, M. I. & Nathanson, N. M. (1987b). Isolation and characterization of monoclonal antibodies specific for the cardiac muscarinic acetylcholine receptor. *Biochemistry*, **26**, 6892–6.

Luthin, G. R., Harkness, J., Artymyshyn, R. P. & Wolfe, B. B. (1988). Antibodies to a synthetic peptide can be used to distinguish between muscarinic acetylcholine receptor binding sites in brain and heart. *Molecular Pharmacology*, **34**, 327–33.

Luthin, G. R. & Wolfe, B. B. (1984). [^{3}H]Pirenzepine and [^{3}H]quinuclidinyl benzilate binding to brain muscarinic cholinergic receptors. Differences in measured receptor density are not explained by differences in receptor isomerization. *Molecular Pharmacology*, **26**, 164–9.

Madison, D. V., Lancaster, B. & Nicoll, R. A. (1987). Voltage clamp analysis of cholinergic action in the hippocampus. *Journal of Neuroscience*, **7**, 733–41.

Malenka, R. C., Madison, D. V., Andrade, R. & Nicoll, R. A. (1986). Phorbol esters mimic some cholinergic actions in hippocampal pyrimidal neurons. *Journal of Neuroscience*, **6**, 475–80.

Martin, J. M., Hunter, D. D. & Nathanson, N. M. (1985). Islet activating protein inhibits physiological responses evoked by cardiac muscarinic acetylcholine receptors. Role of guanosine triphosphate binding proteins in regulation of potassium permeability. *Biochemistry*, **24**, 7521–5.

Martin, M. W., Smith, M. M. & Harden, T. K. (1984). Modulation of muscarinic cholinergic receptor affinity for antagonists in rat heart. *Journal of Pharmacology and Experimental Therapeutics*, **230**, 424–30.

Masters, S. B., Martin, M. W., Harden, T. K. & Brown, J. H. (1985). Pertussis toxin does not inhibit muscarinic-receptor-mediated phosphoinositide hydrolysis or calcium mobilization. *Biochemical Journal*, **227**, 933–7.

Mattera, R., Pitts, B. J. R., Entman, M. L. & Birnbaumer, L. (1985). Guanine nucleotide regulation of a mammalian myocardial muscarinic receptor system. *Journal of Biological Chemistry*, **260**, 7410–21.

McCormick, D. A. & Prince, D. A. (1986). Acetylcholine induces burst firing in thalamic reticular neurones by activating a potassium conductance. *Nature, London*, **319**, 402–5.

McKinney, M. & Richelson, E. (1986*a*). Muscarinic responses and binding in a murine neuroblastoma clone (N1E-115): cyclic GMP formation is mediated by a low affinity agonist-receptor conformation and cyclic AMP reduction is mediated by a high affinity agonist-receptor conformation. *Molecular Pharmacology*, **30**, 207–11.

McKinney, M. & Richelson, E. (1986*b*). Blockage of N1E-115 murine neuroblastoma muscarinic receptor function by agents that affect the metabolism of arachidonic acid. *Biochemical Pharmacology*, **35**, 2389–97.

Meyer, M. R., Gainer, M. W. & Nathanson, N. M. (1982). *In vivo* regulation of muscarinic cholinergic receptors in embryonic chick brain. *Molecular Pharmacology*, **21**, 280–6.

Mochida, S. & Kobayashi, H. (1986). Three types of muscarinic conductance changes in sympathetic neurons discriminately evoked by the different concentrations of acetylcholine. *Brain Research*, **383**, 299–304.

Moriarty, T. M., Gillo, B., Carty, D. J., Premont, R. T., Landau, E. M. & Iyengar, R. (1988). $\beta\gamma$ subunits of GTP-binding proteins inhibit muscarinic receptor stimulation of phospholipase C. *Proceedings of the National Academy of Sciences, USA*, **85**, 8865–9.

Murad, F., Chi, Y.-M., Rall, T. W. & Sutherland, E. W. (1962). The effect of catecholamines and choline esters on the formation of adenosine 3′,5′-phosphate by preparations from cardiac muscle and liver. *Journal of Biological Chemistry*, **237**, 1233–8.

Nakamura, T. & Ui, M. (1985). Simultaneous inhibitions of inositol phospholipid breakdown, arachidonic acid release, and histamine secretion in mast cells by islet-activating protein, pertussis toxin. *Journal of Biological Chemistry*, **260**, 3584–95.

Nakashima, S., Nagata, K.-I., Ueeda, K. & Nozawa, Y. (1988). Stimulation of arachidonic acid release by guanine nucleotides in saponin-permeabilized neutrophils: evidence for involvement of GTP-binding protein in phospholipase A_2 activation. *Archives of Biochemistry and Biophysics*, **261**, 375–83.

Nishizuka, Y. (1988). The molecular heterogeneity of protein kinase C and its implications for cellular regulation. *Nature*, London, **334**, 661–5.

North, R. A., Slack, B. E. & Surprenant, A. (1985). Muscarinic M_1 and M_2 receptors mediate depolarization and presynaptic inhibition in guinea pig enteric nervous system. *Journal of Physiology*, **368**, 435–52.

Nowak, L. M. & MacDonald, R. L. (1983). Ionic mechanism of muscarinic cholinergic depolarization of mouse spinal cord neurons in cell culture. *Journal of Neurophysiology*, **49**, 792–803.

Ohako, S. & Deguchi, T. (1984). Receptor-mediated regulation of calcium mobilization and cyclic GMP synthesis in neuroblastoma cells. *Biochemical and Biophysical Research Communications*, **122**, 333–9.

Okajima, F., Katada, T. & Ui, M. (1985). Coupling of the guanine nucleotide regulatory protein to chemotactic peptide receptors in neutrophil membranes and its uncoupling by islet-activating protein, pertussis toxin. *Journal of Biological Chemistry*, **260**, 6761–8.

Orellana, S., Solski, P. A. & Brown, J. H. (1987). Guanosine 5′-O-(thiotriphosphate)-dependent inositol trisphosphate formation in membrane is inhibited by phorbol ester and protein kinase C. *Journal of Biological Chemistry*, **262**, 1638–43.

Oron, Y., Dascal, N., Nadler, E. & Lupu, M. (1985). Inositol 1,4,5-trisphosphate mimics muscarinic response in *Xenopus* oocytes. *Nature, London*, **313**, 141–3.

Ovchinnikov, Y. A. (1982). Rhodopsin and bacteriorhodopsin: structure-function relationships. *Federation of European Biochemical Societies Letters*, **148**, 179–91.

Parker, I. & Miledi, R. (1986). Changes in intracellular calcium and in membrane currents evoked by injection of inositol trisphosphate into *Xenopus* oocytes. *Proceedings of the Royal Society London B*, **228**, 307–15.

Peralta, E. G., Winslow, J. W., Peterson, G. L., Smith, D. H., Ashkenazi, A., Ramachandran, J., Schimerlik, M. I. & Capon, D. J. (1987*a*). Primary structure and biochemical properties and an M_2 muscarinic receptor. *Science*, **236**, 600–5.

Peralta, E. G., Ashkenazi, A., Winslow, J. W., Smith, D. H., Ramachandran, J. & Capon, D. J. (1987*b*). Distinct primary structures, ligand binding properties and tissue-specific expression of four human muscarinic acetylcholine receptors. *EMBO Journal*, **6**, 3923–9.

Peralta, E. G., Ashkenazi, A., Winslow, J. W., Ramachandran, J. & Capon, D. J. (1988). Differential regulation of PI hydrolysis and adenylyl cyclase by muscarinic receptor subtypes. *Nature*, London, **334**, 434–7.

Peterson, G. L. (1989). Consensus residues at the acetylcholine binding site of cholinergic proteins. *Journal of Neuroscience Research*, **22**, 488–503.

Peterson, G. L., Herron, G. S., Yamaki, M., Fullerton, D. S. & Schimerlik, M. I. (1984). Purification of the muscarinic acetylcholine receptor from porcine atria. *Proceedings of the National Academy of Sciences, USA*, **81**, 4993–7.

Peterson, G. L., Rosenbaum, L. C., Broderick, D. J. & Schimerlik, M. I. (1986). Physical properties of the purified cardiac muscarinic receptors. *Biochemistry*, **25**, 3189–202.

Peterson, G. L., Rosenbaum, L. C. & Schimerlik, M. I. (1988). Solubilization and hydrodynamic properties of pig atrial muscarinic acetylcholine receptor in dodecyl β-D-maltoside. *Biochemical Journal*, **255**, 553–60.

Peterson, O. H. & Maruyama, Y. (1984). Calcium activated potassium channels and their role in secretion. *Nature*, London, **307**, 693–6.

Pfaffinger, P. J., Martin, J. M., Hunter, D. D., Nathanson, N. M. & Hille, B. (1985). GTP-binding proteins couple cardiac muscarinic receptors to a K channel. *Nature*, London, **317**, 536–8.

Raposo, G., Dunia, I., Marullo, S., André, C., Guillet, J.-G., Strosberg, A. D., Benedetti, E. L. & Hoebeke, J. (1987). Redistribution of muscarinic acetylcholine receptors on human fibroblasts induced by regulatory ligands. *Biology of the Cell*, **60**, 117–24.

Rauh, J. J., Lambert, M. P., Cho, N. J., Chin, H. & Klein, W. L. (1986). Glycoprotein properties of muscarinic acetylcholine receptors from bovine cerebral cortex. *Journal of Neurochemistry*, **46**, 23–32.

Reichman, M., Nen, W. & Hokin, L. E. (1987). Highly sensitive muscarinic receptors in the cerebellum are coupled to prostaglandin formation. *Biochemical and Biophysical Research Communications*, **146**, 1256–61.

Repke, H. (1987). Muscarinic receptor-detergent complexes with different biochemical properties: selective solubilization, lectin affinity chromatography and ligand binding studies. *Biochimica et Biophysica Acta*, **929**, 47–61.

Repke, H. & Matthies, H. (1980). Biochemical characterization of solubilized muscarinic acetylcholine receptors. *Brain Research Bulletin*, **5**, 703–9.

Repke, H. & Schmitt, M. (1987). Electrophoretic characterization of muscarinic receptors under denaturing and non-denaturing conditions: computer-assisted Ferguson plot analysis. *Biochimica et Biophysica Acta*, **929**, 62–73.

Richelson, E., Prendergast, F. G. & Divinetz-Romero, S. (1978). Muscarinic receptor-mediated cyclic GMP formation by cultured nerve cells – ionic dependence and effects of local anesthetics. *Biochemical Pharmacology*, **27**, 2039–48.

Rodbell, M. (1980). The role of hormone receptors and GTP-regulatory proteins in membrane transduction. *Nature*, London, 284, 17–22.

Roeske, W. R. & Venter, J. C. (1984). The differential loss of [^{3}H]pirenzepine *vs.* [^{3}H](–)quinuclidinyl-benzilate binding to soluble rat brain muscarinic receptors indicates that pirenzepine binds to an allosteric state of the muscarinic receptor. *Biochemical and Biophysical Research Communications*, **118**, 950–7.

Rosenbaum, L. C., Malencik, D. A., Anderson, S. R., Tota, M. R. & Schimerlik, M. I. (1987). Phosphorylation of the porcine atrial muscarinic acetylcholine receptor by cyclic AMP dependent protein kinase. *Biochemistry*, **26**, 8183–8.

Rosenberger, L. B., Yamamura, H. I. & Roeske, W. R. (1980). Cardiac muscarinic cholinergic receptor binding is regulated by Na^+ and guanyl nucleotides. *Journal of Biological Chemistry*, **255**, 820–3.

Schimerlik, M. I., Miller, S., Peterson, G. L., Rosenbaum, L. C. & Tota, M. R. (1986). Biochemical studies on muscarinic receptors in porcine atrium. *Trends in Pharmacological Sciences*, Feb. Suppl., 2–7.

Schimerlik, M. I. & Searles, R. P. (1980). Ligand interactions with membrane-bound porcine atrial muscarinic receptor(s). *Biochemistry*, **19**, 3407–13.

Schreiber, G., Henis, Y. I. & Sokolovsky, M. (1985*a*). Analysis of ligand binding to receptors by competition kinetics. Application to muscarinic antagonists in rat brain cortex. *Journal of Biological Chemistry*, **260**, 8789–94.

Schreiber, G., Henis, Y. I. & Sokolovsky, M. (1985*b*). Rate constants of agonist binding to muscarinic receptor in rat brain medulla. *Journal of Biological Chemistry*, **260**, 8795–802.

Serra, M., Smith, T. L. & Yamamura, H. I. (1986). Phorbol esters alter muscarinic receptor binding and inhibit polyphosphoinositide breakdown in human neuroblastoma (SH-SY5Y) cells. *Biochemical and Biophysical Research Communications*, **140**, 160–6.

Shirakawa, O., Kuno, T. & Tanaka, C. (1983). The glycoprotein nature of solubilized muscarinic acetylcholine receptors from bovine cerebral cortex. *Biochemical and Biophysical Research Communications*, **115**, 814–19.

Shirakawa, O. & Tanaka, C. (1985). Molecular characterization of muscarinic receptor subtypes in bovine cerebral cortex by radiation inactivation and molecular exclusion h.p.l.c. *British Journal of Pharmacology*, **86**, 375–83.

Silva, W. I., Andres, A., Schook, W. & Puszkin, S. (1986). Evidence for the presence of muscarinic acetylcholine receptors in bovine brain coated vesicles. *Journal of Biological Chemistry*, **261**, 14788–96.

Simmons, M. A. & Hartzell, H. C. (1988). Role of phosphodiesterase in regulation of calcium current in isolated cardiac myocytes. *Molecular Pharmacology*, **33**, 664–71.

Sims, S. M., Singer, J. J. & Walsh, J. V. Jr. (1985). Cholinergic agonists suppress a potassium current in freshly dissociated smooth muscle cells of the toad. *Journal of Physiology*, **367**, 503–29.

Sims, S. M., Singer, J. J. & Walsh, J. V. Jr. (1988). Antagonistic adrenergic-muscarinic regulation of M current in smooth muscle cells. *Science*, **239**, 190–3.

Söling, H-D., Machado-De Domenech, E., Kleineke, J. & Fest, W. (1987). Early effects of β-adrenergic and muscarinic secretagogues on lipid and phospholipid metabolism in guinea pig parotid acinar cells. *Journal of Biological Chemistry*, **262**, 16786–92.

Stein, R., Pinkas-Kramarski & Sokolovsky, M. (1988). Cloned M1 muscarinic receptors mediate both adenylate cyclase inhibition and phosphoinositide turnover. *EMBO Journal*, **7**, 3031–5.

Strader, C. D., Sigal, I. S., Register, R. B., Candelore, M. R., Rands, E. & Dixon, R. A. F. (1987). Identification of residues required for ligand binding to the β-adrenergic receptor. *Proceedings of the National Academy of Sciences, USA*, **84**, 4384–8.

Streb, H., Irvine, R. F., Berridge, M. J. & Schultz, I. (1983). Release of Ca^{2+} from a non-mitochondrial intracellular store in pancreatic acinar cells by inositol-1,4,5-trisphosphate. *Nature, London*, **306**, 67–8.

Tanner, L. I., Harden, T. K., Wells, J. N. & Martin, M. W. (1986). Identification of the phosphodiesterase regulated by muscarinic cholinergic receptors of 1321N1 human astrocytoma cells. *Molecular Pharmacology*, **29**, 455–60.

Tota, M. R., Kahler, K. R. & Schimerlik, M. I. (1987). Reconstitution of the purified porcine atrial muscarinic acetylcholine receptor with purified porcine atrial inhibitory guanine nucleotide binding protein. *Biochemistry*, **26**, 8175–82.

Venter, J. C. (1983). Muscarinic cholinergic receptor structure. Receptor size, membrane orientation, and absence of major phylogenetic structural diversity. *Journal of Biological Chemistry*, **258**, 4842–8.

Venter, J. C., Eddy, B., Hall, L. M. & Fraser, C. M. (1984). Monoclonal antibodies detect the conservation of muscarinic cholinergic receptor structure from *Drosophila* to human brain and detect possible structural homology with α_1-adrenergic receptors. *Proceedings of the National Academy of Sciences, USA*, **81**, 272–6.

Wastek, G. J., Lopez, J. R. & Richelson, E. (1980). Demonstration of a muscarinic receptor-mediated cyclic GMP-dependent hyperpolarization of the membrane potential of mouse neuroblastoma cells using [^{3}H]tetraphenylphosphonium. *Molecular Pharmacology*, **19**, 15–20.

Wheatley, M., Birdsall, N. J. M., Curtis, C., Eveleigh, P., Pedder, E. K., Poyner, D., Stockton, J. M. & Hulme, E. C. (1987). The structure and properties of the purified muscarinic acetylcholine receptor from rat forebrain. *Biochemical Society Transactions*, **15**, 113–16.

Wong, H-M. S., Sole, M. J. & Wells, J. W. (1986). Assessment of mechanistic proposals for the binding of agonists to cardiac muscarinic receptors. *Biochemistry*, **25**, 6995–7008.

Worley, P. F., Baraban, J. M., McCarren, M., Snyder, S. H. & Alger, B. E. (1987). Cholinergic phosphatidylinositol modulation of inhibitory, G-protein linked, neurotransmitter actions: electrophysiological studies in rat hippocampus. *Proceedings of the National Academy of Sciences, USA*, **84**, 3467–71.

Worley, P. F., Heller, W. A., Snyder, S. H. & Baraban, J. M. (1988). Lithium blocks a phosphoinositide-mediated cholinergic response in hippocampal slices. *Science*, **239**, 1428–9.

Yamamura, H. I. & Snyder, S. H. (1974). Muscarinic cholinergic binding in rat brain. *Proceedings of the National Academy of Sciences, USA*, **71**, 1725–9.

Yatani, A., Codina, J., Brown, A. M. & Birnbaumer, L. (1987). Direct activation of mammalian atrial muscarinic potassium channels by GTP regulatory protein G_K. *Science*, **235**, 207–11.

Yatani, A., Hamm, H., Codina, J., Mazzoni, M. R., Birnbaumer, L. & Brown, A. M. (1988*a*). A monoclonal antibody to the α subunit of G_K blocks muscarinic activation of atrial K^+ channels. *Science*, **241**, 828–31.

Yatani, A., Mattera, R., Codina, J., Graf, R., Okabe, K., Padrell, E., Iyengar, R., Brown, A. M. & Birnbaumer, L. (1988*b*). The G protein-gated atrial K^+ channel is stimulated by three distinct $G_{i\alpha}$-subunits. *Nature*, London, **336**, 680–2.

10

The insulin receptor

KENNETH SIDDLE

Introduction

Insulin is the major anabolic hormone in mammals. It is concerned in the regulation of carbohydrate, lipid and protein metabolism and can also function as a growth factor. Lack of insulin or resistance to its action in diabetes mellitus results in serious metabolic imbalance (Taylor & Agius, 1988). Interest in the mechanism of insulin action is therefore long-standing, and the insulin receptor was one of the first receptors to be characterised in biochemical, as opposed to pharmacological, terms. Indeed, studies of the insulin receptor have served as models for much subsequent work on receptors for a wide variety of polypeptide hormones and growth factors.

The insulin receptor

The pioneering work of Sutherland and colleagues on the action of glucagon and catecholamines, culminating in the second messenger hypothesis of hormone action (Sutherland, Oye & Butcher, 1965), was instrumental in focussing attention on the plasma membrane as the primary site of interaction of hormones with target cells. The rapid reversal of insulin action on adipocytes by washing or addition of anti-insulin serum (Crofford, 1968), and the loss of insulin sensitivity after tryp-sinisation of adipocytes (Kono, 1969) suggested that insulin also interacted with a cell surface receptor. The activity of insulin covalently bound to Sepharose beads appeared to confirm that insulin, once bound to its receptor, did not enter cells to exert its metabolic effects (Cuatrecasas, 1969), although the validity of these studies was subsequently questioned (Butcher et al., 1973). The demonstration that antibodies to the insulin receptor elicited a wide range of insulin-mimetic metabolic effects provided strong evidence that the mechanisms responsible for regulating intracellular pathways were intrinsic to the receptor, and did not require the participation of insulin itself (Kahn *et al.*, 1981). The overwhelming weight of evidence now indicates that a single type of receptor is responsible for initiating all the effects of insulin on target cells, although many details of the signalling pathways which are activated by insulin binding remain obscure. The structure and function of the insulin receptor has been the subject of several excellent reviews in recent years (Czech, 1985*a*; Goldfine, 1987; Rosen, 1987; Kahn & White, 1988; Sale, 1988; Houslay & Siddle, 1989; Zick, 1989).

Studies on the insulin receptor over the last two decades can be roughly divided into four phases, dictated largely by the availability of appropriate experimental techniques. The initial phase of characterisation in the 1970s involved detailed studies of binding properties of the receptor, in membrane preparations or isolated cells, using $[^{125}I]$-insulin as a tracer. These studies indicated that receptors for insulin were generally similar in all target tissues and vertebrate species. The binding to membranes and isolated receptors was readily reversible, although the kinetics were complex and suggested site heterogeneity or cooperative interactions (Kahn, 1976). It later became apparent that, in intact cells at physiological temperatures, the insulin/receptor complex was internalised rapidly and much of the insulin was degraded intracellularly (Gorden *et al.*, 1980). Receptor-mediated endocytosis was discovered to be a general phenomenon for a wide range of polypeptide hormones, nutrients and other proteins (Goldstein, Anderson & Brown, 1979). In the case of insulin, the process suggests a mechanism not only for hormone degradation but also for the regulation of expression of receptor at the cell surface, by ligand-induced redistribution and/or degradation, resulting in receptor down-regulation (Kahn, 1976, see also Sonne, 1988).

In the late 1970s, several techniques combined to permit advances in the structural characterisation of insulin receptors. These involved affinity chromatography for receptor purification, covalent affinity labelling of receptors with radioactive ligands, and the use of specific antibodies for receptor immunoprecipitation. Two distinct subunits, termed α and β, were identified, of which α was predominantly responsible for insulin binding and β was a transmembrane protein. The structure of the native receptor was established as a disulphide-linked heterotetramer β-α-α-β (Czech, Massague & Pilch, 1981; Jacobs & Cuatrecasas, 1983). It was shown that α and β subunits were derived from a single polypeptide precursor by

REGULATION OF KEY ENZYMES BY COVALENT MODIFICATION
Decreased serine phosphorylation
glycogen synthase (muscle), pyruvate dehydrogenase (fat) – activated
triacylglycerol lipase (fat) – inhibited

Increased serine phosphorylation
acetyl CoA carboxylase (fat) – activated
insulin receptor tyrosyl kinase (various) – inhibited?
ATP citrate lyase (liver, fat), ribosomal S6 (various) – effect unknown

PROTEIN TRANSLOCATION VIA MEMBRANE VESICLES
To plasma membrane: glucose transporters (muscle, fat)
From plasma membrane: insulin receptors (various)

REGULATION OF GENE TRANSCRIPTION
Increased expression: glucokinase (liver)
Decreased expression: PEP carboxykinase (liver)

Fig. 10.1. *Insulin action.* Examples of insulin effects, together with the tissues in which they are prominent, are listed according to what is currently understood to be the underlying molecular mechanism. This is by no means a complete list but rather is intended to illustrate the complexity of insulin action.

proteolysis (Hedo & Gorden, 1985), an unexpected echo of the generation of A and B chains of insulin itself. In this context it might have been more appropriate to regard the insulin receptor as a 'homodimer' of two $(\alpha\beta)$ units, since the α and β 'subunits' do not have an independent existence any more than the A and B chains of insulin.

In the early 1980s progress was made finally in understanding the activity of insulin receptors. It was found that the β-subunit possessed intrinsic tyrosine-specific protein kinase activity which catalysed an insulin-stimulated autophosphorylation reaction (Gammeltoft & Van Obberghen, 1986). This followed the earlier discovery of tyrosine kinase activity in the product of the v-*src* oncogene and in the receptor for epidermal growth factor (Hunter & Cooper, 1985). However, with all these kinases and others discovered subsequently, it has proved difficult to identify physiologically important substrates, although the effort to do so continues.

Recent years have been dominated by the influence of recombinant DNA techniques. Full-length cDNAs for the insulin receptor precursor have been cloned and sequenced, providing new insights into receptor structure (Ullrich *et al.*, 1985; Ebina *et al.*, 1985 *b*). Manipulation of cloned sequences has allowed the construction and expression of chimaeric and mutant receptors as a powerful approach to understanding the relationship between receptor structure and function. Such studies have provided strong evidence for an essential role of the receptor kinase in mediating insulin action (Rosen, 1987; Kahn & White, 1988; Zick, 1989). In the future it is to be expected that expression of the extracellular and kinase domains as discrete soluble proteins will provide material suitable for detailed conformational analysis by n.m.r. spectroscopy and X-ray crystallography.

The mechanism of insulin action

In parallel with studies of the insulin receptor itself a large body of knowledge has accumulated concerning the pathways which are ultimately regulated by insulin and the mechanisms involved in this regulation (see Czech, 1985b; Kahn, 1985; Denton, 1986). The targets include regulatory enzymes, membrane transporters and receptors, and specific nuclear genes (Fig. 10.1). Many regulatory enzymes are modulated by reversible phosphorylation on serine (or threonine) residues, but in some cases the effect of insulin is to decrease phosphorylation and in others to promote an increase. Some of these proteins are susceptible to multisite phosphorylation in response to a variety of stimuli and reflecting the action of different protein kinases. A second category of insulin effects, including the stimulation of glucose transport, involves the translocation of membrane proteins between different subcellular compartments. All these effects of insulin are clearly very rapid, but other actions, on protein synthesis, become apparent more slowly. Insulin stimulates protein synthesis at the translational level, and inhibits proteolysis, in muscle. It also regulates the expression of specific genes, both positively and negatively, particularly in liver. The longer-term nature of these latter effects most probably reflects the period over which changes are easily detected experimentally rather than any inherent difference in the initial signalling pathways. It is not yet clear whether the membrane translocation and gene regulatory effects of insulin depend ultimately on phosphorylation/dephosphorylation reactions, as the full details of these processes remain to be elucidated. The mechanism of the mitogenic effect of insulin is even more obscure, and, although insulin can certainly act as a growth factor in tissue culture, its physiological role in this regard is uncertain.

It is astonishing, considering the effort devoted to studying both the insulin receptor and the intracellular targets regulated by insulin, that there is still no general agreement as to the signalling pathways involved. Ideas were initially dominated by the second messenger hypothesis. Various candidates for the role of messenger (cAMP decrease, cGMP, Ca, peptides) were proposed but not substantiated (Czech, 1977), although effects on cAMP concentration may contribute to some actions of insulin in liver and adipose tissue. More recently, inositol phosphate glycans have been proposed as diffusible mediators of insulin action (Saltiel & Cuatrecasas, 1988). These polar head groups of membrane glycolipids are apparently released as a consequence of the activation of a specific phospholipase C, but evidence for a physiological role in regulating the activity of intracellular enzymes is not yet convincing. Relevant to this hypothesis are suggestions that protein kinase C (Saltiel & Cuatrecasas, 1988) and G proteins (Houslay & Siddle, 1989) are involved in some aspects of insulin action. Increases in diacylglycerol, which activates protein kinase C, could result, at least in part, from the action of an insulin-stimulated phospholipase, and G proteins might be expected to act as transducers of the regulatory effect on phospholipase as in other receptor–effector systems (Gilman, 1987).

As indicated earlier, there is now good evidence that activation of the receptor tyrosine kinase is an essential first step in signalling pathways (Rosen, 1987; White & Kahn, 1989; Zick, 1989) but the nature of its involvement is unclear. The dominant hypothesis in recent years has been that the tyrosine kinase, acting on one or more intracellular substrates, initiates a phosphorylation cascade which ultimately leads to the regulation of serine-specific kinases or phosphatases. Attempts have been made to demonstrate the regulation of known enzymes by tyrosine phosphorylation, and to identify proteins phosphorylated on tyrosine in intact cells, but so far without conclusive results.

The different hypotheses regarding the mechanism of insulin action are not necessarily mutually exclusive. It may be necessary to consider multiple signalling pathways, acting in sequence or in parallel, to obtain a complete picture of insulin action. This review will concentrate on the structure and activity of the insulin receptor itself, emphasising, where possible, aspects of potential importance for the understanding of signalling mechanisms.

Receptor structure

The insulin receptor is a minor intrinsic membrane glycoprotein, representing less than 0.01 % of cell protein even in those tissues where it is most abundant. Rat liver and human placenta have been used generally as sources of receptor for bulk purification and detailed characterisation.

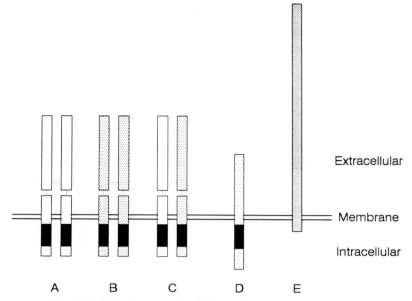

Extracellular

Membrane

Intracellular

A   B   C   D   E

Fig. 10.2. *Subunit structure of the insulin receptor and other growth factor receptors.* The relative sizes of receptors and subunits are drawn approximately in proportion to the corresponding length of peptide sequence. Solid areas indicate tyrosine-specific kinase domains, and dotted lines between subunits indicate putative disulphide bonds. A: insulin receptor; B: type I IGF receptor; C: hybrid insulin/IGF receptor; D: EGF receptor; E: type II IGF receptor.

Isolated cell preparations, particularly rat adipocytes and cultured cell lines such as human IM-9 lymphocytes, also have been widely used for analytical experiments.

Purification methods

Insulin receptors are solubilised readily by using non-ionic detergents such as Triton X-100. The solubilised receptor retains its capacity for insulin binding and its activity as a protein kinase, although both aspects of receptor function, and especially the kinase, are sensitive to denaturing conditions and proteolysis (Petruzzelli *et al.*, 1982; Fujita-Yamaguchi *et al.*, 1983; Kathuria *et al.*, 1986). Purification procedures almost invariably have involved three stages: isolation and solubilisation of a microsomal membrane fraction; partial purification by DEAE-cellulose or lectin-affinity chromatography; purification to near homogeneity by insulin-affinity or immuno-affinity chromatography (Cuatrecasas, 1972; Jacobs *et al.*, 1977; Siegel *et al.*, 1981; Fujita-Yamaguchi *et al.*, 1983; Roth *et al.*, 1986; O'Brien, Soos & Siddle, 1986). The preferred method for elution from insulin-Sepharose columns is by use of acetate buffer pH5 (Fujita-Yamaguchi *et al.*, 1983), taking advantage of the marked pH sensitivity of insulin binding. Elution from antibody affinity columns must be tailored to meet the properties of individual antibodies but in suitable cases may be achieved by use of solutions of insulin (Roth & Cassell, 1983) or synthetic peptides corresponding to known receptor epitopes (Ganderton, R. H., Taylor, C. E. and Siddle, K., unpublished observations).

Antibodies have been particularly important reagents for receptor isolation on an analytical scale by immunoprecipitation. In this context, autoantibodies from the sera of patients with a rare syndrome of extreme insulin resistance have been most valuable (Kahn *et al.*, 1981). Polyclonal antisera have also been produced in rabbits using purified receptor as immunogen (Jacobs *et al.*, 1980; Jacobs & Cuatrecasas, 1981) and more recently by using synthetic peptide immunogens (Herrera *et al.*, 1985; Herrera & Rosen, 1986; Grunfeld *et al.*, 1987). A substantial number of mouse monoclonal antibodies have been described (Roth *et al.*, 1982; Kull *et al.*, 1983; Soos *et al.*, 1986; Morgan & Roth, 1986; Forsayeth *et al.*, 1987*b*) and found many uses (Siddle *et al.*, 1987, 1988).

Subunit composition

The subunit composition of the insulin receptor was investigated initially by SDS–PAGE analysis of purified receptor (Jacobs *et al.*, 1977; Harrison & Itin, 1980; Siegel *et al.*, 1981) or of material which had been affinity-labelled by using photoactivatable [^{125}I]-insulin analogues (Yip, Yeung & Moule, 1978; Jacobs *et al.*, 1979; Wisher *et al.*, 1980) or [^{125}I]-insulin together with cross-linking reagents (Pilch & Czech, 1979). Such studies revealed predominantly an insulin binding component of M_r 130–135 kD which was subsequently termed the α-subunit. A second major subunit was not clearly revealed in early studies although minor components of various sizes (M_r 40–90 kD) were described. The presence of a second type of subunit was deduced by analysing the mobility of affinity-labelled receptor on SDS–PAGE under non-reducing conditions and in the presence of varying concentrations of dithiothreitol (DTT) (Pilch & Czech, 1980*a*; Massague, Pilch & Czech, 1980; Jacobs *et al.*, 1980). These experiments indicated an intact native receptor of structure β-α-α-β with apparent M_r 300–350 kD (Fig. 10.2). Low concentrations of DTT reduced the α-α cross-links (so-called class I disulphides) to generate ($\alpha\beta$) structures, and higher concentrations of DTT under denaturing conditions also reduced α-β links (class II disulphides).

Subunits of approx M_r 135 kD and 95 kD both were clearly seen in material specifically immunoprecipitated from cells labelled by biosynthetic incorporation of [^{35}S]-methionine (Van Obberghen et al., 1981) and [^{3}H]-sugars (Hedo et al., 1981). Both subunits also were detected by cell-surface radioiodination (Lang et al., 1980; Kasuga et al., 1982d). The β-subunit was revealed as a phosphoprotein in cells labelled with [^{32}P]-phosphate (Kasuga, Karlsson & Kahn, 1982a) as discussed in more detail in subsequent sections. Biosynthetic labelling experiments further indicated that α and β subunits were derived from a common precursor, providing confirmation of the proposed stoichiometry. In some cell lines a small amount of uncleaved but fully glycosylated proreceptor has been detected (Hedo & Gorden, 1985) although its cellular location is uncertain and it does not seem to be a significant component in normal tissues.

It is now apparent that the β-subunit is particularly susceptible to proteolysis (Massague, Pilch & Czech, 1981; Fujita-Yamaguchi, 1984), reacts only with some photoaffinity analogues of insulin (Wang et al., 1982; Yip et al., 1982) and stains poorly on polyacrylamide gels (Fujita-Yamaguchi et al., 1983). These properties account for the failure to observe β-subunit as readily as α in early studies. The proteolysed β' subunit probably accounts for the lower M_r species observed in some experiments. Because some degree of proteolysis during extraction is difficult to avoid, receptor as normally isolated appears as a mixture of $\alpha_2\beta_2$, $\alpha_2\beta\beta'$ and $\alpha_2\beta_2'$ species (Massague et al., 1980; Fujita-Yamaguchi, 1984; Boyle et al., 1985; Hofmann et al., 1987; O'Hare & Pilch, 1988). Whether or not a significant fraction of receptor pre-exists in a proteolysed form in intact cells is unclear.

Alternative states of assembly

Several studies have raised the question of whether the $(\alpha\beta)_2$ species is the only one of functional importance. Other species including $(\alpha\beta)$ and free subunits have been reported (Yip & Moule, 1983; Maturo et al., 1983; Chvatchko et al., 1984; Crettaz et al., 1984; Velicelebi & Aiyer, 1984), although the thiol–disulphide exchange which generates these may occur artefactually on boiling samples in SDS in preparation for electrophoresis (Boyle et al., 1985; Helmerhorst et al., 1986). In addition $(\alpha\beta)$ units have sometimes been resolved by gel filtration under native conditions (Koch et al., 1986; Fujita-Yamaguchi & Harmon, 1988), though these have not been apparent in other studies (Boni-Schnetzler, Rubin & Pilch, 1986; Morrison *et al.*, 1988). Some authors have speculated that insulin might influence thiol–disulphide exchange (Maturo, Hollenberg & Aglio, 1983; Chvatchko *et al.*, 1984; Fujita-Yamaguchi & Harmon, 1988) even to the extent of direct involvement and formation of covalent insulin–receptor complexes (Clark & Harrison, 1985). However, when class I (α-α) disulphides in soluble receptors are reduced experimentally to generate ($\alpha\beta$) halves, these remain non-covalently associated as ($\alpha\beta$)$_2$ species under native conditions (Sweet *et al.*, 1985, 1987*a*) and insulin promotes covalent reassociation (Boni-Schnetzler *et al.*, 1988; Morrison *et al.*, 1988). Furthermore, reduction of class I disulphides in intact cells does not impair the ability of the receptor to transmit a regulatory signal (Massague & Czech, 1982). At the present time there is no convincing evidence that interconversion between the ($\alpha\beta$)$_2$ receptor and smaller species under physiological conditions is of any significance in insulin action. The ($\alpha\beta$)$_2$ hetero-tetrameric complex appears to be the only active species in intact cells (Le Marchand-Brustel *et al.*, 1989; Treadway *et al.*, 1990).

Other studies have suggested that higher-order oligomeric forms of the receptor may be significant in signalling. This idea has its origins in the observation that receptors sometimes appear clustered on the surface of intact cells (Jarett & Smith, 1974; Schlessinger *et al.*, 1978) and the demonstration that the insulin-mimetic effects of anti-receptor antibodies are dependent on receptor cross-linking (Kahn *et al.*, 1981). Disulphide-linked oligomeric forms have been isolated under some conditions (Crettaz *et al.*, 1984; Chen *et al.*, 1986) although, as previously mentioned, such analysis is prone to artefacts generated during sample preparation for electrophoresis. Solubilised and partially purified insulin receptors have been resolved into three distinct fractions, apparently representing different non-covalently associated oligomeric states, by non-denaturing electro-phoresis (Kubar & Van Obberghen, 1989). The largest species displayed increased capacity for autophosphorylation relative to the presumed ($\alpha\beta$)$_2$ species. These observations were interpreted as evidence that receptor oligomerisation may play an important role in kinase activation and thus in insulin action. However, other lines of evidence reviewed below suggest strongly that receptor autophosphorylation is an intramolecular event. The significance of receptor oligomers as functional species within intact cells therefore remains to be proved.

The suggestion from earlier studies that the native receptor structure *in situ* may involve components other than the α and β subunits (Baron & Sonksen, 1983; Yip & Moule, 1983) has not been substantiated. No such components co-purify with the receptor in stoichiometric amounts (Fujita-Yamaguchi, 1984). The fact that insulin sensitivity can be conferred on cells which are otherwise unresponsive either by transfer of purified receptor via phospholipid vesicles (Hofmann *et al.*, 1988) or by transfection with cDNA coding for the ($\alpha\beta$) precursor (Hofmann, White & Whittaker, 1989) also argues strongly against a stoichiometric involvement of other components. However, it is obvious that the receptor must interact with other proteins to transmit a regulatory signal, which might be reflected in significant associations. Under some conditions, serine-specific protein kinase activity (Smith, King & Sale, 1988; Lewis *et al.*, 1990*b*) and phosphatidyl inositol kinase activity (Sale, Fujita-Yamaguchi & Kahn, 1986) are present in purified receptor preparations. The insulin receptor may also interact with calmodulin (Graves, Goewert & McDonald, 1985). Other proteins might associate with the receptor in order to modulate its activity. For instance, in some tissues a significant fraction of insulin receptors appears to be associated with MHC class I antigens (Fehlmann *et al.*, 1985; Samson, Cousin & Fehlmann, 1986; Phillips *et al.*, 1986; Due, Simonsen & Olsson, 1986), and insulin binding affinity varies as a function of HLA haplotype (Kittur *et al.*, 1987). Though intriguing, none of these associations has yet been clearly shown to be functionally important and their significance remains to be explored.

Structural heterogeneity

Detailed structural analysis has been carried out largely in rat and human tissues but the basic structure and binding properties of the receptor appear to have been conserved substantially during evolution among vertebrate species (see Gammeltoft, 1984; Czech, 1985*a, b*). The primary sequence of the receptor is conserved very highly between rodents and man (Flores-Riveros *et al.*, 1989; Goldstein & Dudley, 1990). It has been suggested that the structure of the receptor binding site may have been conserved more than that of insulin itself (Muggeo *et al.*, 1979). Not surprisingly, there is substantial immunochemical cross-reactivity among receptors from different species (Kahn *et al.*, 1981) particularly for the kinase domain of the β-subunit (Morgan *et al.*, 1986). However, mono-clonal antibodies for epitopes on the extracellular portion of the human receptor were fairly species-specific, none reacting with rat and only some with rabbit and bovine receptors (Soos *et al.*, 1986). These species differences might reflect variations in amino acid sequence or in glyco-sylation.

Only a single insulin receptor gene has been identified in the human genome (Ullrich *et al.*, 1985; Ebina *et al.*, 1985*b*). Nevertheless, tissue-specific differences in receptors have been suggested by analysis of molecular size, reactivity with antibodies, and binding properties (for review see Gammeltoft, 1984). The brain (neuronal) receptor is somewhat smaller than that in most tissues, the α-subunit having an apparent M_r 115–120 kD (Heidenreich *et al.*, 1983; Hendricks *et al.*, 1984). Minor differences have been noted also in other tissues, the liver receptor being slightly larger than that in muscle and adipose tissue (Burant *et al.*, 1986; Caro *et al.*, 1988). A larger form of receptor is expressed also in some cell lines (McElduff, Grunberger & Gorden, 1985; Whittaker *et al.*, 1987). Immunochemical differences are less obvious and most monoclonal antibodies do not discriminate between receptors in different tissues (Roth *et al.*, 1982; Soos *et al.*, 1986). However, monoclonal antibody MC51 which inhibits binding of insulin in most tissues is relatively unreactive with brain receptor (Roth *et al.*, 1986). There have been conflicting reports regarding another antibody MA10, which appeared unreactive with liver receptor under some conditions (Cordera *et al.*, 1987; Caro *et al.*, 1988).

Tissue-specific glycosylation patterns are responsible for at least some heterogeneity, including the difference between neuronal and other receptors (Heidenreich & Brandenburg, 1986; McElduff *et al.*, 1988; Ota *et al.*, 1988; Heidenreich *et al.*, 1988). It also has been shown that variant forms of receptor, differing in the presence or absence of a 12 amino acid sequence at the carboxyterminus of the α-subunit, arise by alternative splicing of the primary RNA transcript of the receptor gene (Seino *et al.*, 1989). The larger form predominates in liver and the smaller form in brain and adipose tissue, while kidney and placenta express roughly equal proportions of both (Seino & Bell, 1989; Moller *et al.*, 1989). The larger form has a significantly (approx. two-fold) lower binding affinity for insulin than the smaller (Mosthaf *et al.*, 1990). Tissue differences in binding affinity for insulin and certain insulin analogues have been noted (Gammeltoft, 1984) though these are not all as would be predicted from the properties and distribution of splice variants. Glycosylation differences therefore also may play a role in determining binding affinity.

It is unclear whether post-binding aspects of receptor function also might be affected by structural heterogeneity of the extracellular portion. Differences in insulin-stimulated kinase activity and autophosphorylation have been reported for liver and muscle receptors (Burant, Treutelaar & Buse, 1988). It seems most likely that these are related to modulation of the kinase domain *per se* (see below). However, the possibility cannot be ruled out yet that structural variations within extracellular domains could influence the rate or extent of kinase activation, and thus be involved in tissue specificity of signalling and insulin action.

Hybrid and atypical receptors

It has been suggested recently that insulin receptors and the very similar IGF-1 receptors may form hybrid β-α-α^*-β^* structures (Fig. 10.2), as well as the respective homomeric receptors, in tissues where they are co-expressed (Soos & Siddle, 1989; Moxham, Duronio & Jacobs, 1989; Soos *et al.*, 1990). This novel proposal arises from the demonstration that a variety of antibodies specific for the insulin receptor react with a subpopulation of IGF-1 receptors. The size of the immunoreactive (hybrid) subfraction varies, from approximately 30% of total IGF-1 binding in placental membranes and HepG2 hepatoma cells to 80% in IM-9 cells. The hybrids represent a smaller fraction of total insulin binding because of the excess of insulin receptors over IGF receptors in these cells. Hybrid receptors can also be assembled *in vitro* from the separated ($\alpha\beta$) halves of insulin- and IGF-receptors (Treadway *et al.*, 1989*a*). The physiological role of hybrid receptors remains to be investigated. It appears that the hybrids are functional in terms of ligand-stimulated autophosphorylation, and it is tempting to speculate that they provide a mechanism whereby insulin activates signalling pathways normally associated with IGF and vice versa. However, the binding properties of hybrid receptors for the two ligands still have to be documented.

'Atypical' insulin receptors which bind IGFs with relatively high affinity have been demonstrated in placenta (Jonas, Newman & Harrison, 1986; Jonas, Cox & Harrison, 1989), IM-9 cells (Misra, Hintz & Rosenfeld, 1986) and L6 myocytes (Burant *et al.*, 1987). At least in the placenta, these have very similar immunological properties to classical insulin receptors (Jonas *et al.*, 1989) but represent an immunologically distinct subpopulation of IGF receptors (Jonas & Harrison, 1985, 1986). It is possible that these 'atypical' receptors are in fact hybrid structures. However, it appeared that 'atypical' insulin receptors in the placenta were unreactive with the IGF-receptor specific monoclonal antibody αIR-3 (Jonas *et al.*, 1986) and, by this criterion, they would be distinct from hybrid receptors (Soos & Siddle, 1989; Moxham *et al.*, 1989). The structural basis and physiological role of 'atypical' insulin receptors therefore remains unclear.

Information from cDNA cloning

The cloning of full-length cDNA for the human insulin receptor precursor was reported almost simultaneously by two groups (Ullrich *et al.*, 1985; Ebina *et al.*, 1985*b*). In both cases, placental cDNA libraries were screened with oligonucleotide probes constructed on the basis of N-terminal amino acid sequence from purified receptor. The complete amino acid sequence deduced from the cDNA sequence was entirely consistent with the model of receptor structure derived from earlier biochemical studies. The sequence information provided much additional detail, but no immediate insights into conformation or signalling mechanism. However, the potential for manipulation of receptor cDNA *in vitro*, and the expression of mutant receptors in transfected cells, offers a powerful new approach to the study of structure–function relationships. Human insulin receptor cDNA has been expressed in various cells, including hamster, mouse and rat fibroblasts, and rat hepatoma cells (Ebina *et al.*, 1985*a*; Whittaker *et al.*, 1987; McClain *et al.*, 1987; Hofmann *et al.*, 1989; Hawley *et al.*, 1989). The increased sensitivity to insulin was, in general, directly related to the number of human receptors expressed, suggesting that the receptors were fully functional.

Primary sequence

The two reported human receptor cDNA sequences are almost identical except that the Ebina sequence contains an additional 12 amino acids at the C-terminus of the α-subunit, reflecting alternative splicing of the receptor mRNA (see above). Other minor amino acid differences may represent individual polymorphisms and/or cloning artefacts. The notation used here will be based on the Ebina sequence, numbering from the N-terminus of the mature α-subunit. There is a full open reading frame of 1382 (1370) amino acids, including a typical N-terminal signal sequence commencing at a methionine residue (i.e. -27), which is not present in the mature protein (Fig. 10.3). A tetrabasic Arg–Lys–Arg–Arg sequence at residues 732–735 is the site of cleavage to generate α- and β-subunits. The mature α-subunit therefore consists of 732 (720) amino acids and the β-subunit of 620 amino acids, with peptide M_r of approximately 84000 and 70000 respectively. The differences from the apparent subunit M_r values of 135000 and 95000 are mainly due to glycosylation. The α-subunit has 13, and the β-subunit 4, potential N-linked (asparagine) glycosylation sites. It is presumed that most of these are utilised, though the M_r estimates are not

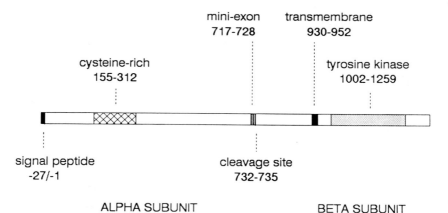

Fig. 10.3. *Primary sequence of the insulin receptor.* The relative position of distinct domains and cleavage sites within the primary sequence is indicated. The numbers refer to amino acid positions according to Ebina *et al.* (1985 b), in which the receptor precursor is a polypeptide of 1355 amino acids plus a 27 amino acid signal sequence.

sufficiently reliable, and the carbohydrate composition is not known in sufficient detail, to permit accurate estimates.

There is a single potential membrane-spanning sequence in the β-subunit and none in the α-subunit. The presumed transmembrane domain is a sequence of 23 uncharged and predominantly hydrophobic amino acids (930–952) flanked by basic residues. Thus it is deduced that the whole of the α-subunit and 194 residues of the β-subunit are extracellular. This is consistent with the pattern of labelling of receptor subunits by radio-iodination, in right-side out and inside-out membrane vesicles (Hedo & Simpson, 1984). It also has been demonstrated that intact α-subunits may be released from the membrane by reduction under denaturing conditions (Grunfeld Shigenagu & Ramachandran, 1985), and that the class I disulphides linking α-subunits are extracellular (Chiacchia, 1988). A model for the arrangement of domains within the receptor, based on sequence information, is given in Fig. 10.4.

The extracellular domain

A conspicuous feature in the sequence of the α-subunit is a cysteine-rich region (155–312), comprising 26 cysteine residues and 72% hydrophilic residues in all, as well as 15% proline and glycine. This domain is homologous to two cysteine-rich regions in the EGF receptor, but distinct from high-cysteine motifs in a variety of other proteins (Doolittle, 1985;

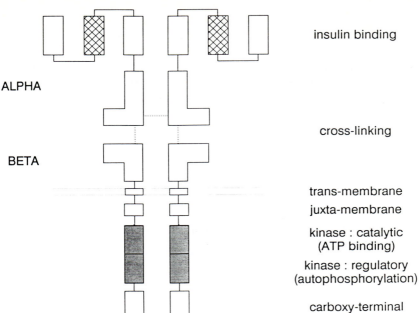

ALPHA

BETA

insulin binding

cross-linking

trans-membrane

juxta-membrane

kinase : catalytic
(ATP binding)

kinase : regulatory
(autophosphorylation)

carboxy-terminal

Fig. 10.4. *Functional domains of the insulin receptor*. This hypothetical model is based in part on that of Bajaj *et al.* (1987). The relative size of domains are drawn approximately in proportion to the corresponding lengths of primary sequence. The cross-hatched area represents the cysteine-rich region and the shaded area the tyrosine kinase domain. The precise locations of the insulin binding site and inter-subunit disulphides are unclear.

Bajaj *et al.*, 1987). The respective cysteine-rich domains of the insulin and EGF receptors have only 23% conserved residues, but these include the majority of the cysteines. The significance of this strongly hydrophilic domain, which presumably includes multiple disulphide bridges, is unclear although recent evidence indicates it contributes to the insulin binding site (see below). Either side of the cysteine-rich region are domains of approximately 120 amino acids with some (approximately 30%) homology to each other and to sequences in the EGF receptor. It is postulated that these may assemble from α-helix/β-sheet motifs in a pseudo-symmetrical dimer (Bajaj *et al.*, 1987), but, as yet, there is no experimental evidence for such structural detail. The C-terminal portion of the α-subunit (440–732) and extracellular portion of the β-subunit (736–929) have no obvious structural features, but contain seven and four cysteines respectively, some of which are likely to be responsible for both class I (α–α) and class II (α–β) inter-subunit disulphide bonds.

The intracellular domain

The intracellular portion of the β-subunit (953–1355) contains a major domain (1002–1257) with clear sequence similarity to other tyrosine kinases, including both receptors and oncoproteins. A consensus ATP binding site Gly–X–Gly–X–X–Gly is found 50 residues from the membrane spanning region in both insulin and EGF receptors. A further 25 residues downstream is a conserved lysine (1030 in the insulin receptor) which is essential for kinase activity, and which may be affinity-labelled with ATP analogues. The C-terminal portion of the β-subunit (1258–1355) has little homology with other proteins. It contains 58 % hydrophilic residues, mostly acidic, and 16 % proline and glycine, suggesting a mobile structure which might have a regulatory function. The activity and regulation of the intracellular domain is discussed in more detail in subsequent sections.

The receptor gene

The insulin receptor gene has been localised to chromosome 19. Multiple mRNA transcripts of 7–11 kbp are detected in most cell types (Ebina *et al.*, 1985*b*), predominantly reflecting varying amounts of 3′ untranslated sequence, the significance of which is unclear. The gene itself is very large, extending over 130 kbp, with 22 exons of various sizes (36–552 bp) and introns which are even more variable (0.5–25 kbp) (Seino *et al.*, 1989; Seino, Seino & Bell, 1990). Some of the exons correspond approximately to recognisable structural units such as the signal peptide (exon 1), cysteine-rich domain (exon 3) alternatively spliced miniexon (exon 11), transmembrane domain (exon 15) or C-terminal tail (exon 22). However, five separate exons contribute to the tyrosine kinase domain.

IGF receptors

The insulin-like growth factors are related structurally to (pro)insulin, and display a spectrum of growth-promoting and metabolic effects which overlap with those of insulin (Froesch *et al.*, 1985; Rechler & Nissley, 1985). Two distinct types of receptor for insulin-like growth factors have been identified and characterised, by affinity labelling (Kasuga *et al.*, 1981), purification (Fujita-Yamaguchi *et al.*, 1986; Kiess *et al.*, 1988) and cloning of the respective cDNAs (Ullrich *et al.*, 1986; Morgan *et al.*, 1987*a*) (Fig. 10.2). The type I receptor is a molecule closely related to the insulin receptor in all respects. It has the same $(\alpha\beta)_2$ structure and displays a low affinity for insulin as well as binding both IGF-I and IGF-II with high affinity (Steele-Perkins *et al.*, 1988). The cloned cDNA (Ullrich *et al.*, 1986) encodes a protein of 1367 amino acids, which is similar in length, overall organisation and sequence to the insulin receptor. The sequence identity is greatest (84%) in the kinase domain, and also considerable (approx 65%) in the domains flanking the cysteine rich region (L1 and L2 of Bajaj *et al.*, 1987), with somewhat lower homology (40–50%) throughout the rest of the molecule. There is strong conservation of critical residues including many cysteines, both within and outside the cysteine-rich region, *N*-linked glycosylation sites, and potential tyrosine phosphorylation sites of the kinase domain. It is clear from the structure that the signalling mechanism of the type I IGF receptor must be at least very similar to that of the insulin receptor. When these receptors are expressed in the same cellular background, by cDNA transfection, they mediate the same spectrum of responses (Steele-Perkins *et al.*, 1988). However, the type I IGF receptor cytoplasmic domain appears more efficient than the insulin receptor in mediating 'growth-promoting' effects (Lammers *et al.*, 1989). It remains a matter of debate whether differences in the physiological effects of insulin and IGF-I *in vivo* reflect differences in inherent signalling capacity of the respective receptors, or differences in receptor distribution among tissues with distinct metabolic patterns.

The type II IGF receptor is a molecule of totally different structure, which is identical to a previously described mannose 6-phosphate receptor (Morgan *et al.*, 1987*a*; Kiess *et al.*, 1988; MacDonald *et al.*, 1988). The cloned cDNA encodes a polypeptide of approximately 275 kD, with 19 potential *N*-linked glycosylation sites. There is a single predicted transmembrane sequence dividing a large extracellular portion (2264 amino acids) and a relatively small cytosolic domain (164 amino acids). The extracellular portion consists almost entirely of 15 repeats based on a 150-residue sequence, with approximately 20% overall sequence identity between repeats but a highly conserved pattern of eight cysteines in each repeat. The cytoplasmic domain is relatively hydrophilic (34%) and unrelated to known kinases. It is, as yet, unclear whether any signalling function is associated with the type II receptor. This receptor may serve merely as a binding protein for IGFs and other ligands (Roth, 1988; Czech, 1989). Indeed, the domain of IGFs which interacts with the type II receptor appears to be similar to that which interacts with IGF binding proteins in serum, and distinct from the binding region for type I IGF receptors and insulin receptors (Froesch *et al.*, 1985).

Very recently, a novel insulin receptor-related gene has been identified by probing genomic DNA libraries with insulin receptor cDNA at low stringency (Shier & Watt, 1989). The pattern of exons corresponds very closely to that of the insulin receptor, and indicates a protein with sequence similarity to both the insulin and type I IGF receptors comparable to that which these have to each other. It will be of great interest to identify the tissues in which this novel receptor is expressed, and the ligand with which it interacts. An attractive possibility is that this is a receptor for IGF-II.

A possible homologue of the insulin receptor has also been identified in *Drosophila*, and the cDNA cloned, again using human insulin receptor cDNA as a probe (Petruzzelli *et al.*, 1986; Nishida *et al.*, 1986). However, although the overall architecture of this molecule is comparable to that of the human insulin receptor (Fernandez-Almonacid & Rosen, 1987), and the presumptive β-subunit showed 63% homology, the α-subunit is only weakly related to the human insulin receptor. The ligand for this receptor is unknown, and the biological function unclear. Studies of the expression of the *Drosophila* gene suggest a possible role in embryogenesis (Garofalo & Rosen, 1988; Rosen, 1988).

Insulin binding

The use of [^{125}I]-insulin as a tracer for quantitative receptor binding studies was pioneered by Cuatrecasas (1971) and Freychet, Roth & Neville (1971). Early studies aimed to characterise the number and affinity of receptors in different tissues and pathophysiological states, by Scatchard analysis of binding data obtained with membrane preparations or isolated cells. Problems arising from tracer heterogeneity and degradation, and receptor internalisation in intact cells, were not fully appreciated at first (Cuatrecasas, 1974) and the interpretation of binding studies, to some extent, remains an area of controversy even now.

Equilibrium and kinetic studies

The criteria to be satisfied for valid binding studies have been discussed by Gammeltoft (1984), who also reviewed the principal results of such studies. Values reported for the association constant (K_a) are generally in the range $0.3–3 \times 10^9$ M^{-1}. Thus, at physiological insulin concentrations (typically of the order 10^{-10} M), the fraction of receptors occupied is small. However, the number of receptors per cell is such that maximal responses are normally obtained at low fractional occupancy. In rat adipocytes for instance, it appears that the full metabolic effect of insulin is observed when only approximately 5% of receptors are occupied, although the relationship may not be the same in other tissues. This finding gave rise to the concept of 'spare receptors' although, in fact, receptor concentration directly influences the number of occupied receptors by mass action, and thus determines cellular insulin sensitivity. The binding of insulin shows a sharp pH optimum near neutrality and decreases markedly at even mildly acid pH. Binding also depends appreciably on temperature, affinity increasing as temperature decreases due mainly to a decreased dissociation rate. An important finding is that almost without exception the biological potency of different insulin analogues directly parallels binding affinity, and competitive antagonists have not been described (for review see Gammeltoft, 1984). Studies with analogues served to identify key residues in the insulin molecule required for receptor binding, involving various portions of the A and B chains which are close together on one face of the native insulin molecule (Pullen *et al.*, 1976; Gammeltoft, 1984).

Many studies of insulin binding to cell membranes of solubilised receptors have produced curvilinear Scatchard plots. These have been interpreted variously as evidence of receptor heterogeneity, the existence of cooperative interactions between binding sites or an experimental artefact. Surprisingly, this is an issue which still has not been fully resolved. The degree of departure from idealised behaviour for reversible binding to a single class of non-interacting sites varies considerably between receptor preparations (Gammeltoft, 1984) and some studies have demonstrated linear Scatchard plots (Kohanski & Lane, 1983; Lipkin, Teller & de Haen, 1986).

It is possible that receptor heterogeneity could result from alternative splicing of receptor mRNA (Seino *et al.*, 1989), although any affinity difference between the isoforms is small (Mosthaf *et al.*, 1990). Type I IGF receptors might be responsible for low affinity insulin binding in some tissues. Substantial evidence has been provided, however, for the existence of negative cooperativity between binding sites. The dissociation of receptor-bound [^{125}I]-insulin is accelerated by addition of excess unlabelled insulin (De Meyts *et al.*, 1976), indicating that the off-rate increases, and affinity decreases, as fractional occupancy increases. The capacity to induce negatively cooperative interactions appears to involve specific amino acid residues within the receptor-binding domain of the insulin molecule (De Meyts *et al.*, 1978). The interpretation of these, and similar experiments, as evidence for the existence of negative cooperativity has been much debated (Pollet, Standaert & Haase, 1977; Levitzki, 1981; Gammeltoft, 1984; Helmerhorst, 1987; de Vries *et al.*, 1988). However, a structural basis for the phenomenon has been demonstrated, in comparing the binding properties of solubilised $(\alpha\beta)_2$ and $(\alpha\beta)$ forms of the receptor. Thus, in spite of an apparently symmetrical bivalent structure, it appears that each $(\alpha\beta)_2$ heterotetramer binds only one molecule of insulin with high affinity (Pang & Shafer, 1983, 1984; Sweet *et al.*, 1987*b*).Moreover, when receptor is reduced to produce $(\alpha\beta)$ units, these have lower affinity for insulin than the intact receptor, and display a linear Scatchard plot (Deger *et al.*, 1986; Boni-Schnetzler *et al.*, 1987; Sweet *et al.*, 1987*b*; Swanson & Pessin, 1989), each $(\alpha\beta)$ unit then binding one molecule of insulin (Sweet *et al.*, 1987*b*). Thus it appears that binding of insulin at one site in the native heterotetramer induces conformational changes which dramatically reduce the affinity of the other site, while isolated $(\alpha\beta)$ units display a binding affinity intermediate between these two states. It is unclear whether type I IGF receptors behave similarly (Feltz *et al.*, 1988; Tollefsen & Thompson, 1988).

The existence of an even higher affinity state of the insulin receptor also has been deduced, on the basis of dissociation kinetics which reveal a decrease in off-rate with increasing association time (Corin & Donner, 1982; Gu *et al.*, 1988; Lipson *et al.*, 1989), and from the apparent presence of a positive cooperative effect at low levels of occupancy (Marsh, Westley & Steiner, 1984). A three-state model of insulin receptor kinetics has been proposed in which both higher affinity, slower dissociation (K super) and lower affinity, faster dissociation (K filled) states are accessible from the initial state (K empty) (Gu *et al.*, 1988). Further evidence for this model is provided by anti-receptor monoclonal antibodies which accelerate or inhibit the dissociation of receptor-bound insulin, apparently by inducing or stabilising different conformation states (Forsayeth *et al.*, 1987*b*; Gu *et al.*, 1988; Wang *et al.*, 1988; Siddle *et al.*, 1988). The kinetic characteristics of the receptor are very similar for intact cells and solubilised, purified preparations (Wang *et al.*, 1988). It is likely, therefore, that these are intrinsic properties of the receptor and reflect ligand-induced conformational changes within the native heterotetramer (although additional cooperative interactions between heterotetramers in intact cells have not been ruled out). Effects of insulin on receptor conformation have been inferred from changes in sensitivity to proteolysis (Pilch & Czech, 1980*b*; Donner & Yonkers, 1983; Lipson *et al.*, 1986), reactivity with thiol reagents (Maturo *et al.*, 1983; Wilden & Pessin, 1987; Chiacchia, 1988), and immunological reactivity (Siddle *et al.*, 1987, 1988).

Insulin binding site

The ectodomain of the insulin receptor has been generated as a secreted soluble protein in both eukaryotic cells (Whittaker & Okamoto, 1988; Ellis, Sissom & Levitan, 1988 *b*; Johnson *et al.*, 1988) and a baculovirus/ insect cell system (Sissom & Ellis, 1989), by using truncated cDNA constructs. The secreted material was a glycosylated, disulphide-linked heterotetramer $(\alpha\beta')_2$, which bound insulin with high affinity. One study demonstrated aggregation of globular structures to form linear macro-arrays in response to insulin binding, though the specificity of this effect was in doubt (Johnson, Wong & Rutter, 1988). The efficiency and kinetics of processing of the truncated proreceptor differed somewhat from those of wild-type receptor. Attempts to generate free α-subunit or truncated α-subunits have been less successful as these proteins were unstable (Johnson *et al.*, 1988). Although the complete ectodomain is necessary for efficient folding and secretion, discrete subdomains can fold in a way which is recognised by conformation-specific monoclonal antibodies (Schaefer, Siddle & Ellis, 1990).

Attempts have been made to define the insulin binding region within the α-subunit. These studies have utilised various approaches including affinity cross linking, site-directed mutagenesis and construction of chimaeric receptors. There is a general consensus that binding involves primarily the N-terminal portion of the α-subunit encoded by exons 2 and 3 (Waugh, DiBella & Pilch, 1989; Andersen *et al.*, 1990). Evidence has been presented for involvement of both the LI domain (Wedekind *et al.*, 1989; Toyoshige *et al.*, 1989; De Meyts *et al.*, 1990) and the cysteine-rich domain (Yip *et al.*, 1988; Rafaeloff *et al.*, 1989; Gustafson & Rutter, 1990). Immunological studies also have suggested a role for the C-terminal half of the α-subunit (Prigent, Stanley & Siddle, 1990; Gustafson & Rutter, 1990). It is possible, and indeed likely, that the binding site will involve elements of different regions of primary sequence which are brought together in the native conformation, just as it does in the insulin molecule. Further studies obviously are required to identify the specific residues which interact with insulin.

Insulin receptor tyrosine kinase

Tyrosine phosphorylation represents a very small fraction of total phosphorylation of cellular proteins, which occurs overwhelmingly on serine, and to a lesser extent threonine, residues. The first tyrosine-specific protein kinases to be identified were the pp60 product of the v-*src* oncogene (Hunter & Sefton, 1980) and the receptor for epidermal growth factor (Ushiro & Cohen, 1980). A common property of both these kinases was that of autophosphorylation and it was a similar property of the insulin receptor which led to its identification as a protein kinase.

Characterisation of the kinase

The first indication that the insulin receptor was a tyrosine kinase came from the demonstration of insulin-induced tyrosine phosphorylation of the β-subunit in intact cells (Kasuga *et al.*, 1982 *a, b*; Van Obberghen & Kowalski, 1982) and in cell-free systems (Kasuga *et al.*, 1982 *c*; Van Obberghen & Kowalski, 1982). These observations soon were confirmed widely with receptor from various tissues (Petruzzelli *et al.*, 1982; Avruch *et al.*, 1982; Van Obberghen *et al.*, 1983; Zick *et al.*, 1983 *b*; Roth & Cassell, 1983; Shia & Pilch, 1983).

Several lines of evidence established that the tyrosine kinase was indeed intrinsic to the receptor. The β-subunit possesses an ATP-binding site, as demonstrated by affinity labelling with ATP analogues (Roth & Cassell, 1983; Shia & Pilch, 1983; Van Obberghen *et al.*, 1983). The insulin-binding and tyrosine kinase activities of the receptor co-purify to homogeneity (Kasuga *et al.*, 1983; Nemenoff *et al.*, 1984; Petruzzelli, Herrera & Rosen, 1984). Finally, the sequence of the presumed intracellular portion of the insulin receptor β-subunit shows extensive similarity to other known tyrosine kinases (Ullrich *et al.*, 1985; Ebina *et al.*, 1985 *b*).

The full extent of insulin-stimulated tyrosine phosphorylation in intact cells probably was underestimated in early experiments because of the action of phosphatases during extraction (see Zick, 1989). Inhibitors such as orthovanadate, fluoride and pyrophosphate are used now to prevent this. It was observed that insulin also stimulated receptor serine and threonine phosphorylation in intact cells, whereas in most experiments the reaction of solubilised receptor appeared specific for tyrosine. This was interpreted as indicating the existence of separate activities, a tyrosine kinase intrinsic to the receptor and a serine/threonine kinase more loosely associated with it (Gazzano *et al.*, 1983). The significance of receptor serine/threonine phosphorylation, and the kinase(s) responsible for this reaction, will be discussed in more detail below.

The kinetic properties of the insulin receptor tyrosine kinase have been reviewed in detail elsewhere (Gammeltoft & Van Obberghen, 1986; Zick, 1989). The enzyme phosphorylates a variety of naturally occurring and synthetic peptide substrates *in vitro*, including histones, casein, angiotensin, reduced-carboxymethylated lysozyme, and copolymers of tyrosine with glutamic acid. Such peptides are not considered to be physiologically relevant substrates but are used conveniently to investigate and compare the properties of tyrosine kinases. The K_m and V_{max} values vary enormously for different substrates. A consensus sequence for efficient phosphorylation has not been established although tyrosines in an acidic environment, with nearby glutamic or aspartic acid residues, are preferred (Hunter & Cooper, 1985). The enzyme has an absolute requirement for ATP as phosphate donor, with a K_m of 30–150 μM, and will not utilise other nucleoside triphosphates. Bivalent metal ions (Mg or Mn) also are required, apparently for a specific regulatory role additional to formation of complexes with ATP. At low ATP concentrations commonly used for assays *in vitro*, Mn^{2+} is most effective, but, at higher ATP concentrations, Mg^{2+} is preferred, although the effects of Mn and Mg appear to synergise for optimal activity. Insulin increases the V_{max} of the enzyme without affecting the K_m for ATP or peptide substrate. The extent of the insulin stimulation reported is variable, depending apparently on the quality of the receptor preparation and possibly on the substrate used, but an increase of at least ten-fold can be observed. The concentration-dependence for this stimulation closely parallels that for insulin binding, with half maximal effects at 2–20 nM under the conditions of *in vitro* phosphorylation assays.

Regulation by autophosphorylation

Insulin-stimulated receptor autophosphorylation is extremely rapid, reaching a steady state within 20 s in intact cells and having a half-time of approximately 30 s for solubilised receptor at 22 °C (White *et al.*, 1984, White, Takayama & Kahn, 1985*b*). The reaction is apparently an intramolecular one, at least for solubilised receptor, as the rate of autophosphorylation is independent of concentration for a given quantity of receptor (Petruzzelli *et al.*, 1984; White *et al.*, 1984). The possibility that some intermolecular phosphorylation may occur in intact cells cannot be ruled out (Beguinot *et al.*, 1988; Ballotti *et al.*, 1989) although the close parallelism between receptor occupancy and extent of autophosphorylation suggests that this is not a major mechanism.

Autophosphorylation plays a key role in regulating the soluble receptor kinase *in vitro*, producing a state which remains activated even if insulin is then removed (Rosen *et al.*, 1983; Yu & Czech, 1984; Kwok *et al.*, 1986).

Activation of the receptor kinase by autophosphorylation similarly occurs in intact cells, as demonstrated by exposure to insulin and subsequent assay of cell extracts for kinase activity towards exogenous substrates (Klein *et al.*, 1986; Yu & Czech, 1986). The receptor kinase also can be activated *in vitro* by insulin-independent phosphorylation by the pp60 *src* kinase (Yu *et al.*, 1985), raising the interesting possibility that the receptor might become constitutively activated *in vivo* under certain conditions, such as cellular transformation.

Autophosphorylation of receptor is accompanied by conformational changes in the β-subunit, as detected by anti-peptide antibodies (Herrera & Rosen, 1986; Perlmann *et al.*, 1989; Baron *et al.*, 1990) and susceptibility to cross-linking (Schenker & Kohanski, 1988). However, these conformational changes do not appear to extend to extracellular domains of the receptor, and there is no evidence that autophosphorylation influences insulin binding (Rosen *et al.*, 1983).

In intact cells, it is likely that receptors remain transiently in an autophosphorylated and activated state even after internalisation into the endosomal system (Khan *et al.*, 1986; Klein *et al.*, 1987; Backer, Kahn & White, 1989*a*). The eventual termination of activation, subsequent to the dissociation of insulin, must require dephosphorylation by phospho-tyrosine-specific protein phosphatases. A multiplicity of such enzymes has been identified recently (Lau, Farley & Baylink, 1989; Hunter, 1989; Tonks & Charbonneau, 1989). However, the phosphatases which act specifically on the insulin receptor are as yet poorly characterised (Tonks, Diltz & Fischer, 1988; King & Sale, 1988; Roome *et al.*, 1988; Mooney & Anderson, 1989; King & Sale, 1990). Both the extent of receptor activation and the rate of deactivation may be critically dependent on the activity and subcellular location of phosphatases. The regulation of phosphatases therefore may provide an important mechanism for control of receptor activity.

Autophosphorylation sites

The initial approach to identifying autophosphorylation sites (Fig. 10.5) was to inspect the insulin receptor sequence for tyrosines which were conserved relative to corresponding sequences of other tyrosine kinases, and to test the ability of relevant synthetic peptides to act as substrates (Stadtmauer & Rosen, 1986*a*). This focussed attention particularly on the region around tyrosine 1162 (1150). (Numbers in parenthesis refer to sequence of Ullrich *et al.*, 1985.) Direct analysis of sites was based on autophosphorylation of receptor *in vitro*, fragmentation by limited

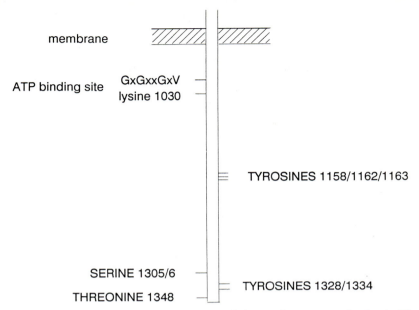

Fig. 10.5. *Phosphorylation sites of the insulin receptor β-subunit.* The diagram illustrates the relative position, within the primary sequence of the β-subunit intracellular domain, of the major tyrosine autophosphorylation sites and of serine/threonine phosphorylation sites, which have so far been identified. The position of the major residues thought to be involved in ATP binding is also shown.

proteolysis, and separation of phosphopeptides by reverse-phase hplc or two-dimensional thin layer chromatography. Phosphopeptides were identified from their reactivity with anti-peptide antibodies (Herrera & Rosen, 1986), and phosphorylation sites defined conclusively by amino acid sequencing (Tornqvist *et al.*, 1987). It was shown that phosphorylation *in vitro* occurred on at least five and possibly six tyrosines in three major domains, the principal sites being 1158/1162/1163 (1146/1150/1151) and 1328/1334 (1316/1322) while a further phosphopeptide was identified tentatively as originating from the juxtamembrane region including 965/972/984 (953/960/972). Similar conclusions were reached on the basis of radiosequencing (White *et al.*, 1988 *b*) and consideration of the charge properties of phosphopeptides (Tavare & Denton, 1988). The complex pattern of phosphopeptides resulted, in part, from the possibilities for multiply-phosphorylated forms of a given peptide and also from alternative points of tryptic cleavage as a consequence of adjacent phosphorylation.

The rapidity of the autophosphorylation events makes it difficult to establish a precise sequence of reactions. It appears that doubly phosphorylated forms of the 1158/1162/1163 domain are generated first but that phosphorylation of all three of these tyrosines is necessary for full kinase activation (White *et al.*, 1988*b*; Tornqvist & Avruch, 1988; Flores-Riveros *et al.*, 1989). Phosphorylation of the C-terminal sites 1328/1334 occurs more slowly and is not required for stimulation of kinase activity (Goren, White & Kahn, 1987; White *et al.*, 1988*b*; McClain *et al.*, 1988). The remaining phosphorylation, putatively in the juxtamembrane domain, occurs most slowly of all (Tavare & Denton, 1988).

Autophosphorylation of receptor in intact cells, in response to insulin, involves essentially the same tyrosines as *in vitro* though with significant quantitative differences (White *et al.*, 1985*b*; White, Takayama & Kahn, 1988*b*; Tornqvist *et al.*, 1988; Tavare *et al.*, 1988). In at least some cell types, the 1158/1162/1163 domain achieves predominantly a bis- rather than tris-phosphorylated state *in situ*. It is not clear whether this is a consequence of specific regulatory mechanisms operating in intact cells, or simply a reflection of a different balance of kinase to phosphatase activity. The C-terminal and juxtamembrane domains also appear considerably less phosphorylated *in situ* than *in vitro*. Most conspicuously, phosphoserine- and phosphothreonine-containing peptides are observed clearly when receptor is phosphorylated in intact cells (see below). These observations indicate that the overall phosphorylation of receptor, and even the net tyrosine phosphorylation, will not necessarily provide a simple measure of kinase activation when comparing receptor under different conditions and from different cell types.

Activation mechanism

The mechanism whereby binding of insulin to the α-subunit activates the tyrosyl kinase is unclear. This must involve some change in conformation of the extracellular domains which is transmitted, via single membrane spanning helices within each (αβ) unit, to influence the conformation or relative positioning of the intracellular domains. A vectorial motion, influencing the proximity of domains to the lipid bilayer, seems energetically problematical and inherently unlikely, given the charged residues which precisely flank the transmembrane domain. More likely, there is lateral or rotational movement of the transmembrane segments which changes the contacts between β-subunits within the native heterotetramer. This allows intramolecular autophosphorylation (Petruzzelli *et al.*, 1984; White *et al.*, 1984; Sweet *et al.*, 1985) and results in a kinase which is activated towards exogenous substrates. Indeed, several studies have suggested that autophosphorylation is essential for kinase activation (Kwok *et al.*, 1986; White *et al.*, 1988*b*). However, others have observed stimulation of kinase activity by insulin (Morrison & Pessin, 1987) or anti-peptide antibodies (Baron *et al.*, 1989) in the absence of autophosphorylation. It is possible, therefore, that conformational change alone is sufficient for some kinase activation. It is interesting to note that, in the case of the structurally related EGF receptor kinase, autophosphorylation occurs only within the C-terminal portion of the molecule and is not required for kinase activation (Downward, Waterfield & Parker, 1985; Clark *et al.*, 1988).

It has been suggested that the function of the α-subunit in the native receptor is to permit a state of assembly which imposes inhibitory control on the β-subunit. Proteolytic truncation of the α-subunit releases the β-subunit from this inhibitory control, resulting in autophosphorylation and constitutive (insulin-independent) kinase activity (Shoelson, White & Kahn, 1988; Hsuan *et al.*, 1989). Consistent with these experiments, a membrane-anchored cytoplasmic domain of the insulin receptor mediates a constitutive metabolic stimulation when expressed in CHO cells (Ellis *et al.*, 1987*a*) and has transforming potential when encoded within the genome of a retrovirus (Wang *et al.*, 1987). However, data on truncated constructs present a complex picture overall. Thus, autophosphorylation of the membrane-anchored domain was not detectable *in situ*, and it was 20-fold less active in terms of autophosphorylation than a soluble kinase construct when assayed *in vitro* (Ellis *et al.*, 1987*a*). Soluble kinase domains generated using a Baculovirus expression system undergo autophosphorylation and phosphorylate exogenous substrates (Herrera *et al.*, 1988; Ellis *et al.*, 1988*a*). However, there is some doubt as to whether the pattern of autophosphorylation is the same as that observed in wild-type receptor stimulated by insulin. It is unclear in any of these experiments whether autophosphorylation of truncated, and therefore monomeric, kinase domains is an intra- or inter-molecular reaction, and it is difficult to compare the 'constitutive' kinase activity of these constructs quantitatively with native insulin receptor.

A different model of kinase activation, requiring β–β interaction, is suggested by other experiments. Thus, isolated (αβ) half-receptors, produced by reduction of holoreceptors, show only basal levels of autophosphorylation and kinase activity (Boni-Schnetzler *et al.*, 1986; Sweet *et al.*, 1987*c*). Insulin stimulates autophosphorylation and kinase activity of such preparations only as a consequence of inducing their covalent reassembly into the $(\alpha\beta)_2$ state (Boni-Schnetzler *et al.*, 1988; Morrison *et al.*, 1988; Wilden, Morrison & Pessin, 1989 *a*). Wheat-germ agglutinin stimulation similarly parallels $(\alpha\beta)_2$ reassembly but in this case by a non-covalent association (Wilden *et al.*, 1989 *b*). Consistent with these observations, the stimulation of autophosphorylation and kinase activity *in vitro* by anti-receptor antibodies appears to result from an intra-molecular 'microaggregation' or cross-linking within the native hetero-tetramer (Heffetz & Zick, 1986; O'Brien, Soos & Siddle, 1987 *c*). An analogous clustering mechanism is possibly involved in the activation of the EGF receptor, which consists of a single polypeptide chain (Schles-singer, 1988).

A rationalisation of these diverse observations on the requirements for receptor activation may be that autophosphorylation within the native heterotetramer normally occurs *trans* (two β-subunits phosphorylating each other) rather than *cis* (each β-subunit phosphorylating itself). The autophosphorylation of 'monomeric' kinase domains (truncated con-structs or $(\alpha\beta)$ units) then might depend very much on concentration, for both membrane-anchored or soluble forms, as well as the presence or absence of α-subunit. Such a model is supported by the observation that only intact $(\alpha\beta)_2$ receptors, and not the partially proteolysed $(\alpha\beta)(\alpha\beta')$ form, demonstrate kinase activity (O'Hare & Pilch, 1988). Studies with mutant/wild-type hybrid receptors assembled *in vitro* also indicate that receptor activation occurs by intramolecular trans-phosphorylation (Treadway *et al.*, 1991).

The role of the extracellular domain in regulating kinase activity has been demonstrated further in studies of chimaeric receptors. Constructs in which the extracellular portion of the insulin receptor is linked to kinase domain derived from the EGF receptor (Riedel *et al.*, 1986), the v-*ros* oncogene (Ellis *et al.*, 1987 *b*) or the type I IGF receptor (Lammers *et al.*, 1989), all show insulin-induced autophosphorylation and kinase stimu-lation. All these chimaeras are assembled as heterotetramers comparable in overall structure to the wild-type insulin receptor. Much more surprising and provocative is a recent description of a chimaera consisting of the *Escherichia coli* aspartate receptor ligand binding domain and the insulin receptor cytoplasmic domain, as a single chain polypeptide. This ap-parently displayed aspartate-stimulated kinase activity towards a limited range of substrates, in the absence of detectable autophosphorylation (Moe, Bollag & Koshland, 1989). The specific activity was not compared to wild-type receptor and may have been low, but the results were interpreted as indicating a possible role of autophosphorylation in substrate discrimination.

Regulation by serine phosphorylation

Many examples are known of the regulation of enzyme activity by multisite phosphorylation (Cohen, 1985; Hunter, 1987). In the case of the insulin receptor, there is evidence that phosphorylation on serine and threonine residues has an inhibitory effect on tyrosine autophosphorylation, and consequently on the kinase activity towards exogenous substrates. The serine/threonine phosphorylation of the receptor in intact cells is stimulated by insulin, and by activation of protein kinase C and possibly also cyclic AMP-dependent protein kinase. Thus, the receptor may be subject to feedback regulation by insulin itself, and also to transmodulation by hormones acting through other receptor–effector systems.

Insulin-stimulated serine kinase

In the basal state, the insulin receptor contains phosphoserine and phosphothreonine, but not phosphotyrosine. Insulin added to intact cells increases the content of phosphoserine as well as phosphotyrosine (Kasuga *et al.*, 1982*b*; Gazzano *et al.*, 1983; Pang *et al.*, 1985; Jacobs & Cuatrecasas, 1986; Ballotti *et al.*, 1987; Duronio & Jacobs, 1990). Increases in phosphothreonine have not been observed consistently, but are detected in transfected cells which express high levels of receptors (Tavare *et al.*, 1988). The increase in phosphotyrosine takes place predominantly in receptors which do not contain phosphoserine, and the increase in phosphoserine occurs more slowly than that in phosphotyrosine (Pang *et al.*, 1985; Ballotti *et al.*, 1987). These results are compatible with a regulatory role for serine phosphorylation, although the possibility cannot be ruled out that differential phosphorylation reflects the responses of receptor subpopulations in different cellular compartments. The observations have also been interpreted as evidence for the activation of a specific serine kinase which might participate more generally in insulin action by phosphorylating a variety of cellular substrates (Sale, 1988; Czech *et al.*, 1988).

Characterisation of the insulin-stimulated receptor serine kinase (IRSK) has proved difficult. It has been suggested that a serine kinase may associate specifically with the receptor. Although solubilised receptors frequently have been found to show only tyrosine autophosphorylation in response to insulin, phosphorylation on serine has been observed sometimes (Gazzano *et al.*, 1983; Zick *et al.*, 1983*a, b*; Yu & Czech, 1984). Conditions now have been described which permit the retention of insulin-stimulated serine kinase activity in solubilised and even purified receptor preparations (Ballotti *et al.*, 1986; Smith *et al.*, 1988; Czech *et al.*, 1988; Lewis *et al.*, 1990*b*). The sites of receptor serine phosphorylation *in vitro* are similar to those in intact cells (Smith & Sale, 1989), and include serine 1305 and/or 1306 (Lewis *et al.*, 1990*b*). The receptor-associated serine kinase is distinct from known enzymes and its activation depends on the tyrosine kinase activity of the receptor (Smith & Sale, 1988) but otherwise little is known of its properties. Inhibition or down-regulation of protein kinase C does not diminish insulin-induced receptor serine phosphorylation in intact cells (Duronio & Jacobs, 1990), indicating that this enzyme is not IRSK.

Protein kinase C

The receptor is a substrate for several serine/threonine kinases both in intact cells and *in vitro*. Phorbol esters, presumed to be acting by stimulation of protein kinase C, increase receptor phosphorylation in various cell types (Jacobs *et al.*, 1983; Takayama *et al.*, 1984; Jacobs & Cuatrecasas, 1986; Hachiya *et al.*, 1987; Duronio & Jacobs, 1990). These observations do not prove that protein kinase C acts directly on the insulin receptor, rather than via intermediate kinases, although phosphorylation of receptor *in vitro* by protein kinase C has been reported (Bollag *et al.*, 1986). Several distinct sites are involved, some of which are identical to those phosphorylated in response to insulin (Lewis *et al.*, 1990*a, b*; T. S. Pillay & K. Siddle, unpublished observations). A significant difference may be the more prominent increase in phosphothreonine in response to phorbol esters compared to insulin (Takayama *et al.*, 1984; Jacobs & Cuatrecasas, 1986). The threonine phosphorylation site has been identified as residue 1348 (1336) (Koshio *et al.*, 1989; Lewis *et al.*, 1990*a*). This C-terminal site is very different from the major site of phorbol ester-induced threonine phosphorylation of the EGF receptor, which is in the juxta-membrane region (Hunter, Ling & Cooper, 1984; Davis & Czech, 1985; Downward *et al.*, 1985). The sites of phorbol ester-induced serine phosphorylation of the insulin receptor have not been defined clearly but probably include residues 1305 and/or 1306 as for insulin-induced phosphorylation (Lewis *et al.*, 1990*a, b*). As discussed previously, the phorbol ester- and insulin-induced receptor serine/threonine phosphoryl-ations appear to depend on distinct enzymes, although involving at least some common sites (Duronio & Jacobs, 1990).

It has been proposed that protein kinase C may participate in regulation of insulin receptor function. Phosphorylation of receptor by protein kinase C *in vitro* decreases insulin-stimulated tyrosine kinase activity (Bollag *et al.*, 1986). Similar effects have been reported for phorbol esters added to intact cells. Thus pretreatment of hepatoma cells with phorbol ester decreases tyrosine autophosphorylation in response to insulin (Takayama *et al.*, 1984, 1988*b*). In both hepatoma cells and adipocytes, phorbol ester pretreatment inhibited insulin-stimulated autophosphorylation and tyrosine kinase activity of receptor assayed *in vitro* (Haring *et al.*, 1986*a*; Obermaier *et al.*, 1987; Takayama *et al.*, 1988*b*). However, in other cell types, including IM-9, Hep G2, vascular endothelium and transfected NIH 3T3 fibroblasts, the effects of phorbol esters and insulin on receptor phosphorylation appeared to be additive (Jacobs *et al.*, 1983; Jacobs & Cuatrecasas, 1986; Hachiya *et al.*, 1987; Pillay, Whittaker & Siddle, 1990). The reasons for these differences are unclear, but might relate to the extent of sites of serine/threonine phosphorylation in the different cell types, possibly reflecting the action of different isoforms of protein kinase C (Nishizuka, 1988). There have been similar inconsistencies in relation to the effects of phorbol esters on insulin binding, a decrease in binding being reported in some cases (Grunberger & Gorden, 1982; Thomopoulos *et al.*, 1982; Haring *et al.*, 1986*a*), but no effect in others (Takayama *et al.*, 1984; Jacobs & Cuatrecasas, 1986; Bollag *et al.*, 1986). There is evidence that phorbol esters modulate the internalisation and recycling of insulin receptors (Blake & Strader, 1986; Hachiya *et al.*, 1987; Bottaro *et al.*, 1989). It is worth noting in relation to all of this work that acute effects on insulin receptor phosphorylation and function comparable to those induced by phorbol esters as yet have not been demonstrated with naturally occurring agonists, which are believed to act via the phospholipase–diacyglycerol–protein kinase C pathway. The physiological relevance of the receptor phosphorylation by protein kinase C therefore remains uncertain, although impairment of insulin receptor kinase has been noted in various pathological states (see below).

Cyclic AMP-dependent protein kinase

The data regarding possible effects of cyclic AMP-dependent protein kinase on insulin receptor function are, if anything, even less clear than those for protein kinase C. Phosphorylation of insulin receptor by cyclic AMP-dependent kinase *in vitro*, with a small concomitant inhibition of tyrosine kinase activity, has been reported (Roth & Beaudoin, 1987). Other studies have failed to find a functional interaction between cyclic AMP-dependent kinase and insulin receptors *in vitro* (Joost *et al.*, 1986), or have described an inhibition of receptor tyrosine kinase which is not dependent on phosphorylation of serine residues (Tanti *et al.*, 1987). In intact IM-9 cells, cyclic AMP analogues or forskolin (an adenylate cyclase stimulator) increased receptor phosphorylation on serine and threonine residues and decreased insulin-induced autophosphorylation *in situ* and tyrosine kinase activity assayed *in vitro* (Stadtmauer & Rosen, 1986*b*). Significantly, effects of cyclic AMP on receptor phosphorylation have not been reported for other cell types. However, in adipocytes, pretreatment with catecholamines decreased receptor autophosphorylation and tyrosine kinase activity assayed *in vitro*, an effect which was only partially explained by an apparent decrease in insulin binding affinity (Haring *et al.*, 1986*b*; Obermaier *et al.*, 1987). However, in these experiments the phosphorylation state of the receptor following catecholamine treatment was not examined directly. Therefore, although there are some data consistent with an inhibitory action of cyclic AMP-dependent kinase on insulin receptor function, the general significance of such a mechanism remains in doubt.

Studies of the phosphorylation of insulin receptor *in vitro* by other serine-specific protein kinases, including casein kinases I and II, similarly have produced conflicting and inconclusive results (Tuazon *et al.*, 1985; Haring *et al.*, 1985).

Other regulatory mechanisms

Regulation of receptor tyrosine kinase activity may occur also by mechanisms other than serine phosphorylation. Peptides derived from MHC class I antigens inhibit receptor autophosphorylation *in vitro*, apparently by an allosteric mechanism (Hansen *et al.*, 1989). It is not clear whether this observation is related to the demonstration that insulin receptors in some cells interact with MHC class I molecules (Fehlmann *et al.*, 1985; Samson *et al.*, 1986; Phillips *et al.*, 1986; Due *et al.*, 1986). Basic proteins and polycations are potent stimulators of kinase activity under some conditions (Morrison *et al.*, 1989; Kohanski, 1989; Fujita-Yamaguchi *et al.*, 1989*a*). Receptor kinase activity is also influenced by phospholipid environment (Lewis & Czech, 1987; Sweet *et al.*, 1987*a*). The physiological importance of such effects remains to be demonstrated. Possible regulatory mechanisms influencing substrate selection, or the activity of the receptor kinase in patho-physiological states, are discussed in subsequent sections.

Involvement of receptor tyrosine kinase in signalling

Before the discovery of the tyrosine kinase activity of the receptor, no single hypothesis on the mechanism of insulin action commanded widespread support (Czech, 1977; Denton, Brownsey & Belsham, 1981). The demonstration of insulin-stimulated kinase activity, intrinsic to the receptor, immediately suggested a signalling mechanism. It was obvious that the receptor might regulate directly intracellular enzyme activity by tyrosine phosphorylation, initiating a cascade of reactions which would lead to the metabolic and growth promoting effects. The time course, concentration dependence and other properties of the kinase stimulation, as measured by autophosphorylation, were entirely consistent with such a role. Moreover, tyrosine kinase activity was found in a variety of growth factor receptors and retroviral transforming proteins (Hunter & Cooper, 1985) suggesting a common involvement of this activity in growth control. All of these considerations are circumstantial, however. Direct experimental evidence was required for an obligatory involvement of the tyrosine kinase in the metabolic effects of insulin, and for the nature of this involvement. The possibility had to be ruled out that the tyrosine kinase was involved only in growth-promoting effects of the hormone, or in some aspect of receptor function aside from signalling, such as endocytosis and turnover.

Evidence relating to the role of the kinase has come from several sources. It has been shown that some states of insulin resistance are associated with decreased receptor autophosphorylation and kinase activity, whether induced acutely by phorbol esters or catecholamines (Takayama *et al.*, 1984, 1988*b*; Haring *et al.*, 1986*a, b*; see above) or as a feature of pathological states of extreme insulin resistance, non-insulin dependent diabetes and obesity (Reddy & Kahn, 1988; see below). More convincingly, the importance of receptor kinase activity has been shown in experiments with insulin-mimetic agents, kinase-inhibitory antibodies and mutant receptors.

Insulin-mimetic agents

A number of chemically diverse agents with insulin-like activity have been shown to stimulate receptor autophosphorylation in parallel with their metabolic effects. These include trypsin (Leef & Larner, 1987; Hsuan *et al.*, 1989), hydrogen peroxide (Hayes & Lockwood, 1987; Kadota *et al.*, 1987;

Koshio *et al.*, 1988), and lectins (Roth *et al.*, 1983; Wilden *et al.*, 1989*b*). The mechanism of action of anti-receptor antibodies and vanadate, as insulin-mimetic agents, has been more controversial.

Receptor antibodies

Stimulation of receptor kinase has been demonstrated with some insulin-mimetic anti-receptor antibodies (Roth *et al.*, 1983; Gherzi *et al.*, 1987; Ponzio *et al.*, 1987; Takayama-Hasumi *et al.*, 1989). However, other studies have failed to find evidence of kinase activation by antibodies, both polyclonal (Simpson & Hedo, 1984; Zick *et al.*, 1984; Ponzio *et al.*, 1988) and monoclonal (Forsayeth *et al.*, 1987*a*; Hawley *et al.*, 1989; Soos *et al.*, 1989; Sung *et al.*, 1989). There are several problems in interpreting these data. First, it is clear that the action of antibodies, unlike that of insulin, does not depend on recognition of specific sites on the receptor, but can result from binding to multiple epitopes on both α- and β-subunits (Taylor *et al.*, 1987; O'Brien *et al.*, 1987*c*). The effects of antibodies depend on their bivalency, and ability to cross-link receptor subunits (Kahn *et al.*, 1981; Heffetz & Zick, 1986; O'Brien *et al.*, 1987*c*). It is possible that this cross-linking induces conformational changes in the receptor which mimic those accompanying insulin-induced autophosphorylation. A second problem arises from the 'spare receptor' phenomenon, in that, in most cells, it is necessary only to activate a small fraction of receptors to initiate maximum metabolic effects. It may be difficult to detect the small amount of autophosphorylation which is adequate to initiate a significant metabolic stimulation. However, in detailed studies with monoclonal anti-receptor antibodies and transfected cells expressing high levels of receptor, it was shown that, whereas the occupancy/metabolic response relationship was very similar for insulin and antibodies, the occupancy/autophosphorylation response was very different. Indeed, no antibody-induced autophosphorylation was detectable by [^{32}P]-phosphate labelling (Soos *et al.*, 1989) although paradoxically the same antibodies clearly stimulated autophosphorylation of solubilised receptor (O'Brien *et al.*, 1987*c*). More recently, it has been concluded that antibodies do induce phosphorylation in intact cells of both the receptor and other cellular substrates, when assayed by sensitive techniques (Brindle *et al.*, 1990; Steele-Perkins & Roth, 1990). These effects of antibodies are slow in onset and small in extent relative to those of insulin, and the sites of autophosphorylation are unknown. It is therefore difficult to say whether or not the autophosphorylation is commensurate with the metabolic activity of the antibodies. However, it would certainly not be appropriate at the present time to cite experiments with antibodies as evidence against a role of the tyrosine kinase in the actions of insulin itself.

Vanadate

Some confusion has also surrounded experiments on the insulin-like effects of vanadate. This compound is an inhibitor of phosphotyrosyl phosphatases (Swarup, Cohen & Garbers, 1982), but may also have direct effects on the receptor kinase (Tamura *et al.*, 1984; Gherzi *et al.*, 1988). Although vanadate has been demonstrated to increase receptor autophosphorylation in intact cells (Tamura *et al.*, 1983; Kadota *et al.*, 1987), it has been claimed that the magnitude of this effect is not appropriate for the accompanying metabolic responses (Mooney *et al.*, 1989; Strout *et al.*, 1989). However, it is likely that a relatively small activation of the receptor kinase would suffice to induce a critical level of phosphorylation of key (unidentified) intracellular substrates, if phosphatases are simultaneously inhibited.

The overall conclusion from experiments with insulin-mimetic agents is that these compounds do induce receptor autophosphorylation *in situ*, and therefore presumably stimulate kinase activity towards other substrates. Detailed comparison of the relative magnitude of effects of insulin and insulin-mimetic agents on metabolism and receptor autophosphorylation may not be appropriate in view of possible differences in time course and phosphorylation sites, or additional effects on other aspects of the signalling pathway.

Kinase inhibition

A different approach to analysing the role of the receptor kinase used a kinase-inhibitory monoclonal antibody directed against the cytoplasmic domain of the β-subunit. When microinjected into *Xenopus* oocytes, this antibody specifically blocked the ability of insulin, but not progesterone, to induce a maturation response (Morgan *et al.*, 1986). The same antibody, when introduced into fibroblasts, adipocytes or hepatoma cells by osmotic lysis of pinocytotic vesicles, inhibited acute metabolic responses to insulin, including deoxyglucose uptake, ribosomal protein S6 phosphorylation and glycogen synthesis (Morgan & Roth, 1987). The major reservation concerning these data is that immunoglobulins are very large molecules relative to the receptor kinase domain and therefore could have effects on receptor function additional to inhibition of autophosphorylation. However, a non-inhibiting antibody to the kinase domain had no effect on insulin action. These studies, therefore, provide evidence for an obligatory role of the kinase in mediating a wide range of insulin effects.

Mutant receptors with defective kinase activity

The strongest evidence, to date, for an obligatory role of the kinase in signalling has come from the study of receptor mutants. The kinase activity of the receptor depends critically on lysine residue 1030 (1018), which occupies an analogous position in a variety of kinases and which can be affinity-labelled with ATP analogues. This residue can be substituted by *in vitro* site-directed mutagenesis of receptor cDNA. The resulting mutant receptors, which are devoid of kinase activity, fail to mediate metabolic or growth-promoting effects of insulin when expressed, by transfection, in a variety of fibroblast cell lines (Ebina *et al.*, 1987; Chou *et al.*, 1987; McClain *et al.*, 1987). The mutation also abolishes the capacity of receptors to support insulin-like metabolic effects of anti-receptor antibodies (Ebina *et al.*, 1987). Replacement of tyrosine residues 1162 and 1163, which are critical sites of autophosphorylation, similarly results in a receptor with compromised metabolic signalling potential (Ellis *et al.*, 1986). It is difficult to rule out completely the possibility that changing even a single amino acid residue, however conservatively, may have effects on receptor conformation and function which extend beyond the specific changes anticipated. Further, it is unfortunate that the cell lines available for transfection are not typical of physiologically important target tissues, and permit the examination of only a limited range of rather small effects of insulin. It is likely, for instance, that the 1.5–2-fold stimulation of deoxyglucose uptake in fibroblasts is not representative of the much greater stimulation observed in muscle and fat, which probably involves a different insulin-responsive isoform of the glucose transporter (James *et al.*, 1988; James, Strube & Mueckler, 1989). However, these reservations aside, the studies of receptor mutants are entirely consistent with an obligatory role of the tyrosine kinase in all the effects of insulin.

Role of the kinase

It has proved difficult so far to identify precisely how the receptor tyrosine kinase is involved in signalling (Fig. 10.6). Two major questions arise. First, it is unclear whether the kinase activity is required only for autophosphorylation or also for phosphorylation of other key substrates. In the former case, it could be envisaged that autophosphorylation induces

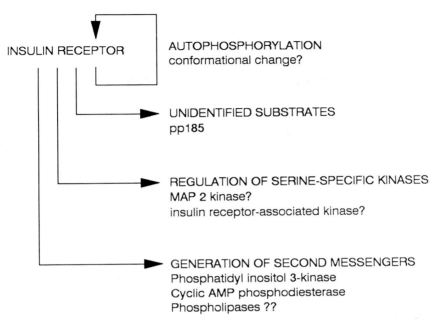

Fig. 10.6. *Mechanism of involvement of receptor tyrosine kinase in insulin action.* The diagram indicates different reactions by which the receptor tyrosine kinase might be involved in mediating insulin action. These all remain more or less speculative at present, and which if any are physiologically important has still to be established.

conformational change which modulates interaction with other proteins. In the latter, autophosphorylation is seen merely as a stage in activation of the kinase towards exogenous substrates. This issue is difficult to resolve because of the reciprocal interdependence of kinase activity and auto-phosphorylation. Mutations which abolish kinase activity must necessarily prevent autophosphorylation, while mutation of autophosphorylation sites impairs kinase activation towards exogenous substrates. The capacity of anti-receptor antibodies to induce insulin-mimetic metabolic effects without conspicuous receptor autophosphorylation, perhaps argues in favour of a key role for conformational changes, but, as discussed previously, the data are open to several interpretations. The issue will most obviously be settled if, and when, physiologically relevant substrates for the receptor kinase are identifed.

A second question is whether the receptor generates only one or multiple signals, given the diverse effects of insulin on different metabolic pathways. Attention has focussed particularly on whether the signalling pathways for acute metabolic regulation, and for so-called growth-promoting effects, might diverge at the level of the receptor. Some receptor mutations appear differentially to affect 'metabolic' and 'mitogenic' effects of insulin. Thus a mutant receptor in which tyrosines 1162/1163 have been replaced by phenylalanines will not mediate insulin stimulation of glucose uptake (Ellis *et al.*, 1986), while the mitogenic effect of the hormone is apparently unaffected (Debant *et al.*, 1988). A metabolic response was restored by cross-linking the mutant receptor with an anti-receptor antibody (Debant *et al.*, 1989). The same antibody mimicked the metabolic but not the mitogenic effects of insulin in hepatoma cells (Ponzio *et al.*, 1988). It was suggested, on the basis of these experiments, that acute metabolic effects are dependent on receptor aggregation, which may be induced either by phosphorylation of tyrosines 1162/1163 or by cross-linking with antibodies (Debant *et al.*, 1989). Mitogenic signalling was proposed to follow a different pathway, perhaps dependent on autophosphorylation of different tyrosines, or on a 'cryptic' kinase activity which is not detected with conventional artificial substrates (Debant *et al.*, 1988).

Differential effects on insulin responses have also been seen with a truncated receptor lacking 43 amino acids from the C-terminus of the β-subunit. When expressed in rat fibroblasts, this mutant receptor showed impaired metabolic signalling, but supported an augmented mitogenic response relative to the wild-type receptor, although tyrosine kinase activity appeared normal (Maegawa *et al.*, 1988*a*; Thies, Ullrich & McClain, 1989). The converse situation also has been reported, in that replacement of tyrosine 1158 (1146) produced a mutant receptor in which tyrosine kinase activity was decreased and the capacity for mitogenic signalling was impaired, but metabolic responses were normal (Wilden *et al.*, 1990). However, it appears that the properties of mutant receptors may depend on the cell type in which they are expressed, and/or the assay methods used. It has been reported recently that the C-terminally truncated receptor mediates normal responses when expressed in Chinese hamster ovary cells (Myers *et al.*, 1991), while mutation of tyrosine 1158 did not affect the kinase activity of receptor expressed in COS cells (Zhang *et al.*, 1991). Thus, it may be premature to draw general conclusions from any single set of data pertaining to the signalling potential of mutant receptors.

Still further complexity is indicated by studies in adipocytes using a series of substrate analogues as competitive inhibitors of the receptor kinase. These compounds generally blocked insulin-stimulated lipogenesis, but not antilipolysis, suggesting distinct signalling pathways even among acute metabolic effects (Schechter *et al.*, 1989). Some of these observations may be explained by requirements for different degrees of phosphorylation of substrates essential to different metabolic pathways.

Consideration has also been given to the possible role in substrate selection of the C-terminal and juxtamembrane regions, which flank the conserved tyrosine kinase domain. The former includes potential sites for both tyrosine autophosphorylation and serine/threonine phosphorylation, although, as presently understood, the latter does not. The properties of the C-terminally truncated receptor (Maegawa *et al.*, 1988*a*) indicate that tyrosine kinase activity *per se* is not sufficient to guarantee normal signalling, and suggest involvement of the C-terminal 'tail' in substrate binding. The role of this region has also been investigated by limited proteolysis, although, in this case, it is only possible to test the consequences with artificial substrates *in vitro*. Removal of a 10 kD fragment with trypsin did not affect activity in terms of autophosphorylation of the kinase domain (Goren *et al.*, 1987). However, removal of a similar-sized fragment with a membranal protease impaired both autophosphorylation and kinase activity, suggesting an important contribution of the 'tail' in maintaining affinity of the kinase for its substrates (Seger, Zick & Shaltiel, 1989). Mutations affecting tyrosine 972 (960) in the juxtamembrane region, which is not an autophosphorylation site, also inhibit signal transmission but do not affect autophosphorylation or kinase activity as measured *in vitro* (White *et al.*, 1988*a*). These observations suggest a role for the juxtamembrane region in substrate recognition *in vivo*. An anti-peptide antibody directed against this region inhibits autophosphorylation, but not the kinase activity of receptor which is already autophosphorylated (Herrera *et al.*, 1985). Involvement of the juxtamembrane region in substrate recognition may therefore be limited to specific endogenous cellular substrates.

Several studies have reported that over-expression of mutant, kinase-defective receptors in transfected cells impairs the signalling capacity of endogenous receptors (Ellis *et al.*, 1986; Ebina *et al.*, 1987; McClain *et al.*, 1987). It is generally assumed that this reflects the incorporation of wild-type, endogenous receptors into inactive heteromeric hybrids (Ebina *et al.*, 1987; Whittaker, Soos & Siddle, 1990). However, others have proposed that the phenomenon reflects a competition between receptors for binding of potential substrates (Maegawa *et al.*, 1988*b*). Such competition would not depend on kinase activity *per se*, but the regions of the receptor involved have not been identified.

In addition to the possible role of factors intrinsic to the receptor, it has been suggested that substrate specificity may be influenced by other

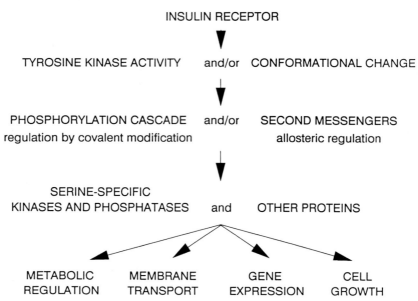

Fig. 10.7. *Mechanism of insulin action.* The diagram summarizes the likely major steps in the intracellular signalling pathway(s), as currently envisaged. Many details remain unclear as discussed in the text.

cellular proteins (Fujita-Yamaguchi *et al.*, 1989*b*; Yonezawa & Roth, 1990). It is clear that many aspects of the insulin receptor kinase remain to be clarified before its role in signalling is understood.

Post-receptor signalling pathways

It is outside the scope of this chapter to discuss post-receptor events in any detail, and it has to be admitted that our understanding of these events is still poor (Fig. 10.7). Given the diversity of effects elicited by insulin, it is possible that more than one type of signalling pathway exists, or that the pathway involves a complex interplay of various factors.

Substrates for the receptor kinase

It remains an attractive hypothesis that the receptor kinase phosphorylates one or more substrates on tyrosine residues as a first step in signalling (Fig. 10.6). However, identifying specific intracellular substrates of physiological importance for any tyrosine-specific protein kinase has proved surprisingly difficult.

Phosphorylation in vitro

Various proteins implicated in other pathways of cellular regulation have been shown to be phosphorylated by insulin receptors *in vitro*. Prominent among these are the guanine nucleotide-binding regulatory proteins (Gilman, 1987) including Gi/Go (O'Brien *et al.*, 1987*a*; Krupinski *et al.*, 1988) and transducin (Zick *et al.*, 1986). In all cases, the inactive (GDP-bound) form was the preferred substrate. However, evidence that the activity of G-proteins is modulated by tyrosine phosphorylation, or that phosphorylation occurs in intact cells, is lacking. Calmodulin (Klee, Crouch & Richman, 1980) is phosphorylated by insulin receptors *in vitro* (Sacks & McDonald, 1988; Wong *et al.*, 1988). Again the functional consequences of this phosphorylation are unknown, and it is disputed whether or not phosphorylation occurs in intact cells (Colca *et al.*, 1987; Blackshear & Haupt, 1989). Other proteins which serve as substrates for the receptor kinase *in vitro* include tubulin and microtubule-associated proteins (Kadowaki *et al.*, 1985), the progesterone receptor (Woo *et al.*, 1986) and Class I histocompatibility antigens (Peyron & Fehlmann, 1988). There is little or no evidence that these reactions are physiologically important, although it is intriguing that Class I antigens have also been implicated in modulation of receptor kinase activity (Hansen *et al.*, 1989) and insulin binding (Kittur *et al.*, 1987).

Endogenous cellular substrates

An alternative approach to identifying substrates has been to look for components which show insulin-stimulated tyrosine phosphorylation in intact cells. This is achieved by incubating cells with $[^{32}P]$-phosphate and by using anti-phosphotyrosine antibodies to immunoprecipitate labelled proteins from cell extracts. Three proteins revealed in this way have received particular attention. The smallest of these, pp15, is a cytosolic protein found in adipocytes, which was implicated initially in regulation of glucose uptake (Bernier, Laird & Lane, 1987, 1988). However, it has been identified as a homologue of myelin P2 and related fatty acid binding proteins (Hresko *et al.*, 1988), casting doubt on its role in signalling. In hepatocytes and hepatoma cells, a membrane glycoprotein pp120 was the most prominent endogenous substrate (Rees-Jones & Taylor, 1985; Sadoul *et al.*, 1985; Accili *et al.*, 1986; Perrotti *et al.*, 1987). This protein is also phosphorylated in response to EGF (Phillips, Perrotti & Taylor, 1987). It, too, now has been identified, as an integral membrane protein of the bile canalicular domain (Margolis *et al.*, 1988), which seems unlikely to have a primary role in signalling. The most widely distributed endogenous substrate is a cytoplasmic protein, pp185, which is rapidly phosphorylated in response to both insulin and IGF-1 (White, Maron & Kahn, 1985 *a*; White *et al.*, 1987; Izumi *et al.*, 1987; Tashiro-Hashimoto *et al.*, 1989). Although this meets several criteria for a role in signalling, assessment of its significance is difficult until its identity is known. A wide range of other endogenous substrates have been described in membrane preparations or intact cells, and await further characterisation (Haring *et al.*, 1987; Yu, Khalaf & Czech, 1987 *a*; Machicao *et al.*, 1987; Kwok & Yip, 1987; Caro *et al.*, 1987 *a*; Madoff, Martensen & Lane, 1988; Momomura *et al.*, 1988; O'Brien *et al.*, 1989).

A limited number of studies have demonstrated tyrosine phosphorylation of known cellular components which might participate in signalling pathways. Evidence has been presented that two different insulin-stimulated serine-specific protein kinases are phosphorylated or tyrosine residues (Yu, Khalaf & Czech, 1987 *b*; Ray & Sturgill, 1988), although in neither case was the stoichiometry of tyrosine phosphorylation demonstrated. In the case of the so-called MAP kinase, it appears that tyrosine phosphorylation directly regulates serine kinase activity (Andersen *et al.*, 1990) though there is, as yet, no evidence that the enzyme is directly phosphorylated by the insulin receptor. However, it remains an attractive hypothesis that insulin controls the activity of a cascade of serine-specific protein kinases via the tyrosine phosphorylation of key 'switch-kinases' by the receptor itself.

In intact hepatocytes, insulin induced tyrosine phosphorylation of a plasma membrane cyclic AMP phosphodiesterase, in parallel with stimulation of the enzyme activity (Pyne *et al.*, 1989). If it is confirmed that this phosphorylation directly modulates the phosphodiesterase, this could be part of the mechanism whereby insulin lowers cyclic AMP concentration, which, in turn, probably contributes to some of its metabolic effects in liver.

Substrates for other kinases

Comparative studies of the EGF- and insulin-receptor kinases *in vitro*, using synthetic peptide and other artificial substrates, have shown very similar but distinct specificities (Blackshear, Nemenoff & Avruch, 1984; Pike *et al.*, 1984; Zick *et al.*, 1985; Klein *et al.*, 1985). Likewise, both common and specific sets of proteins are phosphorylated in response to EGF and insulin in intact cells (Kadowaki *et al.*, 1987; Yarden & Ullrich, 1988), which may relate to their overlapping biological effects (Haystead &

Hardie, 1986; Bosch *et al.*, 1986). It has been shown that one form of phospholipase C is a substrate for EGF- and PDGF-induced tyrosine phosphorylation, and that phosphorylation of phosphatidylinositol 3-kinase is induced by PDGF (Ullrich & Schlessinger, 1990; Cantley *et al.*, 1991). The phosphorylated substrates associate in complexes with the respective receptors, although their precise roles in signalling are unclear. However, phospholipase C is not a substrate for the insulin receptor kinase (Nishibe *et al.*, 1990). Phosphatidylinositol 3-kinase is phosphorylated by the insulin receptor but does not associate with it (Endemann, Yonezawa & Roth, 1990). The significance of this phenomenon for insulin signalling pathways remains to be established.

Insulin mediators

Much effort has been devoted to the hypothesis that insulin action requires the mediation of a diffusible, low molecular weight second messenger. Evidence for such compounds was sought by preparing extracts from insulin-treated cells or membranes and testing these for regulatory activity on defined enzyme systems *in vitro*. The effects observed in this way implicated multiple 'peptide-like' insulin mediators (for reviews see Cheng & Larner, 1985; Jarett *et al.*, 1985). These compounds were characterised eventually as inositol phosphate glycans (phospho-oligosaccharides), formed by hydrolysis of glycosyl-phosphatidyl-inositol in the plasma membrane (Saltiel & Cuatrecasas, 1986; Saltiel *et al.*, 1986). Subsequent work on these potential insulin mediators has been reviewed (Low & Saltiel, 1988; Saltiel & Cuatrecasas, 1988; Saltiel, 1990). Production of inositol phosphate glycans is most readily achieved using exogenously added phospholipase C but an endogenous phosphatidylinositol-glycan specific phosphospholipase C has been identified in liver membranes (Fox *et al.*, 1987). The mechanism whereby such an enzyme is activated by insulin is unclear, although direct tyrosine phosphorylation, or the participation of G-proteins (Houslay, Wakelam & Pyne, 1986) are obvious possibilities.

A proper assessment of the physiological role of inositol phosphate glycans will require their detailed structural characterisation, and the availability of purified material for studies of the mechanism and specificity of effects on intact cells and isolated enzymes. The glycosyl-phosphatidyl-inositol precursors are structurally related to membrane protein anchors (Low, 1987; Low & Saltiel, 1988). There is evidence of considerable heterogeneity in the structure of inositol phosphate glycans, involving both myo- or chiro-inositol moieties, glucosamine or galactosamine, and variable phosphate content (Mato *et al.*, 1987; Larner *et al.*, 1988; Merida *et al.*, 1988; Saltiel & Cuatrecasas, 1988). The topological distribution of insulin-sensitive glycolipid within the plasma membrane is uncertain (Saltiel & Cuatrecasas, 1988). Some evidence suggests the predominant location may be at the outer surface of the cell (Alvarez *et al.*, 1988), and insulin-mimetic oligosaccharides have been isolated from the conditioned medium of insulin-treated cells (Witters & Watts, 1988). If inositol phosphate glycans were, indeed, released extracellularly, a specific uptake mechanism almost certainly would be required to explain their intracellular effects.

Preparations of 'mediator' added to intact cells mimic many, but not all, of the actions of insulin, and effects have also been demonstrated on a variety of enzymes in cell-free systems (for review see Saltiel & Cuatrecases, 1988; Witters *et al.*, 1988). However, the specificity and physiological relevance of these effects must remain in doubt while studies depend on impure preparations of 'mediator' added at unknown concentrations. Although the release of inositol phosphate glycans as an acute effect of insulin now seems reasonably well established, some laboratories have found the actions of mediator preparations difficult to reproduce. Even those working actively in this area are cautious about assigning a second messenger role to inositol phosphate glycans based on current data (Saltiel & Cuatrecasas, 1988) and certainly many questions remain to be answered regarding the physiological role of these compounds.

Diacylglycerol and protein kinase C

Several observations have suggested that some effects of insulin may be mediated in part by activation of protein kinase C (Farese & Cooper, 1989), although this has been another area of considerable controversy. It seems clear that the well-established inositol trisphosphate/diacylglycerol signalling system (Berridge, 1987) is not activated by insulin (Farese *et al.*, 1985; Pennington & Martin, 1985; Taylor *et al.*, 1985; Sakai & Wells, 1986; Augert & Exton, 1988). However, the concentration of diacylglycerol is increased by insulin in some cells (Farese *et al.*, 1985) apparently reflecting both increased phosphatidic acid synthesis and phospholipid hydrolysis (Farese *et al.*, 1987; Farese *et al.*, 1988), as well as the hydrolysis of glycolipids (Saltiel, Sherline & Fox, 1987). Diacylglycerol is considered to be the physiological activator of protein kinase C, and stimulation of protein kinase C activity by insulin has been reported (Walaas *et al.*, 1987;

Cooper *et al.*, 1987; Draznin *et al.*, 1988). Further, some effects of insulin, including stimulation of glucose transport, are mimicked by phorbol esters, which are thought to exert their effects by activating protein kinase C (see Denton 1986; Saltiel & Cuatrecasas, 1988; Luttrell *et al.*, 1989 for refs). However, there is also considerable evidence that protein kinase C does not have a significant role in insulin action. Several studies have been unable to demonstrate an effect of insulin on protein kinase C activity or distribution, or have found insulin action to be unaffected by depletion of protein kinase C following prolonged exposure to phorbol esters (Spach, Nemenoff & Blackshear, 1986; Klip & Ramlal, 1987; Corps & Brown, 1988; Caron *et al.*, 1988). Phorbol esters added acutely do not mimic the full effect of insulin even on glucose transport (Muhlbacher *et al.*, 1988; Tanti *et al.*, 1989) and fail to mimic some effects of insulin or even antagonise others (see Denton, 1986; Saltiel & Cuatrecasas, 1988). Insulin and phorbol esters also induce distinct patterns of protein phosphorylation (Blackshear *et al.*, 1985; Spach *et al.*, 1986; Gibbs, Allard & Lienhard, 1986; Luttrell *et al.*, 1989).

The most likely explanation for these conflicting data is that insulin action does not directly involve protein kinase C, but that there is some overlap in the activities of protein kinase C and insulin-stimulated kinases. However, it is now known that multiple subspecies of protein kinase C exist, which may be subtly different in their patterns of tissue expression, sensitivity to activators and substrate specificity (Nishizuka, 1988; Kikkawa, Kishimoto & Nishizuka, 1989). The possibility cannot be ruled out, therefore, that insulin selectively activates one of these subspecies, perhaps by acting on a distinct source of diacylglycerol which could be related to the generation of inositol phosphate glycan mediators (Saltiel & Cuatrecasas, 1988). It has also been reported that diacylglycerol may stimulate directly glucose transport in adipocytes, without activation of protein kinase C (Stralfors, 1988).

Other serine-specific kinases and phosphatases

Many of the metabolic effects of insulin depend ultimately on either increases or decreases in the level of serine/threonine phosphorylation of key regulatory enzymes (for reviews see Denton *et al.*, 1981; Cohen 1985; Denton 1986). The relevant signalling pathways must therefore involve modulation of the activity of protein kinases and/or phosphatases. Several distinct serine-specific kinases have been described which are stimulated by insulin in intact cells in a manner which persists after extraction and, in some cases, purification (reviewed by Czech *et al.*, 1988). It has also been reported that insulin decreases the activity of a multifunctional protein kinase in adipose tissue (Ramakrishna & Benjamin, 1988). Particular attention has been paid to the possibility that insulin-responsive kinases are themselves regulated by covalent modification, and especially by tyrosine phosphorylation (Stefanovic *et al.*, 1986; Yu *et al.*, 1987*b*; Sommercorn *et al.*, 1987; Ray & Sturgill, 1988; Smith & Sale, 1988). Though suggestive, these results do not show clearly yet that serine kinases are primary substrates for the insulin receptor kinase. Further work is required to establish the mechanism by which insulin regulates their activity, and to identify the range of physiological substrates on which they in turn act.

Protein phosphatases are if anything even less well characterised (Cohen, 1989). However, there is evidence that insulin activates protein phosphatases (Olivier, Ballou & Thomas, 1988; Toth, Bollen & Stalmans, 1988; Chan *et al.*, 1988; Cohen, 1989). There has also been speculation that insulin might modulate the activity of protein phosphatases in a way which alters their substrate specificity (Cohen, 1985). It is possible that phosphatases could be regulated either by covalent modification, or by allosteric mechanisms involving second messengers, and these hypotheses remain to be explored.

G-proteins

A growing family of guanine nucleotide-binding regulatory proteins (G-proteins) has been shown to participate in signal transduction between membrane receptors and effector systems, and especially in regulation of adenylate cyclase and phosphatidylinositol-specific phospholipase C (Gilman, 1987). The evidence that insulin interacts with one or more G-proteins has been reviewed (Houslay & Siddle, 1989; Zick, 1989). Pertussis toxin, which blocks the function of some G-proteins by catalysing their ADP-ribosylation, attenuates some effects of insulin (Elks *et al.*, 1983; Heyworth *et al.*, 1986; Hesketh & Campbell, 1987; Luttrell *et al.*, 1988, 1990). Conversely, insulin inhibits toxin-catalysed ADP-ribosylation of putative G-proteins (Heyworth *et al.*, 1985; Irvine & Houslay, 1988; Rothenberg & Kahn, 1988). Reference was made earlier to the phosphorylation of G-proteins by insulin receptor *in vitro* (Zick *et al.*, 1986; O'Brien *et al.*, 1987*a*; Krupinski *et al.*, 1988). The related guanine nucleotide-binding p21 protein, the product of the *ras* oncogene, is also phosphorylated in response to insulin (Kamata *et al.*, 1987; Korn *et al.*,

1987), and under some conditions p21 inhibits insulin receptor tyrosyl kinase activity (O'Brien *et al.*, 1987*b*). Further, p21 has been implicated in mediating insulin-induced maturation of *Xenopus* oocytes (Korn *et al.*, 1987).

The hypothesis that G-proteins transduce the effects of insulin on membrane-bound enzymes such as adenylate cyclase, cyclic AMP phosphodiesterase and phospholipases is attractive. Such effects could influence the production of second messengers (cyclic AMP, inositol phosphate glycans and diacylglycerol) which have been implicated in insulin action. The interaction of the insulin receptor with G-proteins might be influenced either by receptor autophosphorylation or by phosphorylation of the G-proteins themselves. However, at the present time, evidence for the involvement of G-proteins in insulin action is by no means strong compared to their well-established role in other signalling systems, and the importance for insulin action of the effector systems on which they might act is still uncertain.

Receptor biosynthesis and turnover

Although the insulin receptor is widely distributed among mammalian tissues, the number of receptors varies greatly in different cell types. Rather little is known of the factors which control receptor synthesis, assembly and turnover, although these factors must be ultimately important determinants of cellular sensitivity to insulin. The major steps of receptor biosynthesis and turnover are summarised in Fig. 10.8.

Biosynthesis

Studies of the regulation of insulin receptor biosynthesis have been carried out both by labelling of the receptor itself, and by investigation of mRNA levels with specific cDNA probes (Mamula *et al.*, 1990). Receptor biosynthesis is enhanced by glucocorticoids in several cell types. In the IM-9 lymphocyte cell line, this has been shown to be due to increased levels of receptor mRNA (McDonald *et al.*, 1987; Shibasaki *et al.*, 1988), which, in turn, reflect increased rates of transcription (McDonald & Goldfine, 1988; Rouiller *et al.*, 1988). Variable effects of insulin on receptor biosynthesis have been reported. In IM-9 lymphocytes there was a modest stimulation of biosynthesis that was not accompanied by any change in mRNA levels

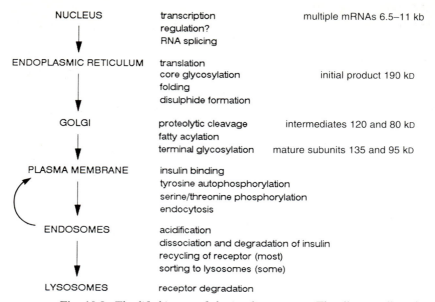

Fig. 10.8. *The life history of the insulin receptor.* The diagram lists the major steps in the biosynthesis and turnover of insulin receptors, according to their subcellular location.

(Rouiller & Gorden, 1987; Rouiller *et al.*, 1988), while in cultured pancreatic acinar cells there was a decrease in biosynthesis and mRNA levels (Okabayashi *et al.*, 1989). In Hep G2 hepatoma cells insulin had no effect on receptor mRNA levels although serum promoted an increase (Hatada *et al.*, 1989). It is unclear to what extent regulation in these cultured cell lines will mirror that in target tissues *in vivo*.

Studies of the insulin receptor gene have produced some information on the promoter region. This has features in common with the regulatory regions of constitutively expressed 'housekeeping' genes, containing neither TATA nor CAAT boxes but possessing several potential binding sites for the transcription factor Sp1 (Araki *et al.*, 1987; Mamula *et al.*, 1988; Seino *et al.*, 1989). Multiple receptor mRNA species of different sizes have been identified (Ullrich *et al.*, 1985; Ebina *et al.*, 1985*b*). The significance of these is unknown, though the differences almost certainly lie in the length of 3'-untranslated sequence. The alternative splicing of a miniexon was discussed above.

Limited studies of receptor biosynthesis have been carried out using an *in vitro* translation system, confirming a primary translation species of approximately 160 kD, which is cotranslationally glycosylated to give a product of approximately 180 kD in the presence of microsomal membranes (Goldstein & Kahn, 1988). These results are in agreement with earlier biosynthetic labelling experiments in a variety of cultured cells including human IM-9 and Hep G2 cells and rodent adipocytes (Hedo *et al.*, 1983; Ronnett *et al.*, 1984; Hedo & Simpson, 1985; Hedo & Gorden, 1985; Forsayeth, Maddux & Goldfine, 1986). The initial translation product *in vivo* is normally the high mannose proreceptor of approximately 190 kD, cotranslational glycosylation being essential for subsequent processing (Ronnett *et al.*, 1984). Formation of disulphide-linked proreceptor dimers probably also takes place in the endoplasmic reticulum (Olson, Bamberger & Lane, 1988) before transfer to the Golgi. There, the proreceptor is proteolytically cleaved to generate the α- and β-subunits, and further processing of the oligosaccharide takes place to produce both high-mannose and complex-type chains (McElduff *et al.*, 1986; Duronio *et al.*, 1988). In at least some cell types, the receptor undergoes an additional covalent modification at an early stage of biosynthesis, by fatty acylation of the β-subunit, probably on a cysteine residue (Magee & Siddle, 1988; Hedo, Collier & Watkinson, 1987). The functional significance of this modification is unknown. The mature receptor is finally transferred from the Golgi to the plasma membrane. The targetting mechanisms which direct this process are unclear. A small fraction of the receptor may escape proteolysis to appear on the cell surface as a fully glycosylated 210 kD form (Hedo & Gorden, 1985).

Endocytosis and down regulation

Mature insulin receptors are located predominantly in the plasma membrane under basal conditions (absence of hormone), and have a half-life of 10–12 h. However, binding of insulin induces a rapid endocytosis of the hormone–receptor complex, which can influence both the distribution and degradation of receptors (for reviews see Heidenreich & Olefsky, 1985; Sonne, 1988; Carpentier, 1989 for refs). Receptor-mediated endocytosis also serves as an important mechanism for degradation of insulin, particularly in the liver (Duckworth, 1988). Not surprisingly, there are significant differences in the details of these complex processes as studied with different analytical techniques and, especially, in different cell types. In general, the internalised insulin-receptor complex is rapidly transferred to an endosomal compartment where acidification results in the dissociation of insulin. Thereafter, the ligand and receptor follow separate paths. The insulin is largely degraded, probably in part in the endosomal system itself, and also after transfer to lysosomes. The receptor is largely recycled to the plasma membrane. However, the new steady state established in the presence of hormone normally results in a redistribution of receptors with an increase in the size of the intracellular pool. This redistribution is rapidly reversible on removal of hormone, but may, in turn, result in increased receptor degradation and net loss of receptors which is then only slowly reversed by *de novo* synthesis. Prolonged exposure of cells to high insulin concentrations, sufficient to occupy a significant fraction of receptors, therefore results in down-regulation of receptor numbers. Thus the internalisation of insulin–receptor complexes may provide a mechanism not only for degradation of the hormone, but also for regulation of receptor numbers in response to ambient hormone concentrations. In endothelial tissues, internalisation of insulin–receptor complexes may, instead, be part of a process of transcytosis, which is important for delivery of insulin to subendothelial target tissues (muscle, adipose tissue) (King & Johnson, 1985; Dernovsek & Bar, 1985). Additionally, receptor endocytosis may serve to deliver activated receptor or insulin itself to intracellular sites (Khan *et al.*, 1986; Klein *et al.*, 1987; Backer *et al.*, 1989*a*), perhaps including specific intracellular compartments such as the nucleus (Soler *et al.*, 1989). A role of internalisation in signalling therefore cannot be ruled out.

The molecular mechanisms which regulate receptor internalisation and subsequent trafficking are unclear. Studies with mutant receptors have suggested that tyrosine kinase activity is necessary to mediate insulin-induced receptor endocytosis and down regulation, as for metabolic signalling (Russell *et al.*, 1987; McClain *et al.*, 1987; Hari & Roth, 1987). However, autophosphorylation and receptor internalisation can apparently be uncoupled in ATP-depleted cells, casting doubt on the role of the kinase (Backer, Kahn & White, 1989*b*). The juxtamembrane region of the receptor, and specifically a sequence motif NPXY, appears to be involved in the interactions which mediate insulin-induced receptor internalisation (Chen, Goldstein & Brown, 1990; Backer *et al.*, 1990; Thies, Webster & McClain, 1990). In vascular endothelial cells, receptor phosphorylation on serine residues may significantly influence internalisation and externalisation rates (Hachiya *et al.*, 1987; Bottaro, Bonner-Weir & King, 1989). It is possible that this is a tissue-specific mechanism involved in the transcytosis of receptors in this cell type.

Anti-receptor antibodies induce rapid internalisation and degradation of receptors by mechanisms which depend on antibody bivalency, but not on receptor kinase activity (Morgan *et al.*, 1987*b*; Russell *et al.*, 1987; Ganderton, Whittaker & Siddle, 1989. Trischitta *et al.*, 1989). It is likely that both the signal for internalisation, and the pathway and fate of internalised receptor, are different for the antibody-induced and insulin-induced processes. Down-regulation of receptors induced by insulin or anti-receptor antibodies may be clinically significant as processes contributing to insulin resistance (see below).

The insulin receptor and insulin resistance

Insulin resistance is a feature of both obesity and non-insulin dependent diabetes mellitus (NIDDM), and indeed is implicated in the aetiology of NIDDM (DeFronzo, 1988; Reaven 1988; Taylor, 1989). Extreme insulin resistance also underlies several much rarer syndromes (Roth & Taylor, 1982; Reddy & Kahn, 1988). Resistance to the action of insulin, either in terms of sensitivity or maximal effect, may be apparent either from whole body glucose utilisation *in vivo*, or from responses of isolated cells or tissues incubated *in vitro* (Olefsky, 1981). Insulin resistance which is demonstrable *in vitro* is unlikely to be a reflection of acutely acting antagonists (metabolites or hormones) which may exert an influence *in vivo*. However, target tissue defects might well depend on the slowly reversible effects of chronic exposure to a metabolic or endocrine imbalance. These effects could include changes in the concentration, distribution, or covalent modification of key proteins, or alterations in membrane lipids, perhaps influenced by the level of glucose or insulin itself. The normally late onset of NIDDM indeed does imply that the associated insulin resistance is an acquired defect. The factors which precipitate this are unclear, although there is ultimately a strong genetic basis (Taylor, 1989). Alternatively, but more rarely, target cell resistance may result directly from a genetic defect at an identifiable locus (Kahn & Goldstein, 1989).

In principle, target cell resistance could reflect defects at the level of the receptor itself, or at post-receptor sites (signalling or metabolic pathways). It is very likely that post-receptor events, including glucose transport, are important sites of insulin resistance, perhaps in part secondary to abnormal patterns of insulin secretion (Berger *et al.*, 1989; Sivitz *et al.*, 1989). However, the broad spectrum of resistance affecting multiple pathways also implicates common early steps including the receptor itself. It is beyond the scope of this chapter to consider in detail the role of receptor abnormalities in insulin resistance, and the topic has been well covered in recent reviews (Reddy & Kahn, 1988; Zick, 1989).

Non-insulin-dependent diabetes mellitus

It has been observed, both in humans and in animal models, that insulin receptor numbers on target cells may be decreased in obesity and NIDDM (Roth & Taylor, 1982; Reddy & Kahn, 1988). However, this appears to be a secondary defect resulting from down-regulation induced by prevailing hyperinsulinaemia. As such, it is reversible by dietary restriction and weight loss. Interest more recently has focussed on the possibility that the signalling capacity of the receptor, as reflected in its tyrosine kinase activity, also is impaired in NIDDM and obesity (Haring & Obermaier-Kusser, 1989). Impaired kinase activity has been described in erythrocytes (Comi, Grunberger & Gorden, 1987), liver (Caro *et al.*, 1986) and adipose tissue (Friedenberg *et al.*, 1987; Sinha *et al.*, 1987; Takayama *et al.*, 1988*a*) of obese NIDDM subjects. Although the defect was apparently absent in obese non-diabetics it was reversed by weight reduction (Friedenberg *et al.*, 1988). Slightly different results were obtained for skeletal muscle, which showed impaired kinase activity in obesity and no additional effect of NIDDM (Caro *et al.*, 1987*b*; Arner *et al.*, 1987). This defect was specific for the insulin receptor and did not affect the Type I IGF receptor (Livingston *et al.*, 1988). Impaired receptor kinase activity also has been reported in animal models of obesity (Le Marchand-Brustel *et al.*, 1985; Shargill *et al.*, 1986), but not in other insulin-resistant states (Truglia, Hayes & Lockwood, 1988). The mechanism of kinase impairment observed in such studies is unclear. In all cases, results were normalised for equal amounts of insulin binding. One study reported that the pattern of autophosphorylation was altered in receptor from diabetic muscle, with a decreased amount of the 1158/1162/1163 tris-phosphorylated state required for maximal kinase activity (Obermaier-Kusser *et al.*, 1989). Other work, on adipose tissue, suggested that the decreased kinase activity reflected an increased subpopulation of receptors which bound insulin but did not undergo autophosphorylation (Brillon *et al.*, 1989). Mechanisms which regulate the autophosphorylation and thus the kinase activity of receptors might include phosphorylation on serine/threonine residues, as discussed previously. Alternatively, partial proteolysis of receptors, perhaps secondary to increased internalisation, might be responsible in part for loss of kinase activity (O'Hare & Pilch, 1988). Chronic hyperinsulinism itself may lead to desensitisation of kinase activity by mechanisms which

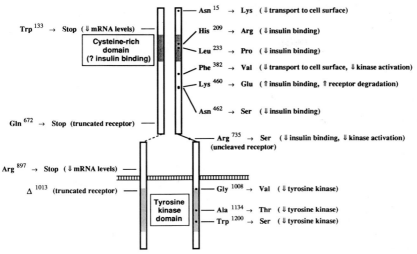

Deletions   **Missense mutations**

Trp ^{133} → Stop (⇓ mRNA levels)

Cysteine-rich domain (? insulin binding)

Gln ^{672} → Stop (truncated receptor)

Arg ^{897} → Stop (⇓ mRNA levels)

Δ ^{1013} (truncated receptor)

Tyrosine kinase domain

Asn ^{15} → Lys (⇓ transport to cell surface)

His ^{209} → Arg (⇓ insulin binding)

Leu ^{233} → Pro (⇓ insulin binding)

Phe ^{382} → Val (⇓ transport to cell surface, ⇓ kinase activation)

Lys ^{460} → Glu (⇑ insulin binding, ⇑ receptor degradation)

Asn ^{462} → Ser (⇓ insulin binding)

Arg ^{735} → Ser (⇓ insulin binding, ⇓ kinase activation) (uncleaved receptor)

Gly ^{1008} → Val (⇓ tyrosine kinase)

Ala ^{1134} → Thr (⇓ tyrosine kinase)

Trp ^{1200} → Ser (⇓ tyrosine kinase)

Fig. 10.9. *Insulin receptor mutants.* The diagram indicates the relative position in the primary sequence of receptor mutations identified in patients with insulin resistance due to defective receptor function. These mutations are discussed and referenced in the text.

do not depend on increased serine/threonine phosphorylation of receptor β-subunit (Treadway, Whittaker & Pesin, 1989 *b*).

Extreme insulin resistance

Genetic defects of the insulin receptor apparently underly some rare forms of severe insulin resistance, including the type A syndrome, leprechaunism and lipoatrophic diabetes (see Kahn & White, 1988; Reddy & Kahn, 1988). These defects may result in severe impairment of insulin binding and/or receptor kinase activity (Grigorescu, Flier & Kahn, 1986; Reddy, Lauris & Kahn, 1988; Kriauciunas *et al.*, 1988). Progress has been made recently in identifying the precise nature of the mutations in some of these patients (Fig. 10.9). These have turned out to be very diverse (Taylor *et al.*, 1990; Moller *et al.*, 1990; Kadowaki *et al.*, 1990). Gross rearrangements of the receptor gene appear to be uncommon (Muller-Wieland *et al.*, 1989), although in one patient a deletion of the most of the kinase domain occurred (Taira *et al.*, 1989). A gene encoding a truncated receptor as a result of a nonsense mutation causing premature chain termination also has been reported (Kadowaki *et al.*, 1988). Kinase-defective receptors also arise as a result of point mutations, and corresponding single amino acid substitutions, within different regions of the kinase domain itself (Moller & Flier, 1988; Odawara *et al.*, 1989; Goldstein & Kahn, 1989; Moller *et al.*, 1990). Impaired insulin binding has been seen to result from single amino acid substitutions within the extracellular domain (Kadowaki *et al.*, 1988; Klinkhamer *et al.*, 1989). In one case, a point mutation at the tetrabasic proteolytic processing site results in the production of uncleaved receptor which has a very low affinity for insulin (Yoshimasa *et al.*, 1988; Kobayashi *et al.*, 1988). Another amino acid substitution, within the α-subunit, leads to inefficient processing and transport of receptors and a consequent low level of surface expression (Accili *et al.*, 1989). In other patients, decreased receptor numbers appear to reflect low levels of mRNA (Ojamaa *et al.*, 1988). As no two patients have yet been found with the same mutation, it is likely that the study of further subjects will reveal many more deleterious mutations at different sites.

Extreme insulin resistance, in some cases, results from a homozygous mutation in a consanguineous family (Accili *et al.*, 1989; Klinkhamer *et al.*, 1989), and in others from compound heterozygous mutations (Kadowaki *et al.*, 1988). In some instances, the mutation behaves as an autosomal dominant (Taira *et al.*, 1989; Moller *et al.*, 1990) and in other cases heterozygotes display significant insulin resistance (Kadowaki *et al.*, 1988; Klinkhamer *et al.*, 1989). It remains to be determined whether genetic abnormalities resulting in minor impairment of receptor function contribute more generally to the insulin resistance associated with NIDDM. At present, it seems likely that receptor mutations at best will be found in only a small subset of diabetic patients (Taylor *et al.*, 1990).

Extreme insulin resistance may also be caused by autoantibodies directed against the insulin receptor (type B syndrome) (Flier, Kahn & Roth, 1979). A spectrum of antibodies with different properties has been described, which may acutely mimic insulin action, although generally inhibiting insulin binding and blocking insulin action in the long term. The chronic effects of these anti-receptor antibodies probably reflect, at least in part, the down regulation of insulin receptor numbers. Very low titres of anti-receptor antibodies have also been reported in a small proportion of subjects with NIDDM (Batarseh *et al.*, 1988), in an animal model of NIDDM (Harrison & Itin, 1979), and in newly diagnosed IDDM (Maron *et al.*, 1983). It has been suggested that these antibodies may arise in part as anti-idiotypes of anti-insulin antibodies (Schecter *et al.*, 1984; Ludwig, Faiman & Dean, 1987).

A full description of the mechanisms underlying insulin resistance will obviously depend on a better understanding of the mechanism of insulin action in normal subjects. However, studies of insulin resistance themselves have made important contributions to research on basic mechanisms. Autoantibodies from patients with type B severe insulin resistance have been invaluable reagents in studies of insulin receptor structure and function. It is likely also that characterisation of receptor mutations in other syndromes of insulin resistance will help to identify critically important residues, and thus provide new insights into relationships between receptor structure and function.

Conclusion

The structure of the insulin receptor is well established as disulphide-linked β–α–α–β heterotetramer, in which the extracellular α-subunit binds insulin, and the transmembrane β-subunit is an insulin-stimulated tyrosine-specific protein kinase. There is no convincing evidence for functionally significant disulphide interchange within the mature receptor. However, the detailed mechanism of transmembrane activation is unclear and requires further study. The result of this activation is an intramolecular autophosphorylation reaction, which significantly modifies the conformation of the cytosolic domain and activates the kinase towards other substrates. The evidence that the kinase activity is an essential component of signalling mechanisms for most, if not all, of insulin's metabolic and growth promoting effects is very strong. However, this cannot be quite conclusive until subsequent steps are clarified. Post-receptor signalling events may require phosphorylation of cellular substrates which have yet to be identified or might depend only on autophosphorylation of the receptor itself. These events may take the form of a cascade of covalent modifications, or involve the production of second messengers which act as allosteric regulators of enzyme activity. The increasingly detailed knowledge of the receptor itself so far has failed to shed much light on these elusive details of the signalling mechanism.

Receptor function may be regulated at various levels, apart from acute activation by insulin. Binding affinity may be subtly dependent on receptor isoforms arising by alternative mRNA splicing, or on post-translational modifications including formation of hybrids with IGF-1 receptors. Cell surface receptor numbers may be influenced by modulation of endocytosis, recycling and degradation in response to insulin itself and other factors. The receptor tyrosine kinase is probably subject to both acute and chronic regulation, involving serine/threonine phosphorylation and other mechanisms. It remains to be determined to what extent impaired receptor function contributes to states of insulin resistance and the development of non-insulin dependent diabetes mellitus.

It is sobering to reflect, while writing these concluding remarks, that there are still so many issues relating to receptor function on which it is difficult to make confident assertions. However, there has been dramatic progress over the last decade in structural analysis and in characterisation of the intrinsic kinase activity. With the increasingly intensive application of molecular genetic and cell biological techniques, to supplement traditional biochemical approaches, the receptor surely cannot withold its remaining secrets for much longer.

Acknowledgements

Work in my laboratory is supported by the Wellcome Trust, Medical Research Council, British Diabetic Association and Serono Diagnostics. I thank many colleagues and collaborators, whose work is cited, for stimulating discussions. I am particularly grateful to Lee Creswell and Jackie Sharpe for their excellent typing of a difficult manuscript.

References

Accili, D., Frapier, C., Mosthaf, L., McKeon, C., Elbein, S. C., Permutt, M. A., Ramos, E., Lander, E., Ullrich, A. & Taylor, S. I. (1989). A mutation in the insulin receptor gene that impairs transport of the receptor to the plasma membrane and causes insulin-resistant diabetes. *EMBO Journal*, **8**, 2509–17.

Accili, D., Perrotti, N., Rees-Jones, R. W. & Taylor, S. I. (1986). Tissue distribution and subcellular localization of an endogenous substrate (pp120) for the insulin receptor-associated tyrosine kinase. *Endocrinology*, **119**, 1274–80.

Alvarez, J. F., Varela, I., Ruiz-Albusac, J. M. & Mato, J. M. (1988). Localisation of the insulin-sensitive phosphatidylinositol glycan at the outer surface of the cell membrane. *Biochemical and Biophysical Research Communications*, **152**, 1455–62.

Andersen, A. S., Kjeldsen, T., Wiberg, F. C., Christensen, P. M., Rasmussen, J. S., Norris, K., Moller, K. B. & Moller, N. P. H. (1990). Changing the insulin receptor to possess insulin-like growth factor I ligand specificity. *Biochemistry*, **29**, 7363–6.

Anderson, N. G., Maller, J. L., Tonks, N. K. & Sturgill, T. W. (1990). Requirement for integration of signals from two distinct phosphorylation pathways for activation of MAP kinase. *Nature*, London, **343**, 651–3.

Araki, E., Shimada, F., Uzawa, H., Mori, M. & Ebina, Y. (1987). Characterization of the promoter region of the human insulin receptor gene. *Journal of Biological Chemistry*, **262**, 16186–91.

Arner, P., Pollare, T., Lithell, H. & Livingston, J. N. (1987). Defective insulin receptor tyrosine kinase in human skeletal muscle in obesity and type 2 (non-insulin-dependent) diabetes mellitus. *Diabetologia*, **30**, 437–40.

Augert, G. & Exton, J. H. (1988). Insulin and oxytocin effects on phosphoinositide metabolism in adipocytes. *Journal of Biological Chemistry*, **264**, 3600–9.

Avruch, J., Nemenoff, R. A., Blackshear, P. J., Pierce, M. W. & Osathanondh, R. (1982). Insulin-stimulated tyrosine phosphorylation of the insulin receptor in detergent extracts of human placental membranes. *Journal of Biological Chemistry*, **257**, 15162–6.

Backer, J. M., Kahn, C. R., Cahill, D. A., Ullrich, A. & White, M. F. (1990). Receptor-mediated internalization of insulin requires a 12-amino acid sequence in the juxtamembrane region of the insulin receptor β-subunit. *Journal of Biological Chemistry*, **265**, 16450–4.

Backer, J. M., Kahn, C. R. & White, M. F. (1989 *a*). Tyrosine phosphorylation of the insulin receptor during insulin-stimulated internalization in rat hepatoma cells. *Journal of Biological Chemistry*, **264**, 1694–701.

Backer, J. M., Kahn, C. R. & White, M. F. (1989 *b*). Tyrosine phosphorylation of the insulin receptor is not required for receptor internalization: studies in 2,4-dinitrophenol-treated cells. *Proceedings of the National Academy of Sciences, USA*, **86**, 3209–13.

Bajaj, M., Waterfield, M. D., Schlessinger, J., Taylor, W. R. & Blundell, T. (1987). On the tertiary structure of the extracellular domains of the epidermal growth factor and insulin receptors. *Biochimica et Biophysica Acta*, **916**, 220–6.

Ballotti, R., Kowalski, R., Le Marchand-Brustel, Y. & Van Obberghen, E. (1986). Presence of an insulin-stimulated serine kinase in cell extracts from IM-9 cells. *Biochemical and Biophysical Research Communications*, **139**, 179–85.

Ballotti, R., Kowalski, A., White, M. F., Le Marchand-Brustel, Y. & Van Obberghen, E. (1987). Insulin stimulates tyrosine phosphorylation of its receptor β-subunit in intact rat hepatocytes. *Biochemical Journal*, **241**, 99–104.

Ballotti, R., Lammers, R., Scimeca, J. C., Dull, T., Schlessinger, J., Ullrich, A. & Van Obberghen, E. (1989). Intermolecular transphosphorylation between insulin receptors and EGF-insulin receptor chimerae. *EMBO Journal*, **8**, 3303–9.

Baron, M. D. & Sonksen, P. H. (1983). Elucidation of the quaternary structure of the insulin receptor. *Biochemical Journal*, **212**, 79–84.

Baron, V., Gautier, N., Rochet, N., Ballotti, R., Rossi, B., Saint-Pierre, S., Van Obberghen, E. & Dolais-Kitabgi, J. (1989). Antibodies to insulin receptor tyrosine kinase stimulate its activity towards exogenous substrates without inducing receptor autophosphorylation. *Biochemical Journal*, **260**, 749–56.

Baron, V., Gautier, N., Komoriya, A., Hainaut, P., Scimeca, J. C., Mervic, M., Lavielle, S., Dolais-Kitabgi, J., & Van Obberghen, E. (1990). Insulin binding to its receptor induces a conformational change in the receptor C-terminus. *Biochemistry*, **29**, 4634–41.

Batarseh, H., Thompson, R. A., Odugbesan, O. & Barnett, A. H. (1988). Insulin receptor antibodies in diabetes mellitus. *Clinical and Experimental Immunology*, **71**, 85–90.

Beguinot, F., Smith, R. J., Kahn, C. R., Maron, R., Moses, A. C. & White, M. F. (1988). Phosphorylation of insulin-like growth factor I receptor by insulin receptor tyrosine kinase in intact cultured skeletal muscle cells. *Biochemistry*, **27**, 3222–8.

Berger, J., Biswas, C., Vicario, P. P., Strout, H. V., Saperstein, R. & Pilch, P. F. (1989). Decreased expression of the insulin-responsive glucose transporter in diabetes and fasting. *Nature*, London, **40**, 70–2.

Bernier, M., Laird, D. M. & Lane, M. D. (1987). Insulin-activated tyrosine phosphorylation of a 15-kilodalton protein in intact 3T3-L1 adipocytes. *Proceedings of the National Academy of Sciences, USA*, **84**, 1844–8.

Bernier, M., Laird, D. M. & Lane, M. D. (1988). Effect of vanadate on the cellular accumulation of pp15, an apparent product of insulin receptor tyrosine kinase action. *Journal of Biological Chemistry*, **263**, 13626–34.

Berridge, M. J. (1987). Inositol trisphosphate and diacylglycerol: two interacting second messengers. *Annual Review of Biochemistry*, **56**, 159–93.

Blackshear, P. J. & Haupt, D. M. (1989). Evidence against insulin-stimulated phosphorylation of calmodulin in 3T3-L1 adipocytes. *Journal of Biological Chemistry*, **64**, 3854–8.

Blackshear, P. J., Nemenoff, R. A. & Avruch, J. (1984). Characteristics of insulin and epidermal growth factor stimulation of receptor autophosphorylation in detergent extracts of rat liver and transplantable rat hepatomas. *Endocrinology*, **114**, 141–8.

Blackshear, P. J., Witters, L. A., Girard, P., Kuo, J. F. & Quamo, S. N. (1985). Growth factor-stimulated protein phosphorylation in 3T3-L1 cells: evidence for protein kinase C-dependent and -independent pathways. *Journal of Biological Chemistry*, **260**, 13304–15.

Blake, A. D. & Strader, C. D. (1986). Potentiation of specific association of insulin with Hep G2 cells by phorbol esters. *Biochemical Journal*, **236**, 227–34.

Bollag, G. E., Roth, R. A., Beaudoin, J., Mochly-Rosen, D. & Koshland, D. E. (1986). Protein kinase C directly phosphorylates the insulin receptor *in vitro* and reduces its protein–tyrosine kinase activity. *Proceedings of the National Academy of Sciences, USA*, **83**, 5822–4.

Boni-Schnetzler, M., Kaligian, A., DelVecchio, R. & Pilch, P. F. (1988). Ligand-dependent intersubunit association within the insulin receptor complex activates its intrinsic kinase activity. *Journal of Biological Chemistry*, **263**, 6822–8.

Boni-Schnetzler, M., Rubin, J. B. & Pilch, P. F. (1986). Structural requirements for the transmembrane activation of the insulin receptor kinase. *Journal of Biological Chemistry*, **261**, 15281–7.

Boni-Schnetzler, M., Scott, W., Waugh, S. E., DiBella, E. & Pilch, P. F. (1987). The insulin receptor: structural basis for high affinity ligand binding. *Journal of Biological Chemistry*, **262**, 8395–401.

Bosch, F., Bouscarel, B., Slaton, J., Blackmore, P. F. & Exton, J. H. (1986). Epidermal growth factor mimics insulin effects in rat hepatocytes. *Biochemical Journal*, **239**, 523–30.

Bottaro, D. P., Bonner-Weir, S. & King, G. L. (1989). Insulin receptor recycling in vascular endothelial cells: regulation by insulin and phorbol ester. *Journal of Biological Chemistry*, **264**, 5916–23.

Boyle, T. R., Campana, J., Sweet, L. J. & Pessin, J. E. (1985). Subunit structure of the purified human placental insulin receptor. *Journal of Biological Chemistry*, **260**, 8593–600.

Brillon, D. J., Friedenberg, G. R., Henry, R. R. & Olefsky, J. M. (1989). Mechanism of defective insulin-receptor kinase activity in NIDDM: evidence for two receptor populations. *Diabetes*, **38**, 397–403.

Brindle, N. P. J., Tavare, J. M., Dickens, M., Whittaker, J. & Siddle, K. (1990). Anti-(insulin receptor) monoclonal antibody-stimulated tyrosine phosphorylation in cells transfected with human insulin receptor cDNA. *Biochemical Journal*, **268**, 615–20.

Burant, C. F., Treutelaar, M. K., Allen, K. D., Sens, D. A. & Buse, M. G.

(1987). Comparison of insulin and insulin-like growth factor I receptors from rat skeletal muscle and L6 myocytes. *Biochemical and Biophysical Research Communications*, **147**, 100–7.

Burant, C. F., Treutelaar, M. K., Block, N. E. & Buse, M. G. (1986). Structural differences between liver- and muscle-derived insulin receptors in rats. *Journal of Biological Chemistry*, **261**, 14361–4.

Burant, C. F., Treutelaar, M. K. & Buse, M. G. (1988). Tissue-specific differences in the insulin receptor kinase activated *in vitro* and *in vivo*. *Endocrinology*, **122**, 427–37.

Butcher, R. W., Crofford, O. B., Gammeltoft, S., Gliemann, J., Gavin, J. R., Goldfine, I. D., Kahn, C. R., Rodbell, M., Roth, J., Jarett, L., Larner, J., Lefkowitz, R. J., Levine, R. & Marinetti, G. V. (1973). Insulin activity: the solid matrix. *Science*, **182**, 396–7.

Cantley, L. C., Auger, K. R., Carpenter, C., Duckworth, B., Graziani, A., Kapeller, R. & Soltoff, S. (1991). Oncogenes and signal transduction. *Cell*, **64**, 281–302.

Caro, J. F., Ittoop, O., Pories, W. J., Meelheim, D., Flickinger, E. G., Thomas, F., Jenquin, M., Silverman, J. F., Khazanie, P. G. & Sinha, M. K. (1986). Studies on the mechanism of insulin resistance in the liver from humans with noninsulin-dependent diabetes. *Journal of Clinical Investigation*, **78**, 249–58.

Caro, J. F., Raju, S. M., Sinha, M. K., Goldfine, I. D. & Dohm, G. L. (1988). Heterogeneity of human liver, muscle and adipose tissue insulin receptor. *Biochemical and Biophysical Research Communications*, **151**, 123–32.

Caro, J. F., Shafer, J. A., Taylor, S. I., Raju, S. M., Perrotti, N. & Sinha, M. K. (1987*a*). Insulin stimulated protein phosphorylation in human plasma liver membranes: detection of endogenous or plasma membrane associated substrates for the insulin receptor kinase. *Biochemical and Biophysical Research Communications*, **149**, 1008–16.

Caro, J. F., Sinha, M. K., Raju, S. M., Ittoop, D., Pories, W. J., Flickinger, E. G., Meelheim, D. & Dohm, G. L. (1987*b*). Insulin receptor kinase in human skeletal muscle from obese subjects with and without noninsulin dependent diabetes. *Journal of Clinical Investigation*, **79**, 1330–7.

Caron, M., Cherqui, G., Wicek, D., Capeau, J., Bertrand, J. & Picard, J. (1988). Effect of protein kinase C activation and depletion on insulin stimulation of glycogen synthesis in cultured hepatoma cells. *Experientia*, **44**, 34–7.

Carpentier, J. L. (1989). The cell biology of the insulin receptor. *Diabetologia*, **32**, 627–35.

Chan, C. P., McNall, S. J., Krebs, E. G. & Fischer, E. H. (1988). Stimulation of protein phosphatase activity by insulin and growth factors in 3T3 cells. *Proceedings of the National Academy of Sciences, USA*, **85**, 6257–61.

Chen, J. J., Kosower, N. S., Petryshyn, R. & London, I. M. (1986). The effects of *N*-ethylmaleimide on the phosphorylation and aggregation of insulin receptors in the isolated plasma membranes of 3T3-F442A adipocytes. *Journal of Biological Chemistry*, **261**, 902–8.

Chen, W.-J., Goldstein, J. L. & Brown, M. S. (1990). NPXY, a sequence often found in cytoplasmic tails, is required for coated pit-mediated internalization of the low density lipoprotein receptor. *Journal of Biological Chemistry*, **265**, 3116–23.

Cheng, K. & Larner, J. (1985). Intracellular mediators of insulin action. *Annual Review of Physiology*, **47**, 405–24.

Chiacchia, K. B. (1988). Reoxidation of the class I disulfides of the rat adipocyte insulin receptor is dependent on the presence of insulin: the class I disulfide of the insulin receptor is extracellular. *Biochemistry*, **27**, 4894–902.

Chou, C. K., Dull, T. J., Russell, D. S., Gherzi, R., Lebwohl, D., Ullrich, A. & Rosen, O. M. (1987). Human insulin receptors mutated at the ATP-binding site lack protein tyrosine kinase activity and fail to mediate postreceptor effects of insulin. *Journal of Biological Chemistry*, **262**, 1842–7.

Chvatchko, Y., Gazzano, H., Van Obberghen, E. & Fehlmann, M. (1984). Subunit arrangement of insulin receptors in hepatoma cells. *Molecular and Cellular Endocrinology*, **36**, 59–65.

Clark, S., Cheng, D. J., Hsuan, J. J., Haley, J. D. & Waterfield, M. D. (1988). Loss of three major autophosphorylation sites in the EGF receptor does not block the mitogenic action of EGF. *Journal of Cellular Physiology*, **134**, 421–8.

Clark, S. & Harrison, L. C. (1985). Structure of covalent insulin–receptor complexes (I-S-S-R) in isolated rat adipocytes and human placental membranes. *Biochemical Journal*, **229**, 513–19.

Cohen, P. (1985). The role of protein phosphorylation in the hormonal control of enzyme activity. *European Journal of Biochemistry*, **151**, 439–48.

Cohen, P. (1989). The structure and regulation of protein phosphatases. *Annual Review of Biochemistry*, **58**, 453–508.

Colca, J. R., DeWald, D. B., Pearson, J. D., Palazuk, B. J., Laurino, J. P. & McDonald, J. M. (1987). Insulin stimulates the phosphorylation of calmodulin in intact adipocytes. *Journal of Biological Chemistry*, **262**, 11399–402.

Comi, R. J., Grunberger, G. & Gorden, P. (1987). Relationship of insulin binding and insulin-stimulated tyrosine kinase activity is altered in type II diabetes. *Journal of Clinical Investigation*, **79**, 453–62.

Cooper, D. R., Konda, T. S., Standaert, M. L., Davis, J. S., Pollett, R. J. & Farese, R. J. (1987). Insulin increases membrane and cytosolic protein kinase C activity in BC3H-1 myocytes. *Journal of Biological Chemistry*, **262**, 3633–9.

Cordera, R., Andraghetti, G., Gherzi, R., Adezati, L., Montemurro, A., Lauro, R., Goldfine, I. D. & De Pirro, R. (1987). Species specificity of insulin binding and insulin receptor protein tyrosine kinase activity. *Endocrinology*, **121**, 2007–10.

Corin, R. E. & Donner, D. B. (1982). Insulin receptors convert to a higher affinity state subsequent to insulin binding. *Journal of Biological Chemistry*, **257**, 104–10.

Corps, A. N. & Brown, K. D. (1988). Insulin-like growth factor 1 and insulin reduce epidermal growth factor binding to Swiss 3T3 cells by an indirect mechanism that is apparently independent of protein kinase C. *FEBS Letters*, **233**, 303–6.

Crettaz, M., Jialal, I., Kasuga, M. & Kahn, C. R. (1984). Insulin receptor regulation and desensitization in rat hepatoma cells. *Journal of Biological Chemistry*, **259**, 11543–9.

Crofford, O. B. (1968). The uptake and inactivation of native insulin by isolated fat cells. *Journal of Biological Chemistry*, **243**, 362–9.

Cuatrecasas, P. (1969). Interaction of insulin with the cell membrane: the primary action of insulin. *Proceedings of the National Academy of Sciences, USA*, **63**, 450–7.

Cuatrecasas, P. (1971). Properties of the insulin receptor of isolated fat cell membranes. *Journal of Biological Chemistry*, **246**, 7265–74.

Cuatrecasas, P. (1972). Affinity chromatography and purification of the insulin receptor of liver cell membranes. *Proceedings of the National Academy of Sciences, USA*, **69**, 1277–81.

Cuatrecasas, P. (1974). Membrane receptors. *Annual Review of Biochemistry*, **43**, 169–214.

Czech, M. P. (1977). Molecular basis of insulin action. *Annual Review of Biochemistry*, **46**, 359–84.

Czech, M. P. (1985 a). The nature and regulation of the insulin receptor: structure and function. *Annual Review of Physiology*, **47**, 357–81.

Czech, M. P. (ed.) (1985 b). *Molecular Basis of Insulin Action*. New York: Plenum Press.

Czech, M. P. (1989). Signal transmission by the insulin-like growth factors. *Cell*, **59**, 235–8.

Czech, M. P., Klarlund, J. K., Yagaloff, K. A., Bradford, A. P. & Lewis, R. E. (1988). Insulin receptor signaling: activation of multiple serine kinases. *Journal of Biological Chemistry*, **263**, 11017–20.

Czech, M. P., Massague, J. & Pilch, P. F. (1981). The insulin receptor: structural features. *Trends in Biochemical Sciences*, **6**, 222–5.

Davis, R. J. & Czech, M. P. (1985). Tumor-promoting phorbol diesters cause the phosphorylation of epidermal growth factor receptors in normal human fibroblasts at threonine-654. *Proceedings of the National Academy of Sciences, USA*, **82**, 1974–8.

Debant, A., Clauser, E., Ponzio, G., Filloux, C., Auzan, C., Contreres, J. O. & Rossi, B. (1988). Replacement of insulin receptor tyrosine residues 1162 and 1163 does not alter the mitogenic effect of the hormone. *Proceedings of the National Academy of Sciences, USA*, **85**, 8032–6.

Debant, A., Ponzio, G., Clauser, E., Contreres, J. O. & Rossi, B. (1989). Receptor cross-linking restores an insulin metabolic effect altered by mutation on tyrosine 1162 and tyrosine 1163. *Biochemistry*, **28**, 14–17.

DeFronzo, R. A. (1988). The triumvirate: β-cell, muscle, liver. A collusion responsible for NIDDM. *Diabetes*, **37**, 667–87.

Deger, A., Kramer, H., Rapp, R., Koch, R. & Weber, U. (1986). The non-classical insulin binding of insulin receptors from rat liver is due to the presence of two interacting α-subunits in the receptor complex. *Biochemical and Biophysical Research Communications*, **135**, 458–64.

De Meyts, P., Bianco, A. R. & Roth, J. (1976). Site–site interactions among insulin receptors: characterization of the negative cooperativity. *Journal of Biological Chemistry*, **251**, 1877–88.

De Meyts, P., Gu, J. L., Shymko, R. M., Kaplan, B. E., Bell, G. I. & Whittaker, J. (1990). Identification of a ligand-binding region of the human insulin receptor encoded by the second exon of the gene. *Molecular Endocrinology*, **4**, 409–16.

De Meyts, P., Van Obberghen, E., Roth, J., Wollmer, A. & Brandenburg, D. (1978). Mapping of the residues responsible for the negative cooperativity of the receptor-binding region of insulin. *Nature*, London, **273**, 504–9.

Denton, R. M. (1986). Early events in insulin actions. *Advances in Cyclic Nucleotide and Protein Phosphorylation Research*, **20**, 293–341.

Denton, R. M., Brownsey, R. W. & Belsham, G. J. (1981). A partial view of the mechanism of insulin action. *Diabetologia*, **21**, 347–62.

Dernovsek, K. & Bar, R. S. (1985). Processing of cell-bound insulin by capillary and macrovascular endothelial cells in culture. *American Journal of Physiology*, **284**, E244–51.

De Vries, C., van Haeften, T. W. & van der Veen, E. A. (1988). A short incubation time in insulin binding experiments leads to disappearance of negative cooperativity in H35 hepatoma cells. *Biochemical and Biophysical Research Communications*, **157**, 1390–5.

Donner, D. B. & Yonkers, K. (1983). Hormone-induced conformational changes in the hepatic insulin receptor. *Journal of Biological Chemistry*, **258**, 9413–18.

Doolittle, R. F. (1985). The genealogy of some recently evolved vertebrate proteins. *Trends in Biochemical Sciences*, **10**, 233–7.

Downward, J., Waterfield, M. D. & Parker, P. J. (1985). Autophosphorylation and protein kinase C phosphorylation of the epidermal growth factor receptor: effect on tyrosine kinase activity and ligand binding affinity. *Journal of Biological Chemistry*, **260**, 14538–46.

Draznin, B., Leitner, J. W., Sussman, K. E. & Sherman, N. A. (1988). Insulin and glucose modulate protein kinase C activity in rat adipocytes. *Biochemical and Biophysical Research Communications*, **156**, 570–5.

Duckworth, W. C. (1988). Insulin degradation: mechanisms, products and significance. *Endocrine Reviews*, **9**, 319–45.

Due, C., Simonsen, M. & Olsson, L. (1986). The major histocompatibility complex class I heavy chain as a structural subunit of the human cell membrane insulin receptor: implications for the range of biological functions of histocompatibility antigens. *Proceedings of the National Academy of Sciences, USA*, **83**, 6007–11.

Duronio, V. & Jacobs, S. (1990). The effect of protein kinase-C inhibition on insulin receptor phosphorylation. *Endocrinology*, **127**, 481–7.

Duronio, V., Jacobs, S., Romero, P. A. & Herscovics, A. (1988). Effects of inhibitors of N-linked oligosaccharide processing on the biosynthesis and function of insulin and insulin-like growth factor 1 receptors. *Journal of Biological Chemistry*, **263**, 5436–45.

Ebina, Y., Araki, E., Taira, M., Shimada, F., Mori, M., Craik, C. S., Siddle, K., Pierce, S. B., Roth, R. A. & Rutter, W. J. (1987). Replacement of lysine residue 1030 in the putative ATP-binding region of the insulin receptor abolishes insulin- and antibody-stimulated glucose uptake and receptor kinase activity. *Proceedings of the National Academy of Sciences, USA*, **84**, 704–8.

Ebina, Y., Eldery, M., Ellis, L., Standring, D., Beaudoin, J., Roth, R. A. & Rutter, W. J. (1985a). Expression of a functional human insulin receptor from a cloned cDNA in Chinese hamster ovary cells. *Proceedings of the National Academy of Sciences, USA*, **82**, 8014–8.

Ebina, Y., Ellis, L., Jarnagin, K., Edery, M., Graf, L., Clauser, E., Ou, J. H., Masiarz, F., Kan, Y. W., Goldfine, I. D., Roth, R. A. & Rutter, W. J. (1985b). The human insulin receptor cDNA: the structural basis for hormone-activated transmembrane signalling. *Cell*, **40**, 747–58.

Elks, M. L., Watkins, P. A., Manganiello, V. C., Moss, J., Hewlett, E. & Vaughan, M. (1983). Selective regulation by pertussis toxin of insulin-induced activation of particulate cAMP phosphodiesterase activity in 3T3-L1 adipocytes. *Biochemical and Biophysical Research Communications*, **116**, 593–8.

Ellis, L., Clauser, E., Morgan, D. O., Edery, M., Roth, R. A. & Rutter, W. J. (1986). Replacement of insulin receptor tyrosine residues 1162 and 1163 compromises insulin-stimulated kinase activity and uptake of 2-deoxyglucose. *Cell*, **45**, 721–32.

Ellis, L., Levitan, A., Cobb, M. H. & Ramos, P. (1988*a*). Efficient expression in insect cells of a soluble, active human insulin receptor protein-tyrosine kinase domain by use of a baculovirus vector. *Journal of Virology*, **62**, 1634–9.

Ellis, L., Morgan, D. O., Clauser, E., Roth, R. A. & Rutter, W. J. (1987*a*). A membrane-anchored cytoplasmic domain of the human insulin receptor mediates a constitutively elevated insulin-independent uptake of 2-deoxyglucose. *Molecular Endocrinology*, **1**, 15–24.

Ellis, L., Morgan, D. O., Jong, S. M., Wang, L. H., Roth, R. A. & Rutter, W. J. (1987*b*). Heterologous transmembrane signalling by a human insulin receptor-v-ros hybrid in Chinese hamster ovary cells. *Proceedings of the National Academy of Sciences, USA*, **84**, 5101–5.

Ellis, L., Sissom, J. & Levitan, A. (1988*b*). Truncation of the ectodomain of the human insulin receptor results in secretion of a soluble insulin binding protein from transfected CHO cells. *Journal of Molecular Recognition*, **1**, 25–31.

Endemann, G., Yonezawa, K. & Roth, R. A. (1990). Phosphatidylinositol kinase or an associated protein is a substrate for the insulin receptor tyrosine kinase. *Journal of Biological Chemistry*, **265**, 396–400.

Farese, R. V. & Cooper, D. R. (1989). Potential role of phospholipid-signaling systems in insulin action and states of clinical insulin resistance. *Diabetes/Metabolism Reviews*, **5**, 455–74.

Farese, R. V., Cooper, D. R., Konda, T. S., Nair, G., Standaert, M. L., Davis, J. S. & Pollett, R. J. (1988). Mechanisms whereby insulin increases diacylglycerol in BC3H-1 myocytes. *Biochemical Journal*, **256**, 175–184.

Farese, R. V., Davis, J. S., Barnes, D. E., Standaert, M. L., Babischkin, J. S., Hock, R., Rosic, N. K. & Pollet, R. J. (1985). The de novo phospholipid effect of insulin is associated with increases in diacylglycerol but not inositol phosphates or cytosolic Ca^{2+}. *Biochemical Journal*, **231**, 269–78.

Farese, R. V., Konda, T. S., Davis, J. S., Standaert, M. L., Pollet, R. J. & Cooper, D. R. (1987). Insulin rapidly increases diacylglycerol by activating *de novo* phosphatidic acid synthesis. *Science*, **236**, 586–9.

Fehlmann, M., Peyron, J. F., Samson, M., Van Obberghen, E., Brandenburg, D. & Brossette, N. (1985). Molecular association between major histocompatibility complex class I antigens and insulin receptors in mouse liver membranes. *Proceedings of the National Academy of Sciences, USA*, **82**, 8634–7.

Feltz, S. M., Swanson, M. L., Wemmie, J. A. & Pessin, J. E. (1988). Functional properties of an isolated $\alpha\beta$ heterodimeric human placenta insulin-like growth factor I receptor complex. *Biochemistry*, **27**, 3234–42.

Fernandez-Almonacid, R. & Rosen, O. M. (1987). Structure and ligand specificity of the *Drosophila melanogaster* insulin receptor. *Molecular and Cellular Biology*, **7**, 2718–27.

Flier, J. S., Kahn, C. R. & Roth, J. (1979). Receptors, anti-receptor antibodies and mechanisms of insulin resistance. *New England Journal of Medicine*, **300**, 413–19.

Flores-Riveros, J. R., Sibley, E., Kastelic, T. & Lane, D. M. (1989). Substrate phosphorylation catalysed by the insulin receptor tyrosine kinase: kinetic correlation to autophosphorylation of specific sites in the β subunit. *Journal of Biological Chemistry*, **264**, 21557–72.

Forsayeth, J. R., Caro, J. F., Sinha, M. K., Maddux, B. A. & Goldfine, I. D. (1987*a*). Monoclonal antibodies to the human insulin receptor that activate glucose transport but not insulin receptor kinase activity. *Proceedings of the National Academy of Sciences, USA*, **84**, 3448–51.

Forsayeth, J., Maddux, B. & Goldfine, I. D. (1986). Biosynthesis and processing of the human insulin receptor. *Diabetes*, **35**, 837–46.

Forsayeth, J. R., Montemurro, A., Maddux, B. A., De Pirro, R. & Goldfine, I. D. (1987*b*). Effect of monoclonal antibodies on human insulin receptor autophosphorylation, negative cooperativity and down-regulation. *Journal of Biological Chemistry*, **262**, 4134–40.

Fox, J. A., Soliz, N. M. & Saltiel, A. R. (1987). Purification of phosphatidylinositol–glycan-specific phospholipase C from liver plasma membranes: a possible target of insulin action. *Proceedings of the National Academy of Sciences, USA*, **84**, 2663–7.

Freychet, P., Roth, J. & Neville, D. M. (1971). Insulin receptors in the liver: specific binding of [^{125}I]-insulin to the plasma membrane and its relation to insulin bioactivity. *Proceedings of the National Academy of Sciences, USA*, **68**, 1833–7.

Friedenberg, G. R., Henry, R. R., Klein, H. H., Reichart, D. R. & Olefsky, J. M. (1987). Decreased kinase activity of insulin receptors from adipocytes of non-insulin-dependent diabetic subjects. *Journal of Clinical Investigation*, **79**, 240–50.

Friedenberg, G. R., Reichart, D., Okefsky, J. M. & Henry, R. R. (1988). Reversibility of defective adipocyte insulin receptor kinase activity in non-insulin-dependent diabetes mellitus: effect of weight loss. *Journal of Clinical Investigation*, **82**, 1398–406.

Froesch, E. R., Schmid, C., Schwander, J. & Zapf, J. (1985). Actions of insulin-like growth factors. *Annual Review of Physiology*, **47**, 443–67.

Fujita-Yamaguchi, Y. (1984). Characterization of purified insulin receptor subunits. *Journal of Biological Chemistry*, **259**, 1206–11.

Fujita-Yamaguchi, Y., Choi, S., Sakamoto, Y. & Itakura, K. (1983). Purification of insulin receptor with full binding activity. *Journal of Biological Chemistry*, **258**, 5045–9.

Fujita-Yamaguchi, Y. & Harmon, J. T. (1988). A monomer-dimer model explains the results of radiation inactivation: binding characteristics of insulin receptor purified from human placenta. *Biochemistry*, **27**, 3252–60.

Fujita-Yamaguchi, Y., Kathuria, S., Xu, Q-Y., McDonald, J. M., Nakano, H. & Kamata, T. (1989*b*). *In vitro* tyrosine phosphorylation studies on RAS proteins and calmodulin suggest that polylysine-like basic peptides or domains may be involved in interactions between insulin receptor kinase and its substrate. *Proceedings of the National Academy of Sciences, USA*, **86**, 7306–10.

Fujita-Yamaguchi, Y., LeBon, T., Tsubokawa, M., Henzel, W., Kathuria, S., Koyal, D. & Ramachandran, J. (1986). Comparison of insulin-like growth factor I receptor and insulin receptor purified from human placental membranes. *Journal of Biological Chemistry*, **261**, 16727–31.

Fujita-Yamaguchi, Y., Sacks, D. B., McDonald, J. M., Sahal, D. & Kathuria, S. (1989*a*). Effect of basic polycations and proteins on purified insulin receptor: insulin-independent activation of the receptor tyrosine-specific protein kinase by poly(L-lysine). *Biochemical Journal*, **263**, 813–22.

Gammeltoft, S. (1984). Insulin receptors: binding kinetics and structure-function relationship of insulin. *Physiological Reviews*, **64**, 1322–78.

Gammeltoft, S. & Van Obberghen, E. (1986). Protein kinase activity of the insulin receptor. *Biochemical Journal*, **235**, 1–11.

Ganderton, R. H., Whittaker, J. & Siddle, K. (1989). Turnover of insulin receptors in cells transfected with human insulin receptor cDNA. *Biochemical Society Transactions*, **17**, 197–8.

Garofalo, R. S. & Rosen, O. M. (1988). Tissue localization of Drosophila melanogaster insulin receptor transcripts during development. *Molecular and Cellular Biology*, **8**, 1638–47.

Gazzano, H., Kowalski, A., Fehlmann, M. & Van Obberghen, E. (1983). Two different protein kinase activities are associated with the insulin receptor. *Biochemical Journal*, **216**, 575–82.

Gherzi, R., Caratti, C., Andraghetti, G., Bertolini, S., Montemurro, A., Sesti, G. & Cordera, R. (1988). Direct modulation of insulin receptor protein tyrosine kinase by vanadate and anti-insulin receptor monoclonal antibodies. *Biochemical and Biophysical Research Communications*, **152**, 1474–80.

Gherzi, R., Russell, D. S., Taylor, S. I. & Rosen, O. M. (1987). Reevaluation of the evidence that an antibody to the insulin receptor is insulinmimetic without activating the protein tyrosine kinase activity of the receptor. *Journal of Biological Chemistry*, **262**, 16900–5.

Gibbs, E. M., Allard, W. J. & Lienhard, G. E. (1986). The glucose transporter in 3T3-L1 adipocytes is phosphorylated in response to phorbol ester but not in response to insulin. *Journal of Biological Chemistry*, **261**, 16597–603.

Gilman, A. G. (1987). G proteins: transducers of receptor-generated signals. *Annual Review of Biochemistry*, **56**, 615–49.

Goldfine, I. D. (1987). The insulin receptor: molecular biology and transmembrane signaling. *Endocrine Reviews*, **8**, 235–55.

Goldstein, B. J. & Dudley, A. L. (1990). The rat insulin receptor: primary structure and conservation of tissue-specific alternative messenger RNA splicing. *Molecular Endocrinology*, **4**, 235–44.

Goldstein, B. J. & Kahn, C. R. (1988). Initial processing of the insulin receptor precursor *in vivo* and *in vitro*. *Journal of Biological Chemistry*, **263**, 12809–12.

Goldstein, B. J. & Kahn, C. R. (1989). Insulin receptor messenger ribonucleic acid sequence alterations detected by ribonuclease cleavage in patients with syndromes of insulin resistance. *Journal of Clinical Endocrinology and Metabolism*, **69**, 15–24.

Goldstein, J. L., Anderson, R. G. W. & Brown, M. S. (1979). Coated pits, coated vesicles and receptor-mediated endocytosis. *Nature, London*, **279**, 679–85.

Gorden, P., Carpentier, J. L., Freychet, P. & Orci, L. (1980). Internalization of polypeptide hormones: mechanism, intracellular localization and significance. *Diabetologia*, **18**, 263–74.

Goren, H. J., White, M. F. & Kahn, C. R. (1987). Separate domains of the insulin receptor contain sites of autophosphorylation and tyrosine kinase activity. *Biochemistry*, **26**, 2374–82.

Grande, J., Perez, M. & Itarto, E. (1988). Phosphorylation of hepatic insulin receptor by casein kinase 2. *FEBS Letters*, **232**, 130–4.

Graves, C. B., Goewert, R. R. & McDonald, J. M. (1985). The insulin receptor contains a calmodulin-binding domain. *Science*, **230**, 827–9.

Grigorescu, F., Flier, J. S. & Kahn, C. R. (1986). Characterization of binding and phosphorylation defects of erythrocyte insulin receptors in the type A syndrome of insulin resistance. *Diabetes*, **35**, 127–38.

Grunberger, G. & Gorden, P. (1982). Affinity alteration of insulin receptor induced by a phorbol ester. *American Journal of Physiology*, **243**, E319–24.

Grunfeld, C., Shigenaga, J. K., Huang, B. J., Fujita-Yamaguchi, Y., McFarland, K. C., Burnier, J. & Ramachandran, J. (1987). Identification of the intact insulin receptor using a sequence-specific antibody directed against the C-terminus of the β-subunit. *Endocrinology*, **121**, 948–57.

Grunfeld, C., Shigenaga, J. K. & Ramachandran, J. (1985). Urea treatment allows dithiothreitol to release the binding subunit of the insulin receptor from the cell membrane: implications for the structural organisation of the insulin receptor. *Biochemical and Biophysical Research Communications*, **133**, 389–96.

Gu, J. L., Goldfine, I. D., Forsayeth, J. R. & DeMeyts, P. (1988). Reversal of insulin-induced negative cooperativity by monoclonal antibodies that stabilise the slowly dissociating (K super) state of the insulin receptor. *Biochemical and Biophysical Research Communications*, **150**, 694–701.

Gustafson, T. A. & Rutter, W. J. (1990). The cysteine-rich domains of the insulin and insulin-like growth factor I receptors are primary determinants of hormone binding specificity: evidence from receptor chimeras. *Journal of Biological Chemistry*, **265**, 18663–7.

Hachiya, H. L., Takayama, S., White, M. F. & King, G. L. (1987). Regulation of insulin receptor internalization in vascular endothelial cells by insulin and phorbol ester. *Journal of Biological Chemistry*, **262**, 6417–24.

Hansen, T., Stagsted, J., Pedersen, L., Roth, R. A., Goldstein, A. & Olsson, L. (1989). Inhibition of insulin receptor phosphorylation by peptides derived from major histocompatibility complex class I antigens. *Proceedings of the National Academy of Sciences, USA*, **86**, 3123–6.

Hari, J. & Roth, R. A. (1987). Defective internalization of insulin and its receptor in cells expressing mutated insulin receptors lacking kinase activity. *Journal of Biological Chemistry*, **262**, 15341–4.

Haring, H., Kirsch, D., Obermaier, B., Ermel, B. & Machicao, F. (1986a). Tumor-promoting phorbol esters increase the K_m of the ATP-binding site of the insulin receptor kinase from rat adipocytes. *Journal of Biological Chemistry*, **261**, 3869–75.

Haring, H., Kirsch, D., Obermaier, B., Ermel, B. & Machicao, F. (1986b). Decreased tyrosine kinase activity of insulin receptor isolated from rat adipocytes rendered insulin-resistant by catecholamine treatment *in vitro*. *Biochemical Journal*, **234**, 59–66.

Haring, H. & Obermaier-Kusser, B. (1989). Insulin receptor kinase defects in insulin-resistant tissues and their role in the pathogenesis of NIDDM. *Diabetes/Metabolism Reviews*, **5**, 431–41.

Haring, H. U., White, M. F., Kahn, C. R., Ahmad, Z., DePaoli-Roach, A. A. & Roach, P. J. (1985). Interaction of the insulin receptor kinase with serine/threonine kinases *in vitro*. *Journal of Cellular Biochemistry*, **28**, 171–82.

Haring, H. U., White, M. F., Machicao, F., Ermel, B., Schleicher, E. & Obermaier, B. (1987). Insulin rapidly stimulates phosphorylation of a 46 kDa membrane protein on tyrosine residues as well as phosphorylation of several soluble proteins in intact fat cells. *Proceedings of the National Academy of Sciences, USA*, **84**, 113–7.

Harrison, L. C. & Itin, A. (1979). A possible mechanism for insulin resistance and hyperglycaemia in NZO mice. *Nature, London*, **279**, 334–6.

Harrison, L. C. & Itin, A. (1980). Purification of the insulin receptor from human placenta by chromatography on immobilised wheat germ lectin and receptor antibody. *Journal of Biological Chemistry*, **255**, 12066–72.

Hatada, E. N., McClain, D. A., Potter, E., Ullrich, A. & Olefsky, J. M. (1989). Effects of growth and insulin treatment on the levels of insulin receptors and their mRNA in HEP G2 cells. *Journal of Biological Chemistry*, **264**, 6741–7.

Hawley, D. M., Maddux, B., Patel, R. G., Wong, K. Y., Mamula, P. W., Firestone, G. L., Brunetti, A., Verspohl, E. & Goldfine, I. D. (1989). Insulin receptor monoclonal antibodies that mimic insulin action without activating tyrosine kinase. *Journal of Biological Chemistry*, **264**, 2438–44.

Hayes, G. R. & Lockwood, D. (1987). Role of insulin receptor phosphorylation in the insulinomimetic effects of hydrogen peroxide. *Proceedings of the National Academy of Sciences, USA*, **84**, 8115–19.

Haystead, T. A. J. & Hardie, D. G. (1986). Both insulin and epidermal growth factor stimulate lipogenesis and acetyl-CoA carboxylase activity in isolated adipocytes. *Biochemical Journal*, **234**, 279–84.

Hedo, J. A., Collier, E. & Watkinson, A. (1987). Myristyl and palmityl acylation of the insulin receptor. *Journal of Biological Chemistry*, **262**, 954–7.

Hedo, J. A. & Gorden, P. (1985). Biosynthesis of the insulin receptor. *Hormone and Metabolic Research*, **17**, 487–90.

Hedo, J. A., Kahn, C. R., Hayashi, M., Yamada, K. & Kasuga, M. (1983). Biosynthesis and glycosylation of the insulin receptor: evidence for a single polypeptide precursor of the two major subunits. *Journal of Biological Chemistry*, **258**, 10020–6.

Hedo, J. A., Kasuga, M., Van Obberghen, E., Roth, J. & Kahn, C. R. (1981). Direct demonstration of glycosylation of insulin receptor subunits by biosynthetic and external labelling: evidence for heterogeneity. *Biochemical and Biophysical Research Communications*, **78**, 4791–5.

Hedo, J. A. & Simpson, I. A. (1984). Internalization of insulin receptors in the isolated rat adipose cell: demonstration of the vectorial disposition of receptor subunits. *Journal of Biological Chemistry*, **259**, 11083–9.

Hedo, J. A. & Simpson, I. A. (1985). Biosynthesis of the insulin receptor in rat adipose cells: intracellular processing of the M_r-190000 pro-receptor. *Biochemical Journal*, **232**, 71–8.

Heffetz, D. & Zick, Y. (1986). Receptor aggregation is necessary for activation of the soluble insulin receptor kinase. *Journal of Biological Chemistry*, **261**, 889–94.

Heidenreich, K. A. & Brandenburg, D. (1986). Oligosaccharide heterogeneity of insulin receptors: comparison of *N*-linked glycosylation of insulin receptors in adipocytes and brain. *Endocrinology*, **118**, 1835–42.

Heidenreich, K. A., Gilmore, P. R., Brandenburg, D. & Hatada, E. (1988). Peptide mapping and Northern blot analyses of insulin receptors in brain and adipocytes. *Molecular and Cellular Endocrinology*, **56**, 255–61.

Heidenreich, K. A. & Olefsky, J. M. (1985). The metabolism of insulin receptors: internalization, degradation, and recycling. In *Molecular Basis of Insulin Action*, ed. M. P. Czech, pp. 45–65. New York: Plenum Press.

Heidenreich, K. A., Zahiser, N. R., Berhanu, P., Brandenburg, D. & Olefsky, J. M. (1983). Structural differences between insulin receptors in the brain and peripheral target tissues. *Journal of Biological Chemistry*, **258**, 8527–30.

Helmerhorst, E. (1987). The insulin-receptor interaction: is the kinetic approach for inferring negative-cooperative site-site interactions valid? *Biochemical and Biophysical Research Communications*, **147**, 399–407.

Helmerhorst, E., Ng, D. S., Moule, M. L. & Yip, C. C. (1986). High molecular weight forms of the insulin receptor. *Biochemistry*, **25**, 2060–5.

Hendricks, S. A., Agardh, C. D., Taylor, S. I. & Roth, J. (1984). Unique features of the insulin receptor in rat brain. *Journal of Neurochemistry*, **43**, 1302–9.

Herrera, R., Lebwohl, D., de Herreros, A. G., Kallen, R. G. & Rosen, O. M. (1988). Synthesis, purification and characterization of the cytoplasmic domain of the human insulin receptor using a baculovirus expression system. *Journal of Biological Chemistry*, **263**, 560–8.

Herrera, R., Petruzzelli, L., Thomas, N., Bramson, H. N., Kaiser, E. T. & Rosen, O. M. (1985). An antipeptide that specifically inhibits insulin receptor autophosphorylation and protein kinase activity. *Proceedings of the National Academy of Sciences, USA*, **82**, 7899–903.

Herrera, R. & Rosen, O. M. (1986). Autophosphorylation of the insulin receptor *in vitro*: designation of phosphorylation sites and correlation with receptor kinase activation. *Journal of Biological Chemistry*, 261, 11980–5.

Hesketh, J. E. & Campbell, G. P. (1987). Effects of insulin, pertussis toxin and cholera toxin on protein synthesis and diacylglycerol production in 3T3 fibroblasts: evidence for a G-protein mediated activation of phospholipase C in the insulin signal mechanism. *Bioscience Reports*, **7**, 533–41.

Heyworth, C. M., Grey, A. M., Wilson, S. R., Hanski, E. & Houslay, M. D. (1986). The action of islet activating protein (pertussis toxin) on insulin's ability to inhibit adenylate cyclase and activate cyclic AMP phosphodiesterase in hepatocytes. *Biochemical Journal*, **235**, 145–9.

Heyworth, C. M., Whetton, A. D., Wong, S., Martin, B. R. & Houslay, M. D. (1985). Insulin inhibits the cholera-toxin-catalysed ribosylation of a M_r-25 000 protein in rat liver plasma membranes. *Biochemical Journal*, **228**, 593–603.

Hofmann, C., Thys, R., Sweet, L. J., Spector, A. A. & Pessin, J. E. (1988). Transfer of functional insulin receptors to receptor-deficient target cells. *Endocrinology*, **122**, 2865–72.

Hofmann, C., White, M. F. & Whittaker, J. (1989). Human insulin receptors expressed in insulin-insensitive mouse fibroblasts couple with extant cellular effector systems to confer insulin sensitivity and responsiveness. *Endocrinology*, **124**, 257–64.

Hofmann, K., Romovacek, H., Titus, G., Ridge, K., Raffensperger, J. A. & Finn, F. M. (1987). The rat liver insulin receptor. *Biochemistry*, **26**, 7384–90.

Houslay, M. D. & Siddle, K. (1989). Molecular basis of insulin receptor function. *British Medical Bulletin*, **45**, 264–84.

Houslay, M. D., Wakelam, M. J. O. & Pyne, N. J. (1986). The mediator is the message: is it part of the answer of insulin's action? *Trends in Biochemical Sciences*, **11**, 393–4.

Hresko, R. C., Bernier, M., Hoffman, R. D., Flores-Riveros, J. R., Liao, K., Laird, D. M. & Lane, M. D. (1988). Identification of phosphorylated 422 (aP2) protein as pp15, the 15-kilodalton target of the insulin receptor tyrosine kinase in 3T3-L1 adipocytes. *Proceedings of the National Academy of Sciences, USA*, **85**, 8835–9.

Hsuan, J. J., Downward, J., Clark, S. & Waterfield, M. D. (1989). Proteolytic generation of constitutive tyrosine kinase activity of the human insulin receptor. *Biochemical Journal*, **259**, 519–27.

Hunter, T. (1987). A thousand and one protein kinases. *Cell*, **50**, 823–9.

Hunter, T. (1989). Protein-tyrosine phosphatases: the other side of the coin. *Cell*, **58**, 1013–16.

Hunter, T. & Cooper, J. A. (1985). Protein-tyrosine kinases. *Annual Review of Biochemistry*, **54**, 897–930.

Hunter, T., Ling, N. & Cooper, J. A. (1984). Protein kinase C phosphorylation of the EGF receptor at a threonine residue close to the cytoplasmic face of the membrane. *Nature*, London, **311**, 480–3.

Hunter, T. & Sefton, B. M. (1980). Transforming gene product of Rous sarcoma virus phosphorylates tyrosine. *Proceedings of the National Academy of Sciences, USA*, **77**, 1311–15.

Irvine, F. J. & Houslay, M. D. (1988). Insulin and glucagon attenuate the ability of cholera toxin to activate adenylate cyclase in intact hepatocytes. *Biochemical Journal*, **251**, 447–52.

Izumi, T., White, M. F., Kadowaki, T., Takaku, F., Akanuma, Y. & Kasuga, M. (1987). Insulin-like growth factor I rapidly stimulates tyrosine phosphorylation of a M_r 185000 protein in intact cells. *Journal of Biological Chemistry*, **262**, 1282–7.

Jacobs, S. & Cuatrecasas, P. (1981). Insulin receptor antibodies. *Methods in Enzymology*, **74**, 471–8.

Jacobs, S. & Cuatrecasas, P. (1983). Insulin receptors. *Annual Review of Pharmacology and Toxicology*, **23**, 461–79.

Jacobs, S. & Cuatrecasas, P. (1986). Phosphorylation of receptors for insulin and insulin-like growth factor I: effects of hormones and phorbol esters. *Journal of Biological Chemistry*, **261**, 934–9.

Jacobs, S., Hazum, E. & Cuatrecasas, P. (1980). The subunit structure of rat liver insulin receptor: antibodies directed against the insulin-binding subunit. *Journal of Biological Chemistry*, **255**, 6937–40.

Jacobs, S., Hazum, E., Schecter, Y. & Cuatrecasas, P. (1979). Insulin receptor: covalent labeling and identification of subunits. *Proceedings of the National Academy of Sciences, USA*, **76**, 4918–21.

Jacobs, S., Sahyoun, N. E., Saltiel, A. R. & Cuatrecasas, P. (1983). Phorbol esters stimulate the phosphorylation of receptors for insulin and somatomedin C. *Proceedings of the National Academy of Sciences, USA*, **80**, 6211–3.

Jacobs, S., Schechter, Y., Bissell, K. & Cuatrecasas, P. (1977). Purification and properties of insulin receptors from rat liver membranes. *Biochemical and Biophysical Research Communications*, **77**, 981–8.

James, D. E., Brown, R., Navarro, J. & Pilch, P. F. (1988). Insulin-regulatable tissues express a unique insulin-sensitive glucose transport protein. *Nature*, London, **333**, 183–5.

James, D. E., Strube, M. & Mueckler, M. (1989). Molecular cloning and characterization of an insulin-regulatable glucose transporter. *Nature*, London, **338**, 83–7.

Jarett, L., Kiechle, F., Macaulay, S. L., Parker, J. C. & Kelly, K. L. (1985). Intracellular mediators of insulin action. In *Molecular Basis of Insulin Action*, ed. M. P. Czech, pp. 183–98. New York: Plenum Press.

Jarett, L. & Smith, R. M. (1974). Electron microscopic demonstration of insulin receptors on adipocyte plasma membranes utilizing a ferritin–insulin conjugate. *Journal of Biological Chemistry*, **249**, 7024–31.

Johnson, J. D., Wong, M. L. & Rutter, W. J. (1988). Properties of the insulin receptor ectodomain. *Proceedings of the National Academy of Sciences, USA*, **85**, 7516–20.

Jonas, H. A., Cox, A. J. & Harrison, L. C. (1989). Delineation of atypical

330   *Kenneth Siddle*

insulin receptors from classical insulin and type 1 insulin-like growth factor receptors in human placenta. *Biochemical Journal*, **257**, 101–7.

Jonas, H. A. & Harrison, L. C. (1985). The human placenta contains two distinct binding and immunoreactive species of insulin-like growth factor I receptors. *Journal of Biological Chemistry*, **260**, 2288–94.

Jonas, H. A. & Harrison, L. C. (1986). Disulphide reduction alters the immunoreactivity and increases the affinity of insulin-like growth factor I receptors in human placenta. *Biochemical Journal*, **236**, 417–23.

Jonas, H. A., Newman, J. D. & Harrison, L. C. (1986). An atypical insulin receptor with high affinity for insulin-like growth factors co-purified with placental insulin receptors. *Proceedings of the National Academy of Sciences, USA*, **83**, 4124–8.

Joost, H. G., Steinfelder, H. J. & Schmitz-Salue, C. (1986). Tyrosine kinase activity of insulin receptors from human placenta: effects of autophosphorylation and cyclic AMP-dependent protein kinase. *Biochemical Journal*, **233**, 677–81.

Kadota, S., Fantus, I. G., Deragon, G., Guyda, H. J. & Posner, B. I. (1987). Stimulation of insulin-like growth factor II receptor binding and insulin receptor kinase activity in rat adipocytes: effects of vanadate and H_2O_2. *Journal of Biological Chemistry*, **262**, 8252–6.

Kadowaki, T., Bevins, C. L., Cama, A., Ojamaa, K., Marcus-Samuels, B., Kadowaki, H., Beitz, L., McKeon, C. & Taylor, S. I. (1988). Two mutant alleles of the insulin receptor gene in a patient with extreme insulin resistance. *Science*, **240**, 787–90.

Kadowaki, T., Fujita-Yamaguchi, Y., Nishida, E., Takaku, F., Akiyama, T., Kathuria, S., Akanuma, Y. & Kasuga, M. (1985). Phosphorylation of tubulin and microtubule-associated proteins by the purified insulin receptor kinase. *Journal of Biological Chemistry*, **260**, 4016–20.

Kadowaki, T., Kadowaki, H., Rechler, M. M., Serrano-Rios, M., Roth, J., Gorden, P. & Taylor, S. I. (1990). Five mutant alleles of the insulin receptor gene in patients with genetic forms of insulin resistance. *Journal of Clinical Investigation*, **86**, 254–64.

Kadowaki, T., Koyasu, S., Nishida, E., Tobe, K., Izumi, T., Takaku, F., Sakai, H., Yahara, I. & Kasuga, M. (1987). Tyrosine phosphorylation of common and specific sets of cellular proteins rapidly induced by insulin, insulin-like growth factor I, and epidermal growth factor in an intact cell. *Journal of Biological Chemistry*, **262**, 7342–50.

Kahn, C. R. (1976). Membrane receptors for hormones and neurotransmitters. *Journal of Cell Biology*, **70**, 261–86.

Kahn, C. R. (1985). The molecular mechanism of insulin action. *Annual Review of Medicine*, **36**, 429–51.

Kahn, C. R., Baird, K. L., Flier, J. S., Grunfeld, C., Harmon, J. T., Harrison, L. C., Karlsson, F. A., Kasuga, M., King, G. L., Lang, U. C., Podskalny, J. M. & Van Obberghen, E. (1981). Insulin receptors, receptor antibodies and the mechanism of insulin action. *Recent Progress in Hormone Research*, **37**, 477–538.

Kahn, C. R., Baird, K. L., Jarrett, D. B. & Flier, J. S. (1978). Direct demonstration that receptor cross-linking or aggregation is important in insulin action. *Proceedings of the National Academy of Sciences, USA*, **75**, 4209–13.

Kahn, C. R. & Goldstein, B. J. (1989). Molecular defects in insulin action. *Science*, **245**, 13.

Kahn, C. R. & White, M. F. (1988). The insulin receptor and the molecular mechanism of insulin action. *Journal of Clinical Investigation*, **82**, 1151–6.

Kamata, T., Kathuria, S. & Fujita-Yamaguchi, Y. (1987). Insulin stimulates the phosphorylation of v-Ha-ras protein in membrane fraction. *Biochemical and Biophysical Research Communications*, **144**, 19–25.

Kasuga, M., Fujita-Yamaguchi, Y., Blithe, D. L. & Kahn, C. R. (1983). Tyrosine-specific protein kinase activity is associated with the purified insulin receptor. *Proceedings of the National Academy of Sciences, USA*, **80**, 2137–41.

Kasuga, M., Hedo, J. A., Yamada, K. M. & Kahn, C. R. (1982*d*). The structure of insulin receptor and its subunits. *Journal of Biological Chemistry*, **257**, 10392–9.

Kasuga, M., Karlsson, F. A. & Kahn, C. R. (1982*a*). Insulin stimulates the phosphorylation of the 95000-Dalton subunit of its own receptor. *Science*, **215**, 185–6.

Kasuga, M., Van Obberghen, E., Nissley, P. & Rechler, M. (1981). Demonstration of two subtypes of insulin-like growth factor receptors by affinity cross-linking. *Journal of Biological Chemistry*, **256**, 5305–8.

Kasuga, M., Zick, Y., Blithe, D. L., Karlsson, F. A., Haring, H. U. & Kahn, C. R. (1982*b*). Insulin stimulation of phosphorylation of the β subunit of the insulin receptor: formation of both phosphoserine and phosphotyrosine. *Journal of Biological Chemistry*, **257**, 9891–4.

Kasuga, M., Zick, Y., Blithe, D. L., Crettaz, M. & Kahn, C. R. (1982*c*). Insulin stimulates tyrosine phosphorylation of the insulin receptor in a cell free system. *Nature*, London, **298**, 667–9.

Kathuria, S., Hartman, S., Grunfeld, C., Ramachandran, J. & Fujita-Yamaguchi, Y. (1986). Differential sensitivity of two functions of the insulin receptor to the associated proteolysis: kinase action and hormone binding. *Proceedings of the National Academy of Sciences, USA*, **83**, 8570–4.

Khan, M. N., Savoie, S., Bergeron, J. J. M. & Posner, B. I. (1986). Characterization of rat liver endosomal fractions: in vivo activation of insulin-stimulable receptor kinase in these structures. *Journal of Biological Chemistry*, **261**, 8462–72.

Kiess, W., Blickenstaff, G. D., Sklar, M. M., Thomas, C. L., Nissley, S. P. & Sahagian, G. G. (1988). Biochemical evidence that the type II insulin-like growth factor receptor is identical to the cation-independent mannose 6-phosphate receptor. *Journal of Biological Chemistry*, **263**, 9339–44.

Kikkawa, U., Kishimoto, A. & Nishizuka, Y. (1989). The protein kinase C family: heterogeneity and its implications. *Annual Review of Biochemistry*, **58**, 31–44.

King, G. L. & Johnson, S. M. (1985). Receptor-mediated transport of insulin across endothelial cells. *Science*, **227**, 1583–6.

King, M. J. & Sale, G. J. (1988). Insulin-receptor phosphotyrosyl-protein phosphatases. *Biochemical Journal*, **256**, 893–902.

King, M. J. & Sale, G. J. (1990). Dephosphorylation of insulin receptor autophosphorylation sites by particulate and soluble phosphotyrosyl-protein phosphatases. *Biochemical Journal*, **266**, 251–9.

Kittur, D., Shimizu, Y., De Mars, R. & Edidin, M. (1987). Insulin binding to human B lymphobasts is a function of HLA haplotype. *Proceedings of the National Academy of Sciences, USA*, **84**, 1351–5.

Klee, C. B., Crouch, T. H. & Richman, P. G. (1980). Calmodulin. *Annual Review of Biochemistry*, **49**, 489–515.

Klein, H. H., Friedenberg, G. R., Cordera, R. & Olefsky, J. M. (1985). Substrate specificities of insulin and epidermal growth factor receptor kinases. *Biochemical and Biophysical Research Communications*, **127**, 254–63.

Klein, H. H., Friedenberg, G., Kladde, M. & Olefsky, J. M. (1986). Insulin activation of insulin receptor tyrosine kinase in intact rat adipocytes. *Journal of Biological Chemistry*, **261**, 4691–7.

Klein, H. H., Friedenberg, G. R., Matthaei, S. & Olefsky, J. M. (1987). Insulin receptor kinase following internalization in isolated rat adipocytes. *Journal of Biological Chemistry*, **262**, 10557–64.

Klinkhamer, M. P., Groen, N. A., van der Zon, G. C. M., Lindhout, D., Sandkuyl, L. A., Krans, H. M. J., Moller, W. & Maassen, J. A. (1989). A leucine-to-proline mutation in the insulin receptor in a family with insulin resistance. *EMBO Journal*, **8**, 2503–7.

Klip, A. & Ramlal, T. (1987). Protein kinase C is not required for insulin stimulation of hexose uptake in muscle cells in culture. *Biochemical Journal*, **242**, 131–6.

Kobayashi, M., Sasoka, T., Takata, Y., Ishibashi, O., Sugibayashi, M., Shigeta, Y., Hisatomi, A., Nakamura, E., Tamaki, M. & Teraoka, H. (1988). Insulin resistance by unprocessed insulin proreceptors: point mutation at the cleavage site. *Biochemical and Biophysical Research Communications*, **153**, 657–63.

Koch, R., Deger, A., Jack, J. M., Klotz, K. N., Schenzle, D., Kramer, H., Kelm, S., Muller, G., Rapp, R. & Weber, U. (1986). Characterization of solubilized insulin receptors from rat liver microsomes: existence of two receptor species with different binding properties. *European Journal of Biochemistry*, **154**, 281–7.

Kohanski, R. A. (1989). Insulin receptor aggregation and autophosphorylation in the presence of cationic polyamino acids. *Journal of Biological Chemistry*, **264**, 20984–91.

Kohanski, R. A. & Lane, M. D. (1983). Binding of insulin to solubilized insulin receptor from human placenta: evidence for a single class of non-interacting binding sites. *Journal of Biological Chemistry*, **258**, 7460–8.

Kono, T. (1969). Destruction and restoration of the insulin effector system of isolated fat cells. *Journal of Biological Chemistry*, **244**, 5777–84.

Korn, L. J., Siebel, C. W., McCormick, F. & Roth, R. A. (1987). Ras p21 as a potential mediator of insulin action in *Xenopus* oocytes. *Science*, **236**, 840–3.

Koshio, O., Akanuma, Y. & Kasuga, M. (1988). Hydrogen peroxide stimulates tyrosine phosphorylation of the insulin receptor and its tyrosine kinase activity in intact cells. *Biochemical Journal*, **250**, 95–101.

Koshio, O., Akanuma, Y. & Kasuga, M. (1989). Identification of a phosphorylation site of the rat insulin receptor catalyzed by protein kinase C in an intact cell. *FEBS Letters*, **254**, 22–4.

Kriauciunas, K. M., Kahn, C. R., Muller-Wieland, D., Reddy, S. S. K. & Taub, R. (1988). Altered expression and function of the insulin receptor in a family with lipoatrophic diabetes. *Journal of Clinical Endocrinology and Metabolism*, **67**, 1284–93.

Krupinski, J., Rajaram, R., Lakonishok, M., Benovic, J. L. & Cerione, R. A. (1988). Insulin-dependent phosphorylation of GTP-binding proteins in phospholipic vesicles. *Journal of Biological Chemistry*, **263**, 12333–41.

Kubar, J. & Van Obberghen, E. (1989). Oligomeric states of the insulin receptor: binding and autophosphorylation properties. *Biochemistry*, **28**, 1068–93.

Kull, F. C., Jacobs, S., Su, Y. K., Svoboda, M. E., Van Wyk, J. J. & Cuatrecasas, P. (1983). Monoclonal antibodies to receptors for insulin and somatomedin-C. *Journal of Biological Chemistry*, **258**, 6561–6.

Kwok, Y. C., Nemenoff, R. A., Powers, A. C. & Avruch, J. (1986). Kinetic properties of the insulin receptor tyrosine protein kinase: activation through an insulin-stimulated tyrosine-specific, intramolecular autophosphorylation. *Archives of Biochemistry and Biophysics*, **244**, 102–13.

Kwok, Y. C. & Yip, C. C. (1987). Tyrosine phosphorylation of two cytosolic proteins of 50 kDa and 35 kDa in rat liver by insulin-receptor kinase *in vitro*. *Biochemical Journal*, **248**, 27–33.

Lammers, R., Gray, A., Schlessinger, J. & Ullrich, A. (1989). Differential signalling potential of insulin- and IGF-1 receptor cytoplasmic domains. *EMBO Journal*, **8**, 1369–75.

Lang, U., Kahn, C. R. & Harrison, L. C. (1980). Subunit structure of the insulin receptor of the human lymphocyte. *Biochemistry*, **19**, 64–70.

Larner, J., Huang, L. C. & Schwartz, C. F. W., Oswald, A. S., Shen, T. Y., Kinter, M., Tang, G. & Zeller, K. (1988). Rat liver insulin mediator which stimulates pyruvate dehydrogenase phosphatase contains galactosamine and D-chiroinositol. *Biochemical and Biophysical Research Communications*, **151**, 1416–26.

Lau, K. H. W., Farley, J. R. & Baylink, D. J. (1989). Phosphotyrosyl protein phosphatases. *Biochemical Journal*, **257**, 23–36.

Leef, J. W. & Larner, J. (1987). Insulin-mimetic effect of trypsin on the insulin receptor tyrosine kinase in intact adipocytes. *Journal of Biological Chemistry*, **262**, 14837–42.

Le Marchand-Brustel, Y., Ballotti, R., Gremeaux, T., Tanti, J-F., Brandenburg, D. & Van Obberghen, E. (1989). Functional labelling of insulin receptor subunits in live cells: $\alpha_2 \beta_2$ species is the major autophosphorylated form. *Journal of Biological Chemistry*, **264**, 21316–21.

Le Marchand-Brustel, Y., Gremeaux, T., Ballotti, R. & Van Obberghen, E. (1985). Insulin receptor tyrosine kinase is defective in skeletal muscle of insulin-resistant obese mice. *Nature*, London, **315**, 676–9.

Levitzki, A. (1981). Negative cooperativity at the insulin receptor. *Nature*, London, **289**, 442–3.

Lewis, R. E., Cao, L., Perregaux, D. & Czech, M. P. (1990*a*). Threonine 1336 of the human insulin receptor is a major target for phosphorylation by protein kinase C. *Biochemistry*, **29**, 1807–13.

Lewis, R. E. & Czech, M. P. (1987). Phospholipid environment alters hormone-sensitivity of the purified insulin receptor kinase. *Biochemical Journal*, **248**, 829–36.

Lewis, R. E., Wu, G. P., MacDonald, R. G. & Czech, M. P. (1990*b*). Insulin-sensitive phosphorylation of serine 1293/1294 on the human insulin receptor by a tightly associated serine kinase. *Journal of Biological Chemistry*, **265**, 947–54.

Lipkin, E. W., Teller, D. C. & de Haen, C. (1986). Equilibrium binding of insulin to rat white fat cells at 15 °C. *Journal of Biological Chemistry*, **261**, 1694–701.

Lipson, K. E., Kolhatkar, A. A. & Donner, D. B. (1989). Alkylation, reduction, solubilization and enrichment of binding activity do not impair the ability of insulin receptors to convert from a rapid- to a slow-dissociating state. *Journal of Biological Chemistry*, **259**, 871–8.

Lipson, K. E., Yamada, K., Kolhatkar, A. A. & Donner, D. B. (1986). Relationship between the affinity and proteolysis of the insulin receptor. *Journal of Biological Chemistry*, **261**, 10833–8.

Livingston, J. N., Pollare, T., Lithell, H. & Arner, P. (1988). Characterization of insulin-like growth factor I receptor in skeletal muscles of normal and insulin-resistant subjects. *Diabetologia*, **31**, 871–7.

Low, M. G. (1987). Biochemistry of the glycosyl-phosphatidylinositol membrane protein anchors. *Biochemical Journal*, **24**, 1–13.

Low, M. G. & Saltiel, A. R. (1988). Structural and functional roles of glycosyl-phosphatidylinositol in membranes. *Science*, **239**, 268–75.

Ludwig, S. M., Faiman, C. & Dean, H. J. (1987). Insulin and insulin-receptor autoantibodies in children with newly diagnosed IDDM before insulin therapy. *Diabetes*, **36**, 420–5.

Luttrell, L. M., Hewlett, E. L., Romero, G. & Rogol, A. D. (1988). Pertussis toxin treatment attenuates some effects of insulin in BC3H-1 myocytes. *Journal of Biological Chemistry*, **263**, 6134–41.

Luttrell, L., Kilgour, E., Larner, J. & Romero, G. (1990). A pertussis toxin-sensitive G protein mediates some aspects of insulin action in BC3H-1 murine myocytes. *Journal of Biological Chemistry*, **265**, 16873–9.

Luttrell, L. M., Luttrell, D. K., Parsons, S. J. & Rogol, A. D. (1989). Insulin and phorbol ester induce distinct phosphorylations of pp60[c-src] in the BC3H-1 murine myocyte cell line. *Oncogene*, **4**, 317–24.

MacDonald, R. G., Pfeffer, S., Coussens, L., Tepper, M., Brocklebank, C. M., Mole, J. E., Anderson, J. K., Chen, E., Czech, M. P. & Ullrich, A. (1988). A single receptor binds both insulin-like growth factor II and mannose 6-phosphate. *Science*, **239**, 1134–7.

Machicao, F., Haring, H., White, M. F., Carrascosa, J. M., Obermaier, B. & Wieland, O. H. (1987). A M_r 180 000 protein is an endogenous substrate for the insulin-receptor-associated tyrosine kinase in human placenta. *Biochemical Journal*, **243**, 797–801.

Madoff, D. H., Martensen, T. M. & Lane, M. D. (1988). Insulin and insulin-like growth factor I stimulate the phosphorylation on tyrosine of a 160 kDa cytosolic protein in 3T3-L1 adipocytes. *Biochemical Journal*, **252**, 7–15.

Maegawa, H., McClain, D. A., Friedenberg, G., Olefsky, J. M., Napier, M., Lipari, T., Dull, T. J., Lee, J. & Ullrich, A. (1988*a*). Properties of a human insulin receptor with a COOH-terminal truncation: II. truncated receptors have normal kinase activity but are defective in signaling metabolic effects. *Journal of Biological Chemistry*, **263**, 8912–7.

Maegawa, H., Olefsky, J. M., Thies, S., Boyd, D., Ullrich, A. & McClain, D. A. (1988*b*). Insulin receptors with defective tyrosine kinase inhibit normal receptor function at the level of substrate phosphorylation. *Journal of Biological Chemistry*, **263**, 12629–37.

Magee, A. I. & Siddle, K. (1988). Insulin and IGF-1 receptors contain covalently bound palmitic acid. *Journal of Cellular Biochemistry*, **37**, 347–57.

Mamula, P. W., McDonald, A. R., Brunetti, A., Okabayashi, Y., Wong, K. Y., Maddux, B. A., Logsdon, C. & Goldfine, I. D. (1990). Regulating insulin receptor gene expression by differentiation and hormones. *Diabetes Care*, **13**, 288–301.

Mamula, P. W., Wong, K. Y., Maddux, B. A., McDonald, A. R. & Goldfine,

I. D. (1988). Sequence and analysis of promoter region of human insulin-receptor gene. *Diabetes*, **37**, 1241–6.

Margolis, R. N., Taylor, S. I., Seminara, D. & Hubbard, A. L. (1988). Identification of pp120, an endogenous substrate for the hepatocyte insulin receptor tyrosine kinase, as an integral membrane glycoprotein of the bile canalicular domain. *Proceedings of the National Academy of Sciences, USA*, **85**, 7256–9.

Maron, R., Elias, D., de Jongh, B. M., Bruining, G. J., van Rood, J. J., Schechter, Y.l & Cohen, I. R. (1983). Autoantibodies to the insulin receptor in juvenile onset insulin-dependent diabetes. *Nature*, London, **303**, 817–8.

Marsh, J. W., Westley, J. & Steiner, D. F. (1984). Insulin-receptor interactions: presence of a positive cooperative effect. *Journal of Biological Chemistry*, **259**, 6641–9.

Massague, J. & Czech, M. P. (1982). Role of disulfides in the subunit structure of the insulin receptor. *Journal of Biological Chemistry*, **257**, 6729–38.

Massague, J., Pilch, P. F. & Czech, M. P. (1980). Electrophoretic resolution of three major insulin receptor structures with unique subunit stoichiometries. *Proceedings of the National Academy of Sciences, USA*, **77**, 7137–41.

Massague, J., Pilch, P. F. & Czech, M. P. (1981). A unique proteolytic cleavage site on the β subunit of the insulin receptor. *Journal of Biological Chemistry*, **256**, 3182–90.

Mato, J. M., Kelly, K. L., Abler, A., Jarett, L., Corkey, B. E., Cashel, J. A. & Zopf, D. (1987). Partial structure of an insulin-sensitive glycophospholipid. *Biochemical and Biophysical Research Communications*, **146**, 764–70.

Maturo, J. M., Hollenberg, M. D. & Aglio, L. S. (1983). Insulin receptor: insulin-modulated interconversion between distinct molecular forms involving disulfide–sulfhydryl exchange. *Biochemistry*, **22**, 2549–86.

McClain, D. A., Maegawa, H., Lee, J., Dull, T. J., Ullrich, A. & Olefsky, J. M. (1987). A mutant insulin receptor with defective tyrosine kinase displays no biologic activity and does not undergo endocytosis. *Journal of Biological Chemistry*, **262**, 14663–71.

McClain, D. A., Maegawa, H., Levy, J., Huecksteadt, T., Dull, T. J., Lee, J., Ullrich, A. & Olefsky, J. M. (1988). Properties of a human insulin receptor with a COOH-terminal truncation: I. insulin binding, autophosphorylation and endocytosis. *Journal of Biological Chemistry*, **263**, 8904–11.

McDonald, A. R. & Goldfine, I. D. (1988). Glucocorticoid regulation of insulin receptor gene transcription in IM-9 cultured lymphocytes. *Journal of Clinical Investigation*, **81**, 499–504.

McDonald, A. R., Maddux, B. A., Okabayashi, Y., Wong, K. Y., Hawley, D. M., Logsdon, C. D. & Goldfine, I. D. (1987). Regulation of insulin-receptor mRNA levels by glucocorticoids. *Diabetes*, **36**, 779–81.

McElduff, A., Grunberger, G. & Gorden, P. (1985). An alteration in apparent molecular weight of the insulin receptor from the human monocyte cell line U-937. *Diabetes*, **34**, 686–90.

McElduff, A., Poronnik, P., Baxter, R. C. & Williams, P. (1988). A comparison of the insulin and insulin-like growth factor I receptors from rat brain and liver. *Endocrinology*, **122**, 1933–9.

McElduff, A., Watkinson, A., Hedo, J. A. & Gorden, P. (1986). Characterization of the *N*-linked high-mannose oligosaccharides of the insulin pro-receptor and mature insulin receptor subunits. *Biochemical Journal*, **239**, 679–83.

Merida, I., Corrales, F. J., Clemente, R., Ruiz-Albusac, J. M., Villalba, M. &

Mato, J. M. (1988). Different phosphorylated forms of an insulin-sensitive glycosylphosphatidylinositol from rat hepatocytes. *FEBS Letters*, **236**, 251–5.

Misra, P., Hintz, R. L. & Rosenfeld, R. (1986). Structural and immunological characterization of insulin-like growth factor II binding to IM-9 cells. *Journal of Clinical Endocrinology and Metabolism*, **63**, 1400–5.

Moe, G. R., Bollag, G. & Koshland, D. E. (1989). Transmembrane signaling by a chimera of the *Escherichia coli* aspartate receptor and the human insulin receptor. *Proceedings of the National Academy of Sciences, USA*, **86**, 5683–7.

Moller, D. E. & Flier, J. S. (1988). Detection of an alteration in the insulin-receptor gene in a patient with insulin resistance, acanthosis nigricans, and the polycystic ovary syndrome (type A insulin resistance). *New England Journal of Medicine*, **319**, 1526–9.

Moller, D. E., Yokota, A., Caro, J. F. & Flier, J. S. (1989). Tissue-specific expression of two alternatively spliced insulin receptor mRNAs in man. *Molecular Endocrinology*, **3**, 1263–9.

Moller, D. E., Yokota, A., White, M. F., Pazianos, A. G. & Flier, J. S. (1990). A naturally occurring mutation of insulin receptor alanine 1134 impairs tyrosine kinase function and is associated with dominantly inherited insulin resistance. *Journal of Biological Chemistry*, **265**, 14979–85.

Momomura, K., Tobe, K., Seyama, Y., Takaku, F. & Kasuga, M. (1988). Insulin-induced tyrosine-phosphorylation in intact rat adipocytes. *Biochemical and Biophysical Research Communications*, **155**, 1181–6.

Mooney, R. A. & Anderson, D. L. (1989). Phosphorylation of the insulin receptor in permeabilized adipocytes is coupled to a rapid dephosphorylation reaction. *Journal of Biological Chemistry*, **264**, 6850–7.

Mooney, R. A., Bordwell, K. L., Luhowskyj, S. & Casnellie, J. E. (1989). The insulin-like effect of vanadate on lipolysis in rat adipocytes is not accompanied by an insulin-like effect on tyrosine phosphorylation. *Endocrinology*, **124**, 422–9.

Morgan, D. O., Edman, J. C., Standring, D. N., Fried, V. A., Smith, M. C., Roth, R. A. & Rutter, W. J. (1987a). Indulin-like growth factor II receptor as a multifunctional binding protein. *Nature*, London, **329**, 301–307.

Morgan, D. O., Ellis, L., Rutter, W. J. & Roth, R. A. (1987b). Antibody-induced down-regulation of a mutated insulin receptor lacking an intact cytoplasmic domain. *Biochemistry*, **26**, 2959–63.

Morgan, D. O., Ho, L., Korn, L. J. & Roth, R. A. (1986). Insulin action is blocked by a monoclonal antibody that inhibits the insulin receptor kinase. *Proceedings of the National Academy of Sciences, USA*, **83**, 328–32.

Morgan, D. O. & Roth, R. A. (1986). Mapping surface structures of the human insulin receptor with monoclonal antibodies: localization of main immunogenic regions to the receptor kinase domain. *Biochemistry*, **25**, 1364–71.

Morgan, D. O. & Roth, R. A. (1987). Acute insulin action requires insulin receptor kinase activity: introduction of an inhibitory monoclonal antibody into mammalian cells blocks the rapid effects of insulin. *Proceedings of the National Academy of Sciences, USA*, **84**, 41–5.

Morrison, B. D., Feltz, S. M. & Pessin, J. E. (1989). Polylysine specifically activates the insulin-dependent insulin receptor protein kinase. *Journal of Biological Chemistry*, **264**, 9994–10001.

Morrison, B. D. & Pessin, J. E. (1987). Insulin-stimulation of the insulin receptor kinase can occur in the complete absence of β subunit autophosphorylation. *Journal of Biological Chemistry*, **262**, 2861–8.

Morrison, B. D., Swanson, M. L., Sweet, L. J. & Pessin, J. E. (1988). Insulin-dependent covalent reassociation of isolated αβ heterodimeric insulin receptors into an $α_2β_2$ heterotetrameric disulfide-linked complex. *Journal of Biological Chemistry*, **263**, 7806–13.

Mosthaf, L., Grako, K., Dull, T. J., Coussens, L., Ullrich, A. & McClain, D. A. (1990). Functionally distinct insulin receptors generated by tissue-specific alternative splicing. *EMBO Journal*, **9**, 2409–13.

Moxham, C. P., Duronio, V. & Jacobs, S. (1989). Insulin-like growth factor I receptor β subunit heterogeneity: evidence for hybrid tetramers composed of insulin-like growth factor I and insulin receptor heterodimers. *Journal of Biological Chemistry*, **264**, 13238–44.

Muggeo, M., Ginsberg, B. H., Roth, J., Neville, D. M., DeMeyts, P. & Kahn, C. R. (1979). The insulin receptor in vertebrates is functionally more conserved during evolution than insulin itself. *Endocrinology*, **104**, 1383–92.

Muhlbacher, C., Karnieli, E., Schaff, P., Obermaier, B., Mushack, J., Rattenhuber, E. & Haring, H. U. (1988). Phorbol esters imitate in rat fat-cells the full effect of insulin on glucose carrier translocation, but not on 3-*o*-methylglucose transport activity. *Biochemical Journal*, **249**, 865–70.

Muller-Wieland, D., Taub, R., Tewari, D. S., Kriauciunas, K. M., Reddy, S. S. K. & Kahn, C. R. (1989). Insulin receptor gene and its expression in patients with insulin resistance. *Diabetes*, **38**, 31–8.

Myers, M. G., Backer, J. M., Siddle, K. & White, M. F. (1991). The insulin receptor functions normally in Chinese Hamster ovary cells after truncation of the C-terminus. *Journal of Biological Chemistry* **266**, 10616–23.

Nemenoff, R. A., Kwok, Y. C., Shulman, G. I., Blackshear, P. J., Osathanondh, R. & Avruch, J. (1984). Insulin-stimulated tyrosine protein kinase: characterization and relation to the insulin receptor. *Journal of Biological Chemistry*, **259**, 5058–65.

Nishibe, S., Wahl, M. I., Wedegaertner, P. B., Kim, J. J., Rhee, S. G. & Carpenter, G. (1990). Selectivity of phospholipase C phosphorylation by the epidermal growth factor receptor, the insulin receptor, and their cytoplasmic domains. *Proceedings of the National Academy of Sciences, USA*, **87**, 424–8.

Nishida, Y., Hata, M., Nishizuka, Y., Rutter, W. J. & Ebina, Y. (1986). Cloning of a Drosophila cDNA encoding a polypeptide similar to the human insulin receptor. *Biochemical and Biophysical Research Communications*, **141**, 474–81.

Nishizuka, Y. (1988). The molecular heterogeneity of protein kinase C and its implications for cellular regulation. *Nature*, London, **334**, 661–5.

Obermaier, B., Ermel, B., Kirsch, D., Mushack, J., Rattenhuber, E., Biemer, E., Machicao, F. & Haring, H. U. (1987). Catecholamines and tumour promoting phorbolesters inhibit insulin receptor kinase and induce insulin resistance in isolated human adipocytes. *Diabetologia*, **30**, 93–9.

Obermaier-Kusser, B., White, M. F., Pongratz, D. E., Su, Z., Ermel, B., Muhlbacher, C. & Haring, H. U. (1989). A defective intramolecular autoactivation cascade may cause the reduced receptor kinase activity of the skeletal muscle insulin receptor from patients with non-insulin-dependent diabetes mellitus. *Journal of Biological Chemistry*, **264**, 9497–504.

O'Brien, R. M., Houslay, M. D., Brindle, N. P. J., Milligan, G., Whittaker, J. & Siddle, K. (1989). Binding to GDP-agarose identifies a novel 60 kDa substrate for the insulin receptor tyrosyl kinase in mouse NIH 3T3 cells expressing high concentrations of the human insulin receptor. *Biochemical and Biophysical Research Communications*, **158**, 743–8.

O'Brien, R. M., Houslay, M. D., Milligan, G. & Siddle, K. (1987*a*). The insulin receptor tyrosyl kinase phosphorylates holomeric forms of the guanine nucleotide regulatory proteins Gi and Go. *FEBS Letters*, **212**, 281–8.

O'Brien, R. M., Siddle, K., Houslay, M. D. & Hall, A. (1987*b*). Interaction of the human insulin receptor with the ras oncogene product p21. *FEBS Letters*, **217**, 253–9.

O'Brien, R. M., Soos, M. A. & Siddle, K. (1986). Immunoaffinity purification of insulin receptor by use of monoclonal antibodies. *Biochemical Society Transactions*, **14**, 316–17.

O'Brien, R. M., Soos, M. A. & Siddle, K. (1987*c*). Monoclonal antibodies to the insulin receptor stimulate the intrinsic tyrosine kinase activity by cross-linking receptor molecules. *EMBO Journal*, **6**, 4003–10.

Odawara, M., Kadowaki, T., Yamamoto, R., Shibasaki, Y., Tobe, K., Accili, D., Bevins, C., Mikami, Y., Matsuura, N., Akanuma, Y., Takaku, F., Taylor, S. I. & Kasuga, M. (1989). Human diabetes associated with a mutation in the tyrosine kinase domain of the insulin receptor. *Science*, **245**, 66–8.

O'Hare, T. & Pilch, P. F. (1988). Separation and characterization of three insulin receptor species that differ in subunit composition. *Biochemistry*, **27**, 5693–700.

Ojamaa, K., Hedo, J. A., Roberts, C. T., Moncada, V. Y., Gorden, P., Ullrich, A. & Taylor, S. I. (1988). Defects in human insulin receptor gene expression. *Molecular Endocrinology*, **2**, 242–7.

Okabayashi, Y., Maddux, B. A., McDonald, A. R., Logsdon, C. D., Williams, J. A. & Goldfine, I. D. (1989). Mechanisms of insulin-induced insulin-receptor down regulation: decrease of receptor biosynthesis and mRNA levels. *Diabetes*, **38**, 182–7.

Olefsky, J. M. (1981). Insulin resistance and insulin action: an *in vitro* and *in vivo* perspective. *Diabetes*, **30**, 148–62.

Olivier, A. R., Ballou, L. M. & Thomas, G. (1988). Differential regulation of S6 phosphorylation by insulin and epidermal growth factor in Swiss mouse 3T3 cells: insulin activation of type 1 phosphatase. *Proceedings of the National Academy of Sciences, USA*, **85**, 4720–4.

Olson, T. S., Bamberger, M. J. & Lane, M. D. (1988). Post-translational changes in tertiary and quaternary structure of the insulin proreceptor: correlation with acquisition of function. *Journal of Biological Chemistry*, **263**, 7342–51.

Ota, A., Shemer, J., Pruss, R. M., Lowe, W. L. & LeRoith, D. (1988). Characterization of the altered oligosaccharide composition of the insulin receptor on neural-derived cells. *Brain Research*, **443**, 1–11.

Pang, D. T. & Shafer, J. A. (1983). Stoichiometry for the binding of insulin to insulin receptors in adipocyte membranes. *Journal of Biological Chemistry*, **258**, 2514–18.

Pang, D. T. & Shafer, J. A. (1984). Evidence that insulin receptor from human placenta has a high affinity for only one molecule of insulin. *Journal of Biological Chemistry*, **259**, 8589–96.

Pang, D. T., Sharma, B. R., Shafer, J. A., White, M. F. & Kahn, C. R. (1985). Predominance of tyrosine phosphorylation of insulin receptors during the initial response of intact cells to insulin. *Journal of Biological Chemistry*, **260**, 7131–6.

Pennington, S. R. & Martin, B. R. (1985). Insulin-stimulated phosphoinositide metabolism in isolated fat cells. *Journal of Biological Chemistry*, **260**, 11039–45.

Perlmann, R., Bottaro, D. P., White, M. F. & Kahn, C. R. (1989). Conformational changes in the α- and β-subunits of the insulin receptor identified by anti-peptide antibodies. *Journal of Biological Chemistry*, **264**, 8946–50.

Perrotti, N., Accili, D., Marcus-Samuels, B., Rees-Jones, R. W. & Taylor, S. I. (1987). Insulin stimulates phosphorylation of a 120 kDa glycoprotein substrate (pp120) for the receptor-associated protein kinase in intact H-35 hepatoma cells. *Proceedings of the National Academy of Sciences, USA*, **84**, 3137–40.

Petruzzelli, L., Ganguly, S., Smith, C. J., Cobb, M. H., Rubin, C. S. & Rosen, O. M. (1982). Insulin activates a tyrosine-specific protein kinase in extracts of 3T3-L1 adipocytes and human placenta. *Proceedings of the National Academy of Sciences, USA*, **79**, 6792–6.

Petruzzelli, L., Herrera, R., Arenas-Garcia, R., Fernandez, R., Birnbaum, M. J. & Rosen, O. M. (1986). Isolation of a Drosophila genomic sequence homologous to the kinase domain of the human insulin receptor and detection of the phosphorylated Drosophila receptor with an anti-peptide antibody. *Proceedings of the National Academy of Sciences, USA*, **83**, 4710–14.

Petruzzelli, L., Herrera, R. & Rosen, O. M. (1984). Insulin receptor is an insulin-dependent tyrosine protein kinase: copurification of insulin binding activity and protein kinase activity to homogeneity from human placenta. *Proceedings of the National Academy of Sciences, USA*, **81**, 3327–31.

Peyron, J. F. & Fehlmann, M. (1988). Phosphorylation of class I histocompatibility antigens in human B lymphocytes: regulation by phorbol esters and insulin. *Biochemical Journal*, **256**, 763–8.

Phillips, M. L., Moule, M. L., Delovitch, T. L., & Yip, C. C. (1986). Class I histocompatibility antigens and insulin receptors: evidence for interactions. *Proceedings of the National Academy of Sciences, USA*, **83**, 3474–8.

Phillips, S. A., Perrotti, N. & Taylor, S. I. (1987). Rat liver membranes contain a 120 kDa glycoprotein which serves as a substrate for the tyrosine kinases of the receptors for insulin and epidermal growth factor. *FEBS Letters*, **212**, 141–4.

Pike, L. J., Kuenzel, E. A., Casnellie, J. E. & Krebs, E. G. (1984). A comparison of the insulin and epidermal growth factor-stimulated protein kinases from human placenta. *Journal of Biological Chemistry*, **259**, 9913–21.

Pilch, P. F. & Czech, M. P. (1979). Interaction of cross-linking agents with the insulin effector system of isolated fat cells. *Journal of Biological Chemistry*, **254**, 3375–81.

Pilch, P. F. & Czech, M. P. (1980a). The subunit structure of the high affinity insulin receptor. *Journal of Biological Chemistry*, **255**, 1722–1731.

Pilch, P. F. & Czech, M. P. (1980b). Hormone binding alters the conformation of the insulin receptor. *Science*, **210**, 1152–3.

Pillay, T. S., Whittaker, J. & Siddle, K. (1990). Phorbol ester-induced down-regulation of protein kinase C potentiates insulin receptor tyrosine phosphorylation: evidence for a major constitutive role in insulin receptor regulation. *Biochemical Society Transactions*, **18**, 494–5.

Pollet, R. J., Standaert, M. & Haase, B. A. (1977). Insulin binding to the human lymphocyte receptor: evaluation of the negative cooperativity model. *Journal of Biological Chemistry*, **252**, 5828–4.

Ponzio, G., Contreres, J. O., Debant, A., Baron, V., Gautier, N., Dolais-Kitabgi, J. & Rossi, B. (1988). Use of an anti-insulin receptor antibody to discriminate between metabolic and mitogenic effects of insulin: correlation with receptor autophosphorylation. *EMBO Journal*, **7**, 4111–17.

Ponzio, G., Dolais-Kitabgi, J., Louvard, D., Gautier, N. & Rossi, B. (1987). Insulin and rabbit anti-insulin receptor antibodies stimulate additively the intrinsic receptor kinase activity. *EMBO Journal*, **6**, 333–40.

Prigent, S. A., Stanley, K. K. & Siddle, K. (1990). Identification of epitopes on the human insulin receptor reacting with rabbit polyclonal antisera and mouse monoclonal antibodies. *Journal of Biological Chemistry*, **295**, 9970–7.

Pullen, R. A., Lindsay, D. G., Wood, S. P., Tickle, I. J., Blundell, T. L., Woolmer, A., Krail, G., Brandenburg, D., Zahn, H., Gliemann, J. & Gammeltoft, S. (1976). Receptor-binding region of insulin. *Nature, London*, **259**, 369–73.

Pyne, N. J., Cushley, W., Nimmo, H. G. & Houslay, M. D. (1989). Insulin stimulates the tyrosyl phosphorylation and activation of the 52 kDa peripheral plasma-membrane cyclic AMP phosphodiesterase in intact hepatocytes. *Biochemical Journal*, **261**, 897–904.

Rafaeloff, R., Patel, R., Yip, C., Goldfine, I. D. & Hawley, D. M. (1989). Mutation of the high cysteine region of the human insulin receptor α-subunit increases insulin receptor binding affinity and transmembrane signaling. *Journal of Biological Chemistry*, **264**, 15900–4.

Ramakrishna, S. & Benjamin, W. B. (1988). Insulin action rapidly decreases multifunctional protein kinase activity in rat adipose tissue. *Journal of Biological Chemistry*, **263**, 12677–81.

Ray, L. B. & Sturgill, T. W. (1988). Insulin-stimulated microtubule-associated protein kinase is phosphorylated on tyrosine and threonine *in vivo*. *Proceedings of the National Academy of Sciences, USA*, **85**, 3753–7.

Reaven, G. M. (1988). Role of insulin resistance in human disease. *Diabetes*, **37**, 1595–607.

Rechler, M. M. & Nissley, S. P. (1985). The nature and regulation of the receptors for insulin-like growth factors. *Annual Review of Physiology*, **47**, 425–52.

Reddy, S. S. K. & Kahn, C. R. (1988). Insulin resistance: a look at the role of insulin receptor kinase. *Diabetic Medicine*, **5**, 621–9.

Reddy, S. S. K., Lauris, V. & Kahn, C. R. (1988). Insulin receptor function in fibroblasts from patients with leprechaunism. *Journal of Clinical Investigation*, **82**, 1359–65.

Rees-Jones, R. W. & Taylor, S. I. (1985). An endogenous substrate for the insulin receptor-associated tyrosine kinase. *Journal of Biological Chemistry*, **260**, 4461–7.

Riedel, H., Dull, T. J., Schlessinger, J. & Ullrich, A. (1986). A chimaeric receptor allows insulin to stimulate tyrosine kinase activity of epidermal growth factor receptor. *Nature, London*, **324**, 68–70.

Ronnett, G. V., Knutson, V. P., Kohanski, R. A., Simpson, T. L. & Lane, M. D. (1984). Role of glycosylation in the processing of newly translated insulin proreceptor in 3T3-L1 adipocytes. *Journal of Biological Chemistry*, **259**, 4566–75.

Roome, J., O'Hare, T., Pilch, P. F. & Brautigan, D. L. (1988). Protein phosphotyrosine phosphatase purified from the particulate fraction of human placenta dephosphorylates insulin and growth factor receptors. *Biochemical Journal*, **256**, 493–500.

Rosen, O. M. (1987). After insulin binds. *Science*, **237**, 1452–8.

Rosen, O. M. (1988). Protein tyrosine kinases, protein serine kinases, and the mechanism of action of insulin. *Harvey Lectures*, Series 82, 105–22.

Rosen, O. M., Herrera, R., Olowe, Y., Petruzzelli, L. M. & Cobb, M. H. (1983). Phosphorylation activates the insulin receptor tyrosine kinase. *Proceedings of the National Academy of Sciences, USA*, **80**, 3237–40.

Roth, J. & Taylor, S. I. (1982). Receptors for peptide hormones: alterations in diseases of humans. *Annual Review of Physiology*, **44**, 639–51.

Roth, R. A. (1988). Structure of the receptor for insulin-like growth factor II: the puzzle amplified. *Science*, **239**, 1269–71.

Roth, R. A. & Beaudoin, J. (1987). Phosphorylation of purified insulin receptor by cAMP kinase. *Diabetes*, **36**, 123–6.

Roth, R. A. & Cassell, D. J. (1983). Insulin receptor: evidence that it is a protein kinase. *Science*, **219**, 299–301.

Roth, R. A., Cassell, D. J., Maddux, B. A. & Goldfine, I. D. (1983). Regulation of insulin receptor kinase activity by insulin mimickers and an insulin antagonist. *Biochemical and Biophysical Research Communications*, **115**, 245–52.

Roth, R. A., Cassell, D. J., Wong, K. Y., Maddux, B. A. & Goldfine, I. D. (1982). Monoclonal antibodies to the human insulin receptor block insulin binding and inhibit insulin action. *Proceedings of the National Academy of Sciences, USA*, **79**, 7312–16.

Roth, R. A. & Morgan, D. O. (1985). Monoclonal antibodies to the insulin receptor. *Pharmacology and Therapeutics*, **28**, 1–16.

Roth, R. A., Morgan, D. O., Beaudoin, J. & Sara, V. (1986). Purification and characterization of the human brain insulin receptor. *Journal of Biological Chemistry*, **261**, 3753–7.

Rothenberg, P. L. & Kahn, C. R. (1988). Insulin inhibits pertussis toxin-catalyzed ADP-ribosylation of G-proteins: evidence for a novel interaction between insulin receptors and G-proteins. *Journal of Biological Chemistry*, **263**, 15546–52.

Rouiller, D. & Gorden, P. (1987). Homologous down-regulation of the insulin receptor is associated with increased receptor biosynthesis in cultured human lymphocytes (IM-9 line). *Proceedings of the National Academy of Sciences, USA*, **84**, 126–30.

Rouiller, D. G. & McKeon, C., Taylor, S. I. & Gorden, P. (1988). Hormonal regulation of insulin receptor gene expression: hydrocortisone and insulin act by different mechanisms. *Journal of Biological Chemistry*, **263**, 13185–90.

Russell, D. S., Gherzi, R., Johnson, E. L., Chou, C. K. & Rosen, O. M. (1987). The protein–tyrosine kinase activity of the insulin receptor is necessary for insulin-mediated receptor down-regulation. *Journal of Biological Chemistry*, **262**, 11833–40.

Sacks, D. B. & McDonald, J. (1988). Insulin-stimulated phosphorylation of calmodulin by rat liver insulin receptor preparations. *Journal of Biological Chemistry*, **263**, 2377–83.

Sadoul, J. L., Peyron, J. F., Ballotti, R., Debant, A., Fehlmann, M. & Van Obberghen, E. (1985). Identification of a cellular 110000 Da protein substrate for the insulin-receptor kinase. *Biochemical Journal*, **227**, 887–92.

Sakai, M. & Wells, W. W. (1986). Action of insulin on the subcellular metabolism of polyphosphoinositides in isolated rat hepatocytes. *Journal of Biological Chemistry*, **261**, 10058–62.

Sale, G. J. (1988). Recent progress in our understanding of the mechanism of insulin action. *International Journal of Biochemistry*, **20**, 897–908.

Sale, G. J., Fujita-Yamaguchi, Y. & Kahn, C. R. (1986). Characterization of phosphatidylinositol kinase activity associated with the insulin receptor. *European Journal of Biochemistry*, **155**, 345–51.

Saltiel, A. R. (1990). Second messenger of insulin action. *Diabetes Care*, **13**, 244–56.

Saltiel, A. R. & Cuatrecasas, P. (1986). Insulin stimulates the generation from hepatic plasma membranes of modulators derived from an inositol glycolipid. *Proceedings of the National Academy of Sciences, USA*, **83**, 5793–7.

Saltiel, A. R. & Cuatrecasas, P. (1988). In search of a second messenger for insulin. *American Journal of Physiology*, **255**, C1–11.

Saltiel, A. R., Fox, J. A., Sherline, P. & Cuatrecasas, P. (1986). Insulin-stimulated hydrolysis of a novel glycolipid generates modulators of cAMP phosphodiesterase. *Science*, **233**, 967–72.

Saltiel, A. R., Sherline, P. & Fox, J. A. (1987). Insulin-stimulated diacylglycerol production results from the hydrolysis of a novel phosphatidylinositol glycan. *Journal of Biological Chemistry*, **262**, 1116–21.

Samson, M., Cousin, J. L. & Fehlmann, M. (1986). Cross-linking of insulin receptors to MHC antigens in human B lymphocytes: evidence for selective molecular interactions. *Journal of Immunology*, **137**, 2293–8.

Schaefer, E. M., Siddle, K. & Ellis, L. (1990). Deletion analysis of the human insulin receptor ectodomain reveals independently folded soluble subdomains and insulin binding by a monomeric α-subunit. *Journal of Biological Chemistry*, **265**, 13248–53.

Schechter, Y., Elias, D., Maron, R. & Cohen, I. R. (1984). Mouse antibodies to the insulin receptor developing spontaneously as anti-idiotypes. I. Characterization of the antibodies. *Journal of Biological Chemistry*, **259**, 6411–15.

Schechter, Y., Yaish, P., Chorev, M., Gilon, C., Braun, S. & Levitzki, A. (1989). Inhibition of insulin-dependent lipogenesis and anti-lipolysis by protein tyrosine kinase inhibitors. *EMBO Journal*, **8**, 1671–6.

Schenker, E. & Kohanski, R. A. (1988). Conformational states of the insulin receptor. *Biochemical and Biophysical Research Communications*, **157**, 140–5.

Schlessinger, J. (1988). Signal transduction by allosteric receptor oligomerization. *Trends in Biochemical Sciences*, **13**, 443–7.

Schlessinger, J., Schechter, Y., Willingham, M. C. & Pastan, I. (1978). Direct visualization of binding, aggregation and internalization of insulin and epidermal growth factor on living fibroblastic cells. *Proceedings of the National Academy of Sciences, USA*, **75**, 2659–63.

Seger, R., Zick, Y. & Shaltiel, S. (1989). Studying the structure of the intracellular moiety of the insulin receptor with a kinase-splitting membranal protease. *EMBO Journal*, **8**, 435–40.

Seino, S. & Bell, G. I. (1989). Alternative splicing of human insulin receptor messenger RNA. *Biochemical and Biophysical Research Communications*, **159**, 312–16.

Seino, S., Seino, M. & Bell, G. I. (1990). Human insulin-receptor gene. *Diabetes*, **39**, 129–33.

Seino, S., Seino, M., Nishi, S. & Bell, G. I. (1989). Structure of the human insulin receptor gene and characterization of its promoter. *Proceedings of the National Academy of Sciences, USA*, **86**, 114–18.

Shargill, N. S., Tatoyan, A., El-Refai, M. F., Pleta, M. & Chan, T. M. (1986). Impaired insulin receptor phosphorylation in skeletal muscle membranes of db/db mice: the use of a novel skeletal muscle plasma membrane preparation to compare insulin binding and stimulation of receptor phosphorylation. *Biochemical and Biophysical Research Communications*, **137**, 286–94.

Shia, M. A. & Pilch, P. F. (1983). The β subunit of the insulin receptor is an insulin-activated protein kinase. *Biochemistry*, **22**, 717–21.

Shibasaki, Y., Sakura, H., Odawara, M., Shibuya, M., Kanazawa, Y., Akanuma, Y., Takaku, F. & Kasuga, M. (1988). Glucocorticoids increase insulin binding and the amount of insulin-receptor mRNA in human cultured lymphocytes. *Biochemical Journal*, **249**, 715–19.

Shier, P. & Watt, V. M. (1989). Primary structure of a putative receptor for a ligand of the insulin family. *Journal of Biological Chemistry*, **264**, 14605–8.

Shoelson, S. E., White, M. F. & Kahn, C. R. (1988). Tryptic activation of the insulin receptor: proteolytic truncation of the α-subunit releases the β-subunit from inhibitory control. *Journal of Biological Chemistry*, **263**, 4852–60.

Siddle, K., Soos, M. A., O'Brien, R. M., Ganderton, R. H. & Pillay, T. S. (1988). Monoclonal antibodies to the insulin receptor. In *Clinical Application of Monoclonal Antibodies*, ed. R. Hubbard & V. Marks, pp. 87–99. London: Plenum Publishing Corporation.

Siddle, K., Soos, M. A., O'Brien, R. M., Ganderton, R. M. & Taylor, R. (1987). Monoclonal antibodies as probes of the structure and function of insulin receptors. *Biochemical Society Transactions*, **15**, 47–51.

Siegel, T. W., Ganguly, S., Jacobs, S., Rosen, O. M. & Rubin, C. S. (1981). Purification and properties of the human placental insulin receptor. *Journal of Biological Chemistry*, **256**, 9266–73.

Simpson, I. A. & Hedo, J. A. (1984). Insulin receptor phosphorylation may not be a prerequisite for acute insulin action. *Science*, **223**, 1301–4.

Sinha, M. K., Pories, W. J., Flickinger, E. G., Meelheim, D. & Caro, J. F. (1987). Insulin-receptor kinase activity of adipose tissue from morbidly obese humans with and without NIDDM. *Diabetes*, **36**, 620–5.

Sissom, J. & Ellis, L. (1989). Secretion of the extracellular domain of the human insulin receptor from insect cells by use of a baculovirus vector. *Biochemical Journal*, **261**, 119–26.

Sivitz, W. I., DeSautel, S. L., Kayano, T., Bell, G. I. & Pessin, J. E. (1989). Regulation of glucose transporter messenger RNA in insulin-deficient states. *Nature, London*, **340**, 72–4.

Smith, D. M., King, M. J. & Sale, G. J. (1988). Two systems in vitro that show insulin-stimulated serine kinase activity towards the insulin receptor. *Biochemical Journal*, **250**, 509–19.

Smith, D. M. & Sale, G. J. (1988). Evidence that a novel serine kinase catalyses phosphorylation of the insulin receptor in an insulin-dependent and tyrosine kinase-dependent manner. *Biochemical Journal*, **256**, 903–9.

Smith, D. M. & Sale, G. J. (1989). Characterization of sites of serine phosphorylation in human placental insulin receptor copurified with insulin-stimulated serine kinase activity by two dimensional thin-layer peptide mapping. *FEBS Letters*, **242**, 301–4.

Soler, A. P., Thompson, K. A., Smith, R. M. & Jarett, L. (1989). Immunological demonstration of the accumulation of insulin, but not insulin receptors, in nuclei of insulin-treated cells. *Proceedings of the National Academy of Sciences, USA*, **86**, 6640–4.

Sommercorn, J., Mulligan, J. A., Lozeman, f. J. & Krebs, E. G. (1987). Activation of casein kinase II in response to insulin and to epidermal growth factor. *Proceedings of the National Academy of Sciences, USA*, **84**, 8834–8.

Sonne, O. (1988). Receptor-mediated endocytosis and degradation of insulin. *Physiological Reviews*, **68**, 1129–96.

Soos, M. A., O'Brien, R. M., Brindle, N. P. J., Stigter, J. M., Okamoto, A. K., Whittaker, J. & Siddle, K. (1989). Monoclonal antibodies to the insulin receptor mimic metabolic effects of insulin but do not stimulate receptor autophosphorylation in transfected NIH 3T3 fibroblasts. *Proceedings of the National Academy of Sciences, USA*, **86**, 5217–21.

Soos, M. A. & Siddle, K. (1989). Immunological relationships between receptors for insulin and insulin-like growth factor I. Evidence for structural heterogeneity of insulin-like growth factor I receptors involving hybrids with insulin receptors. *Biochemical Journal*, **263**, 553–63.

Soos, M. A., Siddle, K., Baron, M. D., Heward, J. M., Luzio, J. P., Bellatin, J. & Lennox, E. S. (1986). Monoclonal antibodies reacting with multiple epitopes on the human insulin receptor. *Biochemical Journal*, **235**, 199–208.

Soos, M. A., Whittaker, J., Lammers, R., Ullrich, A. & Siddle, K. (1990). Receptors for insulin and insulin-like growth factor I can form hybrid dimers: characterization of hybrid receptors in transfected cells. *Biochemical Journal*, **270**, 383–90.

Spach, D. M., Nemenoff, R. A. & Blackshear, P. J. (1986). Protein phosphorylation and protein kinase activities in BC3H-1 myocytes: differences between the effects of insulin and phorbol esters. *Journal of Biological Chemistry*, **261**, 12750–3.

Stadtmauer, L. & Rosen, O. M. (1986a). Phosphorylation of synthetic insulin receptor peptides by the insulin receptor kinase and evidence that the preferred sequence containing Tyr-1150 is phosphorylated *in vivo*. *Journal of Biological Chemistry*, **261**, 10000–5.

Stadtmauer, L. & Rosen, O. M. (1986b). Increasing the cAMP content of IM-9 cells alters the phosphorylation state and protein kinase activity of the insulin receptor. *Journal of Biological Chemistry*, **261**, 3402–7.

Steele-Perkins, G. & Roth, R. A. (1990). Insulin-mimetic anti-insulin receptor monoclonal antibodies stimulate receptor kinase activity in intact cells. *Journal of Biological Chemistry*, **265**, 9458–63.

Steele-Perkins, G., Turner, J., Edman, J. C., Hari, J., Pierce, S. B., Stover, C., Rutter, W. J. & Roth, R. A. (1988). Expression and characterization of a functional human insulin-like growth factor. *Journal of Biological Chemistry*, **263**, 11486–92.

Stefanovic, D., Erikson, E., Pike, L. J. & Maller, J. L. (1986). Activation of a ribosomal protein S6 protein kinase in *Xenopus* oocytes by insulin and insulin-receptor kinases. *EMBO Journal*, **5**, 157–60.

Stralfors, P. (1988). Insulin stimulation of glucose uptake can be mediated by diacylglycerol in adipocytes. *Nature*, London, **335**, 554–6.

Strout, H. V., Vicario, P. P., Saperstein, R. & Slater, E. E. (1989). The insulin-mimetic effect of vanadate is not correlated with insulin receptor tyrosine kinase activity nor phosphorylation in mouse diaphragm *in vivo*. *Endocrinology*, **124**, 1918–24.

Sung, C. K., Maddux, B. A., Hawley, D. M. & Goldfine, I. D. (1989). Monoclonal antibodies mimic insulin activation of ribosomal protein S6 kinase without activation of insulin receptor tyrosine kinase: studies in cells transfected with normal and mutant insulin receptor. *Journal of Biological Chemistry*, **264**, 18951–9.

Sutherland, E. W., Oye, I. & Butcher, R. W. (1965). The action of epinephrine and the role of the adenyl cyclase system in hormone action. *Recent Progress in Hormone Research*, **21**, 623–46.

Swanson, M. L. & Pessin, J. E. (1989). High affinity insulin binding in the human placenta insulin receptor requires $\alpha\beta$ heterodimeric subunit interactions. *Journal of Membrane Biology*, **108**, 217–25.

Swarup, G., Cohen, S. & Garbers, D. L. (1982). Inhibition of membrane phosphotyrosyl–protein phosphatase activity by vanadate. *Biochemical and Biophysical Research Communications*, **107**, 1104–9.

Sweet, L. J., Dudley, D. T., Pessin, J. E. & Spector, A. A. (1987a). Phospholipid activation of the insulin receptor kinase: regulation by phosphatidylinositol. *FASEB Journal*, **1**, 55–9.

Sweet, L. J., Morrison, B. D. & Pessin, J. E. (1987b). Isolation of functional $\alpha\beta$ heterodimers from the purified human placental $\alpha_2\beta_2$ heterotetrameric insulin receptor complex: a structural basis for insulin binding heterogeneity. *Journal of Biological Chemistry*, **262**, 6939–42.

Sweet, L. J., Morrison, B. D., Wilden, P. A. & Pessin, J. E. (1987c). Insulin-dependent intermolecular subunit communication between isolated $\alpha\beta$ heterodimeric insulin receptor complexes. *Journal of Biological Chemistry*, **262**, 16730–8.

Sweet, L. J., Wilden, P. A., Spector, A. A. & Pessin, J. E. (1985). Incorporation of the purified human placental insulin receptor into phospholipid vesicles. *Biochemistry*, **24**, 6571–80.

Taira, M., Taira, M., Hashimoto, N., Shimada, F., Suzuki, Y., Kanatsuka, A., Nakamura, F., Ebina, Y., Tatibana, M., Makino, H. & Yoshida, S. (1989). Human diabetes associated with a deletion of the tyrosine kinase domain of the insulin receptor. *Science*, **245**, 63–6.

Takayama, S., Kahn, C. R., Kubo, K. & Foley, J. E. (1988a). Alterations in insulin receptor autophosphorylation in insulin resistance: correlation with altered sensitivity to glucose transport and antilipolysis to insulin. *Journal of Clinical Endocrinology and Metabolism*, **66**, 992–9.

Takayama, S., White, M. F. & Kahn, C. R. (1988b). Phorbol ester-induced phosphorylation of the insulin receptor decreases its tyrosine kinase activity. *Journal of Biological Chemistry*, **263**, 3440–7.

Takayama, S., White, M. F., Lauris, V. & Kahn, C. R. (1984). Phorbol esters modulate insulin receptor phosphorylation and insulin action in cultured hepatoma cells. *Proceedings of the National Academy of Sciences, USA*, **81**, 7797–801.

Takayama-Hasumi, S., Tobe, K., Momomura, K., Koshio, O., Tashiro-Hashimoto, Y., Akanuma, Y., Hirata, Y., Takaku, F. & Kasuga, M. (1989). Autoantibodies to the insulin receptor (B10) can stimulate tyrosine phosphorylation of the β-subunit of the insulin receptor and a 185000 molecular weight protein in rat hepatoma cells. *Journal of Clinical Endocrinology and Metabolism*, **68**, 787–95.

Tamura, S., Brown, T. A., Dubler, R. E. & Larner, J. (1983). Insulin-like effect of vanadate on adipocyte glycogen synthase and on phosphorylation of

95000 Dalton subunit of insulin receptor. *Biochemical and Biophysical Research Communications*, **113**, 80–6.

Tamura, S., Brown, T. A., Whipple, J. H., Fujita-Yamaguchi, Y., Dubler, R. E., Cheng, K. & Larner, J. (1984). A novel mechanism for the insulin-like effect of vanadate on glycogen synthase in rat adipocytes. *Journal of Biological Chemistry*, **259**, 6650–8.

Tanti, J. F., Gremeaux, T., Rochet, N., Van Obberghen, E. & Le Marchand-Brustel, Y. (1987). Effect of cyclic AMP-dependent protein kinase on insulin receptor tyrosine kinase activity. *Biochemical Journal*, **245**, 19–26.

Tanti, J. F., Rochet, N., Gremeaux, T., Van Obberghen, E. & Le Marchand-Brustel, Y. (1989). Insulin-stimulated glucose transport in muscle: evidence for a protein-kinase-C-dependent component which is unaltered in insulin-resistant mice. *Biochemical Journal*, **258**, 141–6.

Tashiro-Hashimoto, Y., Tobe, K., Koshio, O., Izumi, T., Takaku, F., Akanuma, Y. & Kasuga, M. (1989). Tyrosine phosphorylation of pp185 by insulin receptor kinase in a cell-free system. *Journal of Biological Chemistry*, **264**, 6879–85.

Tavare, J. M. & Denton, R. M. (1988). Studies on the autophosphorylation sites of the insulin receptor from human placenta. *Biochemical Journal*, **252**, 607–15.

Tavare, J. M., O'Brien, R. M., Siddle, K. & Denton, R. M. (1988). Analysis of insulin-receptor phosphorylation sites in intact cells by two-dimensional phosphopeptide mapping. *Biochemical Journal*, **253**, 783–8.

Taylor, D., Uhing, R. J., Blackmore, P. F., Prpic, V. & Exton, J. H. (1985). Insulin and epidermal growth factor do not affect phosphoinositide metabolism in rat liver plasma membranes and hepatocytes. *Journal of Biological Chemistry*, **260**, 2011–14.

Taylor, R. (1989). Aetiology of non-insulin dependent diabetes. *British Medical Bulletin*, **45**, 73–91.

Taylor, R. & Agius, L. (1988). The biochemistry of diabetes. *Biochemical Journal*, **250**, 625–40.

Taylor, R., Soos, M. A., Wells, A., Argyraki, M. & Siddle, K. (1987). Insulin-like and insulin-inhibitory effects of monoclonal antibodies for different epitopes on the structure insulin receptor. *Biochemical Journal*, **242**, 123–9.

Taylor, S. I., Kadowaki, T., Kadowaki, H., Accili, D., Cama, A. & McKeon, C. (1990). Mutations in insulin-receptor gene in insulin-resistant patients. *Diabetes Care*, **13**, 257–79.

Thies, R. S., Ullrich, A. & McClain, D. A. (1989). Augmented mitogenesis and impaired metabolic signaling mediated by a truncated insulin receptor. *Journal of Biological Chemistry*, **264**, 12820–5.

Thies, R. S., Webster, N. J. & McClain, D. A. (1990). A domain of the insulin receptor required for endocytosis in rat fibroblasts. *Journal of Biological Chemistry*, **265**, 10132–7.

Thomopoulos, E., Testa, U., Gourdin, M. F., Hervy, C., Titeux, M. & Vainchenker, W. (1982). Inhibition of insulin receptor binding by phorbol esters. *European Journal of Biochemistry*, **129**, 389–93.

Tollefsen, S. E. & Thompson, K. (1988). The structural basis for insulin-like growth factor I receptor high affinity binding. *Journal of Biological Chemistry*, **263**, 16267–73.

Tonks, N. K. & Charbonneau, H. (1989). Protein tyrosine dephosphorylation and signal transduction. *Trends in Biochemical Sciences*, **14**, 497–500.

Tonks, N. K., Diltz, C. D. & Fischer, E. H. (1988). Characterization of the major protein–tyrosine–phosphatases of human placenta. *Journal of Biological Chemistry*, **263**, 6731–7.

Tornqvist, H. E. & Avruch, J. (1988). Relationship of site-specific β subunit tyrosine autophosphorylation to insulin activation of the insulin receptor (tyrosine) protein kinase activity. *Journal of Biological Chemistry*, **263**, 4593–601.

Tornqvist, H. E., Gunsalus, J. R., Nemenoff, R. A., Frackelton, A. R., Pierce, M. W. & Avruch, J. (1988). Identification of the insulin receptor tyrosine residues undergoing insulin-stimulated phosphorylation in intact rat hepatoma cells. *Journal of Biological Chemistry*, **263**, 350–9.

Tornqvist, H. E., Pierce, M. W., Frackelton, A. R., Nemenoff, R. A. & Avruch, J. (1987). Identification of insulin receptor tyrosine residues autophosphorylated *in vitro*. *Journal of Biological Chemistry*, **262**, 10212–19.

Toth, B., Bollen, M. & Stalmans, W. (1988). Acute regulation of hepatic protein phosphatases by glucagon, insulin and glucose. *Journal of Biological Chemistry*, **263**, 14061–6.

Toyoshige, M., Yanaihara, C., Hoshino, M., Kaneko, T. & Yanaihara, N. (1989). Insulin receptor in human hepatoma PLC/PRF/5 cells and its immunochemical characterization with anti-synthetic peptide antibodies. *Biomedical Research*, **10**, 139–47.

Treadway, J. L., Morrison, B. D., Goldfine, I. D. & Pessin, J. E. (1989a). Assembly of insulin/insulin-like growth factor-1 hybrid receptors *in vitro*. *Journal of Biological Chemistry*, **264**, 21450–3.

Treadway, J. L., Whittaker, J. & Pesin, J. E. (1989b). Regulation of the insulin receptor kinase by hyperinsulinism. *Journal of Biological Chemistry*, **264**, 15136–43.

Treadway, J. L., Morrison, B. D., Wemmie, J. A., Frias, I., O'Hare, T., Pilch, P. F. & Pessin, J. E. (1990). The endogenous functional turkey erythrocyte and rat liver insulin receptor is an $\alpha_2\beta_2$ heterotetrameric complex. *Biochemical Journal*, **271**, 99–105.

Treadway, J. L., Morrison, B. D., Soos, M. A., Siddle, K., Olefsky, J., Ullrich, A., McClain, D. A. & Pessin, J. E. (1991). Transdominant inhibition of tyrosine kinase activity in mutant insulin/insulin-like growth factor I hybrid receptors. *Proceedings of the National Academy of Sciences, USA*, **88**, 214–18.

Trischitta, V., Wong, K. Y., Brunetti, A., Scalisi, R., Vigneri, R. & Goldfine, I. D. (1989). Endocytosis, recycling, and degradation of the insulin receptor: studies with monoclonal antireceptor antibodies that do not activate receptor kinase. *Journal of Biological Chemistry*, **264**, 5041–6.

Truglia, J. A., Hayes, G. R. & Lockwood, D. H. (1988). Intact adipocyte insulin-receptor phosphorylation and *in vitro* tyrosine kinase activity in animal models of insulin resistance. *Diabetes*, **37**, 147–53.

Tuazon, P. T., Pang, D. T., Shafer, J. A. & Traugh, J. A. (1985). Phosphorylation of the insulin receptor by casein kinase 1. *Journal of Cellular Biochemistry*, **28**, 159–70.

Ullrich, A., Bell, J. R., Chen, E. Y., Herrera, R., Petruzzelli, L. M., Dull, T. J., Gray, A., Coussens, L., Liao, Y. C., Tsubokawa, M., Mason, A., Seeburg, P. H., Grunfeld, C., Rosen, O. M. & Ramachandran, J. (1985). Human insulin receptor and its relationship to the tyrosine kinase family of oncogenes. *Nature*, London, **313**, 756–61.

Ullrich, A., Gray, A., Tam, A. W., Yang-Feng, T., Tsubokawa, M., Collins, C.,

Henzel, W., Le Bon, T., Kathuria, S., Chen, E., Jacobs, S., Francke, U., Ramachandran, J. & Fujita-Yamaguchi, Y. (1986). Insulin-like growth factor I receptor primary structure: comparison with insulin receptor suggests structural determinants that define specificity. *EMBO Journal*, **5**, 2503–12.

Ullrich, A. & Schlessinger, J. (1990). Signal transduction by receptors with tyrosine kinase activity. *Cell*, **61**, 203–12.

Ushiro, H. & Cohen, S. (1980). Identification of phosphotyrosine as a product of epidermal growth factor-activated protein kinase in A-431 cell membranes. *Journal of Biological Chemistry*, **255**, 8363–5.

Van Obberghen, E., Kasuga, M., Le Cam, A., Hedo, J. A., Itin, A. & Harrison, L. C. (1981). Biosynthetic labeling of insulin receptor: studies of subunits in cultured human IM-9 lymphocytes. *Proceedings of the National Academy of Sciences, USA*, **78**, 1052–6.

Van Obberghen, E. & Kowalski, A. (1982). Phosphorylation of the hepatic insulin receptor: stimulating effect of insulin on intact cells and in a cell-free system. *FEBS Letters*, **143**, 179–82.

Van Obberghen, E., Rossi, B., Kowalski, A., Gazzano, H. & Ponzio, G. (1983). Receptor-mediated phosphorylation of the hepatic insulin receptor: evidence that the M_r 95000 receptor subunit is its own kinase. *Proceedings of the National Academy of Sciences, USA*, **80**, 945–9.

Velicelebi, G. & Aiyer, R. A. (1984). Identification of the $\alpha\beta$ monomer of the adipocyte insulin receptor by insulin binding and autophosphorylation. *Proceedings of the National Academy of Sciences, USA*, **81**, 7693–7.

Walaas, S. I., Horn, R. S., Adler, A., Albert, K. A. & Walaas, O. (1987). Insulin increases membrane protein kinase C activity in rat diaphragm. *FEBS Letters*, **220**, 311–18.

Wang, C. C., Goldfine, I. D., Fujita-Yamaguchi, Y., Gattner, H. G., Brandenburg, D. & DeMeyts, P. (1988). Negative and positive site-site interactions, and their modulation by pH, insulin analogs, and monoclonal antibodies, are preserved in the purified insulin receptor. *Proceedings of the National Academy of Sciences, USA*, **85**, 8400–4.

Wang, C. C., Hedo, J. A., Kahn, C. R., Saunders, D. T., Thamm, P. & Brandenburg, D. (1982). Photoreactive insulin derivatives: comparison of biologic activity and labeling properties of three analogues in isolated rat adipocytes. *Diabetes*, **31**, 1068–76.

Wang, L. H., Lin, B., Jong, S. J., Dixon, D., Ellis, L., Roth, R. A. & Rutter, W. J. (1987). Activation of the transforming potential of the human insulin receptor gene. *Proceedings of the National Academy of Sciences, USA*, **84**, 5725–9.

Waugh, S. M., DiBella, E. & Pilch, P. F. (1989). Isolation of a proteolytically-derived domain of the insulin receptor containing the major site of cross-linking/binding. *Biochemistry*, **28**, 2448–55.

Wedekind, F., Baer-Pontzen, K., Bala-Mohan, S., Choli, D., Zahn, H. & Brandenburg, D. (1989). Hormone binding site of the insulin receptor: analysis using photoaffinity-mediated avidin complexing. *Biological Chemistry Hoppe-Seyler*, **370**, 251–8.

White, M. F., Haring, H. U., Kasuga, M. & Kahn, C. R. (1984). Kinetic properties and sites of autophosphorylation of the partially purified insulin receptor from hepatoma cells. *Journal of Biological Chemistry*, **259**, 255–64.

White, M. F. & Kahn, C. R. (1989). Cascade of autophosphorylation in the β-subunit of the insulin receptor. *Journal of Cellular Biochemistry*, **39**, 429–41.

White, M. F., Livingston, J. N., Backer, J. M., Lauris, V., Dull, T. J., Ullrich, A. & Kahn, C. R. (1988 a). Mutation of the insulin receptor at tyrosine 960 inhibits signal transmission but does not affect its tyrosine kinase activity. *Cell*, **54**, 641–9.

White, M. F., Maron, R. & Kahn, C. R. (1985 a). Insulin rapidly stimulates tyrosine phosphorylation of a M_r-185000 protein in intact cells. *Nature, London*, **318**, 183–6.

White, M. F., Shoelson, S. E., Keutmann, H. & Kahn, C. R. (1988 b). A cascade of tyrosine autophosphorylation in the β-subunit activates the phosphotransferase of the insulin receptor. *Journal of Biological Chemistry*, **263**, 2969–80.

White, M. F., Stegmann, E. W., Dull, T. J., Ullrich, A. & Kahn, C. R. (1987). Characterization of an endogenous substrate of the insulin receptor in cultured cells. *Journal of Biological Chemistry*, **262**, 9769–77.

White, M. F., Takayama, S. & Kahn, C. R. (1985 b). Differences in the sites of phosphorylation of the insulin receptor *in vivo* and *in vitro*. *Journal of Biological Chemistry*, **260**, 9470–8.

Whittaker, J. & Okamoto, A. (1988). Secretion of soluble functional insulin receptors by transfected NIH 3T3 cells. *Journal of Biological Chemistry*, **263**, 3063–6.

Whittaker, J., Okamoto, A. K., Thys, R., Bell, G. I., Steiner, D. F. & Hofmann, C. A. (1987). High level expression of human insulin receptor cDNA in mouse NIH 3T3 cells. *Proceedings of the National Academy of Sciences, USA*, **84**, 5237–41.

Whittaker, J., Soos, M. A. & Siddle, K. (1990). Hybrid insulin receptors: molecular mechanisms of negative-dominant mutations in receptor-mediated insulin resistance. *Diabetes Care*, **13**, 576–81.

Wilden, P. A., Backer, J. M., Kahn, C. R., Cahill, D. A., Schroeder, G. J. & White, M. F. (1990). Characterization of the insulin receptor with an in vitro mutation at tyrosine 1146: evidence for separate insulin receptor signals regulating cellular metabolism and growth. *Proceedings of the National Academy of Sciences, USA*, **87**, 3358–62.

Wilden, P. A. & Pessin, J. E. (1987). Differential sensitivity of the insulin receptor kinase to thiol and oxidising agents in the absence and presence of insulin. *Biochemical Journal*, **245**, 325–31.

Wilden, P. A., Morrison, B. D. & Pessin, J. E. (1989 a). Relationship between insulin receptor subunit association and protein kinase activation: insulin-dependent covalent and Mn/MgATP-dependent non covalent association of $\alpha\beta$ heterodimeric insulin receptors into an $\alpha_2\beta_2$ heterotetrameric state. *Biochemistry*, **28**, 785–92.

Wilden, P. A., Morrison, B. D. & Pessin, J. E. (1989 b). Wheat germ agglutinin stimulation of $\alpha\beta$ heterodimeric insulin receptor β-subunit autophosphorylation by non-covalent association into an $\alpha_2\beta_2$ heterotetrameric state. *Endocrinology*, **124**, 971–9.

Wisher, M. H., Baron, M. D., Jones, R. H. & Sonksen, P. H. (1980). Photoreactive insulin analogues used to characterise the insulin receptor. *Biochemical and Biophysical Research Communications*, **92**, 492–8.

Witters, L. A. & Watts, T. D. (1988). An autocrine factor from Reuber hepatoma cells that stimulates DNA synthesis and acetyl-CoA carboylase. *Journal of Biological Chemistry*, **263**, 8027–36.

Witters, L. A., Watts, T. D., Gould, G. W., Lienhard, G. E. & Gibbs, E. M. (1988). Regulation of protein phosphorylation by insulin and an insulinomimetic oligosaccharide in 3T3-L1 adipocytes and Fao hepatoma cells. *Biochemical and Biophysical Research Communications*, **153**, 992–8.

Wong, E. C. C., Sacks, D. B., Laurino, J. P. & McDonald, J. M. (1988). Characteristics of calmodulin phosphorylation by the insulin receptor kinase. *Endocrinology*, **123**, 1830–6.

Woo, D. L., Fay, S. P., Griest, R., Coty, W., Goldfine, I. & Fox, C. F. (1986). Differential phosphorylation of the progesterone receptor by insulin, epidermal growth factor, and platelet-derived growth factor receptor tyrosine protein kinases. *Journal of Biological Chemistry*, **261**, 460–7.

Yarden, Y. & Ullrich, A. (1988). Growth factor receptor tyrosine kinases. *Annual Review of Biochemistry*, **57**, 443–78.

Yip, C. C., Hsu, H., Patel, R. G., Hawley, D. M., Maddux, B. A. & Goldfine, I. D. (1988). Localization of the insulin-binding site to the cysteine-rich region of the insulin receptor α-subunit. *Biochemical and Biophysical Research Communications*, **157**, 321–9.

Yip, C. C., Moule, M. L. (1983). Structure of the insulin receptor of rat adipocytes: the three interconvertible redox forms. *Diabetes*, **32**, 760–67.

Yip, C. C., Moule, M. L. & Yeung, C. W. T. (1982). Subunit structure of insulin receptor of rat adipocytes as demonstrated by photoaffinity labelling. *Biochemistry*, **21**, 2940–5.

Yip, C. C., Yeung, C. W. T. & Moule, M. L. (1978). Photoaffinity labelling of insulin receptor of rat adipocyte plasma membrane. *Journal of Biological Biochemistry*, **253**, 1743–5.

Yonezawa, K. & Roth, R. A. (1990). Various proteins modulate the kinase activity of the insulin receptor. *FASEB Journal*, **4**, 194–200.

Yoshimasa, Y., Seino, S., Whittaker, J., Kakehi, T., Kosaki, A., Kuzuya, H., Imura, H., Bell, G. I. & Steiner, D. F. (1988). Insulin-resistant diabetes due to a point mutation that prevents insulin proreceptor processing. *Science*, **240**, 784–7.

Yu, K. T. & Czech, M. P. (1984). Tyrosine phosphorylation of the insulin receptor β subunit activates the receptor associated tyrosine kinase activity. *Journal of Biological Chemistry*, **259**, 5277–86.

Yu, K. T. & Czech, M. P. (1986). Tyrosine phosphorylation of insulin receptor β-subunits activates the receptor tyrosine kinase in intact H-35 hepatoma cells. *Journal of Biological Chemistry*, **261**, 4715–22.

Yu, K. T., Khalaf, N. & Czech, M. P. (1987a). Insulin stimulates the tyrosine phosphorylation of a $M_r = 160000$ glycoprotein in rat adipocyte plasma membranes. *Journal of Biological Chemistry*, **262**, 7865–73.

Yu, K. T., Khalaf, N. & Czech, M. P. (1987b). Insulin stimulates a membrane-bound serine kinase that may be phosphorylated on tyrosine. *Proceedings of the National Academy of Sciences, USA*, **84**, 3972–6.

Yu, K. T., Werth, D. K., Pastan, I. H. & Czech, M. P. (1985). src kinase catalyzes the phosphorylation and activation of the insulin receptor kinase. *Journal of Biological Chemistry*, **260**, 5838–46.

Zhang, B., Tavare, J. M., Ellis, L. & Roth, R. A. (1991). The regulatory role of known tyrosine and autophosphorylation sites of the insulin receptor kinase domain: an assessment by replacement with neutral and negatively charged amino acids. *Journal of Biological Chemistry*, **266**, 990–6.

Zick, Y. (1989). The insulin receptor: structure and function. *Critical Reviews in Biochemistry and Molecular Biology*, **24**, 217–69.

Zick, Y., Grunberger, G., Podskalny, J. M., Moncada, V., Taylor, S. I., Gorden, P. & Roth, J. (1983a). Insulin stimulates phosphorylation of serine residues in soluble insulin receptors. *Biochemical and Biophysical Research Communications*, **116**, 1129–35.

Zick, Y., Grunberger, G., Rees-Jones, R. W. & Comi, R. J. (1985). Use of tyrosine-containing polymers to characterize the substrate specificity of insulin and other hormone-stimulated tyrosine kinases. *European Journal of Biochemistry*, **148**, 177–82.

Zick, Y., Kasuga, M., Kahn, C. R. & Roth, J. (1983 *b*). Characterization of the insulin-mediated phosphorylation of the insulin receptor in a cell-free system. *Journal of Biological Chemistry*, **258**, 75–80.

Zick, Y., Rees-Jones, R. W., Taylor, S. I., Gorden, P. & Roth, J. (1984). The role of antireceptor antibodies in stimulating phosphorylation of the insulin receptor. *Journal of Biological Chemistry*, **59**, 4396–400.

Zick, Y., Sagi-Eisenberg, R., Pines, M., Gierschik, P. & Spiegel, A. M. (1986). Multisite phosphorylation of the α subunit of transducin by the insulin receptor kinase and protein kinase C. *Proceedings of the National Academy of Sciences, USA*, **83**, 9294–7.

11

The ligand-dependent superfamily of transcriptional regulators

CHRISTOPHER K. GLASS,
JEFFREY M. HOLLOWAY AND
MICHAEL G. ROSENFELD

Introduction

Steroid and thyroid hormones coordinate diverse aspects of gene expression central to normal growth, development and homeostasis. These substances regulate patterns of gene expression by binding to specific intracellular receptor proteins that act to increase or decrease the rates of transcription of target genes. The cloning of cDNAs encoding these receptor proteins has revealed that all steroid and thyroid hormone receptors belong to a superfamily of transcription factors that includes nuclear receptors for the substances retinoic acid and vitamin D, as well as novel transcription factors that may be regulated by, as yet, unidentified ligands. The focus of this chapter will be to provide an overview of the scope of this superfamily of transcription factors, describe their domain structures and the mechanisms by which they interact with their respective ligands and target genes, and to delineate current thinking on the mechanisms by which they regulate gene transcription.

A superfamily of nuclear receptors related by similar DNA-binding domains

Fig. 11.1 illustrates representative examples of presently known members of the steroid/thyroid hormone receptor superfamily of transcription factors. Because many of the members of this group do not bind steroid or thyroid hormone ligands, it has also been referred to as the 'nuclear receptor' superfamily. The expression cloning of the glucocorticoid receptor provided the first opportunity to examine the primary structure and functional domains of a steroid hormone receptor (Hollenberg *et al.*, 1985). Its predicted amino acid sequence revealed a striking degree of

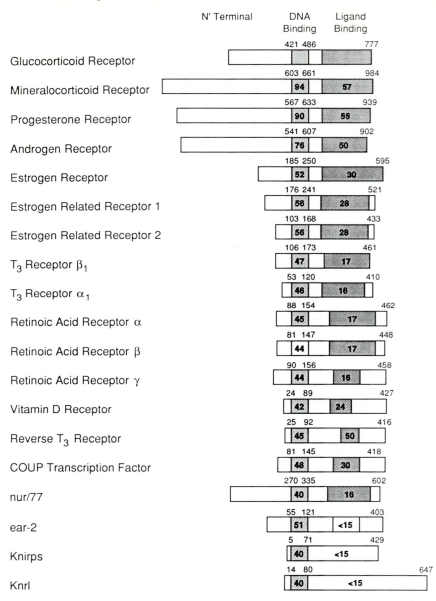

Fig. 11.1. Schematic representation of the structure of the ligand-dependent superfamily of transcriptional activators. Protein alignment is based on the highly-conserved cysteine-rich DNA binding domain (shaded box), with homology of each protein in the DNA binding and the C′ terminal, ligand-binding domains with the glucocorticoid receptor are shown for each receptor. Receptors characterized are the glucocorticoid

similarity to extensive regions of the product of the v-*erb* A protooncogene (Weinberger *et al.*, 1985), particularly in a cysteine-rich region later demonstrated to encode the DNA binding domain of each protein (see below). Subsequent cloning of the estrogen receptor (Green *et al.*, 1986) also demonstrated substantial sequence identity to both the v-*erb* A oncogene and the glucocorticoid receptor, providing the first evidence for the close relationship between receptors for distinct steroid ligands. That all steroid receptors are members of this family was demonstrated eventually by the cloning of cDNAs encoding the progesterone (Gronemyer *et al.*, 1987), mineralocorticoid (Arriza *et al.*, 1987) and androgen (Chang, Kokontis & Liao, 1988*a*, *b*; Lubahn *et al.*, 1988) receptors.

The initial observation of the sequence similarity between the glucocorticoid receptor and the v-*erb* A gene product raised the possibility that the cellular homologue of the v-*erb* A gene also encoded a ligand-dependent transcription factor. This hypothesis was proven to be correct by the identification of chicken and human c-*erb* A homologs that encoded high affinity receptors for thyroid hormones (Weinberger *et al.*, 1986; Sap *et al.*, 1986). The screening of cDNA libraries with fragments containing the DNA binding domain of *erb* A revealed much more diversity of the T3 receptor than had been expected from biochemical studies. Isoforms of c-*erb* A proteins were identified arising from two distinct genes, termed α (Thompson *et al.*, 1987; Benbrook & Pfahl, 1987) and β. The primary transcripts of each of these genes appear to be capable of being alternatively spliced, giving rise to α1, α2 (Izumo & Mahdavi, 1988; Mitsuhashi, Tennyson & Nikodem, 1988; Lazar *et al.*, 1988), and β1, β2 protein products (Hodin *et al.*, 1989) (see below).

The screening of cDNA libraries with DNA fragments containing *erb* A or estrogen receptor DNA binding domains also revealed the existence of novel classes of putative transcription factors. The use of estrogen receptor cDNA probes resulted in the identification of the estrogen receptor-related

---

(Hollenberg *et al.*, 1985), mineralocorticoid (Arriza *et al.*, 1987), progesterone (Gronemeyer *et al.*, 1987; Misrahi *et al.*, 1987), androgen (Chang, Kokontis & Liao, 1988*a*, *b*), estrogen (Green *et al.*, 1986), estrogen related 1 and 2 (Giguère *et al.*, 1988), β thyroid hormone (Weinberger *et al.*, 1986), α thyroid hormone (Sap *et al.*, 1986; Thompson *et al.*, 1987), α retinoic acid (Giguère *et al.*, 1987; Petkovich *et al.*, 1987), β retinoic acid (Brand *et al.*, 1988; Benbrook, Lernhardt & Pfahl, 1988), γ retinoic acid (Zelent *et al.*, 1989), Vit D (Baker *et al.*, 1988), reverse thyroid hormone (Miyajima *et al.*, 1988; Lazar *et al.*, 1989), COUP (Wang *et al.*, 1989; Miyajima *et al.*, 1988), Nu[r]-77 (Hazel, Nathans & Lau 1988; Millbrandt 1988), Ear-2 (Miyajima *et al.*, 1988), Knirps (Nauber *et al.*, 1988), and Knirps-related (Knrl) (Oro *et al.*, 1988*a*,*b*).

clones, ERR1 and ERR2 (Giguère *et al.*, 1988). Although the predicted amino acid sequences of these clones include putative ligand binding domains, specific ligands for the protein products of these clones have not been identified and their cellular functions remain unknown. The use of *erb* A DNA binding domain sequences as probes also resulted in the identification of additional novel members of the nuclear receptor superfamily. These included clones encoding receptors for the morphogen retinoic acid (Giguère *et al.*, 1987; Petkovich *et al.*, 1987), and clones termed *erb* A-related 2,3 and 7 (*ear*-2,3 and 7) (Miyajima *et al.*, 1988). *Ear*-3 has recently been demonstrated to encode COUP transcription factor (Wang *et al.*, 1989), a protein required for efficient expression of the chicken ovalbumin gene. *Ear*-2 and *ear*-7 have not as yet been assigned specific cellular functions.

It is of interest to note that a few members of the nuclear receptor superfamily have been identified by quite different cloning strategies. The second retinoic acid receptor gene to be found (RARβ) was identified on the basis of analysis of the site of an integration event of hepatitis B virus in a hepatocellular carcinoma cell line (de The *et al.*, 1987; Benbrook, Lernhardt & Pfahl, 1988; Brand *et al.*, 1988). A novel nuclear receptor cDNA, termed *nur*-77, was obtained by differential screening of a cDNA library constructed using mRNA isolated from serum-stimulated NIH3T3 cells (Hazel, Nathans & Lau, 1988). This murine clone is homologous to the rat NGF IB cDNA isolated by differential hybridization of a cDNA library constructed using mRNA derived from NGF-treated PC12 cells (Millbrandt, 1988). Although the specific cellular functions of the NGF IB/*nur*-77 protein are currently unknown, the fact that they were obtained from cells that had been stimulated to grow by NGF or serum, respectively, suggests that they play a role in regulatory events leading to cell division.

Although nearly all of the molecular characterization of nuclear receptors has been performed in mammalian and avian species, at least two representatives have been identified in *Drosophila*. These include the proteins encoded by the *knirps* gene (Nauber *et al.*, 1988), which is required for normal abdominal segmentation, and the *knirps*-related gene (Oro *et al.*, 1988a,b), which has not as yet been assigned a specific function. These findings show only that nuclear receptors are widely utilized among complex, multicellular organisms to perform functions similar to those of steroid receptors in vertebrate organisms.

When representatives of each nuclear receptor protein are aligned according to the conserved cysteine-rich DNA binding domain, a number of general features related to their domain structures become apparent

(Fig. 11.2). The DNA binding domain consists of 66 or 68 highly conserved residues that range from 98% to 40% identity with the amino acid sequence of the glucocorticoid receptor DNA binding domain. Nine cysteine residues and fourteen other amino acids are invariant among all members of this family. Examination of the C-terminal sequences reveals significant regions of amino acid identity over a span of approximately 200 amino acids for most of these proteins, but the degree of identity is much less than that observed in the DNA binding domain. Three regions show little or no amino acid similarity, the amino terminus, the extreme carboxyl terminus and the sequence of amino acids connecting the DNA binding and hormone binding domains. The amino terminus is also notable for the extreme variation in length among the various family members, ranging from perhaps as few as five amino acids in the case of the *knirps* protein to as many as 500 amino acids in the case of the mineralocorticoid receptor (Fig. 11.1). It has been proposed that these divergent regions function to mediate specific interactions between each type of receptor and other components of the transcriptional machinery (discussed in further detail below).

Interactions of nuclear receptor proteins with target genes

The structural and functional properties of the DNA binding domain have been most intensively studied in the contexts of the glucocorticoid and estrogen receptors. In the case of the glucocorticoid receptor, mutations within this region, but not in other regions of the receptor, abolish high affinity binding to their respective DNA response elements (Giguère *et al.*, 1986; Hollenberg *et al.*, 1987). Furthermore, deletion of sequences from both the amino and carboxyl terminal regions of either the glucocorticoid or estrogen receptors up to, but not including, the DNA binding domain yielded truncated receptors that continued to bind to DNA, although with somewhat decreased affinities (Kumar & Chambon, 1988; Tsai *et al.*, 1987). Demonstration that this domain was alone responsible for specific DNA binding was demonstrated when the DNA binding domain of the estrogen receptor was replaced with that of the glucocorticoid receptor. This chimeric receptor was then introduced into receptor-deficient cells and tested for its ability to activate transcription from estrogen and glucocorticoid responsive promoters. The chimeric estrogen receptor containing the glucocorticoid receptor DNA binding domain was found to selectively activate transcription from glucocorticoid-responsive promoters.

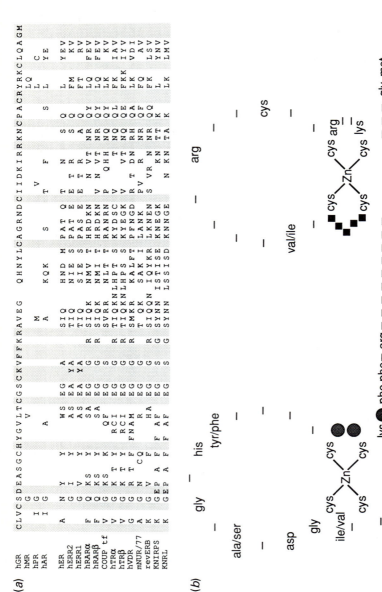

Fig. 11.2. DNA binding domain of the ligand-dependent transcription factor family. Panel (*a*). Comparison of the amino acid sequence within the highly conserved DNA binding domains of the nuclear receptors; shaded amino acids are invariant within the zinc-ligand-dependent finger superfamily of transcription factors. Panel (*b*). Identification diagram of the two zinc-fingers, showing invariant residues. Amino acid positions represented by the circle are implicated in determining sequence-specific DNA-binding of the glucocorticoid and estrogen receptors (Unesomo & Evans, 1989; Danielsen *et al.*, 1989). Those represented by the square are important in sequence-specificity and protein–protein interactions between receptor homodimers (Umesono & Evans, 1989).

An examination of the hormone response elements for the glucocorticoid and estrogen receptors reveals them to be closely related but distinct. A relatively large number of glucocorticoid and estrogen responsive genes have been subjected to sufficiently detailed analysis to allow identification of specific sequences mediating receptor binding and transactivation. Representative examples are illustrated in Fig. 11.3. Based on the numbering of 'consensus' sequences, base pairs at positions 2, 5, 6, 10, 11, 14 and 15 tend to be similar for both classes of response elements (Fig. 11.3). In many, but not all cases, response elements for estrogen and glucocorticoid receptors display a substantial degree of dyad symmetry. In fact, the *Xenopus* vitellogenin 2A estrogen response element exhibits perfect dyad symmetry and the palindromic variants of natural glucocorticoid response elements shown in Fig. 11.3 are capable of conferring glucocorticoid responsiveness to heterologous promoters. Conversion of the T/A base pairs of the vitellogenin ERE at positions 4 and 12 to A/T base pairs is sufficient to convert this element from an estrogen response element to a weak glucocorticoid response element (Martinez, Givel & Wahli, 1987; Klock, Strähle & Schütz, 1987). Consistent with the symmetry of their respective response elements, the glucocorticoid and estrogen receptors have been demonstrated to bind to representative examples of these sequences as dimers (Kumar & Chambon 1988; Tsai *et al.*, 1988).

Consensus response elements have not been established for the remaining members of the nuclear receptor superfamily. However, promoters that are responsive to glucocorticoids have also been demonstrated to be responsive to mineralocorticoids (Cato & Weinmann, 1988), progestins (Cato *et al.*, 1986; Von der Ahe *et al.*, 1985; Otten, Sanders & McKnight, 1988), and androgens (Cato, Henderson & Ponta, 1987). In many cases, the specific sequences mediating these effects have been demonstrated to overlap identically to those interacting with the glucocorticoid receptor, although base pair contacts between receptor and response element are not identical (Chalepakis *et al.*, 1988; Van der Ahe *et al.*, 1986).

Sequences mediating thyroid hormone responsiveness that have been identified in the rat growth hormone (Koenig *et al.*, 1987; Glass *et al.*, 1987) and alpha myosin heavy chain genes (Izumo & Mahdavi, 1988) and in regulatory sequences of the Maloney leukemia virus (MLV) (Sap *et al.*, 1989) are shown in Fig. 11.3. Although a consensus sequence cannot be drawn from this limited number of response elements, all contain at least one copy of a near perfect ERE half site (TGACCT). Furthermore, a variant of the rat growth hormone T3 response element that consists of a

Glucocorticoid/Mineralocorticoid/Progestin/Androgen Response Elements

```
MMTV    (-131)   TTTGGTATCAAATGTTCTGAT  (-105)
MMTV    (-184)   ATGGTTACAAACTGTTCTTAA  (-164)
hGH     (+90)    TTGGGCACAATGTGTTCCTGA  (+110)
hMTIID  (-265)   CCCGGTACACTGTGTCCTCCC  (-245)
TO      (-436)   ATATGCACAGCGAGTTCTAGT  (-416)
TO      (-1171)  TCCCTTTCATGATGTCCTGGC  (-1200)
TAT     (-2417)  GCAGGACTTGTTTGTTCTAGT  (-2449)
TAT     (-2512)  TGCTGTACAGGATGTTCTAGC  (-2492)
MSV     (-248)   AGCTGTTCCCATCTGTCTTGG  (-268)
MSV     (-174)   CTGGGGACCATCTGTTCTTGG  (-194)
```

```
       Position      1   5   10   15
                     |   |   |    |
       Consensus     GGTACAnnnTGTTCT
```

```
       Palindromic   AGAACAnnnTGTTCT
       variants      AGGACAnnnTGTCCT
```

Estrogen Receptor/ERR1/ERR2 Response Elements

```
Vit A2(-335)  GTCAGGTCACAGTGACCTGAT  (-317)
cVit  (-622)  TCCTGGTCAGCGTGACCGGAG  (-609)
rPrl (-1584)  TTTTTGTCACTATGTCCTAGA (-1560)
cOval (-176)  TTCAGGTAACAATGTGTTTTC  (-197)
cOval  (-49)  GGTGGGTCAATTCAGGCTATA  (-27)
      Position     1   5   10   15
                   |   |   |    |
      Consensus    AGGTCACnGTGACCT
                   T         A  TG
```

```
      Palindrome   AGGTCAnnnTGACCT
```

Thyroid Hormone/Retinoic Acid* Response Elements

```
*rGH (-187)GGTAAGATCAGGGACGTGACCGCAGG  (-162)
rαMHC(-124)    TTGGCTCTGGAGGTGACAGGAGG  (-146)
MLV   (331)    CAAGGACCTGAAATGACCCTG    (351)
```

```
      CONSENSUS     Not  established
*Palindromic        AGGTCATGACCT
 variants           AGGTCGCGACCT
```

COUP Transcription Factor

```
cOVAL  (-90)   CTATGGTGTCAAAGGTCAAACTT  (-68)
rINS II(-60)   TCCAGGGGTCAGGGGGGGGGTGG  (-38)
```

```
CONSENSUS     Not established
```

Fig. 11.3. For legend see facing page.

perfect palindrome of the TGACCT half site without the three base-pair spacer present in estrogen response elements binds the T3 receptor with a higher affinity than wild-type elements and confers a more substantial transcriptional response to heterologous promoters *in vivo* (Glass *et al.*, 1988).

Response elements for retinoic acid have not as yet been characterized in naturally occurring target genes, but the retinoic acid receptor has been demonstrated to activate transcription from T3-responsive promoters (Umesono *et al.*, 1988). These include the endogenous rat GH gene in pituitary tumor cells and transfected heterologous promoters containing the palindromic or wild type rat GH TREs. Sequence specificity for transcriptional activation does not completely overlap for T3 and retinoic acid receptors, however, because the retinoic acid receptor does not activate transcription from a heterologous promoter containing the alpha myosin heavy chain TRE, while the T3 receptor does.

COUP transcription factor has been demonstrated to bind to *cis*-active

---

Fig. 11.3. DNA response elements for the ligand-dependent receptor gene family. The glucocorticoid response elements are also response elements for other, related receptors. Some well-defined response elements include those in Maloney mammary tumor virus (MMTV)-LTR (Buetti & Kühnel, 1986; Scheidereit *et al.*, 1983; Ham *et al.*, 1988), the human growth hormone gene, hGH (Slater *et al.*, 1985), the human metallothionein II D gene, hMT II D (Karin *et al.*, 1984), the tryptophan oxygenase gene, TO (Danesch *et al.*, 1987), the tyrosine aminotransferase gene, TAT (Jantzen *et al.*, 1987), and the Maloney murine sarcoma virus, MSV (Miksicek *et al.*, 1986). The glucocorticoid family consensus sequence and palindromic variants are depicted below these response elements. The next panel shows, by comparison, estrogen and estrogen related response elements located in the *Xenopus* vitellogenin A2 gene, cVit (Cato *et al.*, 1988*a*, *b*), the rat prolactin gene, rPrl (Waterman *et al.*, 1988), and the chicken ovalbumin gene, coval (Tora *et al.*, 1988*a*, *b*). The estrogen receptor consensus and palindromic sequences are outlined below. The third panel shows a comparison of the known thyroid hormone/retinoic acid receptor response elements located in the rat growth hormone gene, rGH (Glass *et al.*, 1988; Koenig *et al.*, 1987), the rat alpha myosin heavy chain gene, rmhc (Izumo & Mahdavi, 1988), and the Maloney murine leukemia virus (MLV) (Sap *et al.*, 1989). Two variant palindromic thyroid hormone response elements known to bind thyroid hormone receptors with high affinity are outlined below. Asterisks (*) indicate sequences that do not activate transcription in response to retinoic acid in the presence of retinoic acid receptors. The lowest panel shows the two known binding elements for the COUP transcription factor, located in the chicken ovalbumin gene, cOVAL (Pastorcic *et al.*, 1986) and the rat gene, rINSII (Hwung *et al.*, 1988*a,b*).

elements derived from the chicken ovalbumin promoter (Pastorcic *et al.*, 1986) and the human insulin promoter (Hwung *et al.*, 1988*b*). Methylation and ethylation interference studies indicate that these two binding sequences are quite different from each other (Hwung *et al.*, 1988*a*). However, similar to the case of T3 response elements, they each contain a motif related to an ERE half size (TGACC). The mechanisms by which COUP transcription factor binds to these distinctly different sequences have not been established.

What are the structural features of the DNA binding domain that permit nuclear receptors to specifically recognize their appropriate response elements? The positions of the multiple cysteine residues suggested the possibility that the structure of this domain is coordinated by a metal ion, most likely zinc. This motif is similar to one initially described in the *Xenopus* transcription factor IIIA (TFIIIA), which contains repeating units of cysteine and histidine (Miller, McLachlan & Klug, 1985). This repeat structure has been proposed to fold into 'finger structures' that coordinate zinc ions and which appear to interact with a half turn of DNA. By analogy to the TFIIIA motif, the nuclear receptor DNA binding domain has been proposed to form two fingers in each of which the sulfur atoms of four cysteine residues provide electrons that form a tetrahedral coordination complex with a single zinc ion (Fig. 11.2*b*). That the invariant cysteine residues are critical for the DNA binding of estrogen, glucocorticoid and retinoic acid receptors has been demonstrated by point mutation of these residues (Severne *et al.*, 1988; Green & Chambon 1987; Hollenberg & Evans 1988; Glass *et al.*, 1989). Furthermore, overexpression of the DNA binding domain of the glucocorticoid receptor in bacteria has allowed application of EXAFS to confirm the tetrahedral coordination of zinc by cysteine residues within the finger structures (Freedman *et al.*, 1988).

Individual point mutations of the remaining amino acid residues of the glucocorticoid receptor DNA binding domain to glycine demonstrated that approximately one-third of these mutants were unable to bind GREs with high affinity and were unable to stimulate transcription (Hollenberg & Evans, 1988). One mutation was identified at the base of the first finger that displayed the remarkable property of retaining DNA binding activity but being unable to stimulate transcription. This result implies that the DNA binding domain of the glucocorticoid receptor must play more than a passive role in bringing other functional domains of the receptor into contact with the transcriptional apparatus.

The two hypothetical fingers of nuclear receptors are encoded by separate exons that are apparently evolutionarily distinct (Ponglikitnong-kol, Green & Chambon, 1988) and the individual fingers appear to be unable to bind to DNA with high affinity. The first fingers of the estrogen and glucocorticoid receptors have been demonstrated to be involved in recognizing the base pair differences that distinguish estrogen response elements from glucocorticoid response elements. Mutation of as few as two amino acids at the base of the first finger can alter the response element specificity of the glucocorticoid receptor such that it is capable of activating transcription from estrogen response elements (Danielsen, Hink & Ringold, 1989; Umesono & Evans 1989). Converse mutations of the estrogen receptor allow it to activate transcription from glucocorticoid responsive promoters (Mader *et al.*, 1989). These results suggest that the amino acids necessary for GRE or ERE-specific activation make direct contacts with the base-pairs that distinguish these two classes of response elements. The amino acids that allow GRE-specific interactions are also observed to be conserved among the other members of the steroid receptor superfamily that are capable of activating transcriptional form gluco-corticoid-responsive genes, i.e. the mineralocorticoid, progesterone and androgen receptors. Intriguingly, the amino acids required for specific recognition of EREs are highly conserved in all of the remaining nuclear receptors. Based on these observations, as well as other aspects of sequence divergence and differences in genomic organization, the glucocorticoid and estrogen receptors can be considered to be prototypes of two distinct nuclear receptor subclasses.

Ligands and nuclear receptor ligand binding domains

A detailed discussion of the biochemistry of steroid hormones, thyroid hormones and retinoids is beyond the scope of this chapter and the reader is directed elsewhere for a more extensive treatment of this subject (Wilson & Foster, 1981). It is of interest, however, to examine the diverse structural features of ligands that act to regulate the activities of nuclear receptors. Fig. 11.4 depicts prototypic steroid hormones, retinoic acid and the active forms of vitamin D and thyroid hormone. The compounds that bind to the glucocorticoid receptor subclass of receptors, namely glucocorticoids, mineralocorticoids, progestins and androgens, all contain a four-ring structure derived from cholesterol. The specific patterns of hydroxylations and the presence or absence of carbons 20 and 21 determine the receptor

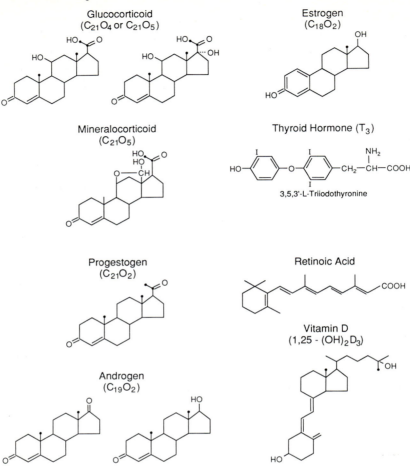

Fig. 11.4. Structures of the active forms of the ligands for the zinc-finger domain nuclear receptor family.

binding properties of these substances. The overall similarity of the structures of these substances is reflected by the high degree of amino acid similarity of the ligand binding domains of their respective receptors, which average 55% identity. Absolute receptor specificity is not observed for this class of ligands, e.g. some glucocorticoids bind with high affinity to both the glucocorticoid and mineralocorticoid receptors.

With the estrogen receptor subclass of nuclear receptors, the structures of known ligands are much more diverse and in the cases of T3 and retinoic acid, are biochemically unrelated. Consistent with this diversity of ligand structure, the ligand binding domains of the respective receptors for these substances exhibit much lower degrees of amino acid similarity with the glucocorticoid receptor class (less than 15 % in the case of the vitamin D receptor) and only 25 to 40 % when compared with other members of the estrogen receptor class. It is of interest to note that all of the putative receptors that have not as yet been identified to bind ligands are members of this class. Because they all retain at least some features of the ligand binding domains of the estrogen, T3, retinoic acid or vitamin D receptors, it seems likely that ligands for many of these proteins exist. It is also noteworthy that no intracellular ligands have been identified that bind to transcription factors that are unrelated to the steroid receptor superfamily. For this reason, it has been tempting to speculate that nuclear receptors mediating the effects of dioxin and negative transcriptional control by intracellular cholesterol metabolites will be members of this superfamily.

The interaction between ligand and receptor controls the receptor's transcriptional properties, presumably by causing a conformational change in the C-terminal domain. In the cases of the glucocorticoid, estrogen and progesterone receptors, regulation of transcriptional activity by ligand at least in part reflects regulation of their subcellular location. In the absence of ligand, these receptors are believed to be complexed to a heat shock protein, hsp 90, and are unable to bind to DNA (Groyer *et al.*, 1987; Joab *et al.*, 1984; Pratt *et al.*, 1988). Upon binding of hormone, the oligomeric complex dissociates, leaving the DNA binding domain of the receptor free to interact with chromatin. Amino acid sequences that mediate association with the hsp 90 are contained within the hormone binding domains, suggesting that the conformational change in receptor structure that occurs upon ligand binding directly results in complex dissociation. Sequences involved in ligand binding span up to 220 to 250 amino acids of the C-terminus of the glucocorticoid, progesterone, estrogen and thyroid hormone receptors (Pratt *et al.*, 1988) (Fig. 11.5).

The ligand binding domain has also been demonstrated to participate in protein–protein interactions that stabilize the formation of receptor dimers in the cases of the estrogen (Kumar & Chambon, 1988), glucocorticoid and thyroid hormone receptors (Holloway, unpublished observations; Tsai *et al.*, 1988) (Fig. 11.5). In addition, the ligand binding domain and the short stretch of amino acids between it and the DNA binding domain (the 'hinge region') have been suggested to contain nuclear localization signals that facilitate the transport of the glucocorticoid receptor to its functional location in the nucleus (Picard & Yamamoto, 1987).

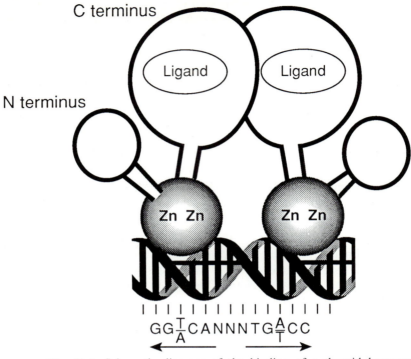

Fig. 11.5. Schematic diagram of the binding of a thyroid hormone receptor or estrogen receptor homodimer to a palindromic DNA response element. Binding of the 'zinc-fingers' (Shaded circles) occurs within the major groove of the DNA. Interactions within the C′ terminal ligand binding domain help to stabilize binding, and assist in determining the ultimate transcriptional effect of the receptor on binding of ligand. The transcriptional outcome depends on ligand spacing between the response element half-sites to which the receptor is bound and interactions at the C′ terminus of the protein.

Perhaps the most intriguing aspect of the ligand binding domain is its apparent ability to regulate directly the transcriptional properties of at least some of the nuclear receptors apart from its ability to regulate subcellular localization. In the case of the estrogen receptor, antiestrogenic compounds have been identified (e.g. TCT 164, 384), that allow dissociation of the receptor from hsp 90 and nuclear localization, but do not confer transcriptional activation. This observation suggests that a very specific conformation is required for transactivation that is not attained with antiestrogens.

Ligand appears to directly control the transcriptional properties of receptor most clearly in the case of the thyroid hormone receptor. Unlike the glucocorticoid, estrogen and progesterone receptors, the thyroid hormone receptor does not seem to interact with heat shock proteins, and biochemical studies have established that it is complexed to chromatin in the presence and absence of hormone (Samuels & Tsai, 1973). Surprisingly, introduction of T3 receptors into otherwise receptor negative cell lines results in the inhibition of transcription from promoters containing palindromic T3 response elements (Damm, Thompson & Evans, 1989). Administration of thyroid hormone dramatically reverses this inhibition and results in a marked transcriptional activation. Because the T3 receptor appears to be bound specifically to T3 response elements at all times, the positive and negative transcriptional effects of ligand presumably reflect alterations in T3 receptor conformation when bound to these sites.

Mechanisms of transcriptional activation by nuclear receptors

The DNA binding domain of nuclear receptors is necessary but usually not sufficient for efficient transcriptional activation of target genes. Removal of both N-terminal amino acids and the ligand binding domain results in truncated receptors that retain the ability to bind to DNA but that are much less active in stimulating transcription than wild-type receptors.

Mutational analysis of the glucocorticoid receptor has identified two amino acid regions that account for the majority of its transcriptional activity on the MMTV promoter. The principal activating region has been localized to an amino terminal sequence extending approximately from amino acids 90 to 290 (Giguère *et al.*, 1986; Hollenberg *et al.*, 1987; Godowski *et al.*, 1988*a*, *b*). Deletion of this sequence results in a loss of more than 90% of the receptor's transcriptional activity. A second region that contributes to active transactivation properties resides on the C-terminal side of the DNA binding domain from amino acids 526 to 556 (Hollenberg & Evans, 1988). These sequences have been demonstrated to be capable of transferring transcriptional activity to the DNA binding domains of other transcription factors (Godowski *et al.*, 1988*a*, *b*; Hollenberg & Evans, 1988).

The major transactivation domain exhibits a significant net negative charge (-20) over the 200 amino acids that comprise this domain. The presence of a high proportion of acidic residues in this region of the glucocorticoid receptor is consistent with the identification of acidic domains within the yeast transcription factors GCN4 and GAL4 indicating that they are essential for their ability to activate transcription (for reviewed see Ptashne, 1988). A specific structure does not appear to be required because the acidic regions of GAL4 and GCN4 can be replaced by other amino acid sequences that have no obvious features in common other than a net negative charge. A number of viral and mammalian transcription factors, such as VP16 and *fos*, have been found to contain highly acidic regions that appear to contribute to their transcriptional effects. These acidic regions have been demonstrated to transfer transcriptional properties to the DNA binding domains of other transcription factors, and the ability of acidic sequences to stimulate transcription in many contexts is well established (Ma & Ptashne, 1987).

Analysis of estrogen (Kumar *et al.*, 1986) and progesterone receptor (Gronemeyer *et al.*, 1987) mutants has revealed the existence of trans-activation domains that, in the case of the estrogen receptor, have been demonstrated to act when transferred to the DNA binding domain of another transcription factor (Webster *et al.*, 1988). In contrast to the location of the major transactivation domain of the glucocorticoid receptor, the principal activation domains of the estrogen and progesterone receptors appear to reside within sequences that overlap with the ligand binding domain. The ability of the C-terminal transcriptional activation domain of the estrogen receptor to function appears to depend in part on promoter context and the cell line in which receptor mutants are tested. While estrogen receptor mutants lacking the C-terminus were relatively inactive on a vitellogenin estrogen response element fused to the thymidine kinase (TK) promoter in HeLa cells, nearly identical mutants were constitutive activators on an estrogen-response prolactin promoter in MB231 cells (Waterman *et al.*, 1988).

Acidic sequences do not appear to account for the transcriptional properties of the transactivation domains of the estrogen and progesterone receptors, as these do not contain regions of significant net negative charge. It is possible that post-translational modifications of these receptors could result in the addition of negatively charged groups. Phosphorylation represents one such mechanism, and several nuclear receptors have been demonstrated to be substrates for various protein kinases. It has not yet been established whether phosphorylation alters the transcriptional properties of these receptors. In the case of the estrogen receptor, tyrosine phosphorylation has been suggested to be necessary for high affinity binding of estrogen (Migliaccio *et al.*, 1989), and would therefore also be implicated in its transactivation properties.

The mechanisms by which the transcriptional activation domains of nuclear receptors function to regulate transcription have not been established. Although substantial progress has been made in understanding the events leading to transcriptional initiation by RNA polymerase II on promoters containing the TATAA recognition motif, no unifying theory has been formulated that provides a general framework for the mechanisms by which sequence-specific DNA binding proteins act to regulate its activity at a distance. In the case of nuclear receptors, *cis*-active regulatory elements necessary for transcriptional activation commonly occur more than a kilobase away from the site of transcriptional initiation. It has been proposed that activation of transcription at a distance results from the formation of loops of DNA in which transcription factors bound to distant enhancer elements mediate protein–protein interactions with transcription factors bound to promoter sequences, resulting in increased transcription rates (Ptashne, 1986). Structures of this type have as yet been difficult to demonstrate in actively transcribed eukaryotic genes, however.

It is likely that the development of a successful model that accounts for the mechanisms by which nuclear receptors function to activate transcription will need to incorporate not only the effects of these proteins on chromatin structure but also their interactions with other transcription factors. Hormonally induced changes in the pattern of DNase I sensitivity have been recognized in regulatory elements of numerous genes, and are correlated with their transcriptional activities (Burch & Weintraub, 1983; Nyborg & Spindler, 1986; Kaye *et al.*, 1986). In the case of the rat growth hormone gene, in which T3 receptors are presumably bound to the gene in both the presence and absence of ligand, the appearance of a T3-dependent hypersensitive site implies that T3 addition results in changes in DNA structure by changing the conformation of the receptor. Whether the changes in DNA structure that result in alterations in the patterns of DNase I sensitivity are directly coupled to changes in transcription rates is not known.

More detailed studies of the potential role of chromatin structure in ligand-dependent transcriptional activation have been performed using episomal copies of MMTV–LTRvc (Richard, Foy & Hager, 1987; Cordingley, Riegel & Hager, 1987; Perlmann & Wrange, 1988). These studies suggest that, in the absence of glucocorticoids, the transcription factor NFI is unable to bind to its cognate recognition element due to the position of a nucleosome-like entity. Following hormonal stimulation, the region acquires DNase I hypersensitivity and the NFI binding site becomes protected as measured by an exonuclease III assay, suggesting that it has become occupied by NF1. Mutation of the NF1 binding site abolishes glucocorticoid dependent activation of transcription. These observations suggest that the glucocorticoid receptor acts to stimulate transcription from the MMTV–LTR in part by effecting a reorganization of nucleosome phasing, permitting the binding of additional transcription factors.

A series of studies indicate that nuclear receptors can interact with a large number of sequence-specific transcription factors resulting in cooperative transcriptional effects. For example, experiments using artificial promoters containing glucocorticoid response elements and various combinations of binding sites for other well-characterized transcription factors demonstrated synergistic activities of these elements as compared to the activities of promoters containing single elements (Jantzen *et al.*, 1987; Schüle *et al.*, 1988*a*,*b*). The synergistic activities were partly dependent on the relative spacing between elements, with optimal effects generally observed when elements were spaced at integrals of helical turns of B-form DNA. Whether these cooperative effects reflect protein–protein interactions between transcription factors or result from changes in chromatin structure have not been established.

Transcriptional inhibition by nuclear receptors

From a physiological standpoint, the function of nuclear receptors as inhibitors of gene expression is at least as important as their stimulatory roles. Studies of genes that are negatively regulated by intracellular ligands, including glucocorticoids and thyroid hormones, indicate that inhibition occurs at the level of transcription. Relatively few model systems have been developed for analysis of the mechanisms responsible for this activity, however, and a general model to account for transcriptional inhibition has not been formulated. Based on currently available data, nuclear receptors appear to be capable of inhibiting transcription by multiple independent mechanisms.

In one model system, inhibition of transcription by the estrogen receptor was demonstrated to occur by a mechanism that did not appear to require DNA binding in that mutant receptors lacking the DNA binding domain retained inhibitory function (Adler *et al.*, 1988). These experiments implicated amino acid sequences in the hinge region and/or ligand binding domain as being responsible for this activity and were interpreted as reflecting protein–protein interactions with limiting transcription factors. Recently, the estrogen receptor has been shown to be capable of inhibiting the activity of the progesterone and glucocorticoid receptors by a mechanism independent of DNA binding (Meyer *et al.*, 1989). Again, sequences in the C-terminus were implicated and suggested that both receptors require interactions with another transcription factor present at limiting concentrations that is required for maximal transcriptional activation.

Studies of glucocorticoid-dependent transcriptional inhibition have demonstrated a requirement for binding of the glucocorticoid receptors to DNA. Model systems have included the bovine prolactin gene (Sakai *et al.*, 1988), the alpha chorionic gonadotropin gene (Akerblom *et al.*, 1988; Oro, Hollenberg & Evans 1988*a*) and the pro-opiomelanocorticoid gene (Drouin *et al.*, 1987), all of which have been demonstrated to be negatively regulated by glucocorticoids *in vivo*. In each case, binding sites for the glucocorticoid receptor can be identified by DNase I footprinting studies of the respective promoters. These binding sites exhibit relatively poor sequence homologies to glucocorticoid response elements that mediate positive regulation by glucocorticoids and in general appear to overlap or reside in close proximity to *cis*-active sequences that bind other positively acting transcription factors. This observation has led to the hypothesis that negative regulation in these contexts results from the 'bumping' of adjacent factors (Fig. 11.6). Experiments utilizing mutant glucocorticoid receptors demonstrate the requirement for an intact DNA binding domain and are consistent with a model in which the binding of adjacent factors is sterically hindered. These experiments do not provide an explanation for why the glucocorticoid receptor is itself transcriptionally inactive in these contexts. The possibility has been raised that the different sequences that comprise 'negative' GREs result in the binding of the glucocorticoid receptor in a transcriptionally inactive form.

In the case of the thyroid hormone receptor, it is difficult to attribute transcriptional inhibition to a 'bumping' model because the receptor appears to be bound to its specific response elements in the presence or absence of ligand. Transcriptional inhibition of the human growth hormone promoter (which lacks the major T3 response element found in the rat growth hormone gene) and the thyroid stimulating hormone gene appears to be mediated by sequences located near the TATA recognition motif that bear resemblance to one-half of the palindromic T3 response element. These observations again suggest that sequences that mediate negative regulation are distinct from those that confer a positive transcriptional effect.

Studies of the DNA binding properties of the T3 receptor indicated that it could bind with high affinity not only to T3 response elements, but also to estrogen response elements (Glass *et al.*, 1988). As illustrated in Fig.

Positive Transcriptional Control

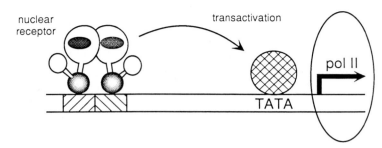

Negative Transcriptional Control

a. Bumping model

b. Silencer model

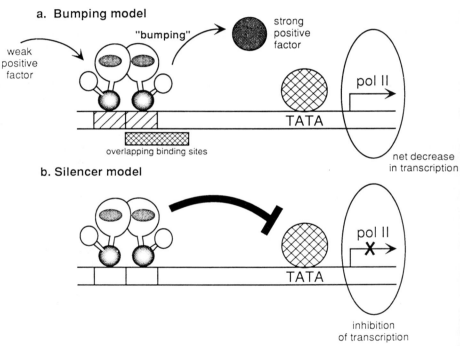

Fig. 11.6. Models for positive and negative transcriptional regulation by ligand-dependent transcription factors. Positive transcriptional control: binding of a nuclear receptor to a positive element leads to activation of transcription by DNA polymerase II. Negative transcription control: a. Bumping model. The binding of a nuclear receptor to a response element results in the displacement of another more active transcription factor binding to the identical or overlapping recognition element, resulting in a decrease in transcription initiation. b. Silencer model. Binding of nuclear receptor to a negative response element results in actual inhibition of polymerase II-initiated transcription. Such negative regulation appears to occur in the presence of ligand, but the effect of ligand is dependent on the sequence of the binding site and the precise identity of the *trans*-acting factors.

11.3, the palindromic versions of T3 and estrogen response elements are identical except for the presence or absence of three base pairs at the center of dyad symmetry. In contrast to the positive transcriptional effects observed on T3 response elements, T3 receptors have been observed to inhibit transcription on promoters containing estrogen response elements (Glass *et al.*, 1988). Because the relative positions of the half sites that form T3 and estrogen response elements, are different spatially and both response elements bind the T3 receptor as a dimer, it has been proposed that inhibition of transcription on estrogen response elements reflects the formation of a specific inhibitory conformation of the receptor on this element. Amino acid sequences responsible for ERE-dependent transcriptional inhibition have been localized to the C-terminal region of the receptor (Holloway *et al.*, 1990).

Further evidence that the T3 receptor contains an intrinsic inhibitory function that is regulated by ligand derives from experiments in which it was introduced into receptor-negative cell lines. In the absence of ligand, the T3 receptor was observed to strongly inhibit the transcription of a promoter containing two copies of the palindromic T3 response element (Damm, Thompson & Evans, 1989). Addition of T3 produced a dramatic reversal of this inhibition, resulting in a net stimulatory effect. The v-*erb* A oncogene, which does not bind thyroid hormone was observed to exert constitutively an inhibitory rather than a stimulatory effect.

Although the inhibition of transcription by the glucocorticoid and thyroid hormone receptors is considered to result from interactions with other sequence-specific DNA binding proteins or the DNA binding site, the possibility remains that these effects are due to alterations in chromatin structure that might be considered to be the inverse of those implicated in positive transcriptional control.

Regulation of tissue specific patterns of gene expression by nuclear receptors

Tissue distribution of nuclear receptors

The observation that nuclear receptors for glucocorticoids, mineralocorticoids, androgens, and progestins are capable of activating transcription from common response elements raises the apparent problem of the mechanisms by which these ligands exert specific effects on patterns of gene expression. For example, the expression of the endogenous genes encoding tyrosine aminotransferase, arginosuccinate synthetase and carbamoyl-phosphate synthetase I is positively regulated by glucocorticoids in the liver and in a rat hepatoma cell line (Fto2B-3 cells) that contains glucocorticoid receptors but lacks progesterone receptors. Stable introduction of the progesterone receptor in Fto2B-3 cells permits positive regulation of these endogenous genes by a specific progesterone agonist (R5020) as well as glucocorticoid (Strähle *et al.* 1989). This result is consistent with the idea that regulation of the tissue distribution of a given receptor represents a major mechanism for defining the target cells of its respective ligand.

In some cases, dynamic regulation of receptor concentration has been demonstrated to be required for the appropriate physiological actions of their respective ligands. The cyclical proliferations and desquamation of the uterine epithelium not only requires the coordination of the relative concentrations of estrogen and progesterone during the menstrual cycle, but also coordinate regulation of the expression of their respective receptors (for review see Clark, Schrader & O'Malley, 1985). In the absence of estrogen, the uterus is relatively insensitive to progesterone because of low progesterone receptor concentrations. As estrogen concentrations increase, estrogen-dependent genes are activated that are required for epithelial proliferation and synthesis of the progesterone receptor is stimulated. Subsequently, the combination of estrogen and rising progestin levels results in the altered pattern of gene expression necessary for the formation of the endometrial secretory phase. Under these conditions, progesterone acts to decrease the synthesis of both the progesterone and estrogen receptors, representing a form of autoregulation.

Regulation of ligand concentration

Although the biosynthesis of most nuclear receptor ligands occurs at sites that are distant from their target issues, local or intracellular metabolism of ligand can be critical to their appropriate physiological function, representing an additional point of regulation. One striking example of this is the presence of a gradient of retinoic acid across the anterior posterior axis of the developing limb bud (Thaller & Eichele, 1987). This gradient appears to provide essential spatial information to developing limb tissues, because its disruption results in aberrant limb formation The mechanisms responsible for the generation and maintenance of this gradient during limb development have not been established.

In the case of thyroid hormones, the major circulating form, L-thyroxine (T4), binds to the T3 receptor with a much lower affinity and is much less active in exerting a thyroid hormone effect than the metabolite, T3 (3,5,3′-L-triiodothyronine). T4 is converted to T3 within the cell by the action of a specific outer ring deiodinase. Alternatively, T4 is inactivated within the cell by an inner ring deiodinase. The relative activities of these two enzymes can thus exert a profound effect on intracellular thyroid hormone economy by regulating T3 levels. Analogous modifications may alter the intracellular concentrations of active steroid receptor metabolites, particularly in the case of androgens (Clark, Schrader & O'Malley, 1985).

Receptor isoforms

A second mechanism that is likely to modify the actions of thyroid hormones, progestins, retinoic acid, and potentially other ligands, is regulation of the pattern of expression of receptor isoforms. As previously mentioned, in the case of the T3 receptor, two distinct genes have been identified that encode high affinity receptors for T3, termed α and β. Although the tissue patterns of expression of these two genes overlap extensively, the β form of the receptor is preferentially expressed in the anterior pituitary and ventromedial nucleus of the hypothalamus (Bradley, Young & Weinberger, 1989). Because thyroid hormones negatively regulate the expression of the TSH and TRH genes in these locations, respectively, it has been proposed that the β T3 receptor is the relevant receptor for this action. Both the α and β forms of the T3 receptor activate transcription from promoters containing the rGH, rMHC and palindromic T3 response elements in transient transfection experiments (Damm et al., 1989; Izumo & Mahdavi, 1988; Forman et al., 1988; Thompson & Evans, 1989; Koenig et al., 1988). Further analysis will be required to determine whether the α and β receptors exert distinct effects on patterns of gene expression. These findings could have potential therapeutic implications if thyroid hormone analogues could be identified that were bound preferentially by one of the two receptors. Clinical situations in which manipulation of thyroid hormone action could be useful include management of cardiac arrhythmias and inotropic state, hyperlipidemic states, and obesity.

In addition to the presence of two distinct genes for T3 receptors, the primary transcripts of each of these genes have been demonstrated to be alternatively spliced. The α receptor message has been shown to be alternatively spliced in the C-terminal domain, giving rise to a protein which, in most reports, fails to bind T3 with high affinity (Izumo & Mahdavi, 1988; Mitsuhashi et al., 1988; Lazar et al., 1988; see also

Benbrook & Pfahl, 1987). In contrast to the actions of the hormone binding forms of the T3 receptor, the α2 variant of the T3 receptor fails to activate transcription from T3-responsive receptors in transient transfection experiments. The α2 variant retains the ability to bind to T3 response elements *in vitro* however, and when co-transfected with the hormone binding forms of the T3 receptor, blocks the transcriptional response to T3 (Koenig *et al.*, 1989). This effect is presumably due to competition for binding to T3 response elements and is similar to the actions of the v-*erb* A proto-oncogene. These observations suggest that hormone binding functions to regulate the transcriptional properties of the receptor following binding to DNA. Although the physiological role of the variant α receptor is not known, its inhibitory effect on T3 dependent transcription suggests that its functions is to modify thyroid hormone action. The alternatively spliced variant of the α T3 receptor is coexpressed with the hormone binding form in virtually all tissues examined, suggesting that the alternative splicing event does not exhibit tissue specificity (Bradley, Young & Weinberger, 1989).

At least three distinct genes have been identified that encode retinoic acid receptors, termed α, β and γ (Zelent *et al.*, 1989). Distinct functional activities for each of these receptors have not as yet been identified. All appear to activate transcription from promoters containing the rGH TRE or a palindromic variant TRE in transient transfection experiments. The tissue distributions of the α and β retinoic acid receptors overlap considerably, while the γ receptor is preferentially expressed in epithelial tissues (Zelent *et al.*, 1989). Based on this observation, it has been suggested that the γ form of the retinoic acid receptor mediates the effects of retinoids on the skin and other epithelial tissues.

Two isoforms have been identified for the chicken and human progesterone receptor, a 'B' form of 109 kD and an 'A' form of 79 kD. In contrast to the mechanisms responsible for the generation of receptor diversity of T3 and retinoic acid receptors, the A form of the progesterone receptor appears to arise from proteolysis or utilization of an alternative initiation site within coding sequence for the amino terminus (Gronemeyer *et al.*, 1987). Both isoforms are present in approximately equal concentrations in the cytosol of chicken oviduct tubular gland cells and human breast cancer T-47D cells (Gronemeyer, Harry & Chambon, 1983; Wei *et al.*, 1987). The B form of the progesterone receptor efficiently stimulates transcription from a promoter containing a GRE in transient transfection experiments, but not the A form (Tora *et al.*, 1988*a, b*). These observations suggest the possibility that the cell can regulate the transcriptional activity of the progesterone receptor by conversion of an active isoform to an inactive isoform. Whether or not this process is regulated and its physiologic significance remains to be established.

Modification of DNA binding properties of receptor via protein–protein interactions

There is a substantial body of evidence accumulating indicating that additional cellular factors play important roles in modifying the DNA binding properties of several members of the nuclear receptor gene family. Two non-histone proteins of 70 kD (NHP1) and 60 kD (NHP2) have been identified in a variety of cell types that increase the binding efficiency of the estrogen receptor to the vitellogenin ERE (Feavers *et al.*, 1987). The mechanisms responsible for this effect and the physiological importance of these proteins in estrogen action remain to be established.

Further evidence that cell-specific factors can influence the target gene specificity of the estrogen receptor has been provided by experiments with the chicken ovalbumin promoter. An estrogen response element has been identified in the vicinity of the TATA box that binds an estrogen receptor fusion protein (Tora *et al.*, 1988*a, b*). This element conferred a functional response to estrogen in chicken embryo fibroblasts but not HeLa cells, suggesting a requirement for a cell-specific transcription factor. Examination of the sequence of this response element revealed it to contain one half of the consensus palindromic estrogen response element (Fig. 11.3), because analogous 'half-sites' bind the estrogen receptor in vitro with much lower affinity than complete palindromic sequences, these results suggest that the cell specific factor present in chicken embryo fibroblasts operates at least in part by altering the affinity of the receptor to this sequence.

The chicken ovalbumin gene recognition sequence for COUP transcription factor also does not exhibit dyad symmetry, but contains two direct repeats of a sequence element (GTCAA) that is similar to estrogen or T3 receptor half sites (GGTCAn) (Pastorcic *et al.*, 1986). Binding of COUP transcription factor to the ovalbumin promoter is necessary, but insufficient to direct transcription from this promoter *in vitro*. Transcriptional activity of COUP *in vitro* is absolutely dependent on an additional factor, termed S300-II, that has been partially purified from HeLa cells (Tsai *et al.*, 1987). Although S300-II appears not to have intrinsic DNA binding properties, it stabilizes the interaction between COUP transcription factor and the ovalbumin promoter recognition sequence. The mechanisms for this interaction have not as yet been established. It will also be of interest to determine whether S300-II interacts with other members of the nuclear receptor superfamily.

As previously stated, the retinoic acid receptor has been demonstrated to activate transcription from thyroid hormone responsive promoters in transient experiments (Umesono *et al.*, 1988). Surprisingly, high affinity binding of retinoic acid receptors to T3 response elements *in vitro* required the presence of additional cellular factors. One of these factors has been identified to be the beta form of the T3 receptor (Glass *et al.*, 1989). Incubation of the β T3 receptor with the α retinoic acid receptor resulted in a cooperative increase in the binding of the retinoic acid receptor to palindromic and myosin heavy chain TREs but not a GH TRE. Cross-linking experiments indicate that this effect results from the formation of a heterodimer between the β T3 receptor and the α retinoic acid receptor. Experiments using retinoic acid and T3 receptor mutants indicate that heterodimer formation requires interactions within a conserved C-terminal domain of the two proteins (Fig. 11.7). The transcriptional properties of the heterodimer were observed to be sequence dependent; the two receptors were weakly cooperative on a palindromic TRE but the retinoic acid receptor inhibited T3 dependent transcription on the myosin heavy chain TRE. Additional factors also influence the DNA binding properties of the retinoic acid receptor because these activities can be purified from cellular extracts that do not contain T3 receptors.

Summary and future directions

The molecular cloning of cDNAs encoding steroid hormone receptors has resulted in the identification of a superfamily of related transcription factors. Many of these newly identified proteins bind ligands that are structurally and biochemically unrelated to steroid hormones and appear to play critical roles in cellular proliferation and development. Structure/function analysis of the estrogen and glucocorticoid receptors, and to a lesser extent other nuclear receptors, has allowed a description of the basic architecture of this class of transcription factors. Amino acid sequences that are responsible for ligand binding, sequence-specific DNA binding, interaction with heat shock proteins, nuclear localization and transcriptional activation have been characterized.

To understand the structural basis for the functional properties of nuclear receptors, it will be necessary to apply technologies that are capable of providing high resolution, three-dimensional information, i.e.

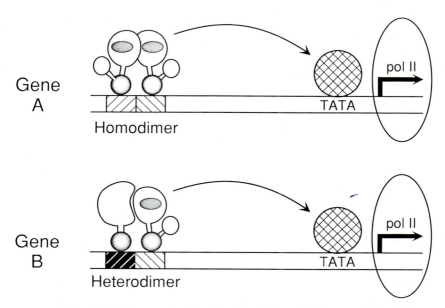

Gene
A

Gene
B

Fig. 11.7. Diagram showing that ligand-dependent transcription factors can bind as homo- or heterodimers to DNA recognition elements. Binding can occur as a homodimer leading to activation or inhibition of transcripts: information within the C′ terminal ligand binding domain assists in stabilization of the DNA–protein complex. DNA binding can also take place between two different nuclear receptors to form a heterodimer based on interaction of specific sequences in the C′ terminal ligand binding domain. The resulting heterodimers can positively or negatively regulate transcription dependent on the response element to which they are bound, the identity of the nuclear receptors in the protein complex, and the presence of ligand.

X-ray diffraction and nuclear magnetic resonance. In particular, it will be of great interest to determine the nature of conformational changes produced by ligand binding. In this regard, structural studies using ligands that produce distinct transcriptional effects may provide new insights into the mechanisms responsible for transcriptional activation. Studies of nuclear receptors bound to DNA response element sequences will provide information on the structural determinants that allow sequence-specific recognition. In addition, use of DNA sequences that confer different transcriptional responses to ligand may provide insights on how these sequences influence receptor structure.

An understanding of the mechanisms by which nuclear receptors act to increase or decrease transcription will require the development of *in vitro* transcription systems that are ligand-dependent. Recent developments in this area using COUP transcription factor and the glucocorticoid and progesterone receptors are encouraging, although ligand dependence has not as yet been established. Assay systems of this type are likely to yield new insights into the role of chromatin structure, the functional importance of receptor modifications (such as phosphorylation) and the biochemical basis for the effects of ligand binding. In addition, these assays will allow the evaluation of interactions with other factors that are necessary for transcriptional activity. The demonstration that steroid hormone receptors function in yeast (Metzger, White & Chambon, 1988; Schena & Yamamoto, 1988) should allow a complementary approach to this problem by permitting the application of the powerful techniques of yeast genetics.

The identification of novel members of the nuclear receptor superfamily that are implicated in fundamental aspects of cellular proliferation and differentiation provides an opportunity to increase our understanding of the regulatory hierarchies that control mammalian development. This pursuit will require the identification of the target genes regulated by nuclear receptor molecules and the determination of their cellular functions.

These directions pose significant technical and intellectual challenges. However, based on the intensity of effort and rapidity of progress that this field has witnessed in recent years, it is likely that many of these objectives will be accomplished in the not too distant future.

References

Adler, S., Waterman, M. L., He, X. & Rosenfeld, M. G. (1988). Steroid receptor-mediated inhibition of rat prolactin gene expression does not require DNA binding domain. *Cell*, **52**, 685–95.

Akerblom, I. W., Slater, E. P., Beato, J. D. & Mellon, P. L. (1988). Negative regulation by glucocorticoids through interference with a cAMP responsive enhancer. *Science*, **241**, 350–3.

Arriza, J. L., Weinberger, C., Cerelli, G., Glaser, T. M., Handelin, B. L., Housemann, D. E. & Evans, R. M. (1987). Cloning of human mineralo-corticoid receptor cDNA: structural and functional kinship with the glucocorticoid receptor. *Science*, **237**, 268–75.

Baker, A. R., McDonnell, D. P., Hughes, M., Crisp, T. M., Manglesdorf, D. J., Haussler, M. R., Pike, J. W., Shine, J. & O'Malley, B. W. (1988).Cloning and expression of full-length cDNA encoding human vitamin D receptor. *Proceedings of the National Academy of Sciences, USA*, **85**, 3294–8.

Benbrook, D., Lernhardt, E. & Pfahl, M. (1988). A new retinoic acid receptor identified from a hepatocellular carcinoma. *Nature*, London, **333**, 669–72.

Benbrook, D. & Pfahl, M. (1987). A novel thyroid hormone receptor encoded by a cDNA clone from a human testis library. *Science*, **238**, 788–91.

Bradley, D. J., Young, III, W. S. & Weinberger, C. (1989). Differential expression of α and β thyroid hormone receptor genes in rat brain and pituitary. *Proceedings of the National Academy of Sciences, USA*, **86**, 7250–4.

Brand, N., Petkovich, M., Krust, A., Chambon, P., de The, H., Marchio, A., Tiollais, P. & Dejean, A. (1988). Identification of a second human retinoic acid receptor. *Neuron*, **2**, 850–3.

Beutti, E. & Kühnel, B. (1986). Distinct sequence elements involved in the glucocorticoid regulation of the mouse mammary tumor virus promoter identified by linker scanning mutagenesis. *Journal of Molecular Biology*, **190**, 379–89.

Burch, J. B. E. & Weintraub, H. (1983). Temporal order of chromatin structural changes associated with activation of the major chicken vitellogenin gene. *Cell*, **33**, 65–76.

Carlstedt-Duke, J., Strömstedt, P. E., Rersson, B., Cederlund, E., Gustafsson, J. A. & Jörnvall, H. (1988). Identification of hormone-interacting amino acid residues within the steroid-binding domain of the glucocorticoid receptor in relation to other steroid hormone receptors. *Journal of Biological Chemistry*, **263**, 6842–6.

Carson, M. A., Tsai, M. J., Conneely, O. M., Maxwell, F. L, Clark, J. H., Dobson, A. D. W., Elbrecht, A., Toft, D. O., Schrader, W. T. & O'Malley, B. W. (1987). Structure–function properties of the chicken progesterone receptor A synthesized from complementary deoxyribonucleic acid. *Molecular Endocrinology*, **1**, 791–801.

Cato, A. C. B & Weinmann, J. (1988). Mineralcorticoid regulation of transfected mouse mammary tumor virus DNA in cultured kidney cells. *Journal of Cell Biology*, **106**, 2119–25.

Cato, A. C. B., Geisse, S., Wenz, M., Wesphal, H. M. & Beato, M. (1984). The nucleotide sequence recognized by the glucocorticoid receptor in the rabbit uteroglobin gene region are located far upstream from the initiation of transcription. *EMBO Journal*, **3**, 2731–6.

Cato, A. C. B., Heitlinger, E., Ponta, H., Klein-Hitpass, L., Ryffel, G. U., Bailly, A., Rauch, C. & Milgrom, E. (1988a). Estrogen and progesterone receptor-binding sites on the chicken vitellogenin II gene: synergism of steroid hormone action. *Molecular and Cell Biology*, **8**, 5323–30.

Cato, A. C. B., Miksicek, R., Schultz, G., Arnemann, J. & Beato, M. (1986). The hormone regulatory element of mouse mammary tumor virus mediates progesterone induction. *EMBO Journal*, **5**, 2237–40.

Cato, A. C. B., Henderson, D. & Ponta, H. (1987). The hormone response element of the mouse mammary tumor virus DNA mediates the progestin androgen induction of transcription in the proviral long terminal repeat region. *EMBO Journal*, **6**, 363–8.

Cato, A. C. B., Skroch, P., Weinmann, J., Butkeraitis, P. & Ponta, H. (1988b). DNA sequences outside the receptor binding sites differentially modulate the responsiveness of the mouse mammary tumor virus promoter to various steroid hormones. *EMBO Journal*, **7**, 1403–7.

Chalepakis, G., Arnemann, J., Slater, E., Brüller, H. -J., Gross, B. & Beato, M. (1988). Differential gene activation by glucocorticoids and progestins through the hormone regulatory element of mouse mammary tumor virus. *Cell*, **53**, 371–82.

Chang, C., Kokontis, J. & Liao, S. (1988*a*). Molecular cloning of the human and rat cDNA encoding androgen receptors. *Science*, **240**, 324–6.

Chang, C., Kokontis, J. & Liao, S. (1988*b*). Structural analysis of complementary DNA and amino acid sequences of human and rat androgen receptors. *Proceedings of the National Academy of Sciences, USA*, **85**, 7211–15.

Clark, J. H., Schrader, W. T. & O'Malley, B. W. (1985). Mechanisms of steroid hormone action. In: *Williams Textbook of Endocrinology*. Wilson, J. D. & Foster, D. W., eds. W. B. Saunders Co., Philadelphia.

Cook, R. G., Conneely, O. M., Sullivan, W. P., Toft, D. O., Birnbaumer, M., Schrader, W. T. & O'Malley, B. W. (1986). Molecular cloning of the chicken estrogen receptors. *Science*, **233**, 767–70.

Cordingley, M. G. & Hager, G. L. (1988). Binding of multiple factors to the MMTV promoter in crude and fractionated nuclear extracts. *Nucleic Acids Research*, **16**, 609–30.

Cordingley, M. G., Riegel, A. T. & Hager, G. L. (1987). Steroid-dependent interaction of transcription factors with the inducible promoter of mouse mammary tumor virus *in vivo*. *Cell*, **48**, 261–70.

Damm, K., Thompson, C. C. & Evans, R. M. (1989). Protein encoded by v-*erb*A functions as a thyroid-hormone receptor antagonist. *Nature*, London, **339**, 593–7.

Danesch, U., Gloss, B., Schmid, W., Schütz, G. & Renkawitz, R. (1987). Glucocorticoid induction of the rat tryptophan oxygenase gene is mediated by two widely separated glucocorticoid-responsive elements. *EMBO Journal*, **6**, 625–30.

Danielsen, M., Hink, L. & Ringold, G. M. (1989). Two amino acids within the knuckle of the first zinc finger specify DNA response element activation by the glucocorticoid receptor. *Cell*, **57**, 1131–8.

Darbré, P., Page, M. & King, R. J. B. (1986). Androgen regulation by the long terminal repeat of mouse mammary tumor virus. *Molecular and Cell Biology*, **6**, 2847–54.

de The, H., Marchio, A., Tiollais, P. & Dejean, A. (1987). A novel steroid thyroid hormone receptor-related gene inappropriately expressed in human hepatocellular carcinoma. *Nature*, London, **330**, 667–70.

Drouin, J., Charron, J., Gagner, J. P., Jeannotte, L., Nemer, M., Plante, R. K. & Wrange, Ö. (1987). The pro-opiomelanocortin gene: a model for negative regulation of transcription by glucocorticoids. *Journal of Cell Biochemistry*, **35**, 293–304.

Feavers, I. M., Jiricny, J., Moncharmont, B., Saluz, H. P. & Jost, J. P. (1987). Interaction of two non-histone proteins with the estradiol response element of the avian vitellogenin gene modulates the binding of estradiol–receptor complex. *Proceedings of the National Academy of Sciences, USA*, **84**, 7453–7.

Forman, B. M., Yang, C., Stanley, Frederick, Casanova, J. & Samuels, H. H. (1988). c-*erb*A protooncogenes mediate thyroid hormone-dependent and independent regulation of the rat growth hormone and prolactin genes. *Molecular Endocrinology*, **2**, 902–11.

Freedman, L. P., Luisi, B. F., Korszun, Z. R., Basavappa, R., Sigler, P. B. & Yamamoto, K. R. (1988). The function and structure of the metal coordination sites within the glucocorticoid receptor DNA binding domain. *Nature*, London, **311**, 543–6.

Giguère, V., Hollenberg, S. M., Rosenfeld, M. G. & Evans, R. M. (1986). Functional domains of the human glucocorticoid receptor. *Cell*, **46**, 645–52.

Giguère, V., Ong, E. S., Segui, P. & Evans, R. M. (1987). Identification of a receptor for the morphogen retinoic acid. *Nature*, London, **330**, 624–9.

Giguère, V., Yang, N., Segui, P. & Evans, R. M. (1988). Identification of a new class of steroid hormone receptors. *Nature*, London, **331**, 91–4.

Glass, C. K., Franco, R., Weinberger, C., Albert, V. R., Evans, R. M. & Rosenfeld, M. G. (1987). A c*erb*-A binding site in rat growth hormone gene mediates transactivation by thyroid hormone. *Nature*, London, **329**, 738–41.

Glass, C. K., Holloway, J. M., Devary, O. V. & Rosenfeld, M. G. (1988). The thyroid hormone receptor binds with opposite transcriptional effects to a common sequence motif in thyroid hormone and estrogen response elements. *Cell*, **54**, 313–23.

Glass, C. K., Lipkin, S. M., Devary, O. V., Umesono, K., Evans, R. M. & Rosenfeld, M. G. (1989). Positive and negative regulation of gene transcription by a retinoic acid/thyroid hormone receptor heterodimer. *Cell*, in press.

Godowski, P. J., Picard, D. & Yamamoto, K. R. (1988*a*). Signal transduction and transcriptional regulation by glucocorticoid receptor–LexA fusion proteins. *Science*, **241**, 812–16.

Godowski, P. J., Rusconi, S., Miesfeld, R. & Yamamoto K. R. (1988*b*). Glucocorticoid receptor mutants that are constitutive activators of transcriptional enhancement. *Nature*, London, **325**, 365–8.

Green, S. & Chambon, P. (1987). Oestradiol induction of a glucocorticoid-responsive gene by chimeric receptor. *Nature*, London, **325**, 75–8.

Green, S., Kumar, B., Thenlaz, I., Wahli, W. & Chambon. P. (1988). The N-terminal DNA binding zinc finger of the oestrogen and glucocorticoid receptors determine target gene specificity. *EMBO Journal*, **7**, 3037–44.

Green, S., Walter, P., Kumar, V., Krust, A., Bornet, J. M., Argos, P. & Chambon, P. (1986). Human estrogen receptor cDNA: sequence, expression and homology to v-*erb*A. *Nature*, London, **320**, 134–9.

Gronemeyer, H., Harry, P. & Chambon, P. (1983). Evidence for two structurally related progesterone receptors in chick oviduct cytosol. *FEBS Letters*, **156**, 287–92.

Gronemeyer, H., Turcotte, B., Quirin-Stricker, C., Bocquel, M. T., Meyer, M. E., Krozowski, Z., Jeltsch, J. M., Lerouge, T., Garnier, J. M. & Chambon, P. (1987). The chicken progesterone receptor: sequence, expression and functional analysis. *EMBO Journal*, **6**, 3985–94.

Groyer, A., Schweizer-Groyer, G., Cadepond, F., Mariller, M. & Baulieu, E. E. (1987). Antiglucocorticosteroid effects suggest why steroid hormone is required for receptors to bind DNA *in vivo* but not *in vitro*. *Nature*, London, **328**, 624–6.

Ham, J., Thomson, A., Needham, M., Webb, P. & Parker, M. (1988). Characterization of response elements for androgens, glucocorticoids and progestins in mouse mammary tumor virus. *Nucleic Acids Research*, **16**, 5263–77.

Hazel, T. G., Nathans, D. & Lau, L. F. (1988). A gene inducible by serum growth factors encodes a member of the steroid and thyroid hormone receptor superfamily. *Proceedings of the National Academy of Sciences, USA*, **85**, 8444–8.

Hodin, R. A., Lazar, M. A., Wintman, B. I., Darling, D. S., Koenig, R. J., Larsen, P. R., Moore, D. D. & Chin, W. W. (1989). Identification of a thyroid hormone receptor that is pituitary-specific. *Science*, **244**, 76–9.

Hollenberg, S. M. & Evans, R. M. (1988). Multiple and cooperative transactivation domains of the human glucocorticoid receptor. *Cell*, **55**, 899–906.

Hollenberg, S. M., Giguère, V., Segui, P. & Evans, R. M. (1987). Colocalization of DNA binding and transcriptional activation functions in the human glucocorticoid receptor. *Cell*, **49**, 39–46.

Hollenberg, S. M., Weinberger, C., Ong, E. S., Cerelli, G., Oro, A., Lebo, R., Thompson, E. B., Rosenfeld, M. G. & Evans, R. M. (1985). Primary structure and expression of a functional human glucocorticoid receptor cDNA. *Nature*, London, **318**, 635–41.

Holloway, J. M., Glass, C. K., Adler, S., Nelson, C. A. & Rosenfeld, M. G. (1990). The C-terminal interaction domain of the thyroid hormone receptor confers the ability of the DNA site to dictate positive or negative transcriptional activity. *Proceedings of the National Academy of Sciences, USA*, **871**, 8160–4.

Hwung, Y. P., Crowe, D. T., Wang, L. H., Tsai, S. Y. & Tsai, M. J. (1988a). The COUP transcription factor binds to an upstream promoter element of the rat insulin II gene. *Molecular and Cell Biology*, **8**, 2070–7.

Hwung, Y. P., Wang, L. H., Tsai, S. Y. & Tsai, M. J. (1988b). Differential binding of the chicken ovalbumin upstream promoter (COUP) transcription factor to two different promoters. *Journal of Biological Chemistry*, **263**, 13460–4.

Izumo, S. & Mahdavi, V. (1988). Thyroid hormone receptor α isoforms generated by alternative splicing differentially active myosin HC gene transcription. *Nature*, London, **334**, 539–42.

Jantzen, H. M., Strähle, U., Gloss, B., Steward, G., Schmid, W., Boshart, M., Miksicek, R. & Shütz, G. (1987). Cooperativity of glucocorticoid response elements located far upstream of the tyrosine aminotransferase gene. *Cell*, **49**, 29–38.

Joab, I., Radanyi, C., Renoir, M., Buchou, T., Catelli, M. G., Binaut, N., Mester, J. & Baulieu, E. S. (1984). Common non-hormone binding component in non-transformed chick oviduct receptor of four steroid hormones. *Nature*, London, **308**, 850–3.

Karin, M., Haslinger, A., Holtgreve, A., Richards, R. I., Krauter, P., Westphal, H. M. & Beato, M. (1984). Characterization of DNA sequences through which cadmium and glucocorticoid hormones induce human metallothionein IIA. *Nature*, London, **308**, 513–19.

Kaye, J. S., Pratt-Kaye, S., Bellard, M., Dretzen, G., Bellard, F. & Chambon, P. (1986). Steroid hormone dependence of four DNAse I hypersensitive regions located within the 7000 bp5′-flanking segment of the ovalbumin gene. *EMBO Journal*, **5**, 277–85.

Klein-Hitpass, Schorpp, M., Wagner, U., & Ryffel, G. U. (1986). An estrogen-responsive element derived from the 5′flanking region of the *Xenopus* vitellogenin A2 gene functions in transfected human cells. *Cell*, **46**, 1053–61.

Klock, G., Strähle, U. & Schütz, G. (1987). Estrogen and glucocorticoid elements are closely related but distinct. *Nature*, London, **329**, 734–6.

Koenig, R. J., Brent, G. A., Warne, R. L., Larsen, P. R. & Moore, D. D. (1987). Thyroid hormone receptor binds to a site in the rat growth hormone promoter required for induction by thyroid hormones. *Proceedings of the National Academy of Sciences, USA*, **84**, 5670–4.

Koenig, R. J., Lazar, M. A., Hodin, R. A., Brent, G. A., Larsen, P. R., Chin, W. W. & Moore, D. D. (1989). Inhibition of thyroid hormone action by a non-hormone binding c-*erb* A protein generated by alternative mRNA splicing. *Nature*, London, **332**, 659–63.

Koenig, R. J., Warne, R. L., Brent, G. A., Harvey, J. W., Larsen, P. R. & Moore, D. D. (1988). Isolation of a cDNA encoding a biologically active thyroid hormone receptor. *Proceedings of the National Academy of Sciences, USA*, **65**, 5031–5.

Kumar, V. & Chambon, P. (1988). The estrogen receptor binds tightly to its responsive element as a ligand-induced homodimer. *Cell*, **55**, 145–56.

Kumar, V., Green, S., Staub, A. & Chambon, P. (1986). Localization of the estradiol binding and putative DNA binding domains of the human estrogen receptor. *EMBO Journal*, **5**, 2231–6.

Lazar, M. A., Hodin, R. A., Darling, D. S. & Chin, W. W. (1988). Identification of a rat c-*erb*Aα-related protein which binds deoxyribonucleic acid but does not bind thyroid hormone. *Molecular Endocrinology*, **2**, 893–901.

Lazar, M. A., Hodin, R. A., Darling, D. S. & Chin, W. W. (1989). A novel member of the thyroid/steroid hormone receptor family is encoded by the opposite strand of the rat c-*erb*Aα transcriptional unit. *Molecular and Cell Biology*, **9**, 1128–36.

Lubahn, D. B., Joseph, D. R., Sullivan, P. M., Willard, H. F., French, F. S. & Wilson, E. M. (1988). Cloning of the human androgen receptor cDNA and localization of the X-chromosome. *Science*, **340**, 327–30.

Ma, J. & Ptashne, M. (1987). A new class of yeast transcriptional activators. *Cell*, **51**, 113–19.

Mader, S., Kumar, V., de Verneuil, H. & Chambon, P. (1989). Three amino acids of the estrogen receptor are essential to its ability to distinguish an estrogen from a glucocorticoid-responsive element. *Nature*, London, **338**, 271–4.

Martinez, E., Givel, F. & Wahli, W. (1987). The estrogen-responsive element as an inducible enhancer: DNA sequence requirements and conversion to a glucocorticoid-responsive element. *EMBO Journal*, **6**, 3719–27.

McDonnell, D. P., Mangelsdorf, D. J., Pike, J. W., Haussler, M. R. & O'Malley, B. W. (1987). Molecular cloning of cDNA encoding the avian receptor for vitamin D. *Science*, **235**, 1214–17.

Metzger, D., White, J. H. & Chambon, P. (1988). The human estrogen receptor functions in yeast. *Nature*, London, **334**, 31–6.

Meyer, M. E., Gronemeyer, H., Turcotte, B., Bocquel, M. T., Tasset, D. & Chambon, P. (1989). Steroid hormone receptors compete for factors that mediate their enhancer function. *Cell*, **57**, 433–42.

Migliaccio, A., Di Domenico, M., Green, S., de Falco, A., Kajtaniak, E. L., Blasi, F., Chambon, P. & Auricchio, F. (1989). Phosphorylation on tyrosine of *in vitro* synthesized human estrogen receptor activates its hormone binding. *Molecular Endocrinology*, **3**, 1061–9.

Miksicek, R., Heber, A., Schmid, W., Danesch, Y., Posseckert, G., Beato, M. & Schütz, G. (1986). Glucocorticoid responsiveness of the transcriptional enhancer of Moloney murine sarcoma virus. *Cell*, **46**, 283–90.

Millbrandt, J. (1988). Nerve growth factor induces a gene homologous to the glucocorticoid receptor gene. *Neuron*, **1**, 183–8.

Miller, J., McLachlan, A. D. & Klug, A. (1985). Repetitive zinc-binding domains in the protein transcription factor IIIA from *Xenopus* oocytes. *EMBO Journal*, **4**, 1609–14.

Misrahi, M., Atger, M., D'Auriol, L., Loosfelt, H., Meriel, C., Friedlansky, F., Guichon-Mantel, A., Galibert, F. & Milgrom, E. (1987). Complete amino acid sequence of the human progesterone receptor deduced from cloned cDNA. *Biochemical and Biophysical Research Communications*, **143**, 740–8.

Mitsuhashi, T., Tennyson, G. E. & Nikodem, V. M. (1988). Alternative splicing generates messages encoding rat c-*erb*A proteins that do not bind thyroid hormone. *Proceedings of the National Academy of Sciences, USA*, **85**, 5804–8.

Miyajima, N., Horiuchi, R., Shibuya, Y., Fukushige, S., Matsubara, K., Toyoshima, K. & Yamamoto, T. (1989). Two *erb*A homologs encoding proteins with different T$_3$binding capacities are transcribed from opposite DNA strands of the same genetic locus. *Cell*, **57**, 31–9.

Miyajima, N., Kadowaki, T., Fukushige, S., Shimizu, S., Semba, K., Yamanashi, Y., Matsubara, K., Toyoshima, K. & Yamamoto, T. (1988). Identification of two novel members of *erb*A superfamily by molecular cloning: the gene products of the two are highly related to each other. *Nucleic Acids Research*, **16**, 11057–74.

Nauber, U., Pankratz, M. J., Kienlin, A., Seifert, E., Klemm, U. & Jäckle, H. (1988). Abdominal segmentation of the *Drosophila* embryo requires a hormone receptor-like protein encoded by the gap gene *knirps*. *Nature, London*, **336**, 489–92.

Nyborg, J. K. & Spindler, S. R. (1986). Alterations in chromatin structure accompany thyroid hormone induction of growth hormone gene transcription. *Journal of Biological Chemistry*, **262**, 5685–8.

Oro, A. E., Hollenberg, S. M. & Evans, R. M. (1988*a*). Transcriptional inhibition by a glucocorticoid receptor-beta-galactogidase fusion protein. *Cell*, **55**, 1109–14.

Oro, A. E., Ong, E. S., Margolis, J. S., Posakony, J. W., McKeown, M. & Evans, R. M. (1988*b*). The *Drosophila* gene *knirps-related* is a member of the steroid-receptor gene superfamily. *Nature, London*, **336**, 493–6.

Otten, A. D., Sanders, M. M. & McKnight, G. S. (1988). The MMTV LTR promoter is induced by progesterone and dihydrotestosterone but not by estrogen. *Molecular Endocrinology*, **2**, 143–7.

Pastorcic, M., Wang, H., Elbrecht, A., Tsai, S. Y., Tsai, M. J. & O'Malley, B. W. (1986). Control of transcription initiation in vitro requires binding of a transcription factor to the distal promoter of the ovalbumin gene. *Molecular and Cell Biology*, **6**, 2784–91.

Payvar, F., DeFranco, D., Firestone, G. L., Edgar, V., Wrange, Ö., Okret, S., Gustafsson, J. A. & Yamamoto, K. R. (1983). Sequence-specific binding of glucocorticoid receptor to MTV–DNA at sites within and upstream of the transcribed region. *Cell*, **35**, 381–92.

Perlmann, T. & Wrange, Ö. (1988). Specific glucocorticoid receptor binding to DNA reconstituted in a nucleosome. *EMBO Journal*, **7**, 3073–9.

Petkovich, M., Brand, N. J., Krust, A. & Chambon, P. (1987). A human retinoid acid receptor belongs to the family of nuclear receptors. *Nature, London*, **330**, 444–50.

Picard, D. & Yamamoto, K. R. (1987). Two signals mediate hormone-dependent nuclear localization of the glucocorticoid receptor. *EMBO Journal*, **6**, 333–40.

Ponglikitnongkol, M., Green, S. & Chambon, P. (1988). Genomic organization of the human estrogen receptor gene. *EMBO Journal*, **7**, 3385–8.

Pratt, W. B., Jolly, D. J., Pratt, D. V., Hollenberg, S. M., Giguère, V., Cadepon, F. M., Schweizer-Groyer, G., Cartelli, M. G., Evans, R. M. & Baulieu, E. E. (1988). A region in the steroid-binding domain determines formation of the non-DNA binding, as glucocorticoid receptor complex. *Journal of Biological Chemistry*, **263**, 267–73.

Ptashne, M. (1986). Gene regulation by proteins acting near and at a distance. *Nature*, London, **335**, 683–9.

Ptashne, M. (1988). How eukaryotic transcriptional activators work. *Nature*, London, **335**, 683–9.

Richard, Foy, H. & Hager, G. L. (1987). Sequence-specific positioning of nucleosomes over the steroid-inducible MMTV promoter. *EMBO Journal*, **6**, 2321–8.

Rusconi, S. & Yamamoto, K. R. (1987). Functional dissection of the hormones and DNA binding activities of the glucocorticoid receptor. *EMBO Journal*, **6**, 1309–15.

Sakai, D. D., Helms, S., Carlstedt-Duke, J., Gustafsson, J. A., Rottman, F. M. & Yamamoto, K. R. (1988). Hormone-mediated repression of transcription: a negative glucocorticoid response element from the bovine prolactin gene. *Genes and Development*, **2**, 1144–54.

Samuels, H. H. & Tsai, J. S. (1973). Thyroid hormone action in cell culture: demonstration of nuclear receptors in intact cells and isolated nuclei. *Proceedings of the National Academy of Sciences, USA*, **70**, 3488–92.

Sap, J., Muñoz, A., Schmitt, J., Stunnenberg, H. & Vennström, B. (1989). Repression of transcription mediated at a thyroid hormone response element by the v-*erb*-A oncogene product. *Nature*, London, **340**, 242–4.

Sap, J., Muñoz, A., Damm, K., Goldberg, Y., Ghysdael, J., Leutz, A., Beug, H. & Vennström, B. (1986). The c-*erb*-A protein is a high affinity receptor for thyroid hormone. *Nature*, London, **324**, 635–40.

Scheidereit, C., Geisse, S., Westphal, H. M. & Beato, M. (1983). The glucocorticoid receptor binds to defined nucleotide sequences near the promoter of mouse mammary tumor. *Nature*, London, **304**, 749–52.

Schena, M. & Yamamoto, K. R. (1988). Mammalian glucocorticoid receptor derivatives enhance transcription in yeast. *Science*, **241**, 965–8.

Schüle, R., Muller, M., Kaltschmidt, C. & Renkawitz, R. (1988*a*). Many transcription factors interact synergistically with steroid receptors. *Science*, **242**, 1418–20.

Schüle, R., Muller, M., Otsuka-Murakami, H. & Renkawit, R. (1988*b*). Cooperativity of the glucocorticoid receptor and the CACCC-box binding factor. *Nature*, London, **332**, 87–90.

Severne, Y., Wieland, S., Schaffner, W. & Rusconi, S. (1988). Metal binding 'finger' structures in the glucocorticoid receptor defined by site-directed mutagenesis. *EMBO Journal*, **7**, 2503–8.

Slater, E. P., Cato, A. C. B., Karin, M., Baxter, J. D. & Beato, M. (1988). Progesterone induction of metallothionein-IIA gene expression. *Molecular Endocrinology*, **2**, 485–91.

Slater, E., Rabenau, O., Karin, M., Baxter, J. D. & Beato, M. (1985). Glucocorticoid receptor binding and activation of a heterologous promoter in response to dexamethasone by the first intron of the human growth hormone gene. *Molecular and Cell Biology*, **5**, 2984–92.

Strähle, U., Boshart, M., Klock, G., Stewart, F. & Schütz, G. (1989). Glucocorticoid- and progesterone-specific effects are determined by differential expression of the respective hormone receptors. *Nature*, London, **339**, 629–32.

Strähle, U., Klock, G. & Schütz, G. (1987). A DNA sequence of 15 base pairs is sufficient to mediate both glucocorticoid and progesterone induction of gene expression. *Proceedings of the National Academy of Sciences, USA*, **84**, 7871–5.

Strähle, U., Schmid, W. & Schütz, G. (1988). Synergistic action of the glucocorticoid receptor with transcription factors. *EMBO Journal*, **7**, 3389–95.

Thaller, C. & Eichele, G. (1987). Identification and spatial distribution of retinoids in the developing chick limb bud. *Nature*, London, **327**, 625–8.

Thompson, C. C. & Evans, R. M. (1989). Transactivation by thyroid hormone receptors; functional parallels with steroid hormone receptors. *Proceedings of the National Academy of Sciences, USA*, **86**, 3494–8.

Thompson, C. C., Weinberger, C., Lebo, R. & Evans, R. M. (1987). Identification of a novel thyroid hormone receptor expressed in mammalian central nervous system. *Science*, **237**, 1610–14.

Tora, L., Gaub, M. P., Mader, S., Dierich, A., Bellard, M. & Chambon, P. (1988*a*). Cell-specific activity of a GGTCA half-palindromic estrogen-responsive element in the chicken ovalbumin gene promoter. *EMBO Journal*, **7**, 3771–8.

Tora, L., Gronemeyer, H., Turcotte, B., Gaub, M. P. & Chambon, P. (1988*b*). The N-terminal region of the chicken progesterone receptor specifies target activation. *Nature*, London, **333**, 185–8.

Tsai, S. Y., Carlstedt-Duke, J., Weigel, N. L., Dahlman, K., Gustafsson, J. A., Tsai, M. J. & O'Malley, B. W. (1988). Molecular interactions of steroid hormone receptor with its enhancer element: evidence for receptor dimer formation. *Cell*, **55**, 361–9.

Tsai, S. Y., Sagami, I., Wang, H., Tsai, M. J. & O'Malley, B. W. (1987). Interactions between a DNA binding transcription factor (COUP) and a non-DNA binding factor (S300-II). *Cell*, **50**, 701–9.

Umesono, K. & Evans, R. M. (1989). Determinants of target gene specificity for steroid/thyroid hormone receptors. *Cell*, **57**, 1139–76.

Umesono, K., Giguère, B., Glass, C. K., Rosenfeld, M. G. & Evans, R. M. (1988). Retinoic acid and thyroid hormone induce gene expression through a common element. *Nature London*, **336**, 262–5.

Visser, T. J. (1978). A tentative review of recent *in vitro* observation of the enzymatic deiodination of iodothyronine and its possible physiologic implication. *Molecular and Cell Endocrinology*, **10**, 241–7.

Von der Ahe, D., Janich, S., Scheidereit, C., Renkawitz, R., Schütz, G. & Beato, M. (1985). Glucocorticoid and progesterone receptors bind to the same sites in two hormonally regulated promoters. *Nature*, London, **313**, 706–9.

Von der Ahe, D., Renoir, J. M., Buchou, T., Baulieu, E. E. & Beato, M. (1986). Receptors for glucocorticosteroid and progesterone recognize distinct features of a DNA regulatory element. *Proceedings of the National Academy of Sciences, USA*, **83**, 2817–21.

Walter, P., Green, S., Greene, G., Krust, A., Bornert, J.-M., Jeltsch, J.-M., Staub, A., Jensen, E., Scrace, G., Waterfield, M. & Chambon, P. (1985). Cloning of the human estrogen receptor cDNA. *Proceedings of the National Academy of Sciences, USA*, **82**, 7889–93.

Wang, L. H., Tsai, S. Y., Cook, R. G., Geattie, W. G., Tsai, M. J. & O'Malley, B. W. (1989). COUP transcription factor is a member of the steroid receptor superfamily. *Nature*, London, **340**, 163–6.

Waterman, M. L., Adler, S., Nelson, C., Greene, G. L., Evans, R. M. & Rosenfeld, M. G. (1988). A single domain of the estrogen receptor confers deoxyribonucleic acid binding and transcriptional activation of the rat prolactin gene. *Molecular Endocrinology*, **2**, 14–21.

Webster, N. J., Green, S., Rui Jin, J. & Chambon, P. (1988). The hormone-binding domains of the estrogen and glucocorticoid receptors contain an inducible transcriptional activation function. *Cell*, **54**, 199–207.

Wei, I. I., Sheridan, P. L., Krett, N. I., Francis, K. C., Toft, D. O., Edwards, D. P. & Horwitz, K. (1987). Immunologic analysis of human breast cancer progesterone receptors. 2. Structure, phosphorylation and processing. *Biochemistry*, **265**, 6262–72.

Weinberger, C., Hollenberg, S. M., Rosenfeld, M. G. & Evans, R. M. (1985). Domain structure of human glucocorticoid receptor and its relationship to the v-*erb*-A oncogen. *Nature*, London, **318**, 670–2.

Weinberger, C., Thompson, C. C., Ong, E. S., Lebo, R., Gruol, D. J. & Evans, R. M. (1986). The c-*erb*-A gene encodes a thyroid hormone receptor. *Nature*, London, **324**, 641–6.

Wilson, J. & Foster, D. W. (1981). *Williams Textbook of Endocrinology*. W. B. Saunders Co., Philadelphia.

Wright, P. A., Crew, M. D. & Spindler, S. R. (1987). Discrete positive and negative thyroid hormone responsive transcription regulatory elements of the rat growth hormone gene. *Journal of Biological Chemistry*, **262**, 5659–63.

Yamamoto, K. R. (1985). Steroid receptor regulated transcription of specific genes and gene networks. *Annual Reviews in Genetics*, **19**, 209–15.

Zelent, A., Krust, A., Petkovich, M., Kastner, P. & Chambon, P. (1989). Cloning of murine α and β retinoic acid receptors and a novel receptor predominantly expressed in skin. *Nature*, London, **339**, 714–17.

12

The nerve growth factor receptors

RONALD D. VALE, MONTE J. RADEKE,
THOMAS P. MISKO AND ERIC M. SHOOTER

Introduction

Nerve growth factor (NGF) is the best known example of an important class of proteins, the neurotrophic proteins or factors (Levi-Montalcini, 1987). During development these proteins control the process of neuronal cell death. Those nerve cells that, once their processes reach their targets, establish a retrograde flow of a protein like NGF from target to nerve cell body survive, those that do not, degenerate. Furthermore, once the retrograde flow of NGF has been established, it continues for the rest of the life of that particular neuron and is a key component in supporting long term survival and differentiation, as well as repair after injury where that occurs. The molecular events underlying this retrograde flow are slowly being unraveled. NGF induces a variety of responses in the nerve cells, sensory, sympathetic and CNS cholinergic, which exhibit the retrograde flow of this protein. Some changes in cellular physiology or morphology are apparent after seconds of NGF treatment, while others become evident only after several days of NGF exposure. NGF affects many parameters of cellular function from the level of the plasma membrane to the regulation of gene transcription. Currently, the mechanism whereby NGF produces its multiplicity of effects on cells is not yet understood. However, it has become clear that the first step in NGF action is binding to specific cell surface receptors. Understanding how these receptors function and how information of NGF binding to its receptor is imparted to the rest of the cell is of fundamental importance for deciphering the mode of action of this factor.

NGF binding to sensory and sympathetic neurons

Studies on the NGF receptor were made possible by the development of an [^{125}I]-labeled derivative of nerve growth factor of high specific activity which retained full biological activity (Banerjee *et al.*, 1973; Frazier *et al.*, 1974; Herrup & Shooter, 1973; Vale & Shooter, 1984). Using this derivative, high affinity, specific NGF binding sites were demonstrated on the surface of responsive sensory and sympathetic neurons. Herrup and Shooter (1973) and Banerjee *et al.* (1973) both found a single class of binding sites with an equilibrium dissociation constant (K_d) in the range of 0.2–0.3 nM on intact dorsal root ganglion cells and rabbit cervical sympathetic ganglion membranes respectively. In contrast, Frazier *et al.* (1974) obtained biphasic Scatchard curves for [^{125}I]-NGF binding to membranes from chick embryonic sensory and sympathetic neurons revealing two apparent affinities of 10^{-10} M and 10^{-6} M. Furthermore, unlabeled NGF enhanced by 30-fold the dissociation of cell-bound [^{125}I]-NGF when compared to dilution conditions alone. DeMeyts, Bianco & Roth (1976) observed these same phenomena for insulin binding to lymphocytes and proposed a negative cooperativity model of receptor binding in which increasing receptor occupancy produces a decrease in receptor affinity. On the basis of their similar results, Frazier *et al.* (1974) proposed that negative cooperativity may also explain the type of NGF binding that they observed.

The question of negative cooperative binding was re-examined several years later by Sutter *et al.* (1979*a, b*). In their study, steady state binding of [^{125}I]-NGF to chick embryonic dorsal root ganglion cells revealed the presence of two apparent classes of binding sites, as shown in the Scatchard plot in Fig. 12.1. One receptor class (Site I) was of high affinity and low capacity ($K_d = 2.3 \times 10^{-11}$ M with 3000 sites per cell) while Site II receptors were low affinity and high capacity ($K_d = 1.7 \times 10^{-9}$ with 45000 sites per cell). Both classes of receptors were observed in experiments conducted at 2 °C as well as 37 °C, and were also seen in membrane preparations (Riopelle *et al.*, 1980).

Association rate constants for NGF binding to Site I or Site II receptors were similar. On the other hand, the rates at which NGF was released from Site I and II receptors differed by two orders of magnitude. If [^{125}I]-NGF was incubated with cells at concentrations below 10^{-11} M (where only binding to Site I receptors is predicted) and then excess unlabeled NGF was added so that dissociation of cell-bound ligand could be followed, monoexponential dissociation kinetics ($t_{\frac{1}{2}} = 10$ min) were observed. How-

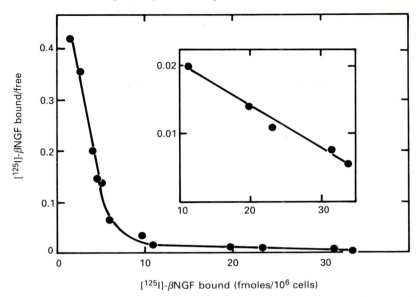

Fig. 12.1. The binding of [¹²⁵I]-NGF to sensory ganglion cells of 8-day old chick embryos. The data are presented as a Scatchard plot of the specific [¹²⁵I]-NGF binding, with the low affinity region expanded in the insert. Cells $(0.6 \times 10^6/\text{ml})$ were incubated at 37 °C for 45 min with various concentrations (3 pM to 3.7 nM) of [¹²⁵I]-NGF and cell bound [¹²⁵I]-NGF determined after a brief (1 min) centrifugation. Non-specific binding was determined in the presence of 10 µg/ml unlabeled NGF and was less than 20% of total binding. Two classes of receptors are distinguished, Site I or high affinity receptors by the section of the graph closest to the ordinate and Site II or low affinity receptors by the data in the insert. (From Sutter *et al.* 1979*a, b.*)

ever, at higher [¹²⁵I]-NGF concentrations where Site II receptors also become occupied, a proportion of the NGF dissociated at a much more rapid rate $(t_{\frac{1}{2}} = 3 \text{ s})$. The rapid and slow phases of dissociation correspond to the release of NGF from Site I and II receptors respectively. Since association rates are very similar, the 100-fold difference in the equilibrium binding constants between Site I and II receptors can be accounted for on the basis of their different rates of dissociation.

Heterogeneity of NGF binding could be interpreted either as the result of multiple receptor species or as the consequence of a single receptor population interacting in a negatively cooperative manner as suggested by Frazier *et al.* (1974). The negative cooperativity model predicts certain kinetic behavior which can be experimentally tested (Delean & Rodbard, 1979). One prediction of this hypothesis is that, when dissociation is initiated by the addition of excess unlabeled ligand, a monoexponential release of NGF corresponding to the low affinity form of the receptor should be obtained. In other words, the kinetics of dissociation are independent of the initial receptor occupancy but are related to the final occupancy after the addition of the unlabeled ligand. However, Sutter *et al.* (1979*a*, *b*) found that the initial concentration of [^{125}I]-NGF incubated with sensory cells determined the subsequent pattern of dissociation kinetics. In fact the relative proportions of rapidly and slowly dissociating binding at various [^{125}I]-NGF concentrations could be predicted on the basis of the theoretical occupancy of Site I and II receptors derived from their equilibrium binding constants.

Another experimental finding which was used to support negative cooperativity was the ability of unlabeled NGF to enhance the dissociation of cell-bound [^{125}I]-NGF when compared to simple dilution conditions. Sutter *et al.* (1979*a*, *b*) also observed this phenomenon; however, the rate of dissociation from Site I receptors in the presence of excess unlabeled ligand did not approach the dissociation rate of Site II receptors and enhanced dissociation was also observed when low concentrations of unlabeled NGF were added which would decrease rather than increase the final receptor occupancy. Both of these results are inconsistent with a model of negative cooperativity. However, the phenomenon of enhanced dissociation is also not compatible with a simple model of two receptor species but can be explained by other mechanisms of hormone-receptor interactions to be discussed later. Very similar characteristics of the NGF receptors on chick embryo sympathetic ganglion cells have been reported (Godfrey & Shooter, 1986).

NGF receptors on PC12 pheochromocytoma cells

Like sensory and sympathetic neurons, NGF receptors have also been documented on the pheochromocytoma PC12 cell line. An initial study by Herrup and Thoenen (1979) documented a single class of receptors with a K_d of 2.9×10^{-9} M when experiments were conducted at 2 °C. This has been confirmed by Woodruff and Neet (1986*a*) and Chandler *et al.* (1984) although the reported number of receptors per cell differs in the three studies. However, subsequent work by Landreth and Shooter (1980) and Schechter and Bothwell (1981) revealed the presence of a heterogeneous population of receptors at 37 °C which displayed rapid and slow dissociation kinetics, very analogous to the results of Sutter *et al.* (1979*a*, *b*). At 37 °C after addition of excess unlabeled NGF, cell-bound [^{125}I]-NGF dissociated from the rapid and slow dissociating receptor subtypes with

Table 12.1. *The characteristics of the high (H-NGFR) and low (L-NGFR) affinity NGF receptors. The K_ds and rates of dissociation refer to NGF receptors on chick sensory or sympathetic neurons; the remainder of the properties are those of NGF receptors on PC12 cells*

Property	H-NGFR (S-NGFR)	L-NGFR (F-NGFR)
K_d ($[^{125}I]$-NGF binding)	10^{-11} M	10^{-9} M
$t_{\frac{1}{2}}$ of dissociation ($[^{125}I]$-NGF release)	~ 10 min	~ 3 s
Trypsin (occupied receptor)	Stable	Labile
Triton X-100 (occupied receptor)	Insoluble	Soluble
M_r (by crosslinking)	140 000	83 000
Mol. wt. of protein (from amino acid seq.)	84 000	42 478

half-times of 30 s and 30 min respectively. If dissociation was conducted at 4 °C instead of 37 °C, $[^{125}I]$-NGF was completely released from rapid dissociating receptor but remained bound to slow dissociating receptors over a period of about 30–60 min. Similar results were also obtained by Woodruff and Neet (1986a). These authors also found that the association kinetics of NGF binding to PC12 cells required two association rates, differing by a factor of approximately four. The correlation of the kinetic data with steady state binding of $[^{125}I]$-NGF to PC12 cells is, however, more complicated than with primary sensory or sympathetic neurons. Steady state binding on PC12 cells at 37 °C yielded Scatchard plots which were slightly curvilinear (Landreth & Shooter, 1980; Schechter & Bothwell, 1981; Woodruff & Neet, 1986a), and K_ds in the 0.2 to 1.0 nM range. Competition assays gave similar results (Cohen *et al.*, 1980). It is reasonable to assume from these results that PC12 cells possess a major population of NGF receptors analogous to the low affinity (Site II) receptors on primary neurons. Interestingly, biphasic binding was observed for solubilized receptors (Lyons, Stach & Perez-Polo, 1983), for cell surface receptors after correction for acid stable NGF (Bernd & Greene, 1984) and for a nuclear fraction (Yankner & Shooter, 1979), providing evidence for a high affinity receptor. The binding data that describe the slowly dissociating component resulted in a concave downward Scatchard plot (Woodruff & Neet, 1986a). These data can be fitted with a model in which part of these receptors have a relatively high K_d of 65 pM and interact in a positively cooperative way, possibly by dimerization or association with another effector (membrane) protein or the cytoskeleton. The differences in the NGF receptor systems between primary neurons and the PC12 cell line is further emphasized by the observation that while the α and γ subunits of 7SNGF inhibit binding to only high affinity receptors on sensory neurons they block all binding to PC12 cells (Woodruff & Neet, 1986b). Nevertheless, PC12 NGF receptors share characteristics of the primary neuron receptors, particularly in showing two different types of receptors as judged by dissociation kinetics.

As shown in Table 12.1, the two receptor subtypes on PC12 can be distinguished by criteria other than their dissociation kinetics. NGF bound to rapidly dissociating receptors is completely degraded by trypsin, while slowly dissociating NGF–receptor complexes are resistant to protease destruction. This protease resistance is not solely the consequence of internalization as protection against trypsin is also observed in binding experiments conducted at 4 °C, a temperature at which endocytosis does not occur. Furthermore, slowly, but not rapidly, dissociating receptors are partially resistant to solubilization by Triton X-100 (Schechter & Bothwell, 1981; Vale & Shooter, 1982). This observation has led to the speculation that slowly dissociating receptors are associated with the cytoskeleton (Schechter & Bothwell, 1981; Vale & Shooter, 1982). The pH optimums of rapidly and slowly dissociating binding are also different (Vale & Shooter, 1983). Furthermore, as will be discussed in more detail later, covalent crosslinking studies have revealed that structural differences may also exist between the two receptor subtypes.

Theoretically, there are a number of ways in which the type of receptor heterogeneity found on sensory and sympathetic neurons and PC12 cells can be generated. Rapidly and slowly dissociating receptors could be (a) independent receptor molecules (either different gene products or derived by differential processing of mRNA or protein precursors), (b) generated one from another by an association of the receptor with a membrane protein which modulates receptor affinity (mobile receptor hypothesis (Boeynaems & Dumont, 1975; Jacobs & Cuatrecasas, 1976), (c) the result of monovalent and divalent NGF–receptor complexes (produced by NGF-induced crossbridging of receptors, NGF being a divalent ligand) (Delisi & Chabay, 1979), and (d) related to one another by a change in the tertiary configuration of a single receptor. The last three hypotheses suppose that

NGF initially binds to a lower affinity form of the receptor and that this ligand–receptor complex subsequently undergoes a change to a higher affinity form.

Evidence in favor of receptor conversion was provided by experiments of Landreth and Shooter (1980) but challenged by Schechter and Bothwell (1981). The latter investigators did not observe a delay in binding of NGF to slowly dissociating receptors, as had Landreth and Shooter, and interpreted the different time courses of binding to the two receptor subtypes as reflecting different ligand association rates. As noted earlier, Woodruff and Neet (1986a) have confirmed this interpretation. Also, when they treated PC12 cells with trypsin prior to adding [^{125}I]-NGF, they observed binding of NGF only to Slow but not Fast receptors (Fast and Slow receptors are terms coined by Schechter and Bothwell (1981) to describe the dissociation kinetics of the two receptors). This finding was used in support of the argument that Fast and Slow receptors exist prior to the addition of NGF and are not formed from a ligand-induced conversion. Yet another alternative has been put forth by Catteneo *et al.* (1983) who have suggested that a homogeneous population of NGF receptors exists in hidden and exposed states and that this situation could give rise to heterogeneous binding kinetics.

More recently, results have been obtained which suggest that rapidly and slowly dissociating receptors are interconvertible. Vale and Shooter (1982) showed that the plant lectin wheat germ agglutinin (WGA) alters the ratios of rapidly and slowly dissociating binding on PC12 cells. Normally, at an [^{125}I]-NGF concentration of 500 pM, 75% of the binding is rapidly dissociating and trypsin sensitive; however, after the addition of WGA, > 90% of the binding becomes slowly dissociating and trypsin resistant. Since total binding remains unaltered after WGA treatment, it appears as though this lectin converts rapidly dissociating receptors into a slowly dissociating form. WGA-induced receptor conversion has also been observed by Grob and Bothwell (1983) and Buxser *et al.* (1983). In the study by Buxser *et al.* (1983), WGA was shown to change receptor affinity in detergent soluble membrane extracts and in soluble extracts reconstituted into phospholipid vesicles. Furthermore, Block and Bothwell (1983) showed that fusion of PC12 cell surface membranes with 3T3 cells (which themselves have no NGF receptors) results in a conversion of the transplanted NGF receptors to a slowly dissociating state. It is still unclear whether the change in receptor affinity induced by WGA or membrane fusion is the same as a naturally occurring receptor conversion process proposed by Landreth and Shooter (1980).

The slowly dissociating NGF receptor is also resistant to Triton X-100 solubilization, as was first noted by Schechter and Bothwell (1981). Vale and Shooter (1982) later discovered that the WGA-induced change in the kinetic state of the receptor was correlated with a conversion of the receptor to a Triton X-100 insoluble form. Both events were rapid (maximal effect within 2 min of WGA addition) and shared identical time courses. NGF receptors also became Triton X-100 insoluble after fusion of PC12 cell membranes with 3T3 cells along with their conversion to a slowly dissociating state (Block & Bothwell, 1983). The Triton X-100 insolubility of the receptor may indicate an association with a cytoskeleton-associated protein or with the cytoskeleton itself. A good deal of evidence has accumulated indicating that some membrane proteins are linked to the cytoskeleton (Bennett, 1982; Mescher, Jose & Balk, 1981), and in some cases, this interaction is enhanced by multivalent antibodies or lectins (Flanagan & Koch, 1978; Painter & Ginsberg, 1982). Association of the receptor with the Triton X-100 insoluble residue of differentiated PC12 can be visualized by autoradiography; however, recent evidence indicates that WGA attaches many cell surface proteins to the cytoskeleton in addition to the NGF receptor (Vale *et al.*, 1985*b*). This lectin may act by crossbridging proteins that are normally free to diffuse in the lipid bilayer to other proteins that are immobile and anchored to the cytoskeleton.

By what mechanism are NGF receptors converted to a slowly dissociating form? Some evidence indicates that clustering of receptors may be involved in producing a high affinity hormone–receptor complex (Kahn *et al.*, 1978; Schechter *et al.* 1979*a*). Since WGA is a multivalent lectin, its ability to crosslink receptors may be important for producing receptor conversion. To further investigate this question, Vale and Shooter (1983) artificially crosslinked NGF–receptor complexes using anti-NGF antibodies. Intact antibodies, but not monovalent F_{ab} fragments, were found to convert rapidly dissociating NGF–receptor complexes to a slowly dissociating, trypsin-resistant form. Furthermore, anti-NGF antibodies also increased the proportion of NGF binding which was resistant to Triton X-100 solubilization.

These results suggest that receptor clustering may be involved in generating a high affinity form of the NGF receptor. In fact, a number of receptors (Schlessinger *et al.*, 1978), including the NGF receptor (Bernd & Greene, 1983; Levi *et al.*, 1980), undergo ligand-induced cell surface clustering, and it has been speculated that clustering may be a critical event in the mechanism of action of several polypeptide hormones (Kahn *et al.*, 1978; Schechter *et al.*, 1979*b*; Schreiber *et al.*, 1983). Interestingly, some component of the intact cell appears to be involved in the generation of slowly dissociating NGF receptors, since purified plasma membranes of PC12 cells (but not sensory neurons) display only rapidly dissociating NGF binding (Block & Bothwell, 1983). The absence of slowly dissociating receptors on these membranes is not the result of degradation, since slowly dissociating binding can be recovered by fusion of PC12 membranes with 3T3 cells (Block & Bothwell, 1983). The correlation of Triton X-100 insolubility and slowly dissociating NGF binding further suggests that the cytoskeleton may be involved in regulating the affinity of the NGF–receptor complex. Therefore, the cytoskeleton or a cytoskeletal-associated protein could act as an effector molecule in the mobile receptor hypothesis of Boeynaems and Dumont (1975) and Jacobs and Cuatrecasas (1976). This model predicts curvilinear Scatchard plots and enhanced dissociation kinetics upon addition of unlabeled ligand, findings that have been observed for NGF. Furthermore, biochemical evidence for the existence of effector molecules involved in the regulation of the β-adrenergic and insulin receptors has been obtained (Gilman, 1987). In addition, recent studies reported association of membrane-bound effector proteins with the cytoskeleton (Jesaitis *et al.*, 1984; Sahyoun *et al.*, 1981).

NGF itself has the potential to cluster receptors. This protein consists of two identical 13 000 molecular weight subunits joined by noncovalent forces (Greene *et al.*, 1971) and therefore is potentially divalent with respect to receptor binding. Clustering of receptors by a divalent ligand can produce the type of heterogeneous binding that has been observed for NGF (Delisi & Chabay, 1979). In accordance with this hypothesis, WGA and anti-NGF antibodies could increase slowly dissociating NGF binding by promoting receptor clustering. However, no evidence for the existence of slowly dissociating NGF receptors have been found on the stably transfected PCNA cell line which expresses very high numbers (2.5×10^6) of rapidly dissociating receptors per cell (Radeke *et al.*, 1987) and where the potential for clustering with the divalent NGF is high.

What is the functional significance of having two receptor subtypes for the mechanism of action of NGF? Two receptor classes are found on a variety of NGF responsive cells as determined either by steady state or dissociation kinetics. Exceptions are the melanoma A875 cell line which primarily displays low affinity receptors (Buxser & Johnson, 1982; Fabricant, De Larco & Todaro, 1977), and the SH-SY5Y neuroblastoma cell line which primarily has high affinity receptors (Sonnenfeld & Ishii, 1982). Interestingly, the A875 line also does not exhibit a classic differentiation response to NGF, even though it is the richest known source of receptors. Pharmacological dose response curves indicate that high affinity NGF receptors mediate long term differentiation responses such as neurite outgrowth. A correlation of high affinity receptor occupancy and survival and process formation by sensory neurons has also been observed (Sutter *et al.*, 1979*b*; Zimmerman *et al.*, 1978). Moreover, high affinity receptors appear developmentally on the chick DRG neurons when the cells become responsive to NGF (Zimmerman *et al.*, 1978). Also, neurite outgrowth can be elicited by NGF in a neuroblastoma cell line with only Slow receptors indicating that the low affinity form of the receptor is apparently not needed for this response (Sonnenfeld & Ishii, 1982). Low affinity receptors may serve functions distinct from high affinity receptors. For example, stimulation of amino acid uptake in PC12 cells requires significantly higher concentrations of NGF than are necessary for promoting neurite outgrowth and thus this response may be mediated by low affinity receptors (Kedes, Gunning & Shooter, 1982; McGuire & Greene, 1979). In addition, glial cells from the chick dorsal root ganglion (Zimmerman & Sutter, 1973; Hosang & Shooter, 1985) and Schwann cells from injured sciatic nerve only exhibit low affinity receptors, but their function is unknown (Taniuchi *et al.*, 1986*a*; Heumann *et al.*, 1987).

Biochemistry of the NGF receptors

Covalent crosslinking of [^{125}I]-NGF has yielded more detailed information on the structure of these receptors. Using a heterobifunctional, photo-activated cross-linking reagent, Massague *et al.* (1981*a*) found NGF cross-linked to two molecules on the surface of rabbit superior cervical ganglion membranes forming complexes with molecular weights of 143000 and 112000. Proteolytic peptide maps of the 143000 and 112000 complexes were similar, indicating a possible precursor–product relationship between these two species, conceivably involving proteolytic degradation. Evidence for proteolytic cleavage of the insulin and the epidermal growth factor receptors by membrane proteases also has been obtained (Lindsley & Fox, 1980; Massague, Pilch & Czech, 1981*b*). Massague *et al.* (1982) later conducted similar cross-linking studies with PC12 cells and only observed the association of NGF with a single polypeptide which migrated in a SDS polyacrylamide gel with an apparent molecular weight of 148000–158000. However, using the same cell line, Hosang and Shooter (1985) obtained different results and found [^{125}I]-NGF crosslinked into two major complexes with molecular weights of 158000 and 100000 with a minor amount present in a complex of 225000 (Fig. 12.2). Subtracting the weight of the βNGF monomer, assuming only one is present in the crosslinked complex, one obtains receptor molecular weights of approximately 85 000, 145 000 and 210 000. Moreover, the association of NGF with the 158 000 and 100 000 molecular weight complexes showed the same characteristics as binding to slowly and rapidly dissociating receptors respectively (Fig. 12.2*a*). When [^{125}I]-NGF was first bound to PC12 cells and excess unlabeled NGF then added at 4 °C to remove rapidly dissociating NGF prior to cross-linking, only the 158 000 species was observed. Massague *et al.* (1982) washed cells to remove unbound [^{125}I]-NGF before crosslinking (which also removes rapidly dissociating binding), and this procedure may account for their failure to detect the 100 000 molecular weight band. Furthermore, binding of NGF to the 158 000 complex was trypsin resistant while binding to the 100 000 complex was not, also arguing that the 158 000 complex is the slowly dissociating form of the receptor (Fig. 12.2*b*). As predicted, in non-neuronal ganglion cells, which only have the low affinity receptor, only the lower molecular weight NGF–receptor complex was observed.

Several laboratories have also crosslinked [^{125}I]-NGF to A875 melanoma cells and have observed bands of about 200 000 and 100 000 molecular weight (Grob, Berlot & Bothwell, 1983; Hosang & Shooter, 1985; Puma *et al.*, 1983), and Hosang and Shooter (1985) also observed an additional minor band of 160 000. Unlike PC12 cells, only a small proportion of the receptors are slowly dissociating on A875 cells. Thus all three molecular weight NGF–receptor complexes are associated with a rapidly dissociating kinetic state (Buxser & Johnson, 1982).

Monoclonal antibodies are powerful reagents for probing structural–functional relationships of receptors. Chandler *et al.* (1984) generated a monoclonal antibody (MC192) which enhances NGF binding on PC12 cells. Interestingly, this antibody specifically affected the low affinity receptor by increasing its affinity by three- to four-fold. Since this antibody had the same number of binding sites on PC12 cells as did NGF, it was likely that this antibody recognized the receptor or a receptor associated protein. This was confirmed by Taniuchi, Schweitzer & Johnson (1986*c*) who demonstrated that MC-192 immunoprecipitated the 100 000 molecular weight [^{125}I]-NGF–crosslinked receptor complex as well as the surface [^{125}I]-labeled 80 000 receptor. Since this antibody selectively affects the low affinity receptor, it appears that the two receptor subtypes are antigenically distinct. Ross *et al.* (1984) examined a series of monoclonal antibodies specific for human melanomas and not other tumor cells and found six antibodies which recognized the NGF receptor and inhibited binding. These antibodies do not cross-react with the receptor in other

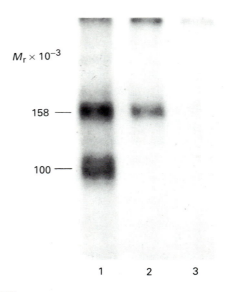

(a) HIGH MOLECULAR WEIGHT NGF-RECEPTORS
ARE SLOW AND LOW MOLECULAR WEIGHT
NGF-RECEPTORS ARE FAST RECEPTORS

(PC12 cells incubated with ligands,
crosslinking with 50 μM HSAB)

$M_r \times 10^{-3}$

158 —

100 —

1   2   3

1 : 0.8 nM [^{125}I]-NGF;
2 : 0.8 nM [^{125}I]-NGF; then plus 0.8 nM NGF (30 min, 0.5°C);
3 : 0.8 nM [^{125}I]-NGF; then plus 0.8 nM NGF (30 min, 37°C).

Fig. 12.2 (*a*). For legend see facing page

species. Anti-receptor antibodies were used to immunoprecipitate the receptor labeled externally with ^{125}I or biosynthetically with radioactive amino acids. These experiments indicate that the receptor has a molecular weight of 85000, in good agreement with the results obtained with crosslinking for the fast or low affinity receptor, and that it may be joined by disulphide bonds to other receptors forming higher molecular weight dimers or oligomers. Using immunoprecipitation, these investigators also showed that the NGF receptor is phosphorylated on two serine residues and not on tyrosine, and that NGF binding inhibits receptor phosphorylation in cells and membranes. The *in vitro* phosphorylation of the receptor by a cAMP-independent kinase has also been demonstrated (Taniuchi *et al.*, 1986*b*).

(b) [^{125}I]-NGF;BOUND TO LOW MOLECULAR WEIGHT
NGF-RECEPTORS IS TRYPSIN SENSITIVE

(PC12 cells incubated with [^{125}I]-NGF, (+) or (−)
trypsin prior to crosslinking with 50 µM HSAB)

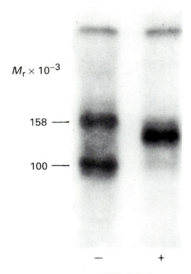

$M_r \times 10^{-3}$

158 —

100 —

−   +

TRYPSIN

Fig. 12.2. The identification of high and low affinity NGF receptors with different molecular weight species. (*a*) After incubation of PC12 cells with [^{125}I]-NGF (0.8 nM for 60 min at 23 °C) the cells were cooled to 0.5 °C and incubated for 30 min at the same temperature in the absence (lane 1) or presence of 0.8 µM NGF (lane 2). After cross-linking with 50 µM HSAB the samples were analysed by electrophoresis and autoradiographed. Lane 1 shows the two different molecular weight [^{125}I]-NGF–receptor complexes and lane 2 identifies the higher molecular weight complex as the slowly dissociating, low affinity NGF–receptor complex. Lane 3 shows that [^{125}I]-NGF is released from both receptors at 37 °C. (*b*) In a similar experiment cells were incubated for 30 min at 0.5 °C in the absence (−) or presence of 0.5 µg/ml trypsin (+). The lower molecular weight receptor complex is completely degraded by trypsin while the higher molecular weight complex is stable but loses an approx. M_r 10000 peptide. This experiment again identifies the high and low molecular weight complexes as the high and low affinity NGF–receptor complexes respectively. (From Hosang & Shooter, 1985.)

The ability of lectins to alter the binding properties of the NGF receptor (Buxser *et al.*, 1983; Costrini & Kogan, 1981; Grob & Bothwell, 1983; Vale & Shooter, 1982) and the binding of the receptor to a WGA affinity column (Grob *et al.*, 1983) suggests that the receptor is a glycoprotein. NGF receptors contain sialic acid as part of their carbohydrate moiety. Removal of terminal sialic acid residues by neuraminidase lowers the molecular weights of the 158 000 and 100 000 NGF–receptor complexes by 10 000 (Vale, Hosang & Shooter, 1985a). This decrease in apparent molecular weight most likely reflects an altered mobility of the receptor in a SDS polyacrylamide gel after removal of highly charged sialic acid residues. Furthermore, neuraminidase treatment of PC12 cells inhibits the WGA conversion of NGF receptors to a slowly dissociating form, indicating that this lectin binds to sialic acid rather than N-acetyl-D-glucosamine residues on the receptor (Vale *et al.*, 1985d). The receptor also undergoes a 10 000 increase in its molecular weight after its initial synthesis on ribosomes due to glycosylation (Grob *et al.*, 1985). The function of carbohydrate residues on the receptor is unknown. After tunicamycin treatment, Grob *et al.* (1983) found that [^{125}I]-NGF could still be crosslinked to cell surface PC12 receptors, indicating that the receptor appears on plasma membranes and binds NGF in the absence of attached carbohydrate. Furthermore, exoglycosidase treatment does not impair NGF binding to its receptor (Vale *et al.*, 1985a).

The NGF receptor has been partially purified using a WGA affinity column (Grob *et al.*, 1983). Puma *et al.* (1983) however, were able to achieve greater purification of the receptor using two sequential NGF–Sepharose affinity columns. These investigators used A875 cell membranes as their starting material because of its abundance of receptors (Fabricant *et al.*, 1977). The NGF binding activity, purified essentially to homogeneity from detergent solubilized membranes, consisted of two polypeptide species with molecular weights of 85 000 and 200 000. Photoaffinity labeling experiments showed that these two binding species were present in intact cells and soluble cell extracts in addition to the purified receptor preparation. Moreover, the purification of a 200 000 molecular weight NGF binding polypeptide indicates that the presence of this high molecular weight band in photoaffinity labeling experiments is not due to cross-linking of two molecular weight receptors. Purification of the receptor has also been achieved using affinity chromatography with anti-receptor antibodies (Marano *et al.*, 1987).

The molecular cloning of the low affinity, fast NGF receptor

The availability of antibodies which recognize either the rat or human low affinity, fast NGF receptor has permitted the isolation of the cDNAs for these two receptors and from their nucleotide sequences the deduced amino acid sequences of the receptors (Johnson *et al.*, 1986; Radeke *et al.*, 1987). In the following discussion, the receptor is referred to as the low affinity receptor, i.e. the one with a K_d of approximately 10^{-9} M. In these studies either the rat (from PC12 cells) or human (from A875 cells) low affinity NGF receptor gene was first transferred to and expressed in mouse L fibroblasts. The respective receptor antibodies were then used to select the mouse L cells expressing the low affinity NGF receptor. Amongst a series of stable transformants obtained in the transfer of the rat NGF receptor gene was one which amplified the expression of the gene through ten selections (each selection involved isolation of the most highly fluorescent cells stained with the MC192 rat NGF receptor antibody plus a fluorescent second antibody and the growth in HAT medium) to give, finally, a mouse L cell with 2.5×10^6 rat low affinity NGF receptors per cell. These receptors bound NGF with the same K_d, showed the same rapid dissociation and had the same apparent molecular weight as the low affinity receptor on PC12 cells. The cDNA for the rat low affinity receptor was rescued from this cell line by hybridizing its total single stranded DNA with mRNA from the host mouse L cells. The proof that the cDNA coded for the low affinity receptor was obtained by transfection of the cDNA, in an appropriate plasmid, into mouse L cells and the demonstration that the sense, but not the anti-sense, cDNA sequence expressed low affinity receptors with all the properties of the parent low affinity receptor in PC12 cells. The isolation of the cDNA of the human low affinity receptor relied on the presence of the repetitive *alu* sequences in the human gene (Johnson *et al.*, 1986).

The nucleotide sequences of the two cDNAs are highly homologous with the open reading frames falling in the 5′ halves of the mRNAs. The genomic DNA is quite large, spanning approximately 20kD with six exons (Sehgal *et al.*, 1988). The amino acid sequences deduced from the nucleotide sequences show peptide chains with a single membrane spanning domain dividing the receptors into roughly equal length intra- and extracellular domains. The rat low affinity NGF receptor contains 425 amino acid residues, 29 of which are in a signal peptide and the protein molecular weight of the mature receptor is 42478 (Fig. 12.3). Although the low affinity receptor shows no overall homology to any other protein, the extracellular domain contains four repeats of segments rich in cysteine residues, similar to segments seen in other receptors which bind proteins or peptides. The recent sequencing of the chick low affinity NGF receptor (Large *et al.*, 1989), which shows less homology to the other two receptors, permits an analysis of the probable functional domains of the receptor. For

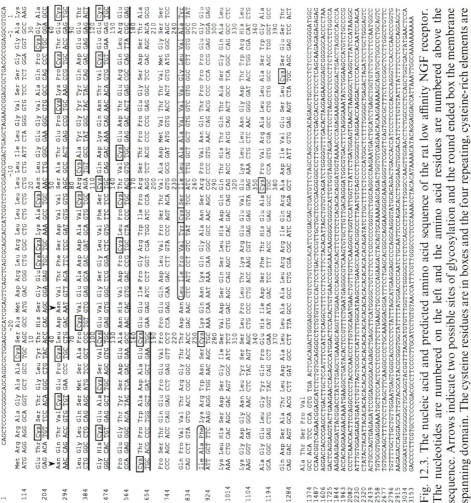

Fig. 12.3. The nucleic acid and predicted amino acid sequence of the rat low affinity NGF receptor. The nucleotides are numbered on the left and the amino acid residues are numbered above the sequence. Arrows indicate two possible sites of glycosylation and the round ended box the membrane spanning domain. The cysteine residues are in boxes and the four repeating, cysteine-rich elements are

Homologies of rat, human and chick NGFR

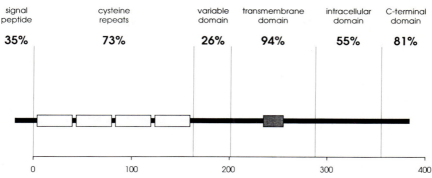

Fig. 12.4. A comparison of the NGF receptors from chicken, rat and human. A schematic model for the chicken low affinity NGF receptor showing the degree of amino acid conservation (per cent identity) for the several domains of the receptor. The four cysteine-rich repeating elements are shown as hatched boxes and the membrane spanning domain as a single darkly hatched box. (From Large *et al.*, 1989.)

example, the homology between all three receptors is high over the region of the cysteine repeats which, coupled with the fact that this region is predicted to be highly contorted and rigid, suggests it comprises at least part of the NGF binding domain (Fig. 12.4). This has recently been confirmed by showing that a cDNA construct comprising only the cysteine-rich domain (the 168 amino acids at the N-terminus), when expressed in L cells, produces a secreted, truncated receptor which still binds NGF (M. Radeke *et al.*, unpublished observations). Similarly, cultured Schwann or Schwannoma cells secrete a truncated receptor which binds NGF (Distefano & Johnson, 1988). Moreover, the cystein-rich domain is highly negatively charged, appropriate for the binding of the basic NGF protein. In contrast, the region immediately next to the cysteine-rich domain shows little homology and the secondary structure here is predicted to be highly flexible (Fig. 12.4). The greatest homology occurs over the membrane spanning domain (predicted to be an α-helix) and restricted regions on either side suggesting that this part of the structure may be very important in the signal transduction mechanism of the receptor. Another region of high homology occurs at the C-terminus of the receptor and, again, this region may be important for signal transduction.

The low affinity NGF receptor undergoes several posttranslational modifications. One N-linked glycosylation consensus sequence is present in the human and chick receptors and two in the rat. The latter result was derived from the rat cDNA sequence. However, the rat genomic sequence shows only one such consensus sequence, the one found in the human and chick receptors. The reason for this difference is not clear. It is possible that the rat cDNA represents a rare message, resulting from alternative splicing or an error in splicing at an exon boundary, or it could result from a cloning artefact. Whatever the reason, it is clear that the second site in the rat sequence coded for by the cDNA is glycosylated without affecting the NGF binding properties of the receptor. There are numerous potential serine and threonine residues for O-linked glycosylation and an analysis of a number of truncated receptors suggests that the region between the membrane spanning domain and the cysteine-rich region contains specific sites of O-linked glycosylation. Glycosylation is probably the major contributor to the mass difference between the mature low affinity NGF receptor (approx. 85000) and its peptide chain (42478). The rat low affinity NGF receptor is also modified by fatty acylation, notably a palmitate group, probably in a thioester-link to the cysteine residue just inside the membrane spanning domain, and also, possibly, a myristate group on the N-terminus (T. Misko *et al.*, unpublished observations). One or both of these residues may help anchor segments of the receptor to the membrane. As noted earlier, phosphorylation on serine and threonine but not on tyrosine residues has been observed.

These analyses of the structural domains of the low affinity NGF receptor reveal little about the signal transduction mechanism used by NGF. The cytoplasmic domain is too short to encompass an intrinsic tyrosine kinase activity and this is emphasized by the lack of the consensus sequence for the binding of ATP. The cDNA for the low affinity receptor only hybridizes to one mRNA species, that coding for the receptor, even in cells which express both receptors (Radeke *et al.*, 1987). One explanation of this result is that the low affinity receptor is an NGF binding protein which can exist by itself or as a high affinity NGF receptor when complexed to an effector protein. Evidence that both low and high affinity receptors contain a common 85000 molecular weight NGF binding protein has been obtained from crosslinking studies (Green & Greene, 1986). It is now known, however, that there is a second NGF receptor, the protooncogene *trk* (Bothwell, 1991). Trk is a single peptide chain with a single membrane spanning domain and an intrinsic tyrosine kinase activity. When cross-linked to NGF, it forms the 158000 molecular weight complex. Some evidence supports the idea that *trk* is the high affinity, biologically active receptor while other evidence suggests that the latter is formed by the interaction of *trk* and the low affinity NGF receptor (Bothwell, 1991). Possible clues to the nature of the signal cascade come from experiments in which the products of oncogenes have been shown to mimic the effects of NGF. Transfection of PC12 cells with the activated N-*ras* gene, but not the normal proto-oncogenic *ras* gene, causes neurite outgrowth in the absence of NGF (Guerrero *et al.*, 1986). The same result was obtained with transfection of the Kirsten murine sarcoma virus (Noda *et al.*, 1985). Microinjection of the *ras* p21 oncogenic, but not the proto-oncogenic protein, also induced morphological differentiation (Bar-Sagi & Fera-misco, 1985) while antibody to *ras* p21 inhibited the differentiated effects of NGF (Hagag *et al.*, 1986). The finding that microinjection of a human proto-oncogenic Ha-*ras* protein complexed with GTPγS is a very efficient inducer of neurite formation in PC12 cells (Satoh, Nakamura & Kaziro, 1987) suggests that GTP binding proteins may play a role in the signal transduction mechanism of NGF.

The regulation of the expression of NGF receptors

Two different modes of receptor regulation have been observed in nerve cells and Schwann cells. Although short-term exposure of PC12 cells to NGF results in NGF receptor down-regulation by internalization of high affinity receptors (Layer & Shooter, 1983; Bernd & Greene, 1984; Green & Greene, 1986; Hosang & Shooter, 1987) long-term exposure leads to an increase in both high and low affinity receptors (Yankner & Shooter, 1979; Bernd & Greene, 1984; Doherty *et al.*, 1988). This last result could be interpreted as demonstrating a close relationship between the two recep-tors. An increase in the amount of the low affinity NGF receptor mRNA was observed in PC12 cells after a one day exposure to NGF, before significant neurite outgrowth occurred. The amount of mRNA continued to increase significantly over the next several days to a maximum 20–30 fold increase over control before declining to near control levels by day eight. Nuclear run-on experiments showed a four-fold increase in mRNA indicating that part of the regulation was at the level of transcription (Radeke, M. *et al.*, unpublished observations). Doherty *et al.* (1988) also observed an increase in NGF receptor mRNA levels in PC12 cells after NGF treatment. Although adult sensory neurons survive in the absence of NGF, they express NGF receptors and the biosynthesis of the neuro-peptides substance P and CGRP is regulated by NGF (Lindsay & Harmer, 1989). Lindsay *et al.* (1990) observed that NGF also up-regulated the expression of the low affinity NGF receptor mRNA in adult sensory

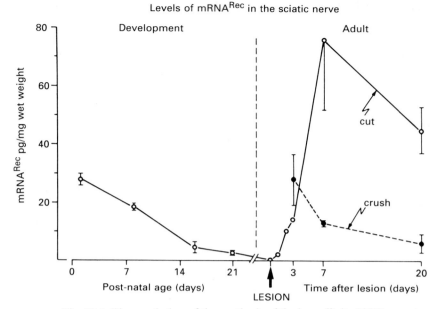

Fig. 12.5. The regulation of the synthesis of the low affinity NGF receptor on Schwann cells in rat sciatic nerve. The figure shows the levels of the mRNA for the NGF receptor (mRNA^{Rec}) in either intact (for post-natal days 1 to 21) or in 4 mm segments of rat sciatic nerve immediately distal to a lesion. In the plot labeled 'cut' regeneration of nerve fibers was inhibited while in that labeled 'crush' regeneration was permitted. (Modified from Heumann *et al.*, 1987.)

neurons leading to a five- to six-fold increase after 5 days' exposure. Since these experiments used very low concentrations of NGF, it is clear that the regulation is mediated by the high affinity NGF receptor. The ability of NGF to up-regulate its own receptor in neurons may be useful in neuronal response to injury.

An entirely different type of regulation is seen in non-neuronal cells. NGF has no effect on the expression of the low affinity receptor in Schwann cells. However, Heumann *et al.* (1987) observed that Schwann cells in developing rat sciatic nerve expressed the mRNA for the low affinity NGF receptor and the receptor itself (Fig. 12.5). Similar results have been reported by Yasuda *et al.* (1987) who confirmed that the receptors were low affinity by binding. The function of this receptor is unknown since the developing nerve fibers are devoid of NGF receptors and grow in an NGF-independent manner (Davies *et al.*, 1987). It has been suggested that NGF produced by the same Schwann cells may regulate expression of cell contact or adhesion molecules through the NGF receptor on Schwann cells (Seilheimer & Schachner, 1987). As the nerve matured and close contact between the axonal and myelin membranes was established the expression of the low affinity NGF receptor decreased (Fig. 12.5) until little or no Schwann cell NGF receptors were seen in the adult nerve (Heumann *et al.* 1987). However, axotomy of the sciatic nerve and degeneration of the distal nerve stump led to a steady rise in Schwann cell NGF receptor mRNA production reaching a maximum three days after injury if regeneration of the nerve fibers was prevented (Fig. 12.5) (Heumann *et al.*, 1987; Lemke & Chao, 1988). At the protein level maximum expression of the receptor occurred by day seven after injury (Taniuchi *et al.*, 1986a). Regeneration of nerve fibers rapidly down-regulated NGF receptor expression indicating, again, that regulation of the NGF receptor gene is achieved by axonal-Schwann cell contacts. Taniuchi *et al.* (1986a) have suggested that the Schwann cell receptors by binding NGF in the distal part of the injured nerve function provide neurotrophic support for the regrowing sensory and sympathetic fibers.

Conclusion

The finding that all the known neurotrophins bind to the low affinity NGF receptor argues either in favor of it being a part of the high affinity receptor or that it serves a unique function (for review see Bothwell, 1991). The generation of appropriate probes for the NGF receptor, namely the antibodies and the cDNAs for the rat and human receptors has permitted detailed studies on the distribution and development of the receptor and confirmed and extended our knowledge of NGF-dependent neuronal systems. Further information will come from studying the distribution of *trk*. NGF or NGF-like receptors appear transiently in development in unexpected locations, none perhaps more surprisingly than on non-neuronal cells (Heumann *et al.*, 1987). Why Schwann cells apparently express only low affinity NGF receptors remains a puzzle but their different mode of regulation by axonal contact, as contrasted to the NGF-dependent regulation of the receptor in neuronal cells, opens new avenues for the exploration of receptor regulation. Finally, the discovery of the expression of the NGF receptor on cells of the immune system (Ernfors *et al.*, 1988) emphasizes that there is still much to be learned about the functions of NGF.

Acknowledgements

Original work described in this review was supported by grants from NINDS (NS 04270), the American Cancer Society and the Isabelle M. Niemala Trust.

References

Banerjee, S. P., Snyder, S. H., Cuatrecasas, P. & Greene, L. A. (1973). Binding of nerve growth factor in sympathetic ganglia. *Proceedings of the National Academy of Sciences, USA*, **70**, 2519–28.

Bar-Sagi, D. & Feramisco, J. R. (1985). Microinjection of the *ras* oncogene protein into PC12 cells induces morphological differentiation. *Cell*, **42**, 841–8.

Bennett, V. (1982). The molecular basis for membrane-cytoskeleton association in human erythrocytes. *Journal of Cell Biochemistry*, **181**, 49–65.

Bernd, P. & Greene, L. A. (1983). Electron-microscopic autoradiographic localization of iodinated nerve growth factor bound to and internalized by PC12 cells. *Journal of Neuroscience*, **3**, 631–43.

Bernd, P. & Greene, L. A. (1984). Association of ^{125}I-nerve growth factor with PC12 pheochromocytoma cells. Evidence for internalization via high affinity receptors only and/or long term regulation by nerve growth factor of both high- and low-affinity receptors. *Journal of Biological Chemistry*, **259**, 15509–16.

Block, T. & Bothwell, M. (1983). The nerve growth factor receptor on PC12 cells: interconversion between two forms with different binding properties. *Journal of Neurochemistry*, **40**, 1654–63.

Boeynaems, J. M. & Dumont, J. E. (1975). Quantitative analysis of the binding of ligands to their receptors. *Journal of Cyclic Nucleotide Research*, **1**, 123–42.

Bothwell, M. (1991). Keeping track of neurotrophin receptors. *Cell*, **65**, 915–18.

Buxser, S. F. & Johnson, G. L. (1982). Multiple states of nerve growth factor receptors on PC12 and A875 cells: a comparison of structure and binding kinetics. *Journal of Cell Biochemistry*. Supplement 6: **165**,

Buxser, S. E., Kelleher, D. J., Watson, L., Puma, P. & Johnson, G. L. (1983). Change in state of nerve growth factor receptor: modification of receptor affinity by wheat germ agglutinin. *Journal of Biological Chemistry*, **258**, 3741–9.

Cattaneo, A., Biocca, S., Nasi, S. & Calissano, P. (1983). Hidden receptors for nerve growth factor in PC12 cells. *European Journal of Biochemistry*, **135**, 285–90.

Chandler, C. E., Parsons, L., Hosang, M. & Shooter, E. M. (1984). A monoclonal antibody modulates the interaction of nerve growth factor with PC12 cells. *Journal of Biological Chemistry*, **259**, 6882–9.

Cohen, P., Sutter, A., Landreth, G., Zimmerman, A. & Shooter, E. M. (1980). Oxidation of tryptophan-21 alters the biological activity and receptor binding characteristics of mouse nerve growth factor. *Journal of Biological Chemistry*, **255**, 2949–54.

Costrini, N. V. & Kogan, M. (1981). Lectin-induced inhibition of nerve growth factor binding by receptors of sympathetic ganglia. *Journal of Neurochemistry*, **36**, 1175–80.

Davies, A. M., Bandtlow, C. E., Heumann, R., Korsching, S., Rohrer, H. & Thoenen, H. (1987). The site and timing of nerve growth factor (NGF) synthesis in developing skin in relation to its innervation by sensory neurons and their expression of NGF receptors. *Nature*, London, **326**, 353–8.

Delean, A. & Rodbard, D. (1979). Kinetics of cooperative binding. In: *The Receptors*, O'Brien, R. D., ed., vol. 1, pp. 143–92, Plenum, New York.

Delisi, C. & Chabay, R. (1979). The influence of cell surface receptor clustering on the thermodynamics of ligand binding and the kinetics of its dissociation. *Cell Biophysics*, **1**, 117–31.

Demeyts, P., Bianco, A. R. & Roth, J. (1976). Site–site interactions among insulin receptors, characterization of the negative cooperativity. *Journal of Biological Chemistry*, **251**, 1877–88.

Distefano, P. S. & Johnson, E. M. Jr. (1988). Identification of a truncated form of the nerve growth factor receptor. *Proceedings of the National Academy of Sciences, USA*, **85**, 270–74.

Doherty, P., Seaton, P., Flanigan, T. P. & Walsh, F. S. (1988). Factors controlling the expression of the NGF receptor in PC12 cells. *Neuroscience Letters*, **92**, 222–7.

Ernfors, P., Hallbook, F., Ebendal, T., Shooter, E. M., Radeke, M. J., Misko, T. P. & Persson, H. (1988). Developmental and regional expression of β-nerve growth factor receptor mRNA in the chick and rat. *Neuron*, **1**, 983–96.

Fabricant, R. N., De Larco, J. E. & Todaro, G. J. (1977). Nerve growth factor receptors on human melanoma cells in culture. *Proceedings of the National Academy of Sciences, USA*, **74**, 565–9.

Flanagan, J. & Koch, G. L. E. (1978). Cross-linked surface Ig attaches to actin. *Nature*, London, **273**, 278–81.

Frazier, W. A., Boyd, L. F. & Bradshaw, R. A. (1974). Properties of the specific binding of [^{125}I] nerve growth factor to responsive peripheral neurons. *Journal of Biological Chemistry*, **249**, 5513–19.

Gilman, A. G. (1987). G proteins: transducers of receptor-generated signals. *Annual Review of Biochemistry*, **56**, 615–49.

Godfrey, E. W. & Shooter, E. M. (1986). Nerve growth factor receptors on chick embryo sympathetic ganglion cells: binding characteristics and development. *Journal of Neuroscience*, **6**, 2543–50.

Green, S. H. & Greene, L. A. (1986). A single M_r approximately 103 000 [^{125}I]-β-nerve growth factor-affinity-labeled species represents both the low and high affinity forms of the nerve growth factor receptor. *Journal of Biological Chemistry*, **261**, 15316–26.

Greene, L. A., Varon, S., Piltch, A. & Shooter, E. M. (1971). Substructure of the β-subunit of mouse 7S nerve growth factor. *Neurobiology*, **1**, 37–48.

Grob, P. M., Berlot, C. H. & Bothwell, M. A. (1983). Affinity labeling and partial purification of nerve growth factor receptors from rat pheochromocytoma and human melanoma cells. *Proceedings of the National Academy of Sciences, USA*, **801**, 6819–23.

Grob, P. M. & Bothwell, M. A. (1983). Modification of nerve growth factor receptor properties by wheat germ agglutinin. *Journal of Biological Chemistry*, **258**, 14136–43.

Grob, P. M., Ross, A. H., Koprowski, H. & Bothwell, M. (1985).

Characterization of the human melanoma nerve growth factor receptor. *Journal of Biological Chemistry*, **260**, 8044–9.

Guerrero, I., Wong, H., Pellicer, A. & Burstein, D. E. (1986). Activated N-*ras* gene induces neuronal differentiation of PC12 rat pheochromocytoma cells. *Journal of Cell Physiology*, **129**, 71–6.

Hagag, N., Halegona, S. & Viola, M. (1986). Inhibition of growth factor-induced differentiation of PC12 cells by microinjection of antibody to *ras* p21. *Nature*, London, **319**, 680–2.

Herrup, K. & Shooter, E. M. (1973). Properties of the β-nerve growth factor receptor of avian dorsal root ganglia. *Proceedings of the National Academy of Sciences, USA*, **70**, 3884–8.

Herrup, K. & Thoenen, H. (1979). Properties of the nerve growth factor receptor of a clonal line of rat pheochromocytoma (PC12) cells. *Experimental Cell Research*, **121**, 71–8.

Heumann, R., Lindholm, D., Bandtlow, C., Meyer, M., Radeke, M. J., Misko, T. P., Shooter, E. M. & Thoenen, H. (1987). Differential regulation of mRNA encoding nerve growth factor and its receptor in rat sciatic nerve during development, degeneration and regeneration; role of macrophage. *Proceedings of the National Academy of Sciences, USA*, **84**, 8735–9.

Hosang, M. & Shooter, E. M. (1985). Molecular characteristics of nerve growth factor receptors in PC12 cells. *Journal of Biological Chemistry*, **260**, 655–62.

Hosang, M. & Shooter, E. M. (1987). The internalization of nerve growth factor receptors on pheochromocytoma PC12 cells. *EMBO Journal*, **6**, 1197–202.

Jacobs, S. & Cuatrecasas, P. (1976). The mobile receptor hypothesis and cooperativity of hormone binding. Application to insulin. *Biochimica et Biophysica Acta*, **433**, 482–95.

Jesaitis, A. J., Naemura, J. R., Sklar, L. A., Cochrane, C. G. & Painter, R. G. (1984). Rapid modulation of N-formyl chemotactic peptide receptors on the surface of human granulocytes: formation of slowly dissociating ligand–receptor complexes in transient association with cell cytoskeleton. *Journal of Cell Biology*, **98**, 1378–87.

Johnson, D., Lanahan, A., Buck, C. R., Sehgal, A., Morgan, C., Mercer, E., Bothwell, M. & Chao, M. (1986). Expression and structure of the human NGF receptor. *Cell*, **47**, 545–54.

Kahn, C. R., Baird, K. L., Jarret, D. B. & Flier, J. S. (1978). Direct demonstration that receptor cross-linking or aggregation is important in insulin action. *Proceedings of the National Academy of Sciences, USA*, **75**, 4209.

Kedes, D. H., Gunning, P. W. & Shooter, E. M. (1982). Nerve growth factor-induced stimulation of alpha-aminoisobutyric acid uptake in PC12 cells: evaluation of its biological significance. *Journal of Neuroscience Research*, **8**, 367–74.

Landreth, G. E. & Shooter, E. M. (1980). Nerve growth factor receptors on PC12 cells: ligand-induced conversion from low- to high-affinity states. *Proceedings of the National Academy of Sciences, USA*, **77**, 4751–5.

Large, T. H., Weskamp, G., Helder, J. C., Radeke, M. J., Misko, T. P., Shooter, E. M. & Reichardt, L. F. (1989). Structure and developmental expression of the nerve growth factor receptor in the chicken central nervous system. *Neuron*, **2**, 1123–34.

Layer, P. & Shooter, E. M. (1983). Binding and degradation of nerve growth factor by PC12 pheochromocytoma cells. *Journal of Biological Chemistry*, **258**, 3012–18.

Lemke, G. & Chao, M. (1988). Axons regulate Schwann cell expression of the major myelin and NGF receptor genes. *Development*, **102**, 499–504.

Levi, A., Schechter, Y., Neufeld, E. J. & Schlessinger, J. (1980). Mobility, clustering, and transport of nerve growth factor in embryonal sensory cells and in a sympathetic neuronal cell line. *Proceedings of the National Academy of Sciences, USA*, **77**, 3469–73.

Levi-Montalcini, R. (1987). The nerve growth factor: thirty five years later. *EMBO Journal* **6**, 1145–54.

Lindsay, R. M. & Harmar, A. J. (1989). Nerve growth factor regulates expression of neuropeptide genes in adult sensory neurons. *Nature, London*, **337**, 362–4.

Lindsay, R. M., Radeke, M. J., Misko, T. P., Shooter, E. M., Dechant, G., Thoenen, H. and Lindholm, D. (1990). Nerve growth factor regulates expression of the nerve growth factor receptor gene in adult sensory neurons. *European Journal of Neuroscience*, **2**, 389–96.

Lindsley, P. & Fox, C. F. (1980). Controlled proteolysis of EGF receptors: evidence for transmembrane distribution of EGF binding and phosphate acceptor sites. *Journal of Supramolecular Structures*, **14**, 461–71.

Lyons, C. R., Stach, R. W. & Perez-Polo, J. R. (1983). Binding constants of isolated NGF-receptors from different species. *Biochemical and Biophysical Research Communications*, **115**, 368–74.

Marano, N., Dretzschold, B., Farley, J. J. Jr., Schatteman, G., Thompson, S., Grob, P., Ross, A. H., Bothwell, M., Atkinson, B. F. & Koprowski, A. (1987). Purification and amino terminal sequencing of human melanoma nerve growth factor receptor. *Journal of Neurochemistry*, **48**, 225–32.

Massague, J., Buxser, S., Johnson, G. L. & Czech, M. P. (1982). Affinity labeling of a nerve growth factor receptor component on rat pheochromocytoma (PC12) cells. *Biochimica et Biophysica Acta*, **693**, 205–12.

Massague, J., Guillette, B. J., Czech, M. P., Morgan, C. J. & Bradshaw, R. A. (1981a). Identification of a nerve growth factor receptor protein in sympathetic ganglia membranes by affinity labeling. *Journal of Biological Chemistry*, **256**, 9419–24.

Massague, J., Pilch, P. F. & Czech, M. P. (1981b). A unique proteolytic cleavage site on the β-subunit of the insulin receptor. *Journal of Biological Chemistry*, **256**, 3182–90.

McGuire, J. C. & Greene, L. A. (1979). Rapid stimulation by nerve growth factor of amino acid uptake by clonal PC12 pheochromocytoma cells. *Journal of Biological Chemistry*, **254**, 3362–7.

Mescher, M. F., Jose, M. J. L. & Balk, S. P. (1981). Actin-containing matrix associated with the plasma membrane of murine tumor and lymphoid cells. *Nature, London*, **289**, 139–44.

Noda, M., Ko, M., Ogura, A., Liu, D., Amano, T., Takano, T. & Ikawa, Y. (1985). Sarcoma viruses carrying *ras* oncogenes induce differentiation-associated properties in a neuronal cell line. *Nature, London*, **318**, 73–5.

Painter, R. G. & Ginsberg, M. (1982). Concanavalin A induces interactions between surface glycoproteins and the platelet cytoskeleton. *Journal of Cell Biology*, **92**, 565–73.

Puma, P., Buxser, S. E., Watson, L., Kelleher, D. J. & Johnson, G. L. (1983). Purification of the receptor for nerve growth factor from A875 melanoma cells by affinity chromatography. *Journal of Biological Chemistry*, **258**, 3370–5.

Radeke, M. J., Misko, T. P., Hsu, C., Herzenberg, L. A. & Shooter, E. M. (1987). Gene transfer and molecular cloning of the rat nerve growth factor receptor. *Nature*, London, **325**, 593–7.

Riopelle, R. J., Klearman, M. & Sutter, A. (1980). Nerve growth factor receptors: analysis of the interaction of β-NGF with membranes of chick embryo dorsal root ganglia. *Brain Research*, **199**, 63–77.

Ross, A. H., Grob, P., Bothwell, M., Elder, D. E., Ernst, C.S., Marano, N., Ghrist, B. F. D., Slemp, C. C., Hermlyn, M., Atkinson, B. & Koprowski, H. (1984). Characterization of nerve growth factor receptor in neural crest tumors using monoclonal antibodies. *Proceedings of the National Academy of Sciences, USA*, **81**, 6681–5.

Sahyoun, N. E., Levine, H., III, Davis, J., Hedon, G. M. & Cuatrecasas, P. (1981). Molecular complexes involved in the regulation of adenylate cyclase. *Proceedings of the National Academy of Sciences, USA*, **78**, 6158–62.

Satoh, T., Nakamura, S. & Kaziro, Y. (1987). Induction of neurite formation in PC12 cells by microinjection of proto-oncogenic Ha-*ras* protein preincubated with guanosine-5'-*o*-(3-thiotriphosphate). *Molecular Cell Biology*, **7**, 4553–6.

Schechter, A. L. & Bothwell, M. A. (1981). Nerve growth factor receptors on PC12 cells: evidence for two receptor classes with differing cytoskeletal association. *Cell*, **24**, 867–74.

Schechter, Y., Chang, K.-J., Jacobs, S. & Cuatrecasas, P. (1979a). Modulation of binding and bioactivity of insulin by anti-insulin antibody: relation to possible role of receptor self-aggregation in hormone action. *Proceedings of the National Academy of Sciences, USA*, **76**, 2720–4.

Schechter, Y., Hernaez, L., Schlessinger, J. & Cuatrecasas, P. (1979b). Local aggregation of hormone–receptor complexes is required for activation by epidermal growth factor. *Nature*, London, **278**, 835–8.

Schlessinger, J., Schechter, Y., Willingham, M. C. & Pastan, I. (1978). Direct visualization of binding, aggregation and internalization of insulin and epidermal growth factor on living fibroblastic cells. *Proceedings of the National Academy of Sciences, USA*, **75**, 2659–63.

Schreiber, A. B., Liberman, T. A., Lax, I., Yarder, Y. & Schlessinger, J. (1983). Biological role of epidermal growth factor receptor clustering: investigation with monoclonal anti-receptor antibodies. *Journal of Biological Chemistry*, **258**, 846–53.

Sehgal, A., Patil, N. & Chao, M. (1988). A constitutive promoter directs expression of the nerve growth factor receptor gene. *Molecular Cell Biology*, **8**, 3160–7.

Seilheimer, B. & Schachner, M. (1987). Regulation of neural cell adhesion molecule expression on cultured mouse Schwann cells by nerve growth factor. *EMBO Journal*, **6**, 1611–16.

Sonnenfeld, H. H. & Ishii, D. N. (1982). Nerve growth factor effects and receptors in cultured human neuroblastoma cell lines. *Journal of Neuroscience Research*, **8**, 375–91.

Sutter, A., Riopelle, R. J., Harris-Warrick, R. M. & Shooter, E. M. (1979a). Nerve growth factor receptors. Characterization of two distinct classes of binding sites on chick embryo sensory ganglia cells. *Journal of Biological Chemistry*, **254**, 5972–82.

Sutter, A., Riopelle, R. J., Harris-Warrick, R. M. & Shooter, E. M. (1979b). The heterogeneity of nerve growth factor receptors. In: *Transmembrane*

Signaling, Collier, R. J., Steiner, D. F. & Fox, C. F, eds., pp. 659–67 New York: Alan Liss Inc.

Taniuchi, M., Clark, H. B. & Johnson, E. M. Jr. (1986a). Induction of nerve growth factor receptor in Schwann cells after axotomy. *Proceedings of the National Academy of Sciences, USA*, **83**, 4094–8.

Taniuchi, M., Johnson, E. M. Jr., Roach, P. J. & Lawrence, J. C. Jr, (1986b). Phosphorylation of nerve growth factor receptor proteins in sympathetic neurons and PC12 cells. *In vitro* phosphorylation by the cAMP-independent protein kinase FA/GSK-3. *Journal of Biological Chemistry*, **261**, 13342–9.

Taniuchi, M., Schweitzer, J. B. & Johnson, E. M. Jr. (1986c). Nerve growth factor receptor molecules in rat brain. *Proceedings of the National Academy of Sciences, USA*, **83**, 1950–4.

Vale, R. D., Hosang, M. & Shooter, E. M. (1985a). Sialic acid residues on NGF receptors on PC12 cells. *Developmental Neuroscience*, **7**, 55–64.

Vale, R. D., Ignatius, M. J. & Shooter, E. M. (1985b). Association of nerve growth factor receptors with the Triton-X-100 cytoskeleton of PC12 cells. *Journal of Neuroscience*, **5**, 2672–770.

Vale, R. D. & Shooter, E. M. (1982). Alteration of binding properties and cytoskeletal attachment of nerve growth factor receptors in PC12 cells by wheat germ agglutinin. *Journal of Cell Biology*, **94**, 710–17.

Vale, R. D. & Shooter, E. M. (1983). Conversion of nerve growth factor receptor complexes to a slowly dissociating, Triton X-100 insoluble state by anti-nerve growth factor antibodies. *Biochemistry*, **22**, 5022–8.

Vale, R. D. & Shooter, E. M. (1984). Assaying binding of nerve growth factor to cell surface receptors. *Methods in Enzymology*, **109**, 21–39.

Woodruff, N. R. & Neet, K. E. (1986a). β-Nerve growth factor binding to PC12 cells. Association kinetics and cooperative interactions. *Biochemistry*, **25**, 7956–66.

Woodruff, N. R. & Neet, K. E. (1986b). Inhibition of β-nerve growth factor binding to PC12 cells by α-nerve growth factor and γ-nerve growth factor. *Biochemistry*, **25**, 7967–74.

Yankner, B. A. & Shooter, E. M. (1979). Nerve growth factor in the nucleus: interaction with receptors on the nuclear membrane. *Proceedings of the National Academy of Sciences, USA*, **76**, 1269–73.

Yasuda, T., Sobue, G., Mokuno, K., Kreider, B. & Pleasure, D. (1987). Cultured rat Schwann cells express low affinity receptors for nerve growth factor. *Brain Research*, **436**, 113–19.

Zimmerman, A. & Sutter, A. (1983). β Nerve growth factor (βNGF) receptors on glial cells. Cell–cell interaction between neurones and Schwann cells in cultures of chick sensory ganglia. *EMBO Journal*, **2**, 879–85.

Zimmerman, A., Sutter, A., Samuelson, J. & Shooter, E. M. (1978). A serological assay for the detection of cell surface receptors of nerve growth factor. *Journal of Supramolecular Structures*, **91**, 351–61.

13

Oncogenes and their relation to growth regulating receptors

MICHAEL J. FRY, GEORGE PANAYOTOU
AND MICHAEL D. WATERFIELD

Introduction

Growth factors exert their mitogenic effects by interacting with specific cell surface receptors to induce a complex cascade of biochemical events which can result in the stimulation of DNA synthesis. A large number of these growth factors interact with a family of receptors that possess intrinsic, ligand-stimulated protein–tyrosine kinase (PTK) activity. These receptors are membrane-inserted proteins which are able to transmit a mitogenic signal following binding of their respective ligand. Effects elicited due to receptor activation include protein phosphorylation, inositol lipid break-down, ion fluxes, changes in gene expression, etc. (for review see Rozengurt, 1986). Although the above responses may be involved in triggering mitogenesis, we are still some way from integrating these effects into a coherent scheme of molecular mechanisms underlying growth stimulation. Even the earliest steps, which must involve phosphorylation of substrate proteins on tyrosine residues, are only just beginning to be elucidated with the identification of phospholipase $C_{\gamma 1}$, phosphatidyl-inositol (PI) 3'-kinase, the GTPase activating protein (GAP), *raf*, and *src*-family PTKs as primary substrates which appear to associate with PTK receptors (for review see Ullrich & Schlessinger, 1990).

Much of our understanding of the important proteins involved in cell growth control has come from the study of oncogenes. These oncogenes, which when expressed in normal cells lead to uncontrolled proliferation, have been identified either as cellular genes transduced by acutely transforming retroviruses or by transfection of tumour cell DNA into NIH 3T3 cells (Bishop, 1987). Two distinct mechanisms of oncogene activation involving either growth factors or receptors have been identified. The first evidence providing a link between oncogenes and components of growth control networks was obtained by determining the primary structure of the PDGF B-chain thus revealing its identity with the v-*sis* oncogene product (Doolittle *et al.*, 1983; Waterfield *et al.*, 1983). Constitutive ligand production is thought to lead to uncontrolled cell proliferation by activating the respective receptor either at the cell surface or in some intracellular compartment. Such autocrine activation of receptors in malignancies is beyond the scope of this review, but has recently been dealt with elsewhere (Heldin & Westermark, 1989).

An increasing number of oncogenes have been shown to be altered versions of growth factor receptors. The structural characteristics which defined several families of PTK receptors will be briefly examined, followed by a review of receptor gene products which have recently been proposed to play a role in cancer.

Receptor structure

Only the basic structural motifs of the major families of receptor PTKs will be described here (for a recent in-depth review of PTK receptor structural characteristics see Yarden & Ullrich (1988)). Primary sequence data provided by the cloning of a number of receptor PTKs has led to the proposal of a common structure and membrane topology for this family of growth factor receptors (see Fig. 13.1). All possess an extracellular domain, which binds to its respective ligand, connected via a hydrophobic membrane-spanning segment to a cytoplasmic domain. This latter domain can be further subdivided. A juxtamembrane domain contains the stop-transfer sequence consisting of several charged amino acids and lies between the membrane and the start of the kinase domain. Next comes the catalytic PTK domain responsible for the generation of an intracellular signal. Finally there is a C-terminal tail of variable length (containing autophosphorylation sites) which may be involved in regulation of kinase activity and in the clearing of activated receptor from the cell surface (see Fig. 13.1; Chen *et al.*, 1989). By comparing known receptor sequences, these general structural characteristics have been used to group the receptor PTKs into four subclasses (Yarden & Ullrich, 1988; Ullrich & Schlessinger, 1990).

Subclass I

This first class is represented by the EGF receptor (c-*erb*B), and its relatives *neu*/c-*erb*B-2 and c-*erb*B-3. This class is defined by the presence of two

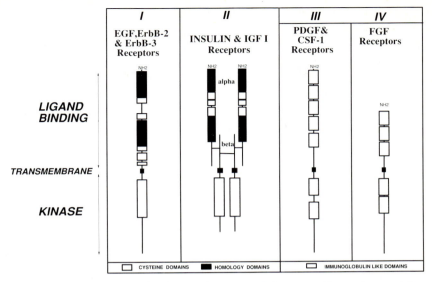

Fig. 13.1. Classification and structural features of receptor protein–tyrosine kinases. Only a single representative schematic for each of the four subclasses is shown. The classification is based on that proposed previously by Yarden & Ullrich (1988) and recently extended by Ullrich & Schlessinger (1990).

cysteine-rich repeat sequences within the extracellular ligand-binding domain (Fig. 13.1).

Subclass II

Subclass II is defined by the insulin and IGF-1 receptors. Unlike class I and III/IV receptors, this family has a heterotetrameric structure (Fig. 13.1). The holoreceptor molecule is composed of two extracellular α-chains, which provide the ligand-binding domain. Each α-chain contains a single cysteine-rich domain. These chains are disulphide bonded to two β-chains which span the plasma membrane and possess intrinsic PTK activity.

Subclass III

The third class is defined by the PDGF receptors (α- and β-subtypes), the CSF-1 receptor (c-*fms*) and c-*kit*. This family is quite distinct from classes I and II. Instead of cysteine-rich repeat motifs in the extracellular domain, this subclass has a different signature which includes regions related to immunoglobulin domains. A distinct feature of this family of receptors is the large kinase insert a domain of 60–100 residues (see Figs. 13.1 and 13.3). The function of this domain is unclear. It has been suggested that it may play a role in catalysis, modulation of the kinase activity, regulate intracellular transport of the receptor or recognition of cellular substrate proteins. Of these possibilities there is some evidence to support the last suggestion in that it has been shown that PI 3′-kinase binds to phosphorylated tyrosine residues within this domain of both the PDGF receptor and c-*fms* (Kazlauskas & Cooper, 1989; Shurtleff *et al.*, 1990).

Subclass IV

This final subclass is a modification to an older classification scheme (Yarden & Ullrich, 1988) which was recently proposed by Ullrich & Schlessinger (1990) in response to the identification and cloning of several fibroblast growth factor (FGF) receptors (Ruta *et al.*, 1989; Pasquale & Singer, 1989; Lee *et al.*, 1989), and the novel PTKs, *flg* (Ruta *et al.*, 1988), and *bek* (Kornbluth, Paulson & Hanafusa, 1988). This family is closely related to the subclass III receptors having immunoglobulin-like repeats defining their extracellular domains and an insert within their kinase domains (Fig. 13.1). In this class however only three immunoglobulin-like repeats, rather than five, are observed and these repeats show weak but significant homology to the external ligand binding domain of the interleukin 1 receptor (Ruta *et al.*, 1989).

With the identification of a number of other PTK receptors for which the ligands are as yet unknown (retroviral gene v-*ros*, the fusion genes isolated during DNA transfection experiments – *trk* A and B, *ret* and *met* (see later) – and the *Drosophila sevenless* gene (Hafen *et al.*, 1987) it is clear that this classification is insufficient. This grouping of receptors is still of use, however, to aid in the comparison of structure-function relationships between receptors.

Oncogenic potential of PTK receptors

As with other key components involved in the generation of a mitogenic signal, PTK receptors possess a latent ability, which if activated in a non-physiological manner, can result in the subversion of normal growth control and establishment of the transformed state. In general, for the oncogenic potential of a receptor molecule to be unmasked two conditions need to be met (1). Constitutive activation of the PTK (2). Release of signal generating activity from the normal regulatory controls. As will be described in the following sections, this can be brought about by a number of mechanisms involving over-expression or mutation of the receptor by deletion, point mutation or fusion with unrelated sequences.

1. EGF receptor family

With the recent identification of c-*erb*B-3 (see below) there are now three members of this family of receptors for which the EGF receptor is the prototype. All three display strikingly similar features including overall size, an extracellular ligand-binding domain with two signature cysteine clusters and a highly related, uninterrupted tyrosine kinase domain. The three major mechanisms of activation of the transforming potential of a gene, addition/deletion of sequences, point mutation and over-expression of the receptor protein, are all found to affect various members of this receptor family and are discussed below.

EGF receptor

The EGF receptor provided the first example of an oncogene derived by alteration of a growth factor receptor and as such has been most extensively studied. The EGF receptor is a 170 kD membrane glycoprotein, consisting of a 621 amino acid long extracellular ligand-binding domain which is connected to the 541 amino acid catalytic, tyrosine kinase domain via a single transmembrane stretch of 24 amino acids. A C-terminal tail contains three major autophosphorylation sites and is connected to the kinase domain by a protease-sensitive region. Sequence analysis of tryptic peptides obtained from purified human EGF receptor were found to show close similarity with the chicken v-*erb*B oncogene of the avian erythroblastosis virus (Downward *et al.*, 1984). These data, followed by the subsequent cloning of the cDNA of the EGF receptor (Ullrich *et al.*, 1984), showed that extensive truncation of most of the extracellular, ligand-binding domain and of 32 amino acids of the C-terminal tail of the receptor, containing the major site of autophosphorylation, resulted in activation of the oncogenic potential (Fig. 13.2). This evidence suggested that a truncated, ligand-independent receptor transmits a continuous mitogenic signal similar to that of a ligand-activated intact receptor and thus contributes to transformation.

Using different experimental approaches and cell types, several labora-

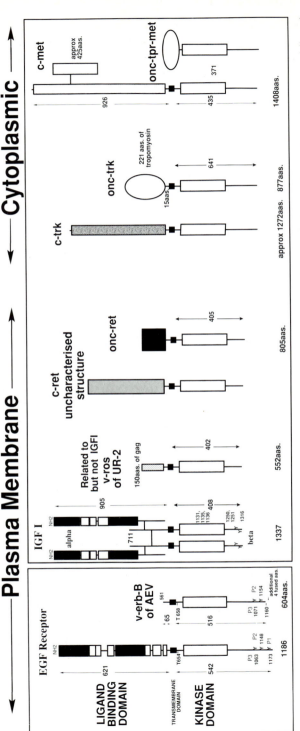

Fig. 13.2. Conversion of receptor protein-tyrosine kinases to oncogenic forms by deletion and/or replacement of their external ligand binding domains.

tories have examined the role of truncations of the EGF receptor in activating its transforming potential. However, despite a number of studies of v-*erb*B and its mutants, it is still not clearly resolved how the various N- and C-terminal mutations contribute to the oncogenic activation of the EGF receptor (Many of these data are reviewed in Beug, Hayman & Vennstrom, 1987.) The next section briefly summarises some of the discrepancies between reports from different laboratories. It would seem that the C-terminal domain is responsible for interaction with specific cellular components in a cell type-specific manner. Transformation by avian leukosis virus requires truncation of the external domain but not the C-terminal domain and results in erythroblastosis but not sarcomas (Nilsen *et al.*, 1985). Moreover, Khazaie *et al.* (1988) have shown using chicken fibroblasts that the transforming potential of an N-terminal truncated EGF receptor is further enhanced by C-terminal deletion of 32 amino acids. The same requirements were found for the constitutive self-renewal of erythroblasts. Wells & Bishop (1988) constructed EGF receptor v-*erb*B chimeras and also found that deletion of the ligand-binding domain alone was sufficient for transformation with further deletions having no added effect. Haley, Hsuan & Waterfield (1989) have analysed the ability of a spectrum of EGF receptor mutants to transform rat fibroblasts. They found that truncation of the ligand-binding domain was sufficient to activate the tyrosine kinase activity and transforming potential in these cells. The transforming potential was enhanced however by further truncation of the C-terminal domain containing the tyrosine autophosphorylation sites P1 and P2 (see Fig. 13.2). However, in contrast to these results, Riedel, Schlessinger & Ullrich (1987) found that the transforming potential of a chimeric molecule, consisting of the EGF-binding domain of the EGF receptor and the cytoplasmic domain of v-*erb*B, was not affected when compared to v-*erb*B. This suggests that the C−terminal deletion may be the most important parameter for transformation. The apparent differences between all of these studies may be accounted for by cell type and construct differences and the measuring of transformation efficiencies. However, they do demonstrate that a range of structural alterations of a growth factor can contribute to tumourigenesis.

Structurally intact growth factor receptors can contribute to transformation when over-expressed at sufficiently high numbers. The EGF receptor and the *neu* proto-oncogene (see below) have been extensively studied in this respect. Over-expression of EGF receptors may sometimes be associated with gene amplification and has been observed in a variety of tumour cell lines. One of the earliest examples was the A431 human carcinoma line expressing more than 2×10^6 receptors per cell and on which most studies of the EGF receptor have been performed (Ullrich *et al.*, 1984). Human tumours have been extensively examined in this respect. Gliomas represent the most common example of primary human tumours where amplification of the EGF receptor has been observed (Libermann *et al.*, 1985). However, no clear correlation with tumour stage or prognosis of the disease has been found yet. In some glioma patients, rearrangements of the EGF receptor gene have also been observed, often resulting in the generation of truncated EGF receptors (Humphrey *et al.*, 1988; Malden *et al.*, 1988). In three cases, an 802 nucleotide in-frame deletion has resulted in the loss of amino acid residues 5–273 of the ligand binding domain and the generation of a glycine residue at the fusion point (Humphrey *et al.*, 1990). An association between EGF receptor-positive bladder tumours and recurrence, progression and death has also recently been demonstrated (Neal *et al.*, 1990). In addition to these observations, *in vitro* experiments suggests that receptor over-expression can contribute to tumourigenic conversion. For example, Velu *et al.* (1987) used retroviral vectors to over-express the EGF receptor in NIH 3T3 cells and found that more than 10^5 receptors per cell resulted in EGF-dependent transformation and tumourigenesis *in vivo*.

ErbB-2/neu

Another member of the EGF receptor family that has been implicated in transformation is c-*erb*B-2 (human gene), also called *neu* (rat gene). The molecular weight of the product of this oncogene is 185 kD and its primary sequence is about 50% identical to that of the EGF receptor (Bargmann, Hung & Weinberg, 1986). Despite this close similarity there is no demonstrable EGF (or TGF-α) binding activity to c-*erb*B-2 and as yet no ligand has been clearly characterised (although see Yarden & Weinberg, 1989). Point mutations have been implicated in the oncogenic subversion of this receptor-like protein. Comparison with the normal rat cellular homologue has shown that a single non-conservative amino acid substitution (Val 664 → Glu) within the transmembrane domain appears to be sufficient for its conversion into a transforming protein (Bargmann *et al.*, 1986). Substitutions in neighbouring positions by site-directed mutagenesis do not have the same effect, suggesting that residue 664 plays a crucial role in signal transduction by the *neu* protein (Bargmann & Weinberg, 1988*a*). Increased autophosphorylation *in vitro* (Bargmann & Weinberg, 1988*b*) as well as phosphorylation of cellular substrates *in vivo* (Stern, Kamps & Cao, 1988) accompanies the activation of the *neu* proto-oncogene. Therefore, apart from the truncations discussed above, specific mutations within the transmembrane domain provide another structural alteration that can result in activating the transforming potential of a receptor.

Moreover, various studies have shown such a correlation for over-expression of the c-*erb*B-2 protein in breast carcinomas, suggesting an important role of this receptor protein in tumour progression (Slamon *et al.*, 1987). The significance of this finding was initially questioned as only small numbers of tumours were examined (Ali *et al.*, 1988). Subsequent studies, however, seem to confirm the prognostic value of the presence of c-*erb*B-2 alterations (Slamon *et al.*, 1989; Wright *et al.*, 1989; Borg *et al.*, 1990).

Over-expression of normal *neu* in NIH 3T3 cells gave similar results to those described above for the EGF receptor. When placed under the control of the Moloney leukemia virus long terminal repeat, normal *neu* contributed to full oncogenic transformation, but had no effect when expressed at low, detectable levels using the SV40 promoter (Di Fiore *et al.*, 1987). Recent studies by Kokai *et al.* (1989) suggest that transformation by over-expressed normal $p185^{neu}$ protein requires the presence of EGF and a synergistic interaction with the endogenous EGF receptor. Whether over-expressed normal *neu* can transform cells lacking endogenous EGF receptors remains to be tested. It seems therefore that the *neu* protein may be involved in transformation in two ways, one involving point mutation (as seen in the case of rat *neu*) and one involving over-expression of the unaltered gene (human c-*erb*B-2).

ErbB-3

A third member of the EGF receptor family, c-*erb*B-3, was recently detected and isolated by low stringency hybridization of v-*erb*B sequences to human genomic DNA (Kraus *et al.*, 1989; Plowman *et al.*, 1990). This gene encodes a 160 kD mature transmembrane protein when expressed in COS cells (Plowman *et al.*, 1990). The ligand for this receptor is not clear although amphiregulin remains a possible candidate (Shoyab *et al.*, 1989). Although less characterized than the EGF receptor or *neu*, initial studies have shown increased levels of expression in a variety of human mammary tumour cells lines (Kraus *et al.*, 1989). This result suggests that c-*erb*B-3, like the EGF receptor and c-*erb*B-2, may play a role in some human malignancies.

2. PDGF receptor family

The PDGF receptor is the prototype of a family of PTK receptors (PDGF receptor α and β subtypes, CSF-1 (c-*fms*) and c-*kit*) that have external domains with regions of homology to the immunoglobulin primordial domain and large inserts in their kinase domains (see Fig. 13.3; Yarden *et al.*, 1986). To date, there are no recorded examples of activation of PDGF receptors resulting in neoplasia. Other members of this family, however, have been found in activated forms. The first to be identified, and the one studied in the most detail, is v-*fms*.

Fms

The v-*fms* oncogene was identified as the transforming gene of two feline retroviruses, the Susan McDonough (Donner *et al.*, 1982) and the Hardy–Zuckerman 5 (Besmer *et al.*, 1986*a*) strains of feline sarcoma virus, where it was probably transduced from feline cellular DNA by recombination events with a feline leukaemia virus (FeLV). Hampe *et al.* (1984) sequenced the v-*fms* gene and demonstrated its homology to the family of PTKs. Sherr *et al.* (1985) have since shown that this oncogene is related to the colony-stimulating factor-1 (CSF-1, also called M-CSF) receptor. The structure, function and transforming ability of this receptor has recently been extensively reviewed by Sherr (1990). The human CSF-1 receptor is a 972 amino acid, 150 kD integral transmembrane glycoprotein. It possesses a 512 amino acid extracellular ligand-binding domain linked by a 25 amino acid membrane spanning segment to a 435 amino acid intracellular domain containing the PTK domain (see Fig. 13.3). It is highly related to other members of the PDGF receptor family and its head to tail localisation with the PDGF β receptor on chromosome 5 suggests that the two genes arose by gene duplication followed by sequence divergence (Roberts *et al.*, 1988).

Transduction of c-*fms* to generate v-*fms* in the isolates described above, resulted in both a C-terminal deletion (replaced by a short stretch of unrelated residues) and the introduction of scattered point mutations (see schematic comparison in Fig. 13.3). This resulted in the generation of a constitutively active, ligand-independent receptor PTK (Sacca *et al.*, 1986). The mature v-*fms* product is a 120 kD glycoprotein (generated by proteolytic removal of *gag* sequences at the N-terminus) which remains membrane-associated and possesses intrinsic PTK activity (Coussens *et al.*, 1986). Both glycosylation and membrane localization have been shown necessary for the transforming activity of v-*fms* (Roussel *et al.*, 1984;

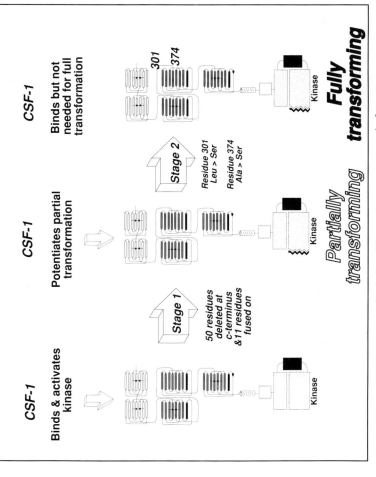

Fig. 13.3. Conversion of subclass III receptors to oncogenic forms.

Nichols *et al.*, 1985). Therefore, in contrast to v-*erb*B described above which is severely truncated in its external domain, the v-*fms* oncogene retains most of the structural features of c-*fms* (cf. Fig. 13.2 and Fig. 13.3). In fact v-*fms* can still bind CSF-1 but this factor is not required for continued growth or the maintenance of the transformed phenotype (Sacca *et al.*, 1986).

The effect of these changes on the normal receptor and their role in transformation have been investigated. Browning *et al.* (1986) demonstrated using v-*fms*/c-*fms* chimeras, replacing the C-terminal truncation of v-*fms* with c-*fms* sequences, that this deletion contributes to the full transforming potential. A tyrosine residue in the C — terminal tail of c-*fms* (Tyr 969; cf., *src* Tyr 527), which is deleted in v-*fms*, was mutated to a phenylalanine (Roussel *et al.*, 1987). Expression of this mutant in NIH 3T3 cells resulted in the transformed phenotype when CSF-1 was added, although not in the absence of ligand, suggesting that additional changes are required to confer ligand-independent activation. However, a v-*fms*/c-*fms* C-terminal tail chimera incorporating the Phe 969 mutation was fully transforming indicating this mutation is able to activate the transforming potential of c-*fms* in conjunction with some of the other N-terminal point mutations.

Activation has been recently shown to occur following the mutation of a single amino acid (Leu 301 → Ser) in the external domain of the human receptor. This results in properties resembling the v-*fms* product, i.e. ligand-independent kinase activity and transformation of NIH 3T3 cells (Roussel *et al.*, 1988). This mutation presumably induces a conformational change in the receptor which constitutively activates its kinase without impairing its ability to bind ligand. However, in other cell types it seems that more alterations may be required for expression of the full transformed phenotype. Woolford *et al.* (1988) have cloned the feline cellular *fms* gene, from which the original v-*fms* oncogene was derived. As well as the 50 residue C-terminal deletion, this allowed nine amino acid substitutions between the two feline genes to be identified. By constructing chimeric genes replacing specific amino acids they found that the transforming potential of the CSF-1 receptor was unmasked by the C-terminal deletion in concert with two point mutations at positions 301 and 374 of the extracellular ligand-binding domain (see Fig. 13.3).

The contribution of kinase insert domain to the enzymatic and transforming activities of both c- and v-*fms* has recently been investigated by Taylor *et al.* (1989). They found that partial or complete deletion of this domain had no effect on the kinase activity or autocrine transforming ability of c-*fms* in NIH 3T3 cells nor on the transforming ability of v-*fms* in the same cell type. However it is possible that this domain may play a specific role in monocytes or trophoblasts where c-*fms* is normally expressed.

There is some preliminary evidence for a role of the *fms* gene in human malignancies. Ridge *et al.* (1990) have investigated mutations at codon 301 and 969 of the *fms* gene (see above). They found that 12.7% of patients with myelodysplasia or acute myeloblastic leukaemia (AML) had mutations at codon 969 and 2% had mutations at codon 301. Furthermore, 20% of chronic myelomonocytic leukaemia and 23% of AML type M4 patients also had *fms* mutations. Further studies are required to determine what function these mutations may have in malignancy or predisposition to a particular malignancy.

Kit

The c-*kit* proto-oncogene encodes a transmembrane PTK receptor whose ligand has been recently identified as a 30 kD product of the *steel* locus (for review see Witte, 1990). This receptor was first identified as the transforming gene of the Hardy–Zuckerman 4 feline sarcoma virus (isolated from a feline sarcoma) where it was expressed as a truncated p80$^{gag-kit}$ fusion protein (Besmer *et al.*, 1986*b*). The extensive nature of the deletions involved in generating p80$^{gag-kit}$ result in a minimal kinase domain being fused to 340 amino acids of FeLV *gag*. That the v-*kit* gene was derived from a novel PTK receptor was demonstrated with the cloning of the human (Yarden *et al.*, 1987) and murine (Qui *et al.*, 1988) c-*kit* genes. c-*kit* has recently been shown to be allelic with the mouse W locus and its product is presumably required for the development of stem cells of the melanocyte, hematopoietic and germ cell lineages (Chabot *et al.*, 1988; Geissler, Ryan & Houseman, 1988).

c-*kit* encodes a 145 kD transmembrane glycoprotein which is most closely related to the CSF-1 receptor and also to a lesser degree the PDGF receptor (Yarden *et al.*, 1987). The whole of the extracellular and transmembrane domain, 17 amino acids of the juxtamembrane domain and 49 amino acids from the C-terminal tail are deleted in v-*kit*. These alterations in structure probably account for the oncogenic activation of v-*kit* (Qui *et al.*, 1988). However it is possible that noted point mutations, rather than being simply species specific sequence differences, also play a role as already described for v-*fms*. The *gag–kit* protein is able to associate with cellular membranes, as are other *gag–onc* fusion proteins, probably involving *gag*-linked myristoylation (Besmer *et al.*, 1986*b*; Majumder, Ray

& Besmer, 1990). Comparison of the resulting juxtamembrane domains of v-*kit* and c-*kit*, which separate the membrane attachment region from the kinase domain, suggest that there may be a qualitatively different interaction between the kinases of c-*kit* and v-*kit* and cellular membranes which might contribute to the transforming potential (see Fig. 13.3). v-*kit* is, therefore, distinct from most other known oncogenic variants of receptor PTK genes where the transmembrane domain is not deleted and so the subcellular localization is not necessarily affected.

3. Other PTK receptor types

The following receptors are not classified according to the three classes defined above. This is because several recently cloned putative receptors possess novel structures which will not readily fit into this classification. This suggests that the above described scheme may be too limited and that other classes will have to be defined to accommodate the receptor molecules discussed below.

Ros

The UR2 virus, an isolate of an acutely transforming retrovirus of chickens (Balduzzi *et al.*, 1981), encodes a fusion protein, p68$^{gag-ros}$, which has intrinsic PTK activity (Feldman *et al.*, 1982), although UR2-transformed cells show no significant increase in phosphotyrosine content (Cooper & Hunter, 1983). Nucleotide sequence analysis revealed that the v-*ros* onco-gene of UR2 encodes a kinase domain related to those present in PTKs (Neckameyer & Wang, 1985). This was later shown to be most highly related to the amino acid sequence of the kinase domain in the human insulin receptor (Ebina *et al.*, 1985; Ullrich *et al.*, 1985). Both chicken and human c-*ros* genes have been isolated which show that the *ros* gene encodes a transmembrane PTK with a large putative ligand binding domain similar to other cell surface receptors for growth or differentiation factors (Birchmeier *et al.*, 1986; Matsushime *et al.*, 1986; Neckameyer *et al.*, 1986). Although related, it is clear that this isolated human *ros* gene is distinct from that encoding the human insulin receptor. A chimera consisting of the ligand binding domain of the insulin receptor fused to the transmembrane and catalytic domains of v-*ros* has been constructed. Although this chimera bound insulin and activated the *ros* kinase no glucose metabolism or mitogenesis was observed, suggesting that *ros* and the insulin receptor interact with distinct intracellular pathways (Ellis *et al.*, 1987).

Details of the mechanism by which the truncations in v-*ros* contribute to the oncogenic conversion have yet to be investigated. That p68$^{gag-ros}$ is a transmembrane protein, with p19^{gag} determinants protruding extracellularly and the PTK domain lying intracellularly, has been demonstrated by Jong & Wang (1987). The relationship between p68$^{gag-ros}$ and c-*ros* is therefore similar to that already described above for gp65^{erbB} and the EGF receptor (see Fig. 13.2).

Ros has also been isolated independently as an oncogene, named *mcf3*, by a combination of a DNA transfection assay on NIH 3T3 cells and a tumourigenicity assay (Fasano *et al.*, 1984). This isolate was later shown to be derived from the human *ros*1 gene by a rearrangement introduced during gene transfer. Partial sequence data suggest that this resulted in the deletion of virtually the entire external domain (only eight amino acids of this domain remain) but left the transmembrane and PTK domains intact (Birchmeier *et al.*, 1986). There are no other apparent gross rearrangements involved in the generation of *mcf3*. Since loss of the external domain is also observed in v-*ros* it seems likely that the deletion of this domain is an important event in activation of the transforming potential of *ros* genes, however, subtle changes such as point mutations in the C-terminal portion of the protein cannot be ruled out.

The human *ros*1 gene has been localized to chromosome 6 region 6q21–q22 (Nagarajan *et al.*, 1986). This region of chromosome 6 is involved in nonrandom chromosomal rearrangement in specific neoplasias including acute lymphoblastic leukaemias, malignant melanomas and ovarian carcinomas. There is some evidence for the specific expression of the *ros*1 gene in about 60% of human glioblastomas analysed, but not in normal glial cell lines or in normal human brain tissue (Birchmeier *et al.*, 1987). A potentially activating mutation of the *ros*1 locus has been described (Sharma *et al.*, 1989). Once again the alteration appears to involve exchange of sequences upstream of the transmembrane domain resulting in the replacement of the putative extracellular ligand-binding domain of *ros*1 with sequences of unknown origin or function. However the role of the *ros* product in glioblastomas remains unclear and further work is required to clarify whether *ros*1 actually contributes to the development of glioblastomas.

Ret

The *ret* oncogene is also a fusion protein which probably resulted from recombination between two unlinked human genes during transfection of NIH 3T3 cells with DNA from a human T-cell lymphoma (Takahashi,

Ritz & Cooper, 1985). The C-terminal portion of the resultant fusion protein is related to the PTK family (Takahashi & Cooper, 1987) and a putative transmembrane domain is observed, while the N-terminal region appears to be a novel, developmentally regulated gene containing a putative metal- and nucleic acid-binding 'finger' domain (Takahashi *et al.*, 1988*b*). Cloning of the c-*ret* gene from the THP-1 human monocyte leukaemia cell line indicates that it encodes a protein which is structurally related to the PTK family of receptors (Takahashi *et al.*, 1988*a*; Takahashi *et al.*, 1989). Interestingly this clone possesses two potential trans-membrane domains. This may point to a novel class of mammalian PTK receptor related to the recently cloned *Drosophila sevenless* gene product which also has two such domains (Hafen *et al.*, 1987; Basler & Hafen, 1988). Alternatively only one of these two hydrophobic regions may function as a membrane spanning domain. Expression studies in human tumour cell lines have shown *ret* to be restricted to neuroblastoma cell lines (11/11 neuroblastoma cell lines tested) suggesting a role for *ret* in cellular functions specific to this cell type (Ikeda *et al.*, 1990).

Comparison of c-*ret* with the transforming *ret* gene revealed that in addition to the N-terminal deletion, the last 51 C-terminal amino acids have been replaced by nine unrelated amino acids. (See Fig. 13.2.) Focus formation assays indicated that *ret* cDNAs containing both the N- and C-terminal deletions scored higher in the transformation assay that *ret* cDNAs containing the N-terminal deletion alone (Takahashi *et al.*, 1988*a*). This suggests a role for both N- and C-terminal changes in the activation of the full transforming potential of *ret* as described for other receptor genes.

Several other activated *ret* genes have been isolated. *Ret*II was obtained by transfection of NIH 3T3 cells with DNA from a human sigmoid colon cancer (Ishizaka *et al.*, 1988). From comparisons with the *ret* proto-oncogene, *ret*II has both 5' and 3' rearrangements but only the former was observed in all transformants. *Ret*II also lacks the transmembrane domain of proto-*ret* and is, therefore, presumed to be cytosolic (Ishizaka *et al.*, 1988). Activated *ret* cDNAs (*ret*TPC) have also recently been isolated from the human papillary thyroid carcinoma cell line TPC-1. Sequence analysis indicated that the recombination had occurred just upstream of the kinase domain (Ishizaka *et al.*, 1990). The position at which this recombination occurred was identical to that previously observed in *ret*II (Ishizaka *et al.*, 1988). Grieco *et al.* (1990) have also identified rearranged *ret* genes in human thyroid papillary carcinomas. They identified a novel activated oncogene by transfection analysis on NIH 3T3 cells in 25% of primary human thyroid papillary carcinomas and also in lymph node metastases. This new oncogene, which they designated PTC for *P*apillary *t*hyroid *c*arcinoma gene, was shown to have resulted from the fusion of unidentified N-terminal sequences to the PTK domain of *ret*. This rearrangement was detected in all the NIH 3T3 transfectants and in all the primary tumour DNAs but not in normal cell DNA from the same patients. This indicates that this genetic lesion probably occurred *in vivo* and is specific to this type of tumour. Finally, the human c-*ret* and *ret*PTC have been mapped to chromosome 10q11–q12 (Donghi *et al.*, 1989; Ishizaka *et al.*, 1989; Grieco *et al.*, 1990). This region has previously been identified as the possible locus responsible for multiple endocrine neoplasia type 2A (MEN2A).

Met

Another member of the PTK family of receptors whose ligand is unknown is *met*. The *met* oncogene was isolated from a *N*-methyl-*N*′-nitronitroso-guanidine-treated human osteogenic sarcoma cell line (Cooper *et al.*, 1984). Preliminary characterization suggested that *met* was related to the insulin receptor and *abl* within the PTK domain, but showing little or no homology in its putative ligand-binding domain (Dean *et al.*, 1985). Further analysis revealed that the *met* oncogene arose as a result of rearrangement of DNA sequences at unlinked loci, during which time a translocated promoter and N-terminus of the *tpr*(*t*ranslocated *p*romoter *r*egion) locus became fused to the *met* kinase domain (Park *et al.*, 1986). This resulted in the expression of a 65 kD$^{tpr-met}$ fusion protein with PTK activity as demonstrated using a C-terminal peptide serum (Gonzatte-Haces *et al.*, 1988). Isolation of the *met* proto-oncogene cDNA confirmed early suspicions of its relationship with previously identified growth factor receptor PTKs (see Fig. 13.2; Park *et al.*, 1987). *Met* encodes a 1408 amino acid protein of predicted molecular weight 157 kD which possesses a 24 amino acid signal sequence followed by a large putative external domain (926 amino acids) with a characteristic array of cysteine residues distinct from both type I and type III/IV PTK receptors (cf., Fig. 13.1 and 13.2). A single small cysteine rich cluster is observed with the majority of the cysteine residues dispersed throughout the external domain. The chromosomal breakpoint which generated the *tpr–met* fusion is located 54 amino acids C-terminal from the 23 amino acid transmembrane domain. Thus unlike v-*erb* and v-*ros*, *tpr–met* lacks not only the external domain but also its transmembrane domain and is therefore cytosolic.

Recently it has become clear that the mature *met* protein is in fact an αβ heterodimer of 190 kD, consisting of the previously identified 140 kD β subunit possessing autophosphorylating activity disulphide linked to a 50 kD α subunit which is exposed at the cell surface (Giordano *et al.*, 1989*b*). Biosynthesis studies strongly suggest that both the 140 and 50 kD chains are derived from a single precursor glycoprotein which is cleaved to generate two subunits during maturation (Giordano *et al.*, 1989*a*). This suggests that p190^{met}, rather than being closely related to the insulin receptor family as proposed in the evolutionary tree of Hanks, Quinn & Hunter (1988), is the prototype for a new class of $\alpha\beta$ dimeric PTK receptors (see Fig. 13.2 and cf. Fig. 13.1).

The human *met* gene has been localized to chromosome 7 band 7q21–q31, a region associated with nonrandom chromosomal deletions in a number of patients with acute nonlymphocytic leukaemia (Dean *et al.*, 1985). However, to date there is no evidence of a role for *met* in human cancer.

Trk A and B (OncD)

Somatic rearrangement between separate genetic loci also appears to be the mechanism of activation of other receptor-like proteins. The *trk* oncogene (originally named *onc*D) was isolated as a gene from a human colon carcinoma which was found to transform transfected NIH 3T3 cells (Pulciani *et al.*, 1982). Molecular cloning of the transforming gene demonstrated that *trk* was the product of fusion between 221 amino acids of a truncated non-muscle tropomyosin gene with the transmembrane domain and a PTK domain of a novel receptor-like PTK (see Fig. 13.2; Martin-Zanca, Hughes & Barbacid, 1986). This rearrangement could be found in the original tumour cell line suggesting that it occurred during the development of the colon carcinoma (Martin-Zanca *et al.*, 1986).

The product of the tropomyosin–*trk* fusion has been identified as a 70 kD cytosolic protein which is able to phosphorylate proteins on tyrosine residues (Mitra, Martin-Zanca & Barbacid, 1987). It was suggested that the N-terminal cytoskeletal domain might play a role in activating the transforming activity of *trk* by altering its subcellular localization and causing it to interact with inappropriate substrates. Also the results of Coulier *et al.* (1989) provided some evidence in support of this, demonstrating that not just any sequence (e.g., actin) could replace the extracellular domain, and that tropomyosin sequences might contribute to malignant activation. However studies from a number of laboratories have shown that the *trk* PTK can fuse with a variety of unrelated sequences (including skeletal muscle tropomyosin, ribosomal subunit protein L7a, etc.) which unmask its transforming potential (Kozma *et al.*, 1988; Oskam *et al.*, 1988; Barnes *et al.*, 1989; Ziemiecki *et al.*, 1990). Some of the above result in non-glycosylated cytoplasmic molecules while others were found to be transmembrane glycoproteins. Interestingly, one group found that three activated *trk* molecules had all fused at exactly the same nucleotide (Kozma *et al.*, 1988). The recombination points in the other activated *trk* species are unknown but must also be in the same region of the *trk* gene. Finally, Coulier *et al.* (1990) have very recently demonstrated that a single point mutation of a conserved cysteine residue (Cys 345) in the external domain to a serine results in the generation of a novel *trk* oncogene illustrating that small mutations within this locus can also lead to its activation. This is reminiscent of a mutation of the v-*fms* gene described earlier. Taken together these results suggest that alteration of the normal extracellular domain, probably resulting in a constitutively active, ligand-independent PTK, is more important than specific structural features of the N-terminal sequences involved in the generation of a transforming *trk* species.

Cloning of the normal allele confirmed that the *trk* proto-oncogene has a molecular structure similar to other known PTK receptor genes (Martin-Zanka *et al.*, 1989). The gene codes for a 790 amino acid protein composed of a 32 amino acid signal peptide, a 375 amino acid external domain (no cysteine-rich regions) linked via a 26 hydrophobic amino acid trans-membrane domain to a PTK domain (with a 14 amino acid kinase insert sequence) and a very short, 15 residue, C-terminal tail. *Trk* shows no particular homology with any of the other receptor PTKs. The best alignment is with members of the insulin receptor family and *ros* (see Fig. 13.2). The mature *trk* glycoprotein has a molecular weight of 140 kD and possesses *in vitro* PTK activity. The rearrangement in the *trk* oncogene resulted in the replacement of the first 392 amino acids with tropomyosin sequences. No other differences were found between the normal and transforming *trk* genes suggesting that alteration of the external domain alone is sufficient to activate the transforming potential of this gene. This external, putative ligand-binding domain shows no homology to other known proteins. A second *trk* gene, *trk*B, has recently been cloned with a suggested role in neurogenesis (Klein *et al.*, 1989). It has been proposed that *trk* and *trk*B genes may code for a new family of cell surface receptors involved in the maintenance and/or development of the nervous system (Klein *et al.*, 1989).

Finally it should be noted that a recent study has documented the activation of *trk* oncogenes in at least 25% of human papillary thyroid carcinomas (Bongarzone *et al.*, 1989). The recent chromosomal localization of human *trk* gene to human chromosome 1 band 1q32–q41 will allow further molecular investigation of the mechanisms activating *trk in vivo* (Miozzo *et al.*, 1990). In this respect it should be noted that a cluster of breakpoints at 1q32 has been reported in various human tumours (reviewed in Olah *et al.*, 1989).

Conclusion

From the above survey of activating mutations of receptor molecules a few general observations can be made. 1. A functional PTK domain is essential for transforming potential of the PTK receptors. 2. Loss or alteration of the external domain appears to be a very common mutation, being observed in most cases documented here. These truncations of extracellular domains may lock receptor molecules into an active state, mimicking constitutive activation by ligand. 3. Truncations, point mutations and other modifications of both extracellular and C-terminal sequences enhance the transforming ability in an often synergistic manner. C-terminal mutations may be involved in either up-regulation of the kinase activity or in the release of receptor from negative regulation. The overall result of these modifications is generally a ligand-independent receptor, with potentially impaired feedback control and reduced substrate specificity which may provide a continuous mitogenic signal and contribute to cellular transformation.

Notes added in proof

1. The c-*met* proto-oncogene product has been recently identified as the receptor for hepatocyte growth factor (Bottaro, D. P., Rubin, J. S., Faletto, D. L., Chan, A. M.-L., Kmiecik, T. E., Van de Woude, G. F. & Aaronson, S. A. (1991). *Science*, **251**, 802–4.)

2. The receptor-like molecules Trk and TrkB have been recently identified as the receptors for nerve growth factor, and brain-derived neurotrophic factor and neurotrophin-3 respectively (reviewed by M. Bothwell (1991). *Cell*, **65**, 915–18). A third member of the Trk receptor family TrkC has recently been described as the receptor for neurotrophin-3 (Lamballe, F., Klein, R. & Barbacid, M. (1991). *Cell*, **66**, 967–79).

References

Ali, I.U., Campbell, G., Liderau, R. (1988). Amplification of c-*erb*B-2 and aggressive human breast tumors? *Science*, **240**, 1795–6.

Balduzzi, P. C., Notter, M. F. D., Morgan, H. R. & Shibuya, M. (1981). Some biological properties of two new avian sarcoma viruses. *Journal of Virology*, **40**, 268–75.

Bargmann, C. I., Hung, M. -C. & Weinberg, R. A. (1986). Multiple independent activations of the *neu* oncogene by a point mutation altering the transmembrane domain of p185. *Cell*, **45**, 649–57.

Bargmann, C. I. & Weinberg, R. A. (1988*a*). Oncogenic activation of the *neu*-encoded receptor protein by point mutation and deletion. *EMBO Journal*, **7**, 2043–52.

Bargmann, C. I. & Weinberg, R. A. (1988*b*). Increased tyrosine kinase activity associated with the protein encoded by the activated *neu* oncogene. *Proceedings of the National Academy of Sciences, USA*, **85**, 5394–8.

Barnes, D., Clayton, L., Chumbley, G. & MacLoed, A. R. (1989). Activation of the *trk* oncogene by alternatively spliced muscle and non-muscle tropomyosin sequences. *Oncogene*, **4**, 259–62.

Basler, K. & Hafen, E. (1988). Control of photoreceptor cell fate by the *sevenless* protein requires functional tyrosine kinase domain. *Cell*, **54**, 299–311.

Besmer, P., Lader, E., George, P. C., Bergold, P. J., Qui, F. -J., Zuckerman, E. E. & Hardy, W. D. (1986*a*). A new acute transforming feline retrovirus with *fms* homology specifies a C-terminally truncated version of the c-*fms* protein that is different from SM-feline sarcoma virus v-*fms* protein. *Journal of Virology*, **60**, 194–203.

Besmer, P., Murphy, J. E., George, P. C., Qui, F., Bergold, P. J., Lederman, L., Snyder, Jr, H. W., Brodeur, D., Zuckerman, E. E. & Hardy, W. D. (1986*b*). A new acute feline retrovirus and relationship of its oncogene v-*kit* with the protein kinase gene family. *Nature*, London, **320**, 415–21.

Beug, H., Hayman, M. J. & Vennstrom, B. (1987). Mutational analysis of v-*erb*B oncogene function. In *Oncogenes and Growth Control*. Kahn, P. & Graf, T. eds. Springer-Verlag, Berlin. Heidelberg, pp. 85–9.

Birchmeier, C., Birnbaum, D., Waitches, G., Fasano, O. & Wigler, M. (1986). Characterisation of an activated human *ros* gene. *Molecular and Cellular Biology*, **6**, 3109–16.

Birchmeier, C., Sharma, S. & Wigler, M. (1987). Expression and rearrangement of the ROS1 gene in human glioblastoma cells. *Proceedings of the National Academy of Sciences, USA*, **84**, 9270–4.

Bishop, J. M. (1987). The molecular genetics of cancer. *Science*, **235**, 305–11.

Bongarzone, I., Pierotti, M. A., Monzini, N., Mondellini, P., Manenti, G., Donghi, R., Pilotti, S., Grieco, M., Santoro, M., Fusco, A. *et al.* (1989). High frequency of activation of tyrosine kinase oncogenes in human papillary thyroid carcinoma. *Oncogene*, **4**, 1457–62.

Borg, A., Tandon, A. K., Sigurdsson, H., Clark, G. M., Ferno, M., Fuqua, S. A. W., Killander, D. & McGuire, W. L. (1990). HER-2/*neu* amplification predicts poor survival in node-positive breast cancer. *Cancer Research*, **50**, 4332–7.

Browning, P. J., Bunn, H. F., Cline, A., Shuman, M. & Nienhuis, A. W. (1986). 'Replacement' of COOH-terminal truncation of v-*fms* with c-*fms* sequences markedly reduces transformation potential. *Proceedings of the National Academy of Sciences, USA*, **83**, 7800–4.

Chabot, B., Stephenson, D. A., Chapman, V. M., Besmer, P. & Bernstein, A. (1988). The proto-oncogene c-*kit* encoding a transmembrane tyrosine kinase receptor maps to the mouse W locus. *Nature*, London, **335**, 88–9.

Chen, W. S., Lazar, C. S., Lund, K. A., Welsh, J. B., Chang, C. -P., Walton, G. M., Der, C. J., Wiley, H. S., Gill, G. N. & Rosenfeld, M. G. (1989). Functional independent of the epidermal growth factor receptor from a domain required for ligand-induced internalization and calcium regulation. *Cell*, **59**, 33–43.

Cooper, J. A. & Hunter, T. (1983). Regulation of cell growth and transformation by tyrosine specific protein kinase: The search for important cellular substrate proteins. *Current Topics in Microbiology and Immunology*, **107**, 125–61.

Cooper, C. S., Park, M., Blair, D. G., Tainsky, M. A., Huebner, K., Croce, C. M. & Van de Woude, G. F. (1984). Molecular cloning of a new transforming gene from a chemically transformed human cell line. *Nature, London*, **311**, 29–33.

Coulier, F., Kumar, R., Ernst, M., Klein, R., Martin-Zanca, D. & Barbacid, M. (1990). Human *trk* oncogenes activated by point mutations, in-frame deletion and duplication of the tyrosine kinase domain. *Molecular and Cellular Biology*, **10**, 4202–10.

Coulier, F., Martin-Zanca, D., Ernst, M. & Barbacid, M. (1989). Mechanism of activation of the human *trk* oncogene. *Molecular and Cellular Biology*, **9**, 15–23.

Coussens, L., Van Beveren, C., SMith, D., Chen, E., Mitchell, R. L., Isacke, C. M., Verma, I. M. & Ullrich, A. (1986). Structural alteration of viral homologue of receptor proto-oncogene *fms* at carboxyl terminus. *Nature, London*, **320**, 277–80.

Dean, M., Park, M., Le Beau, M. M., Robins, T. S., Diaz, M. O., Rowley, J. D., Blair, D. G. & Van Woude, G. F. (1985). The human *met* oncogene is related to tyrosine kinase oncogenes. *Nature, London*, **318**, 385–88.

Di Fiore, P. P., Pierce, J. H., Kraus, M. H., Segatto, O., King, C. R., & Aaronson, S. A. (1987). *erb*B-2 is a potent oncogene when overexpressed in NIH/3T3 cells. *Science*, **237**, 178–82.

Donghi, R., Sozzi, G., Pierotti, M. A., Biunno, I., Miozzo, M., Fusco, A., Grieco, M., Santoro, M., Vecchio, G., Spurr, N. K. & Della Porta, G. (1989). The oncogene associated with human papillary thyroid carcinoma (PTC) is assigned to chromosome 10q11–12 in the same region as multiple endocrine neoplasia type 2A (MEN2A). *Oncogene*, **4**, 521–3.

Donner, L., Fedele, L. A., Garon, C. F., Anderson, S. J. & Sherr, C. J. (1982). McDonough feline sarcoma virus: characterisation of the molecularly cloned provirus and its feline oncogene (v-*fms*). *Journal of Virology*, **41**, 489–500.

Doolittle, R. F., Hunkapillar, M. W., Hood, L. E., Devare, S. G., Robbins, K. C., Aaronson, S. A. & Antoniades, H. N. (1983). Simian sarcoma virus onc-gene, v-*sis*, is derived from the gene (or genes) encoding a platelet-derived growth factor. *Science*, **221**, 275–7.

Downward, J., Yarden, Y., Mayes, E., Scrace, G., Totty, N., Stockwell, P., Ullrich, A., Schlessinger, J. & Waterfield, M. D. (1984). Close similarity of epidermal growth factor receptor and v-*erb-B* oncogene protein sequences. *Nature, London*, **307**, 521–7.

Ebina, Y., Ellis, L., Jarnagin, K., Edery, M., Graf, L., Clauser, E., Ou, J., Masiarz, F., Kan, Y. W., Goldfine, I. D., Roth, R. A. & Rutter, W. J. (1985). The human insulin receptor cDNA: The structural basis for hormone activated transmembrane signalling. *Cell*, **40**, 747–58.

Ellis, L., Morgan, D. O., Jong, S. -M., Wang, L. -H., Roth, R. A. & Rutter,

W. J. (1987). Heterologous transmembrane signalling by a human insulin receptor v-*ros* hybrid in Chinese hamster ovary cells. *Proceedings of the National Academy of Sciences, USA*, **84**, 5101–5.

Fasano, O., Birnbaum, D., Edlund, L., Fogh, J. & Wigler, M. (1984). New human transforming genes detected by a tumourigenicity assay. *Molecular and Cellular Biology*, **4**, 1695–705.

Feldman, R. A., Wang, L. -H., Hanafusa, H. & Balduzzi, P. C. (1982). Avian sarcoma virus UR2 encodes a transforming protein which is associated with a unique protein kinase activity. *Journal of Virology*, **42**, 228–36.

Geissler, E. N., Ryan, M. A. & Houseman, D. E., (1988). The dominant-white spotting locus of the mouse encodes the c-*kit* proto-oncogene. *Cell*, **55**, 185–92.

Giordano, S., Di Renzo, M. F., Narsimhan, R. P., Cooper, C. S., Rosa, C. & Comoglio, P. M. (1989*a*). Biosynthesis of the protein encoded by the c-*met* proto-oncogene. *Oncogene*, **4**, 1383–8.

Giordano, S., Ponzetto, C., Di Renzo, M. F., Cooper, C. S. & Comoglio, P. M. (1989*b*). Tyrosine kinase receptor indistinguishable from the c-*met* protein. *Nature*, London, **339**, 155–6.

Gonzatte-Haces, M., Seth, A., Park, M., Copeland, T., Oroszlan, S. & Van de Woude, G. F. (1988). Characterisation of the TPR–MET oncogene p65 and the MET protooncogene p140 protein-tyrosine kinases. *Proceedings of the National Academy of Sciences, USA*, **85**, 21–5.

Grieco, M., Santoro, M., Berlingieri, M. T., Melillo, R. M., Donghi, R., Bongarzone, I., Pierotti, M. A., Della Porta, G., Fusco, A. & Vecchio, G. (1990). RTC is a novel rearranged form of the *ret* proto-oncogene and is frequently detected *in vivo* in human thyroid papillary carcinomas. *Cell*, **60**, 557–63.

Hafen, E., Basler, K., Edstroem, J.-E. & Rubin, G. M. (1987). *Sevenless*, a cell-specific homeotic gene of *Drosophila*, encodes a putative transmembrane receptor with a tyrosine kinase domain. *Science*, **236**, 55–63.

Haley, J. D., Hsuan, J. J. & Waterfield, M. D. (1989). Analysis of mammalian fibroblast transformation by normal and mutated human EGF receptors. *Oncogene*, **4**, 273–83.

Hampe, A., Gobot, M., Sherr, C. J. & Galibert, F. (1984). The nucleotide sequence of the feline retroviral oncogenes v-*fms* shows unexpected homology with oncogenes encoding tyrosine-specific protein kinases. *Proceedings of the National Academy of Sciences, USA*, **81**, 85–9.

Hanks, S. K., Quinn, A. M. & Hunter, T. (1988). The protein kinase family: conserved features and deduced phylogeny of the catalytic domains. *Science*, **241**, 42–52.

Heldin, C. -H. & Westermark, B. (1989). Growth factors as transforming proteins. *European Journal of Biochemistry*, **184**, 487–96.

Humphrey, P. A., Wong, A. J., Vogelstein, B., Friedman, H. S., Werner, M. H., Bigner, D. D. & Bigner, S. H. (1988). Amplification and expression of the epidermal growth factor receptor gene in human glioma xenografts. *Cancer Research*, **48**, 2231–8.

Humphrey, P. A. Wong, A. J., Vogelstein, B., Zalutsky, M. R., Fuller, G. N., Archer, G. E., Friedman, H. S., Kwatra, M. M., Bigner, S. H. & Bigner, D. D. (1990). Anti-synthetic peptide antibody reacting at the fusion junction of deletion-mutant epidermal growth factor receptors in human glioblastoma. *Proceedings of the National Academy of Sciences, USA*, **87**, 4207–11.

Ikeda, I., Ishizaka, Y., Tahira, T., Suzuki, T., Onda, M., Sugimura, T. &

Nagao, M. (1990). Specific expression of the *ret* proto-oncogene in human neuroblastoma cell lines. *Oncogene*, **5**, 1291–6.

Ishizaka, Y., Itoh, F., Tahira, T., Ikeda, I., Sugimura, T., Tucker, J., Fertitta, A., Carrano, A. V. & Nagao, M. (1989). Human *ret* proto-oncogene mapped to chromosome 10q11.2. *Oncogene*, **4**, 1519–21.

Ishizaka, Y., Tahira, T., Ochiai, M., Ikeda, I., Sugimura, T. & Nagao, M. (1988). Molecular cloning and characterisation of *ret*-II oncogene. *Oncogene Research*, **3**, 193–7.

Ishizaka, Y., Ushijima, T., Sugimura, T. & Nagao, M. (1990). cDNA cloning and characterisation of *ret* activated in a human papillary thyroid carcinoma cell line. *Biochemical and Biophysical Research Communications*, **168**, 402–8.

Jong, S. -M. J. & Wang, L. -H. (1987). The transforming protein p68$^{gag-ros}$ of avian sarcoma virus UR2 is a transmembrane protein with the *gag* portion protruding extracellularly. *Oncogene research*, **1**, 7–21.

Kazlauskas, A. & Cooper, J. A. (1989). Autophosphorylation of the PDGF receptor in the kinase insert region regulates interactions with cell proteins. *Cell*, **58**, 1121–33.

Khazaie, K., Dull, T. J., Graf, T., Schlessinger, J., Ullrich, A., Beug, H. & Vennstrom, B. (1988). Truncation of the human EGF receptor leads to differential transforming potentials in primary avian fibroblasts and erythroblasts. *EMBO Journal*, **7**, 3061–71.

Klein, R., Parada, L., Coulier, F. & Barbacid, M. (1989). *trk*B, a novel tyrosine protein kinase receptor expressed during mouse neural development. *EMBO Journal*, **8**, 3701–9.

Kokai, Y., Myers, J. N., Wada, T., Brown, V. I., LeVea, C. M., Davis, J. G., Dobashi, K. & Greene, M. I. (1989). Synergistic interaction of p185^{c-neu} and the EGF receptor leads to transformation of rodent fibroblasts. *Cell*, **58**, 287–92.

Kornbluth, S., Paulson K. E. & Hanafusa, H. (1988). Novel tyrosine kinase identified by phosphotyrosine antibody screening of cDNA libraries. *Molecular and Cellular Biology*, **8**, 5541–44.

Kozma, S. C., Redmond, S. M. S., Xiao-Chang, F., Saurer, S. M., Groner, B. & Hynes, N. E., (1988). Activation of the receptor kinase domain of the trk oncogene by recombination with two different cellular sequences, *EMBO Journal*, **7**, 147–54.

Kraus, M. H., Issing, W., Miki, T., Popescu, N. C. & Aaronson, S. A. (1989). Isolation and characterisation of *erb*B3, a third member of the *erb*B/epidermal growth factor receptor family: Evidence for over-expression in a subset of human mammary tumours. *Proceedings of the National Academy of Sciences, USA*, **86**, 9193–97.

Lee, P. L., Johnson, D. E., Cousens, L. S., Fried, V. A. & Williams, L. T. (1989). Purification and complementary DNA cloning of a receptor for basic fibroblast growth factor. *Science*, **245**, 57–60.

Libermann, T. A., Nusbaum, H. R., Razon, N., Kris, R., Lax, I., Soreq, H., Whittle, N., Waterfield, M. D., Ullrich, A. & Schlessinger, J. (1985). Amplification, enhanced expression and possible rearrangement of EGF receptor in primary human brain tumours of glial origin. *Nature*, London, **313**, 144–7.

Majumder, S., Ray, P. & Besmer, P. (1990). Tyrosine protein kinase activity of the HZ4-feline sarcoma virus p80$^{gag-kit}$ transforming protein. *Oncogene Research*, **5**, 329–35.

Malden, L. T., Novak, Y., Kaye, A. H. & Burgess, A. W. (1988). Selective amplification of the cytoplasmic domain of the epidermal growth factor receptor gene in glioblastoma multiforme. *Cancer Research*, **48**, 2711–14.

Martin-Zanca, D., Hughes, S. H. & Barbacid, M. (1986). A human oncogene formed by fusion of truncated tropomyosin and protein kinase sequences, *Nature*, London, **319**, 743–8.

Martin-Zanca, D., Oskam, R., Mitra, G., Copeland, T. & Barbacid, M. (1989). Molecular and biochemical characterisation of the human *trk* proto-oncogene. *Molecular and Cellular Biology*, **9**, 24–33.

Matsushime, J., Wang, L. -H. & Shibuya, M. (1986). Human c-*ros*-1 gene homologous to the v-*ros* sequence of UR2 sarcoma virus encodes for a transmembrane receptor-like molecule. *Molecular and Cellular Biology*, **6**, 3000–4.

Miozzo, M., Pierotti, M. A., Sozzi, G., Radice, P., Bongarzone, I., Spurr, N. K. & Della Porta, G. (1990). Human TRK proto-oncogene maps to chromosome 1q32–q41. *Oncogene*, **5**, 1411–14.

Mitra, G., Martin-Zanca, D. & Barbacid, M. (1987). Identification and biochemical characterisation of p70^{trk}, product of the human TRK oncogene. *Proceedings of the National Academy of Sciences, USA*, **84**, 6707–11.

Nagarajan, L., Louie, E., Tsujimoto, Y., Balduzzi, P. C., Huebner, K. & Croce, C. M. (1986). The human c-*ros* gene (ROS) is located at chromosome region 6q16–6q22. *Proceedings of the National Academy of Sciences, USA*, **83**, 6568–72.

Neal, D. E., Sharples, L., Smith, K., Fennelly, J., Hall, R. R. & Harris, A. L. (1990). The epidermal growth factor receptor and prognosis of bladder cancer. *Cancer*, **65**, 1619–25.

Neckameyer, W. S., Shibuya, M., Hsu, M. -T. & Wang, L. -H. (1986). Proto-oncogene c-*ros* codes for a molecule with structural features common to those of growth factor receptors and displays tissue specific and developmentally regulated expression. *Molecular and Cellular Biology*, **6**, 1478–86.

Neckameyer, W. S. & Wang, L. -H. (1985). Nucleotide sequence of avian sarcoma virus UR2 and comparison of its transforming gene with other members of the tyrosine protein kinase oncogene family. *Journal of Virology*, **53**, 879–84.

Nichols, E. J., Manger, R., Hakamori, S. -I., Herscovics, A. & Rohrschneider, L. R. (1985). Transformation by the v-*fms* oncogene product: role of glycosylational processing and cell surface expression. *Molecular and Cellular Biology*, **5**, 3467–75.

Nilsen, T. W., Maroney, P. A., Goodwin, R. G., Rottman, F. M., Crittenden, L. B., Raines, M. A. & Kung, J. -J. (1985). c-*erb*B activation in ALV-induced erythroblastosis: novel RNA processing and promoter insertion result in expression of an amino-truncated EGF receptor. *Cell*, **47**, 719–26.

Olah, E., Balogh, E., Kovacs, I. & Kiss, A. (1989). Abnormalities of chromosome 1 in relation to human malignant diseases. *Cancer Genetics and Cytogenetics*, **43**, 179–94.

Oskam, R., Coulier, F., Ernst, M., Martin-Zanca, D. & Barbacid, M. (1988). Frequent generation of oncogenes by *in vitro* recombination of TRK protooncogenes sequences. *Proceedings of the National Academy of Sciences, USA*, **85**, 2964–8.

Park, M., Dean, M., Cooper, C. S., Schmidt, M., O'Brian, S. J., Blair, D. G. &

Van de Woude, G. F. (1986). Mechanism of *met* oncogene activation. *Cell*, **45**, 895–904.

Park, M., Dean, M., Kaul, K., Braun, M. J., Gonda, M. A. & Van de Woude, G. (1987). Sequence of *met* proto-oncogene cDNA has features characteristic of the tyrosine kinase family of growth factor receptors. *Proceedings of the National Academy of Sciences, USA*, **84**, 6379–83.

Pasquale, E. B. & Singer, S. J. (1989). Identification of a developmentally regulated protein–tyrosine kinase by using anti-phosphotyrosine antibodies to screen a cDNA expression library. *Proceedings of the National Academy of Sciences, USA*, **86**, 5449–53.

Plowman, G. D., Whitney, G. S., Neubauer, M. G., Green, J. M., McDonald, B. L., Todaro, G. J. & Shoyab, M. (1990). Molecular cloning and expression of an additional epidermal growth factor receptor-related gene. *Proceedings of the National Academy of Sciences, USA*, **87**, 4905–9.

Pulciani, S., Santos, E., Lauver, A. V., Long, L. K., Aaronson, S. A. & Barbacid, M. (1982). Oncogenes in solid human tumours. *Nature*, London, **300**, 539–42.

Qui, F., Ray, P., Brown, K., Barker, P. E., Jhanwar, S., Ruddle, F. H. & Besmer, P. (1988). Primary structure of c-*kit*: relationship with the CSF-1/PDGF receptor kinase family–oncogenic activation of v-*kit* involves deletion of extracellular domain and C-terminus. *EMBO Journal*, **7**, 1003–11.

Ridge, S. A., Worwood, M., Oscier, D., Jacobs, A. & Padua, R. A. (1990). FMS mutations in myelodysplastic, leukaemic and normal subjects. *Proceedings of the National Academy of Sciences, USA*, **87**, 1377–80.

Riedel, H., Schlessinger, J. & Ullrich, A. (1987). A chimeric, ligand-binding v-*erb*B/EGF receptor retains transforming potential. *Science*, **236**, 197–200.

Roberts, W. M., Look, A. T., Roussel, M. F. & Sherr, C. J. (1988). Tandem linkage of human CSF-1 receptor (c-*fms*) and PDGF receptor genes. *Cell*, **55**, 655–61.

Roussel, M. F., Rettenmier, C. W., Look, A. T. & Sherr, C. J. (1984). Cell surface expression of v-*fms*-coded glycoproteins is required for transformations. *Molecular and Cellular Biology* **4**, 1999–2009.

Roussel, M. F., Dull, T. J., Rettenmier, C. W., Ralph, P., Ullrich, A. & Sherr, C. J. (1987). Transforming potential of the c-*fms* proto-oncogene (CSF-1 receptor). *Nature*, London, **325**, 549–52.

Roussel, M. F., Downing, J. R., Rettenmier, C. W. & Sherr, C. J. (1988). A point mutation in the extracellular domain of the human CSF-1 receptor (c-*fms* proto-oncogene product) activates its transforming potential. *Cell*, **55**, 979–88.

Rozengurt, E. (1986). Early signals in the mitogenic response. *Science*, **234**, 161–6.

Ruta, M., Burgess, W., Givol, D., Epstein, J., Neiger, N., Kaplow, J., Crumley, G., Dionne, C., Jaye, M. & Schlessinger, J. (1989). Receptor for acidic fibroblast growth factor is related to tyrosine kinase encoded by the *fms*-like gene (FLG). *Proceedings of the National Academy of Sciences, USA*, **86**, 8722–6.

Ruta, M., Howk, R., Ricca, G., Drohan, W., Zabelshansky, M., Laureys, G., Barton, D. E., Francke, U., Schlessinger, J. & Givol, D. (1988). A novel protein tyrosine kinase whose expression is modulated during endothelial cell differentiation. *Oncogene*, **3**, 9–15.

Sacca, R., Stanley, E. R., Sherr, C. J. & Rettenmier, C. W. (1986). Specific binding of the mononuclear phagocyte colony stimulating factor, CSF-1, to the product of v-*fms* oncogene. *Proceedings of the National Academy of Sciences, USA*, **83**, 3331–5.

Sharma, S., Birchmeier, C., Nikawa, J., O'Neill, K., Rodgers, L. & Wigler, M. (1989). Characterisation of the *ros*1-gene products expressed in human glioblastoma cell lines. *Oncogene Research*, **5**, 91–100.

Sherr, C. J. (1990). Colony-stimulating factor-1 receptor. *Blood*, **75**, 1–12.

Sherr, C. J., Rettenmier, C. W., Sacca, R., Roussel, M. F., Look, A. T. & Stanley, E. R. (1985). The c-*fms* proto-oncogene product is related to the receptor for mononuclear phagocyte growth factor, CSF-1. *Cell*, **41**, 665–76.

Shoyab, M., Plowman, G. D., McDonald, V. L., Bradley, J. G. & Todaro, G. J. (1989). Structure and function of human amphiregulin: a member of the epidermal growth factor family. *Science*, **243**, 1074–6.

Shurtleff, S. A., Downing, J. R., Rock, C. O., Hawkins, S. A., Roussel, M. F. & Sherr, C. J. (1990). Structural features of the colony-stimulating factor-1 receptor that affects its association with phosphatidylinositol 3-kinase. *EMBO Journal*, **9**, 2415–21.

Slamon, D. J., Clark, G. M., Wong, S. G., Levin, W. J., Ullrich, A. & McGuire, W. L. (1987). Human breast cancer: correlation of relapse and survival with amplification of the HER-2/*neu* oncogene. *Science*, **235**, 177–82.

Slamon, D. J., Godolphin, W., Jones, L. A., Holt, J. A., Wong, S. G., Keith, D. E., Levin, W. J., Stuart, S. G., Udove, J., Ullrich, A. & Press, M. F. (1989). Studies of the HER-2/*neu* proto-oncogene in human breast and ovarian cancer. *Science*, **244**, 707–12.

Stern, D. F., Kamps, M. P. & Cao, H. (1988). Oncogenic activation of p185[neu] stimulates tyrosine phosphorylation *in vivo*. *Molecular and Cellular Biology*, **8**, 3969–73.

Takahashi, M., Ritz, J. & Cooper, G. M. (1985). Activation of a novel human transforming gene, *ret*, by DNA rearrangement. *Cell*, **42**, 581–8.

Takahashi, M. & Cooper, G. M. (1987). *ret* transforming gene encodes a fusion protein homologous to tyrosine kinases. *Molecular and Cellular Biology*, **7**, 1378–85.

Takahashi, M., Buma, Y., Iwamoto, T., Inaguma, Y., Ikeda, H. & Hiai, H. (1988*a*). Cloning and expression of the *ret* proto-oncogene encoding a tyrosine kinase with two potential transmembrane domains. *Oncogene*, **3**, 571–8.

Takahashi, M., Inaguma, Y., Hiai, H. & Hirose, F. (1988*b*). Developmentally regulated expression of a human 'finger'-containing gene encoded by the 5' half of the *ret* transforming gene. *Molecular and Cellular Biology*, **8**, 1853–6.

Takahashi, M., Buma, Y. & Hiai, H. (1989). Isolation of *ret* proto-oncogene cDNA with an amino-terminal signal sequence. *Oncogene*, **4**, 805–6.

Taylor, G. R., Reedijk, M., Rothwell, V., Rohrschneider, L. & Pawson, T. (1989). The unique insert of cellular and viral *fms* protein tyrosine kinase domains is dispensable for enzymatic and transforming activities. *EMBO Journal*, **8**, 2029–37.

Ullrich, A., Bell, J. R., Chen, E. Y., Herrera, R., Petruzzelli, L. M., Dull, T. J., Gray, A., Coussens, L., Liao, Y. -C., Tsubokawa, M., Mason, A., Seeburg, P. H., Grunfeld, C., Rosen, O. M. & Ramachandran, J. (1985). Human insulin receptor and its relationship to the tyrosine kinase family of oncogenes. *Nature*, London, **313**, 756–61.

Ullrich, A., Coussens, L., Hayflick, J. S., Dull, T. J., Gray, A., Tam, A. W., Lee, T. J., Yarden, Y., Liberman, T. A., Schlessinger, J., Downward, J., Mayes, E. L. V., Whittle, N., Waterfield, M. D. & Seeburg, P. H. (1984). Human epidermal growth factor receptor cDNA sequence and abberant expression of the amplified gene in A431 epidermoid carcinoma cells. *Nature*, London, **309**, 418–25.

Ullrich, A. & Schlessinger, J. (1990). Signal transduction by receptors with tyrosine kinase activity. *Cell*, **61**, 203–12.

Velu, T. J., Beguinot, L., Vass, W. C., Willingham, M. C., Merlino, G. T., Pastan, I. & Lowy, D. R. (1987). Epidermal growth factor-dependent transformation by a human EGF receptor proto-oncogene. *Science*, **238**, 1408–10.

Waterfield, M. D., Scrace, G. T., Whittle, N., Stroobant, P., Johnsson, A., Wasteson, A., Westermark, B., Heldin, C.-H., Huang, J. -S. & Deuel, T. F. (1983). Platelet-derived growth factor is structurally related to the putative transforming protein p28^{sis} of simian sarcoma virus. *Nature*, London, **304**, 35–9.

Wells, A. & Bishop, M. (1988). Genetic determinants of neoplastic transformation by the retroviral oncogene v-*erb*B. *Proceedings of the National Academy of Sciences, USA*, **85**, 7597–601.

Witte, O. N. (1990). Steel locus defines new multipotent growth factor. *Cell*, **63**, 5–6.

Woolford, J., McAuliffe, A. & Rohrschneider, L. R. (1988). Activation of the feline c-*fms* proto-oncogene: multiple alterations are required to generate a fully transformed phenotype. *Cell*, **55**, 965–77.

Wright, C., Angus, B., Nicholson, S., Sainsbury, J. R. C., Cairns, J., Gullick, W. J., Kelly, P., Harris, A. L. & Horne, C. H. W. (1989). Expression of c-*erb*B-2 oncoprotein: a prognostic indicator in human breast cancer. *Cancer Research*, **49**, 2087–90.

Yarden, Y., Escobedo, J. A., Kuang, W.-J., Yang-Feng, T. L., Daniel, T. O., Tremble, P. M., Chen, E. Y., Ando, M. E., Harkins, R. N., Francke, Y., Fried, V. A., Ullrich, A. & Williams, L. T. (1986). Structure of the receptor for the platelet-derived growth factor receptor helps define a family of closely related growth factor receptors. *Nature*, London, **323**, 226–32.

Yarden, Y., Kuang, W. -J., Yang-Feng, T., Coussens, L., Munemitsu, S., Dull, T. J., Chen, E., Schlessinger, J., Francke, U. & Ullrich, A. (1987). Human proto-oncogene c-*kit*: a new cell surface receptor tyrosine kinase for an unidentified ligand. *EMBO Journal*, **6**, 3341–51.

Yarden, Y. & Ullrich, A. (1988). Growth factor receptor tyrosine kinases. *Annual Reviews in Biochemistry*, **57**, 443–78.

Yarden, Y., & Weinberg, R. A. (1989). Experimental approaches to hypothetical hormones: detection of a candidate ligand of the *neu* proto-oncogene. *Proceedings of the National Academy of Sciences, USA*, **86**, 3179–83.

Ziemiecki, A., Muller, R. G., Xiao-Chang, F., Hynes, N. E. & Kozma, S. (1990). Oncogenic activation of the human *trk* proto-oncogene by recombination with the ribosomal large subunit protein L7a. *EMBO Journal*, **9**, 191–6.

14

Structure and function of the T-cell antigen receptor

JUAN S. BONIFACINO,
RICHARD D. KLAUSNER AND
ALLAN M. WEISSMAN

Introduction

T cells recognize and respond to foreign antigens via the T-cell antigen receptor (TCR). This receptor, like immunoglobulin, consists of clonally derived elements that serve the primary purpose of antigen recognition. These components exhibit small intracytoplasmic domains of less than 12 amino acids, which would seem unlikely alone to effect signal transduction. The TCR has been found to contain not only these clonally derived antigen recognition elements, but also a set of invariant subunits, all of which possess more extensive intracellular domains. These invariant components presumably function primarily in signal transduction. The TCR is of great interest to us as a paradigm for the study of the assembly of a multisubunit membrane-bound, cell surface structure. In this chapter, recent work dealing with the structure, assembly and function of this receptor complex will be reviewed.

Structure of the TCR complex

The antigen recognition element on the surface of most T cells consists of disulfide-linked heterodimers $\alpha\beta$. These elements recognize antigen in association with a molecule of the major histocompatibility complex (MHC) on the surface of an antigen-presenting cell. The structure of the DNA encoding these subunits in T cells arises as a result of somatic rearrangement of the germ-like DNA in a fashion similar to that of immunoglobulin (for review see Davis & Bjorkman, 1988; Wilson *et al.*, 1988). These subunits contain both constant and variable domains. The variable regions account for the great diversity of antigens that can be recognized by these receptors. A small fraction of T cells have on their

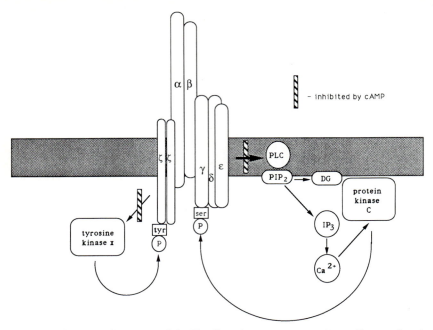

Fig. 14.1. Structure of the T-cell antigen receptor and coupling to signal transduction pathways. PLC: phospholipase C; PIP$_2$: phosphatidyl inositol bisphosphate; DG: diacylglycerol; IP$_3$: inositol trisphosphate.

surface an alternative clonally derived dimer consisting of rearranged subunits termed TCR δ and γ (for review see Brenner, Strominger & Krangel, 1988). The function of these alternative receptor forms is yet to be determined. In all cases, these clonally derived subunits exist on the surface of T cells in association with a set of at least four distinct gene products. These latter receptor components are invariant in their primary structure in all T cell clones. Three of these proteins were initially termed the T3 complex, by virtue of their recognition and co-immunoprecipitation by the monoclonal antibody OKT3. More recently they have been termed, by convention, the CD3 complex. These receptor components are structurally related polypeptides known as CD3-δ, CD3-ε and CD3-γ. They were first identified in human T cells and will be referred to in this review as δ, ε and γ. With the discovery of the invariant components of the murine TCR it became clear that there was an additional receptor component which we have termed ζ. Zeta is always found as part of a disulfide-linked dimer ζ_2. Hence, as a minimal stoichiometry, the TCR consists of $\alpha\beta\delta\varepsilon\gamma\zeta_2$. A number of experimental findings have provided evidence that these subunits are integral components of the receptor. These subunits can be quantitatively co-immunoprecipitated with the $\alpha\beta$ heterodimers from the surface of T cells (Reinherz *et al.*, 1983; Borst *et al.*, 1983*a*; Borst, Prendiville & Terhorst, 1983*b*; Samelson, Harford & Klausner, 1985*a*). Modulation of $\alpha\beta$ from the cell surface by ligands results in specific co-modulation of the CD3 chains (Reinherz *et al.*, 1982; Meuer *et al.*, 1983). Most importantly, for all components evaluated thus far, loss of expression of any of the receptor components results in loss of efficient surface expression of the remaining components (Weiss & Stobo, 1984; Schmitt-Verhulst *et al.*, 1987; Sussman *et al.*, 1988*a*,*b*; Bonifacino *et al.*, 1988*a*, 1989). Thus far cell lines deficient in either of α, β, δ, or ζ have been studied. These will be discussed in greater depth below.

The amino acid sequences of all of the stoichiometric invariant receptor components as well as many rearranged α and β subunits have been determined by cDNA cloning. All of the receptor components are transmembrane structures with single membrane-spanning regions. The extracellular regions of all the clonally derived polypeptides have the predicted β pleated sheet that makes them members of the immunoglobulin gene superfamily. Both α and β contain N-linked carbohydrate chains. These subunits are encoded on human chromosomes 14 and 7, respectively. cDNAs encoding the CD3 components and ζ have been cloned and sequenced from both mouse and man (van den Elsen *et al.*, 1984, 1985; Krissansen *et al.*, 1986, 1987; Haser *et al.*, 1987; Gold *et al.*, 1986, 1987; Weissman *et al.*, 1988*a*,*b*). These structures are lymphoid specific in their tissue distribution. Like the clonally derived subunits, these receptor components are also transmembrane proteins, each containing one membrane-spanning region. The CD3 components are all structurally homologous to each other, especially in the extracellular domains and, like the clonally derived elements, all are members of the immunoglobulin gene superfamily. Delta and γ both contain N-linked sugars while there is no evidence for N-glycosylation of ε. These components have extracellular domains of 79 to 89 amino acids with the intracellular domains being 43–46 amino acids long.

All of the CD3 genes are co-localized to chromosome region 11q23 in the human and to chromosome 9 in the mouse. The δ and γ subunits are located quite close to each other with their promoters arranged in a head-to-head manner 1.6 kb away from each other. The ε subunit has been determined to be from 30 to 300 kb away from the $\delta\gamma$ locus (Tunnacliffe, Buluwela & Rabbits, 1987; Georgopoulos *et al.*, 1988). Although the actual stoichiometry of the receptor has not been determined with certainty, the CD3 components are currently thought to exist as a linked set of $\varepsilon\gamma$ and $\varepsilon\delta$ dimers within the receptor (de la Hera *et al.*, 1991; Manolios *et al.*, 1991).

In contradistinction to the monomeric CD3 components, ζ exists within the receptor solely as a disulfide linked dimer. The majority of ζ (approximately 90%) exists as a homodimer ζ_2. The ζ subunit is not glycosylated (Samelson *et al.*, 1985a). In both mouse and man, it comprises a small, 9-amino acid extracellular domain, a single membrane spanning region and a comparatively long intracytoplasmic tail of 113 amino acids (112 in human). ζ shows no homology to the CD3 components (Weissman *et al.*, 1988a, b). Several structural features of ζ are of note. First, there is a single cysteine located on the extracellular face of the transmembrane region through which dimerization must occur. The intracytoplasmic tail beginning at amino acid 135 contains a consensus motif for a nucleotide binding domain. This motif or a variant of it is found in all proteins which bind nucleotides including kinases. This finding is of particular note as ζ undergoes tyrosine phosphorylation in response to antigenic stimulation (Baniyash *et al.*, 1988b). Like the other invariant receptor components, ζ also contains a single negatively charged amino acid residue in its transmembrane domain (Weissman *et al.*, 1988a, b). The ζ subunit of the receptor is localized to chromosome 1 in human as well as the mouse (Weissman *et al.*, 1988b; Baniyash *et al.*, 1989). The gene encoding the murine ζ gene has recently been cloned and mapped (Baniyash *et al.*, 1989). Several features are of note. The ζ chain gene is organized into 8 exons, spanning at least 31 kilobases of chromosome 1. Like the other invariant components of the receptor, there is no TATA box and several transcription initiation sites are present (Baniyash *et al.*, 1989).

Approximately 10% of the total ζ exists as a heterodimer bound to a distinct subunit which we have termed η (Baniyash *et al.*, 1988a). Unlike the other receptor subunits which would appear to exist on all receptors, the $\zeta\eta$ heterodimers would seem to be present in only $\sim 20\%$ of receptor complexes. Based on immunoblotting with anti-ζ reagents, the η subunit shares some but not all antigenic epitopes with the ζ chain (Orloff *et al.*, 1989). Recent cloning of a cDNA encoding η, has demonstrated that this chain arises from alternative splicing of the ζ mRNA (Jin *et al.*, 1990).

We have recently transfected cDNAs encoding the ζ subunit of the receptor into a ζ deficient cell line MA 5.8 (Weissman *et al.*, 1989). This deficient cell line makes normal amounts of all the stoichiometric receptor components but produces no ζ message or protein. The surface expression of pentameric partial receptors is therefore poor, as described below.

Transfection of ζ results in restoration of normal or supranormal levels of receptor expression. Interestingly, however, these transfectants demonstrate no detectable η, i.e. only the homodimeric form of ζ is present. These data, are supportive of the finding that η is generated from an alternatively spliced form of ζ mRNA.

Assembly and intracellular traffic

Assembly of the TCR complex

The TCR chains are synthesized by membrane-bound ribosomes in the endoplasmic reticulum (ER). They are synthesized with amino terminal signal peptides which are cleaved, presumably co-translationally, on translocation through the ER membrane. The subunits are all configured with their NH_2-termini facing the ER lumen and with their COOH-termini in the cytosol. Modifications of the nascent TCR chains such as glycosylation, disulfide-bond formation and assembly begin either co-translationally or immediately after translocation across the ER membrane.

The TCR complex contains both intrachain and interchain disulfide bonds. The rapid formation of intrachain disulfides in the δ and ε subunits is demonstrated by their migration above the diagonal on non-reducing/reducing gels, even after short pulses. This migration is characteristic of intrachain disulfide linkages, and is presumably due to incomplete protein unfolding under non-reducing conditions. The ζ chain also undergoes rapid interchain disulfide linkage, and in fact dimerization of ζ appears to be an efficient process that is complete at the earliest time points evaluated. The same is true for the formation of $\zeta\eta$ heterodimers. In contrast, the disulfide linkage of the $\alpha\beta$ heterodimer appears to be a considerably slower process. It has been shown that in certain cell lines disulfide-linkage of the α and β chains occurs only after their association with the CD3 chains (Alarcon *et al.*, 1988; Koning *et al.*, 1988). However, we have found that, at least in the case of the α and β chains expressed by the T cell hybridoma line 2B4, α and β are capable of forming disulfide-linked heterodimers when transfected into fibroblasts in the absence of other receptor components (Bonifacino *et al.*, 1988*a*). This apparent discrepancy may be due to the variable nature of the α and β chains. In addition, these studies did not measure the efficiency of disulfide bond formation. It is possible that even though certain $\alpha\beta$ pairs can form disulfide bonds in the absence of CD3 chains, the efficiency of this process is considerably increased by the simultaneous expression of the components of the CD3 complex.

Other interactions that occur shortly after synthesis involve the assembly of the CD3 subunits ($\varepsilon\gamma$ and $\varepsilon\delta$) in association with a protein called ω (Pettey *et al.*, 1987) or T-cell receptor associated protein (TRAP, Bonifacino, *et al.*, 1988*b*). TRAP binds with newly synthesized CD3 chains in the ER, after which it is rapidly dissociated, before transport of assembled TCR complexes into the Golgi system. The CD3 complex appears to behave as an autonomous entity that can assemble even in the absence of other receptor components as demonstrated in mutant T cells and transfected fibroblasts (Bonifacino *et al.*, 1988*a*; Berkhout, Alarcon & Terhorst, 1988). Thus, the first steps in the assembly of the TCR complex appear to be the independent formation of $\varepsilon\gamma$ and $\varepsilon\delta$ complexes and ζ_2 dimers. These partial complexes then come together with the α chains before addition and disulfide-linkage of the β chains. A potential pitfall in these studies is that antibodies may perturb the structure of the complex to some extent so that the order in which subunit interactions are observed may reflect the formation of stable complexes rather than the actual kinetics of subunit assembly.

Analysis of TCR assembly in subunit-deficient T cell mutants and in transfectants reveals that, in many cases, stable interactions between receptor components take place in the absence of the complete receptor. For instance, among the CD3 components, $\gamma\varepsilon$ and $\delta\varepsilon$ pairs are capable of forming in the absence of the third chain (Bonifacino *et al.*, 1988*a*; Berkhout *et al.*, 1988). These dimers can in turn interact with the clonotypic $\alpha\beta$ dimers (Bonifacino *et al.*, 1988*a*). Similarly, the lack of the ζ chain has no detectable effect on the assembly of the rest of the complex (Sussman *et al.*, 1988*a*; Bonifacino *et al.*, 1988*a*). An exception to this ability to assemble stable partial complexes is the failure of the ζ_2 dimer to co-precipitate with the CD3 complex in β deficient cells (Bonifacino *et al.*, 1988*a*; Chen *et al.*, 1988).

Our studies on the antigen-specific murine T cell hybridoma 2B4 have shown that $\alpha, \beta, \gamma, \delta$ and ε chains are synthesized in roughly stoichiometric amounts. In contrast, the ζ chain is synthesized in limiting quantities (Minami *et al.*, 1987*b*). The efficient and stable assembly of the receptor is reflected by the fact that those components in excess of ζ are rapidly and efficiently assembled into stable pentameric complexes. A number of observations suggest that complete assembly of the TCR complex occurs in the ER, even though the structure of the receptor may be further stabilized after its progression to the Golgi apparatus. Evidence for this early association comes from our studies which demonstrated the formation of complete complexes prior to the acquisition of Endo H resistance by α, β, γ and δ (Bonifacino *et al.*, 1988*b*). Additional support comes from studies showing that agents which inhibit transport out of the ER such as monensin do not affect receptor assembly (Alarcon *et al.*, 1988).

Examination of the fate of partial complexes or free TCR chains has shown that their transport out of the ER is related to the quaternary structure. Free chains or partial complexes assembled in T cells that fail to synthesize α, β and δ chains or more than one chain, and in fibroblasts transfected with the TCR genes, all fail to be transported from the ER to the Golgi system (Lippincott-Schwartz *et al.*, 1988; Chen *et al.*, 1988; Bonifacino *et al.*, 1989). This indicates that sorting mechanisms identify incomplete complexes as being structurally abnormal and determine their retention in the ER. In this regard, the TCR complexes behave like other oligomeric proteins for which correct assembly is a requirement for movement out of the ER (for review, see Rose & Doms, 1988). Unlike other oligomers, however, not all subunits of the TCR are necessary for this transport. Indeed, the ζ deficient pentamers which constitute the majority of incomplete complexes in the T cell hybridoma 2B4 are efficiently and rapidly transported out of the ER, processed in the Golgi system and then rapidly degraded in lysosomes (Minami *et al.*, 1987*b*). These findings have been confirmed by the study of ζ deficient variants of 2B4 (Sussman *et al.*, 1988*a*).

Transport to the cell surface

As mentioned above, assembly of the TCR chains in the endoplasmic reticulum results in a mixture of complete $\alpha\beta\gamma\delta\varepsilon\zeta_2$ and incomplete $\alpha\beta\gamma\delta\varepsilon$ complexes. Both species are transported into the Golgi system where the carbohydrate side chains of the glycoprotein subunits are modified. The extent and speed of modification of the carbohydrate chains in the Golgi system varies from chain to chain. All N-linked oligosaccharide chains of 2B4-α and -β are terminally modified, as demonstrated by the acquisition of resistance to Endo H and the addition of sialic acid. The single N-linked oligosaccharide chain of murine γ is also modified to complex carbohydrate. By contrast, only one of the three N-linked oligosaccharides of the mouse δ chain is terminally modified. The other two remain sensitive to Endo H at all times after synthesis. The speed at which the carbohydrate chains are modified in the Golgi system parallels the efficiency of the process. α and β chains are rapidly modified ($t_{\frac{1}{2}} \sim 40$ min) (Bonifacino *et al.*, 1988*b*), processing of γ is slower ($t_{\frac{1}{2}} \sim 1$ h) (Chen *et al.*, 1988) and δ is the slowest ($t_{\frac{1}{2}} = 1$–2 h) (Bonifacino *et al.*, 1988*b*). The different susceptibility of each oligosaccharide chain to Golgi processing may be determined by their location within the complex, so that some may be more readily accessible than others to the modifying enzymes.

After processing in the Golgi system, $\alpha\beta\gamma\delta\varepsilon\zeta_2$ and $\alpha\beta\gamma\delta\varepsilon\zeta\eta$ complexes are delivered to the cell surface. Once the complexes arrive at the plasma membrane, they engage in a process of constitutive recycling between the cell surface and an endosomal compartment (Minami, Samelson & Klausner, 1987*a*). Binding studies with anti-receptor antibodies have shown that at steady state, approximately 70 % of receptors are found on the cell surface and 30 % in an endosomal compartment. Analysis of the time course of disappearance of radioiodinated receptors from the surface of normal spleen T cells or T cell hybridomas has revealed that they possess a long half-life (> 20 h, J. S. Bonifacino, unpublished observations). Addition of phorbol esters, which cause phosphorylation of the ε and γ chains of the receptor by protein kinase C, leads to a reduction of the amount of surface receptor. This loss of receptors from the surface appears to be the result of a redistribution of the same number of receptors between surface and an internal pool (Minami *et al.*, 1987*a*).

Degradation of αβγδε complexes in lysosomes

Several lines of evidence indicate that while complete assembled receptors are delivered to the cell surface, ζ-minus $\alpha\beta\gamma\delta\varepsilon$ complexes, both in the parental T cell hybridoma 2B4 and in the ζ-deficient variant MA 5.8, are targeted to lysosomes for degradation (Minami *et al.*, 1987*b*; Sussman *et al.*, 1988*a*; Chen *et al.*, 1988; Lippincott-Schwartz *et al.*, 1988). In contrast to the surface receptors which survive long term, complexes lacking the ζ chains are rapidly degraded intracellularly after a lag period of 45–60 min. All the receptor chains are degraded at approximately the same rate ($t_{\frac{1}{2}} =$ 1–2 h). This degradation can be inhibited by the addition of agents that interfere with lysosomal proteolysis, such as NH_4Cl, methionine methyl ester and leupeptin (Minami *et al.*, 1987*b*; Chen *et al.*, 1988; Lippincott-Schwartz *et al.*, 1988). In addition, immunofluorescence studies of 2B4 cells using anti-ε antibodies show predominant staining of intracellular vesicles (Bonifacino *et al.*, 1989), most of which can be identified as lysosomes by immunoelectron microscopy (Chen *et al.*, 1988).

These incomplete complexes can be saved from lysosomal degradation by increasing the expression of the limiting ζ chains. A recent study has shown that transfection of a cDNA encoding the mouse ζ chain into the ζ-

deficient line MA 5.8 restores transport of complexes to the cell surface, increasing the proportion of receptor chains that avoid lysosomal degradation and survive long term (Weissman *et al.*, 1989). In addition, increases in the expression of the transfected ζ genes result in proportional increases in the survival of the other chains (Weissman *et al.*, 1989). These observations suggest that the ζ chains have a critical role in determining the sorting of newly synthesized complexes to the cell surface or to lysosomes.

A similar delivery of newly synthesized proteins to lysosomes was observed in studies on the intracellular transport of the envelope glycoprotein precursor of HIV-1, gp160 (Willey *et al.*, 1988). These studies showed that a large proportion of newly synthesized gp160 in infected T cells was degraded over a period of 6–8 h by a process that was sensitive to NH_4Cl. The remaining 5–15 % was cleaved to the mature gp120 and gp41 mature envelope product and transported to the cell surface where it was incorporated into budding viral particles (Willey *et al.*, 1988). Thus targeting of newly synthesized proteins to lysosomes may be a more common process than has been recognized to date. A potential role of this process may be to dispose of proteins that cells recognize as having an 'abnormal' structure. In the case of the ζ-minus TCR complexes, this process prevents the expression of complexes that are not completely functional (Sussman *et al.*, 1988*a*). In addition, this mechanism is an example of how the surface expression of a multisubunit complex can be modulated by simply turning on or off a single gene that encodes a limiting chain.

Selective degradation of TCR chains retained in the endoplasmic reticulum

It has been assumed until recently that the major lumenal site of protein degradation within the secretory system is the lysosome. As discussed above, this is the case for the $\alpha\beta\gamma\delta\varepsilon$ receptor complexes that are transported to lysosomes after traversal of the Golgi system. However, analysis of partial complexes assembled in T cells that fail to synthesize the β or δ chains (Chen *et al.*, 1988; Bonifacino *et al.*, 1989) or in fibroblasts transfected with genes encoding individual TCR chains (Lippincott-Schwartz *et al.*, 1988; Bonifacino *et al.*, 1989) uncovered a completely different situation. By using a combination of carbohydrate analysis, cell fractionation and morphologic techniques, partial complexes in these cells were found to be incapable of being transported into the Golgi system (Lippincott-Schwartz *et al.*, 1988; Chen *et al.*, 1988; Bonifacino *et al.*,

1989). Instead, they were retained within the endoplasmic reticulum system. Surprisingly, we found that whereas some of the receptor chains (ε,ζ) were relatively stable, others (α, β, δ) were rapidly degraded with half times of 30–45 min (Chen *et al.*, 1988; Lippincott-Schwartz *et al.*, 1988; Bonifacino *et al.*, 1989). Degradation of the γ chains depended on the state of oligomerization: free γ chains were degraded whereas γ bound to ε was stable (Bonifacino *et al.*, 1989). Degradation of the sensitive chains differed from lysosomal degradation in that it occurred without transport through the Golgi system and was not inhibited by agents that block lysosomal degradation such as NH_4Cl, chloroquine, methionine methyl ester and leupeptin (Lippincott-Schwartz *et al.*, 1988; Chen *et al.*, 1988). Whether this degradation occurs in the ER or transport to a non-lysosomal degradative organelle is needed, is not known.

The cause of the different fates of TCR chains retained in the ER was examined in COS cells that were transfected with genes encoding TCR chains. These studies showed that, even when expressed alone, TCR chains were degraded at vastly different rates, thus demonstrating that each chain has an intrinsic susceptibility to ER degradation. When transfected in pairs, however, at least one interaction resulted in stabilization of a sensitive chain: the γ chains were more stable when co-expressed with ε than when expressed alone (Bonifacino *et al.*, 1989). Other interactions, characterized in T-cell variants, did not result in dramatic stabilization of sensitive chains (Chen *et al.*, 1988; Bonifacino *et al.*, 1989). Intrinsic susceptibilities and the variable effect of oligomerization combine to generate the pattern of selective degradation described above. A recent study has shown that sensitivity to ER degradation is probably confered by peptide sequences that act as targeting signals for degradation (Bonifacino *et al.*, 1990). Deletion of a 28 amino acid COOH-terminal sequence comprising the transmembrane and cytoplasmic domains of the α chain resulted in a soluble protein that was not degraded, even though it was still retained within the ER system. Fusion of this sequence to the extracellular domain of the Tac antigen, which is normally transported to the cell surface, caused retention and rapid degradation of the chimeric protein within the ER. Additional studies showed that the ability to cause rapid degradation in the ER is localized to a 23 amino acid stretch including the transmembrane domain of the α chain (Bonifacino *et al.*, 1990). This non-lysosomal degradation of TCR chains may be the reflection of a pathway used for the disposal of abnormal proteins retained in the ER, the normal turnover of ER resident proteins, the proteolytic processing of protein precursors or the removal of accessory molecules.

Function

The mechanisms by which stimulation of the T-cell antigen receptor results in the generation of intracellular signals remain largely unknown. However, what is becoming increasingly clear is that multiple and apparently independent signals are generated in response to receptor occupancy (see Fig. 14.1). Stimulation of the T-cell antigen receptor can be affected either by antigen plus MHC on the surface of antigen-presenting cells or by direct cross-linking of anti-receptor antibodies, either using a plastic surface or using Fc bearing B cells (Samelson *et al.*, 1985*b*, 1986, 1987). Additionally, antibodies or combinations or antibodies directed against other cell surface structures such as Thy 1.2 in mouse or CD2 in human result in a pattern of activation and proliferation that by certain criteria mimic the effect of antigen (Klausner *et al.*, 1987; Weissman *et al.*, 1988*c*). Pharmacologic agents such as phorbol myristate acetate (PMA) plus a calcium ionophore also result in a pattern of intracellular events and proliferation that mimic at least some aspects of antigen occupancy with physiologic ligand.

Stimulation of the T-cell antigen receptor results in the serine phosphorylation of the CD3-γ chain and, to a lesser extent, the CD3-ε chain (Samelson *et al.*, 1985*b*). In addition to these serine phosphorylations, the ζ subunit of the receptor undergoes tyrosine phosphorylation (Samelson *et al.*, 1985*b*; Baniyash *et al.*, 1988*b*). These receptor phosphorylation events appear to be the result of the activation of two independent kinase pathways. Protein kinase C (PKC) would appear to be responsible for the serine phosphorylation events. Evidence for this comes from experiments in which protein kinase C is depleted from cells by overnight treatment with the protein kinase C activator PMA (Patel, Samelson & Klausner, 1987). In these experiments the ability of antigen to result in phosphorylation of γ and ε is specifically lost while the ζ tyrosine phosphorylation is maintained. Activation of this pathway is stimulated by the coupling of receptor occupancy to increases in phosphoinositide metabolism leading to the generation of polyphosphoinositols after phospholipase C mediated hydrolysis. It is believed that, as a general mechanism, the generation of these products leads to the release of intracellular Ca^{2+} stores with a later prolonged increase in free Ca^{2+} resulting from an increase in Ca^{2+} flux from outside the cell. This increase in Ca^{2+} in conjunction with increases in free diacylglycerol, the other product of phospholipase C hydrolysis, results in the activation of protein kinase C. The function of this phosphorylation in cellular activation remains unknown. Until recently it was believed that this PI elevation and subsequent PKC activation was required for the subsequent production of the lymphokine interleukin-2 (IL-2). Recent data suggest a dissociation between these phenomena in that in certain hybridoma variants and transfectants normal IL-2 production is seen in the face of disproportionately low phosphoinositol generation (Sussman *et al.*, 1988*b*; Mercep *et al.*, 1988).

The second kinase pathway activated by antigen is one in which occupancy results in the activation of an as yet unknown protein tyrosine kinase. Initially we observed that receptor activation resulted in a tyrosine phosphorylated 21 kD species (Samelson *et al.*, 1986). We have since determined that this apparent heterodimer is actually a ζ homodimer, one partner of which has been phosphorylated on multiple tyrosines resulting in its aberrant migration at 21 kD on SDS/PAGE (Baniyash *et al.* 1988*b*). Interestingly, only a small percentage, $\sim 5\%$ of the $(\zeta)_2$ homodimers, undergo these multiple cooperative phosphorylations. The significance of this subset and what appears to be the role that this phosphorylation plays in receptor function remain largely unknown. Analysis of the deduced amino acid sequence of ζ reveals six intracellular tyrosines all of which might be substrates for this kinase (Weissman *et al.*, 1988*a*). Interestingly, the sequence of ζ reveals a consensus sequence for nucleotide binding; the significance of this remains to be determined (Weissman *et al.*, 1988*a*). In addition, several other intracellular species have been determined to be substrates for this receptor-activated protein tyrosine kinase. The nature of these species and whether one of them is, in fact, the kinase remains to be determined. The TCR is unique among receptors that are tyrosine kinase substrates in that unlike growth factor receptors such as EGF, insulin, etc, the receptor, to the extent that we have defined it, is not itself a kinase. Hence activation via the receptor results in the activation of at least two independent kinase pathways, both of which would appear to feed back on the receptor resulting in phosphorylation of receptor subunits.

References

Alarcon, B., Berkhout, B., Breitmeyer, J. & Terhorst, C. (1988). Assembly of the human T cell receptor–CD3 complex takes place in the endoplasmic reticulum and involves intermediary complexes between the CD3-gamma, delta, epsilon core and single T cell receptor alpha or beta chains. *Journal of Biological Chemistry*, **263**, 2953–61.

Baniyash, M., Garcia-Morales, P., Bonifacino, J. S., Samelson, L. E. & Klausner, R. D. (1988*a*). Disulfide linkage of the zeta and eta chains of the

T cell receptor. Possible identification of two structural classes of receptors. *Journal of Biological Chemistry*, **263**, 9874–8.

Baniyash, M., Garcia-Morales, P., Luong, E., Samelson, L. E. & Klausner, R. D. (1988*b*). The T cell antigen receptor zeta chain is tyrosine phosphorylated upon activation. *Journal of Biological Chemistry*, **263**, 18225–30.

Baniyash, M., Hsu, V. W., Seldin, M. F. & Klausner, R. D. (1989). The isolation and characterization of the murine T cell antigen receptor zeta chain gene. *Journal of Biological Chemistry*, **264**, 13252–7.

Berkhout, B., Alarcon, B. & Terhorst, C. (1988). Transfection of genes encoding the T cell receptor-associated CD3 complex into COS cells results in assembly of the macromolecular structure. *Journal of Biological Chemistry*, **263**, 8528–36.

Bonifacino, J. S., Chen, C., Lippincott-Schwartz, J., Ashwell, J. D. & Klausner, R. D. (1988*a*). Subunit interactions within the T cell antigen receptor: clues from the study of partial complexes. *Proceedings of the National Academy of Sciences, USA*, **85**, 6929–33.

Bonifacino, J. S., Lippincott-Schwartz, J., Chen, C., Antusch, D., Samelson, L. E. & Klausner, R. D. (1988*b*). Association and dissociation of the murine T cell receptor associated protein (TRAP): Early events in the biosynthesis of a multisubunit receptor. *Journal of Biological Chemistry*, **263**, 8965–71.

Bonifacino, J. S., Suzuki, C. K. & Klausner, R. D. (1990). A peptide sequence confers retention and degradation in the endoplasmic reticulum. *Science*, **247**, 79–82.

Bonifacino, J. S., Suzuki, C. K., Lippincott-Schwartz, J., Weissman, A. M. & Klausner, R. D. (1989). Pre-Golgi degradation of newly synthesized T-cell antigen receptor chains: Intrinsic sensitivity and the role of subunit assembly. *Journal of Cell Biology*, **109**, 73–83.

Borst, J., Alexander, S., Elder, J. & Terhorst, C. (1983*a*). The T3 complex on human T lymphocytes involves four structurally distinct glycoproteins. *Journal of Biological Chemistry*, **258**, 5135–41.

Borst, J., Prendiville, M. A. & Terhorst, C. (1983*b*). The T3 complex on human thymus-derived lymphocytes contains two different subunits of 20 kDa. *European Journal of Immunology*, **13**, 576–80.

Brenner, M. B., Strominger, J. L. & Krangel, M. S. (1988). The gamma–delta T cell receptor. *Advances in Immunology*, **43**, 133–92.

Chen, C., Bonifacino, J. S., Yuan, L. & Klausner, R. D. (1988). Selective degradation of T cell antigen receptor chains retained in a pre-Golgi compartment. *Journal of Cell Biology*, **107**, 2149–61.

Davis, M. M. & Bjorkman, P. J. (1988). T-cell antigen receptor genes and T-cell recognition. *Nature*, London, **334**, 395–402.

de la Hera, A., Müller, U., Olsson, C., Isaag, S. & Tunnacliffe, A. (1991). Structure of the T cell antigen receptor (TCR): two CD3ε subunits in a functional TCR/CD3 complex. *Journal of Experimental Medicine*, **177**, 7–17.

Georgopoulos, K., van den Elsen, P., Bier, E., Maxam, A. & Terhorst, C. (1988). A T cell-specific enhancer is located in a DNAse I-hypersensitive area at the B1 end of the CD3-δ gene. *EMBO Journal*, **7**, 2401–7.

Gold, D. P., Clevers, H., Alarcon, B., Dunlap, S., Novotny, J., Williams, A. F. & Terhorst, C. (1987). Evolutionary relationship between the T3 chains of the T-cell receptor complex and the immunoglobulin supergene family. *Proceedings of the National Academy of Sciences, USA*, **84**, 7649–53.

Gold, D. P., Puck, J. M., Pettey, C. L., Cho, J., Coligan, J., Woody, J. N. & Terhorst, C. (1986). Isolation of cDNA clones encoding the 20 K nonglycosylated chain of the human T cell receptor–T3 complex. *Nature, London*, **321**, 431–4.

Haser, W. G., Saito, H., Koyoma, T. & Tonegawa, S. (1987). Cloning and sequencing of murine T3 gamma cDNA from a subtractive cDNA library. *Journal of Experimental Medicine*, **166**, 1186–91.

Jin, Y. J., Clayton, L. K., Howard, F. D., Koyasu, S., Siehs, M., Steinbrich, R., Tarr, G. E. & Reinherz, E. L. (1990). Molecular cloning of the CD3η subunit identifies a CD3η-related product in thymus-derived cells. *Proceedings of the National Academy of Sciences, USA*, **87**, 3319–23.

Klausner, R. D., O'Shea, J. J., Luong, H., Ross, P., Bluestone, J. A. & Samelson, L. E. (1987). T cell receptor tyrosine phosphorylation. Variable coupling for different activating ligands. *Journal of Biological Chemistry*, **262**, 12654–9.

Koning, F., Lew, A. M., Maloy, W. L., Valas, R. & Coligan, J. E. (1988). The biosynthesis and assembly of T cell receptor alpha and beta chains with the CD3 complex. *Journal of Immunology*, **140**, 3126–34.

Krissansen, G. W., Owen, J. J., Fink, P. J. & Crumpton, M. J. (1987). Molecular cloning of the cDNA encoding the T3-gamma subunit of the mouse T3/T cell antigen receptor complex. *Journal of Immunology*, **138**, 3513–18.

Krissansen, G. W., Owen, M. J., Verbi, W. & Crumpton, M. J. (1986). Primary structure of the T3-gamma subunit of the T3/T cell antigen receptor complex deduced from cDNA sequences: evolution of the T3-gamma and delta subunits. *EMBO Journal*, **5**, 1799–808.

Lippincott-Schwartz, J., Bonifacino, J. S., Yuan, L. & Klausner, R. D. (1988). Degradation from the endoplasmic reticulum: disposing of newly synthesized protein. *Cell*, **54**, 209–29.

Manolios, N., Letourner, F., Bonifacino, J. S. & Klausner, R. D. (1991). Pairwise, cooperative and inhibitory interactions describe the assembly and probable structure of the T-cell antigen receptor. *EMBO Journal*, **10**, 1643–51.

Mercep, M., Bonifacino, J. S., Garcia-Morales, P., Samelson, L. E., Klausner, R. D. & Ashwell, J. D. (1988). T cell CD3-$\zeta\eta$ heterodimer expression and coupling to phosphoinositide hydrolysis. *Science*, **242**, 571–4.

Mercep, M., Weissman, A. M., Frank, S. J., Klausner, R. D. & Ashwell, J. D. (1989). Activation-driven programmed cell death and T cell receptor $\zeta\eta$ expression. *Science*, **246**, 1162–5.

Meuer, S., Fitzgerald, K., Hussey, R., Hodgdon, J., Schlossman, S. & Reinherz, E. (1983). Clonotypic structures involved in antigen-specific human T cell function. Relationship to the T3 molecular complex. *Journal of Experimental Medicine*, **157**, 705–19.

Minami, Y., Samelson, L. E. & Klausner, R. D. (1987*a*). Internalization and cycling of the T cell antigen receptor. Role of protein kinase C. *Journal of Biological Chemistry*, **262**, 13342–7.

Minami, Y., Weissman, A. M., Samelson, L. E. & Klausner, R. D. (1987*b*). Binding a multichain receptor. Synthesis, degradation and assembly of the T-cell antigen receptor. *Proceedings of the National Academy of Sciences, USA*, **84**, 2688–92.

Orloff, D. G., Frank, S. F., Robey, F. A., Weissman, A. M. & Klausner, R. D. (1989). Biochemical characterization of the eta chain of the T cell antigen receptor: a unique subunit related to zeta. *Journal of Biological Chemistry*, **264**, 14812–17.

Patel, M. D., Samelson, L. E. & Klausner, R. D. (1987). Multiple kinases and signal transduction: Phosphorylation of the T cell antigen receptor. *Journal of Biological Chemistry*, **262**, 5631–838.

Pettey, C. L., Alarcon, B., Malin, R., Weinberg, K. & Terhorst, C. (1987). T3-p28 is a protein associated with the delta and epsilon chains of the T cell receptor–T3 antigen complex during biosynthesis. *Journal of Biological Chemistry*, **262**, 4854–9.

Reinherz, E., Meuer, S., Fitzgerald, K., Hussey, R. E., Hodgdon, J., Acuto, O. & Schlossman, S. (1983). Comparison of T3-associated 49- and 42-kilodalton cell surface molecules on individual human T-cell clones: evidence for peptide variability in T-cell receptor structures. *Proceedings of the National Academy of Sciences, USA*, **80**, 4104–8.

Reinherz, E. L., Meuer, S., Fitzgerald, K. A., Hussey, R. E., Levine, H. & Schlossman, S. F. (1982). Antigen recognition by human T lymphocytes is linked to surface expression of the T3 molecular complex. *Cell*, **30**, 735–43.

Rose, J. K. & Doms, R. W. (1988). Regulation of protein export from the endoplasmic reticulum. *Annual Reviews in Cell Biology*, **4**, 257–88.

Samelson, L. E., Harford, J. B. & Klausner, R. D. (1985a). Identification of the components of the murine T cell antigen receptor complex. *Cell*, **43**, 223–31.

Samelson, L. E., Harford, J. B., Schwartz, R. H. & Klausner, R. D. (1985b). A 20-kDa protein associated with the murine T-cell antigen receptor is phosphorylated in response to activation by antigen or concanavalin A. *Proceedings of the National Academy of Sciences, USA*, **82**, 1969–73.

Samelson, L. E., O'Shea, J. J., Luong, H., Ross, P., Urdahl, K. V., Klausner, R. D. & Bluestone, J. (1987). T cell antigen receptor phosphorylation induced by an anti-receptor antibody. *Journal of Immunology*, **139**, 2708–14.

Samelson, L. E., Patel, M. D., Weissman, A. M., Harford, J. B. & Klausner, R. D. (1986). Antigen activation of murine T cells induces tyrosine phosphorylation of a polypeptide associated with the T cell antigen receptor. *Cell*, **46**, 1083–90.

Schmitt-Verhulst, A-M., Guimezanes, A., Boyer, D., Peonie, M., Tsien, R., Buferne, M., Hua, C. & Leserman, L. (1987). Pleiotropic loss of activation pathways in T-cell receptor alpha-chain deletion variant of a cytolytic T cell clone. *Nature*, London, **325**, 628–31.

Sussman, J. J., Bonifacino, J. S., Lippincott-Schwartz, J., Weissman, A. M., Saito, T., Klausner, R. D. & Ashwell, J. D. (1988a). Failure to synthesize the T cell CD3-zeta chain: structure and function of a partial T cell receptor complex. *Cell*, **52**, 85–95.

Sussman, J. J., Saito, T., Shevach, E. M., Germain, R. N. & Ashwell, J. D. (1988b). Thy-1 and Ly-6-mediated lymphokine production and growth inhibition of a T cell hybridoma require co-expression of the T cell antigen receptor complex. *Journal of Immunology*, **140**, 2520–6.

Tunnacliffe, A., Buluwela, L. & Rabbits, T. H. (1987). Physical linkage of three CD3 genes on human chromosone 11. *EMBO Journal*, **6**, 2953–7.

van den Elsen, P., Shepley, B-A., Borst, J., Coligan, J. E., Markham, A. F., Orkin, S. & Terhorst, C. (1984). Isolation of cDNA clones encoding the 20 K T3 glycoprotein of human T cell receptor complex. *Nature*, London, **312**, 413–18.

van den Elsen, P., Shepley, B-A., Cho, M. & Terhorst, C. (1985). Isolation and characterization of a cDNA clone encoding the murine homologue of the human 20 K T3/T-cell receptor glycoprotein. *Nature*, London, **314**, 542–4.

Weiss, A., & Stobo, J. D. (1984). Requirement for the co-expression of T3 and the T cell antigen receptor on a malignant human T cell line. *Journal of Experimental Medicine*, **160**, 1284–99.

Weissman, A. M., Baniyash, M., Hou, D., Samelson, L. E., Burgess, W. H. & Klausner, R. D. (1988*a*). Molecular cloning of the zeta chain of the T cell antigen receptor. *Science*, **239**, 1018–21.

Weissman, A. M., Frank, S. J., Orloff, D. G., Mercep, M., Ashwell, J. D. & Klausner, R. D. (1989). Role of the zeta chain in the expression of the T cell receptor: genetic reconstitution studies. *EMBO Journal*, **8**, 3651–6.

Weissman, A. M., Hou, D., Orloff, D. G., Modi, W. S., Seuanez, H., O'Brien, S. & Klausner, R. D. (1988*b*). The molecular cloning and chromosomal localization of the human T cell receptor zeta chain: distinction from the molecular CD3 complex. *Proceedings of the National Academy of Science, USA*, **85**, 9709–13.

Weissman, A. M., Ross, P., Luong, E. T., Garcia-Morales, P., Jelachich, M. L., Biddison, W. E., Klausner, R. D. & Samelson, L. E. (1988*c*). Tyrosine phosphorylation of the human T cell antigen receptor zeta chain: activation via CD3 but not CD2. *Journal of Immunology*, **141**, 3532–6.

Willey, R. L., Bonifacino, J. S., Potts, B. J., Martin, M. A. & Klausner, R. D. (1988). Biosynthesis, processing and degradation of the HIV-1 envelope glycoproteins. *Proceedings of the National Academy of Sciences, USA*, **85**, 9580–4.

Wilson, R. K., Lai, E., Concannon, P., Barth, R. K. & Hood, L. E. (1988). Structure, organization and polymorphism of murine and human T-cell receptor alpha and beta chain gene families. *Immunology Review*, **101**, 149–72.

Index